INTROD
NANOSCIENCE

INTRODUCTORY NANOSCIENCE

Masaru Kuno

LONDON AND NEW YORK

Vice President: Denise Schanck
Editor: Summers Scholl
Editorial Assistant: Kelly O'Connor
Production Editors: Mac Clarke and Georgina Lucas
Cover Design: Andrew Magee
Copyeditor: Mac Clarke
Typesetting: TechSet
Proofreader: Sally Huish

ISBN 978-0-8153-4212-0

Library of Congress Cataloging-in-Publication Data
Kuno, Masaru.
Introductory nanoscience / Masaru Kuno.
p. cm.
Includes bibliographical references.
ISBN 978-0-8153-4424-7 (alk. paper)
1. Nanoscience. I. Title.
QC176.8.N35K86 2011
620'.5–dc23 2011023499

Published by Garland Science, Taylor & Francis Group, LLC, an informa business,
711 Third Avenue, 8th floor, New York, NY 10017, USA, and
2 Park Square, Milton Park, Abingdon, OX14 4RN, UK.

Printed in the United States of America
15 14 13 12 11 10 9 8 7 6 5 4 3 2 1

Visit our web site at http://www.garlandscience.com

To my Parents and to my darling Katya

Preface

This introductory book on nanoscience originated during the spring and summer of 2003. After my initial appointment as an Assistant Professor in Chemistry and Biochemistry at Notre Dame, I agreed to teach an introductory class on nanoscience and nanotechnology for incoming graduate students and upper level undergraduates. However, I quickly realized, after accepting this task, that there were few resources available for teaching such a class, let alone any textbook that one could refer to. So while waiting for equipment to arrive, I undertook upon myself the task to compile a series of lecture notes that would explain to the student some of the underlying concepts behind "nano." These lecture notes have since become the basis of the text that you are now reading.

Philosophy

My underlying motivation has been to describe to you, the student or reader, the physics or at least the quantitative basis behind each concept or assumption commonly encountered in the nanoscience literature. This is as opposed to providing qualitative overviews about developments in the field. In this regard, there now exist many fine texts on nanoscience and nanotechnology that are encyclopedic in scope and that cover a vast range of concepts and observations. My goal has been different and in some ways minimalistic. Rather than illustrate the breadth of nanoscience, I have tried to highlight key quantitative concepts that underlie this new field.

Furthermore, as anyone who has ever tried learning something on his or her own knows, picking up a book and figuring things out is easier said than done. I have often found myself baffled by equations in texts accompanied by the well-known phrase "it can be shown." Usually, this means (for me) a few extra days of work to derive what was apparently obvious. Therefore, to avoid inflicting the same pain onto others, I have tried to be as thorough as possible in my derivations. I hope that the reader out there, diligently trying to figure things out, will see that he or she is not alone.

Scope of the Text

The following describes the pedagogy of this text. Chapter 2 begins with a brief overview of crystalline materials. Chapter 3 follows with a discussion about the length scales that define nano as well as the size regimes where the optical and electrical properties of materials become size- and shape-dependent. In Chapter 4, we highlight three generic types of nanostructures (quantum wells, quantum wires, and quantum dots) that will be the focus of the rest of this text. To aid the reader connect such structures to reality, micrographs of actual systems are provided. In Chapter 5, we introduce the phenomenological absorption and emission properties of these nanostructures and discuss important quantitative concepts such as the absorption cross section. However, a deeper understanding of the physics behind either the absorption or the emission process requires us to know some quantum mechanics. Thus, Chapter 6 starts us on a longer journey towards achieving this goal by briefly reviewing basic quantum mechanics. Chapter 7 introduces representative quantum mechanical models for the above wells, wires, and dots, while Chapter 8 finishes this introduction by describing additional model systems that are commonly encountered. Chapter 9 introduces us to the concept of a density of states as well as the joint density of states, underlying the absorption spectrum of a material. Chapter 10 then describes the actual bands from where these optical transitions originate and end. Chapter 11 revisits time-dependent perturbation theory so that we can complete our theoretical construct for describing absorption probabilities and rates. Finally, Chapter 12 consolidates all of this to describe the interband transitions of quantum wells, wires, and dots. The book concludes with a description about the chemical synthesis of nanostructures (Chapter 13), the approaches by which they are characterized (Chapter 14), and their applications (Chapter 15).

Features of the Text

Throughout this book, I have tried to illustrate and motivate the discussion of concepts with "back of the envelope" calculations. The idea is to get the reader used to running the numbers, since this is often important when carrying out actual experiments. Furthermore, in the spirit of Fermi, I have suggested places where readers can carry out derivations themselves. When possible, the end-of-chapter problems come directly from the literature, or at least refer to it. In this way, I hope that the usefulness of what has been described becomes readily apparent to the reader. My ultimate goal is to provide the reader both a solid introductory foundation and a connection to the literature, from where they can critically analyze, and possibly in the future contribute to, the emerging nanoscience field.

Instructor Resources

Accessible from www.garlandscience.com/nanoscience, the Instructor's Resource Site requires registration and access is available only to qualified instructors. To access the Instructor Resource Site or to report any errors, please contact your local sales representative or email science@garland.com. On the website, the resources may be browsed by individual chapters and there is a search engine. The images from the book are available in two convenient formats: PowerPoint® and JPEG. They have been optimized for display on a computer. Figures are searchable by figure number, by figure name, or by keywords used in the figure legend from the book. You can also access the resources available for other Garland Science titles.

Acknowledgments

I'd like to take this opportunity to thank my colleagues for their technical and nontechnical help over the years. Special thanks go to Pavel Frantsuzov, Greg Hartland, Boldizsar Janko, Prashant Kamat, Tom Kosel, and Dani Meisel (all at the University of Notre Dame). I thank the reviewers of draft material: Wei Chen (University of Texas at Arlington), Cherie Kagun (University of Pennsylvania), Boris Veytsman (George Mason University), Haeyeon Yang (Utah State University), and others who provided feedback. I also want to thank my students and postdocs for their tireless efforts in lab on a daily basis. This reminds me of when, as an assistant professor, I used to go around and give talks. Occasionally, if the talk was going poorly or if I sensed that the audience was bored stiff, I would end with the one and only joke I have. It goes like this: A grad student, a postdoc, and a young assistant professor are walking through the woods. They run across a magic lamp. The grad student runs up and rubs the lamp and, *whoosh*, a genie comes out! The genie says, "I've been trapped in this lamp for a thousand years. You set me free. Now, I usually give three wishes to the person who sets me free, but since there are three of you, I'll give each one of you a wish." The grad student immediately wants to go first. The genie says, "OK, what do you want?" The grad student answers, "I want to be in Hawaii drinking Coronas on the beach." The genie says, "no problem," and, *whoosh*, the student is gone. The postdoc is excited and wants to go next. So the genie asks, "What do you want?" The postdoc says "I too want to be in Hawaii drinking Mai Tais on the beach." The genie says "fine," and, *whoosh*, the postdoc is gone. At this point, the genie turns to the young assistant professor and asks "So, what do you want?" The young assistant professor pauses, thinks about it for a moment, and then turns to the genie, "I want those guys back in lab after lunch."

I want to conclude by thanking the nano subgroup at Notre Dame, who have, in many ways, motivated me to write this text. Those who know me will understand why I'm smiling right now. Finally, I want to thank the University of Notre Dame, the Notre Dame Radiation Laboratory/DOE Office of Basic Energy Sciences, the National Science Foundation, and Research Corporation for supporting my research.

Masaru Kuno
Notre Dame, IN

Contents in Brief

Contents in Detail

Introduction

1.1 PRELIMINARIES

What are nanoscience and nanotechnology? Without providing a definite answer to this question, "nano" is a popular (emerging) area of science and technology. It has attracted the attention of researchers from all walks of life, from physics to chemistry to biology and engineering. Further impetus for this movement comes from the large increase in public and private funding for nano over the last ten years. An example of this is the National Nanotechnology Initiative (NNI) created by former President Bill Clinton in 2001. The NNI coordinates nanoscience and nanotechnology research activities in the United States and has increased funding in these areas by hundreds of millions of dollars yearly, from \$464 million in 2001 to approximately \$1.4 billion in 2007. More information about the NNI can be found on its Web site at www.nano.gov. Private sector contributions have also jumped dramatically, as evidenced by the plethora of small startup firms lining the tech corridors of the East and West.

Nano has even entered popular culture. It has been used as a buzzword in contemporary books, movies, television commercials, and even consumer electronics. For example, in the recent blockbuster movie *Spider-Man*, Willem Dafoe's character (the Green Goblin) is a famous (and wildly wealthy) nanotechnologist whose papers Tobey McGuire's character (Spider-Man) has read and followed. Likewise, in the movie *Minority Report*, Tom Cruise's character undergoes eye surgery to avoid biometric fingerprinting. This involves a retinal eye transplant aided by so-called "nano reconstructors." Furthermore, a scene in the D.C. metro shows him reading a newspaper with the headline "nanotechnology breakthrough." Many other movies and television series have likewise invoked nano at some point. In television commercials, a recent General Electric advertisement for washers and dryers features the storyline of geeky nanotechnologist bumps into supermodel at the laundromat. It's love at first sight. In books, the New York Times bestseller *Prey* by Michael Crighton features nanotechnology run amok, with spawns of tiny nano robots escaping from the laboratory and hunting down people for food. Finally, Apple's iPod nano is clearly a reference to its slim, elegant, styling.

The mantle of nano has also been adopted by various scientific visionaries. Perhaps the most prominent is Eric Drexler, who has founded a center, called the Foresight Institute, devoted to exploring his ideas. Concepts being discussed include the development of tiny nano robots that will "live" inside us and repair our blood vessels when damaged. They will also kill cancer, cure us when we are sick, mend bones when broken, make us smarter, and even give us immortality. These nano robots will further serve as tiny factories, manufacturing anything and everything from food to antibiotics to energy. In turn, nanotechnology will provide a solution to all of mankind's problems whether hunger in developing countries or pollution in developed ones. Drexler therefore envisions a vast industrial revolution of unprecedented size and scale.

At the same time, concurrent with his visions of a utopian future is a darker side, involving themes where such nano robots escape from the laboratory and evolve into sentient creatures completely out of humankind's control. Such beings could then sow the seeds to our own destruction in the spirit of movies and books such as the *Terminator* and *Matrix* series and *Prey*.

Whether such predictions and visions of the future will ever become reality remains to be seen. However, any such developments will ultimately rely on the scientific research of today, which is, on a daily basis, laying down the foundation for tomorrow's nanoscience and nanotechnology.

In today's scientific realm, the prefix "nano" describes physical lengths that are on the order of a billionth of a meter long (i.e., 10^{-9} m). Nanoscale materials therefore lie in a physical size regime between bulk, macroscale, materials (the realm of condensed matter physics) and molecules or atoms (the realm of traditional chemistry and atomic physics). This mesoscopic size regime has previously been unexplored and beckons the researcher with visions of a scientific Wild Wild West where opportunities abound for those willing to pack their wagons and head into this scientific hinterland.

Nanoscale physics, chemistry, biology, and engineering ask basic, yet unanswered, questions such as how the optical and electrical properties of materials evolve from those of individual atoms or molecules. Other questions being asked include the following:

- How does one actually go about making a nanometer-sized object?
- How does one make many such (identical) nanometer-sized objects?
- How do their optical and electrical properties change with "dimensionality"?
- How do charges move in such nanoscale systems?
- Do these materials possess new and previously undiscovered properties?
- Finally, are they useful?

The transition to nanotechnology begins with this last question, where we begin to ask how these new materials might improve our daily lives. Venture-capital-funded startups have therefore taken up this challenge, with many trying to apply nanoscale materials in products ranging from better sunscreen lotions to fluorescent labels for biological imaging applications to next-generation transistors and memory elements able to store the entire content of the Library of Congress on the head of a pin. More established companies, such as General Electric, Hewlett-Packard, Lucent, and IBM, have also started their own in-house nano programs to revolutionize consumer lighting, personal computing, data storage, and so forth. Thus, whether it be for household lighting or consumer electronics, a nano solution exists and there is very likely a company or person pursuing this vision of a nano future. So what is nano? This text tries to answer the question by explaining the underlying physical and quantitative concepts behind why such small, nanoscale, materials are so interesting and so potentially useful.

1.2 OVERVIEW

An outline of the text follows. First in Chapter 2, we discuss the composition of solids to introduce to the reader common crystal structures found in nanomaterials. Solids come in a number of forms, from amorphous to polycrystalline to crystalline. Much of nanoscience and nanotechnology focuses on nanometer-sized crystalline solids—hence the emphasis on understanding crystal structures. **Figure 1.1** illustrates a high-resolution transmission electron micrograph of two isolated cadmium selenide (CdSe) nanocrystals (left). Apparent in the image are lattice fringes associated with the regular columns of atoms making up both particles. On the right is a high-resolution image of a branched CdSe nanowire. Again, the lattice fringes present stem from the regular arrays of atoms making up the wire. Scale bars in both micrographs provide a sense of how small these structures truly are.

Chapter 2 also illustrates the increase in surface-to-volume ratio of nanomaterials over that of comparable bulk systems. This is because in nanostructures up to 50% of atoms lie at the surface. By contrast, such numbers are typically much smaller in counterpart bulk solids. The surface is therefore a potentially important property of nano systems, which helps dictate their optical and electrical properties. Furthermore, this large surface-to-volume ratio is critically important for applications such as catalysis or in photovoltaics. Finally, the concept of a crystal structure and the underlying (regular) periodic potential, due to the ordered arrangement of atoms, is central to the concept of electronic bands, a topic that we will discuss in Chapter 10.

Next, Chapter 3 discusses length scales to put into perspective the actual physical dimensions relevant to nano. Although being nanometer-sized is often considered the essence of nano, relevant lengths (from an optical and electrical standpoint) are actually determined by comparison to the natural dimensions of electrons and holes within bulk solids. These natural length scales can either be referred to by the de Broglie wavelength or by the bulk exciton Bohr radius. Thus, while a given nanometer-sized object of one system may formally qualify as being nano, a similar-sized object of another may not.

Chapter 4 then introduces us to three generic nanostructures, referred to as quantum wells, quantum wires (also called nanowires), and quantum dots (also called nanocrystals). Associated terminology is also

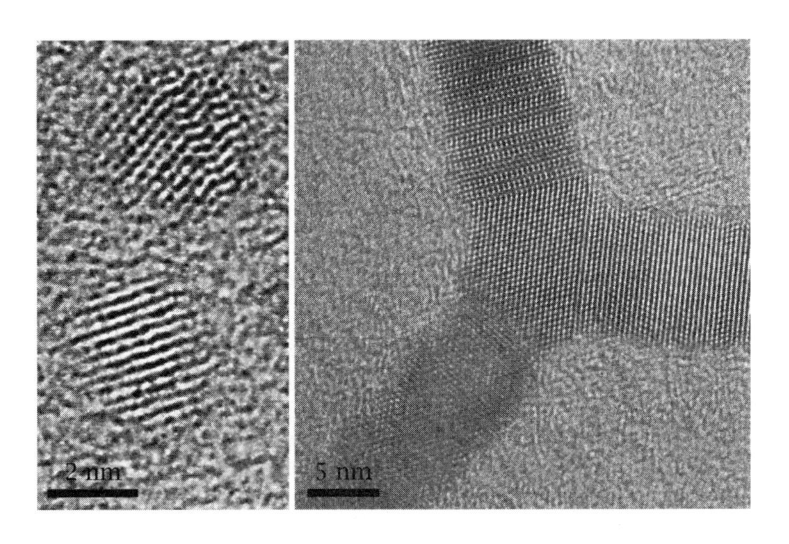

Figure 1.1 Transmission electron micrograph of two CdSe quantum dots (left) and a branched tripod nanowire (right).

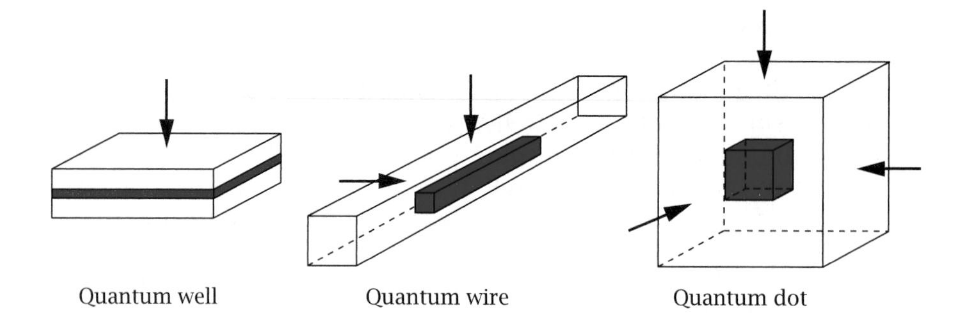

Figure 1.2 Cartoon illustrating carrier confinement along one, two, and three dimensions (left to right). This is representative of quantum wells, quantum wires, and quantum dots.

introduced, since quantum wells are often referred to as 2D (two-dimensional) systems while quantum wires and quantum dots are called 1D and 0D systems, respectively. These numbers refer to the dimensionality of the material and indicate the degrees of "freedom" carriers have. For example, in quantum wells, the motion of electrons and holes is restricted along one direction of the material by its physical size. However, motion along the other two directions remains unhindered. Hence the material is said to have two degrees of freedom while exhibiting one degree of confinement. This same analogy can be carried over to wires and dots, with a corresponding decrease in the available degrees of freedom. **Figure 1.2** provides a cartoon visual of these wells, wires, and dots, with arrows showing the increasing degrees of confinement from one system to the next. The solid regions represent the systems of interest with the surrounding regions representing an insulating host.

Without going into a detailed quantitative description about their absorptive and emissive properties, Chapter 5 introduces simple phenomenological approaches that describe these properties of wells, wires, and dots. The exponential attenuation law, associated absorption coefficients, and corresponding absorption cross sections are introduced. The kinetics of a model two-level system are then described to introduce radiative and nonradiative relaxation processes. Finally, Einstein A and B coefficients are used to relate absorption coefficients to a system's emission lifetime.

In Chapter 6, we review basic concepts underlying quantum mechanics. This ultimately enables us to introduce in Chapter 7 simple quantum mechanical models of a particle in a one-dimensional, two-dimensional, or three-dimensional "box." Such models provide a quantitative basis for what we mean by confined carriers. The importance of quantum confinement and why it is commonly associated with nano stems from the fact that bulk materials generally exhibit continuous absorption spectra. By contrast, when nanoscale materials have physical sizes equivalent to or less than the bulk exciton Bohr radius (or associated de Broglie wavelength), the system's optical and electrical spectra become discrete and atomic-like. In the limiting case of quantum dots, confinement occurs along all three physical dimensions. Their optical and electrical spectra therefore become truly atomic-like. This is one reason why quantum dots are often referred to as artificial atoms.

In all cases, the energies at which confined materials absorb and emit light are size-dependent. As a consequence, one can have tunable absorption and emission energies from a single material by simply varying its physical size. Herein lies the beauty of nanoscale objects.

Figure 1.3 illustrates this property through size-dependent absorption and emission spectra from a small size series of CdSe quantum dots

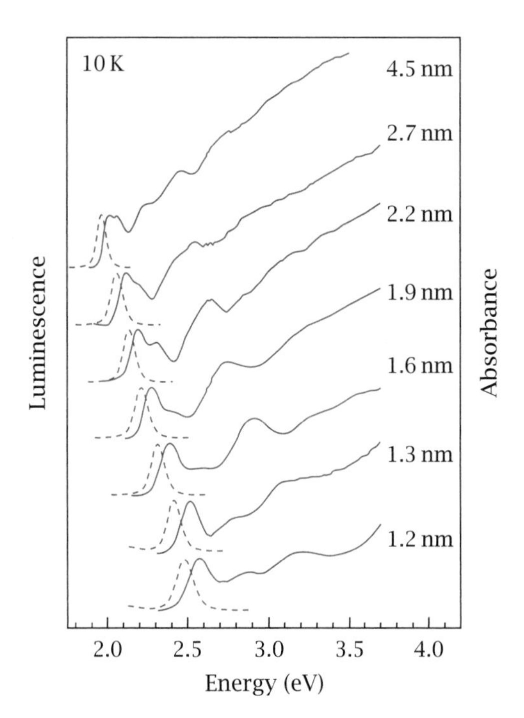

Figure 1.3 Size-dependent absorption and emission spectra of colloidal CdSe quantum dots.

(accompanying numbers represent the radius of the particles). In the plot, several features stand out. First, in the absorption spectra (solid lines), there are peaks that correspond to discrete transitions between unique electronic states. Any broadening of these transitions predominately originates from the finite size distribution of the ensemble (i.e., not all particles are the same size). Next, accompanying the absorption at lower energies is the sample's emission (dashed lines), which tracks the absorption edge. Finally, both the absorption and emission exhibit clear size-dependent trends. Namely, small dots absorb/emit blue or blue/green light while larger dots absorb/emit red light.

Analogies comparing the particle in a one-dimensional box with a quantum well, the particle in a two-dimensional box with a quantum wire, and the particle in a three-dimensional box with a quantum dot are only part of the story. In this regard, our discussion of quantum confinement for the one-dimensional box in Chapter 7 (and extensions of it in Chapter 8) only provides a description of quantum well electronic states and their associated energies along a single confined direction. However, there exist other electronic states associated with the remaining two degrees of freedom. Likewise, in the case of a quantum wire, solving the quantum mechanical problem of a particle in a two-dimensional box only provides descriptions of electronic states associated with the two confined dimensions. The calculation ignores states linked to the remaining degree of freedom.

To better account for these additional states, Chapter 9 introduces the concept of a density of states (DOS). In this manner, one can derive an expression that, while not describing a single state, accounts for all of them at the same time over a given energy range. This DOS argument is subsequently applied to both the valence band and conduction band of a given material, since absorption/emission transitions occur between them. When considered together, this leads to what is called the joint density of states (JDOS). The JDOS is, in turn, related to the absorption coefficient of a material, introduced earlier in Chapter 5.

The concept of a band is more fully developed in Chapter 10. Specifically, bands in metals, semiconductors, and insulators arise from the periodic potential experienced by electrons in a crystal due to the ordered and repeated arrangement of constituent atoms. To describe the evolution of electronic states into a band, the simple quantum mechanical model of a particle in a periodic potential (the Kronig–Penney model) is introduced. It utilizes a number of concepts described earlier in Chapter 6. In addition, two other models—the tight binding model and the nearly free electron model—are introduced. They represent limiting cases for how strongly or loosely bound carriers are to their parent atoms and can also be used to describe the evolution of bands in a solid.

The occupation of these bands by carriers allows us to distinguish metals from semiconductors, insulators, and semimetals. In particular, metals have "full" conduction bands while semiconductors and insulators have "empty" ones. Furthermore, in semiconductors and insulators, there exist a range of energies between the valence and conduction bands that cannot be populated by carriers. This forbidden range of energies is referred to as the band gap and is important for optoelectronic applications. For example, the magnitude of the band gap determines the color of light absorbed and emitted by a given semiconductor.

Next, to provide more quantitative descriptions of these absorption and emission events, Chapter 11 introduces time-dependent quantum mechanics. This subsequently provides us the foundation for calculating transition probabilities as well as transition rates for the absorption of light by semiconductors in Chapter 12. These latter rates can then be connected to the absorption coefficient of a material and allow us to tie together the phenomenological absorption coefficients (or absorption cross sections) discussed earlier in Chapter 5 with the underlying transition probabilities/rates derived in this section.

Finally, we turn to three topics that begin the transition from nanoscience to nanotechnology. Chapter 13 describes current methods for making nanoscale materials. Our particular focus is on the synthesis of colloidal nanostructures. Next, once such nanostructures have been made, tools are needed to study them. Chapter 14 introduces some of the classical techniques used to characterize nanostructures. These include transmission electron microscopy (TEM), scanning electron microscopy (SEM), atomic force microscopy (AFM), and scanning tunneling microscopy (STM). Finally, Chapter 15 goes over some representative applications of quantum dots, quantum wires, and quantum wells.

It is our hope that this text will provide the reader with a coherent, pedagogical, starting point for understanding key *quantitative* aspects of current nanoscience.

1.3 FURTHER READING

There are a number of general references on nanoscience and nanotechnology that the reader may find useful in addition to this text, including the books by Lindsay (2010), Ozin and Arsenault (2008), Poole and Owens (2003), and Ratner and Ratner (2003), as well as the compilation from *Scientific American* (2002).

1.4 THOUGHT PROBLEMS

1.1 Fact, fiction, and hype

Read the following articles:

- Richard Feynman. There's plenty of room at the bottom. www.zyvex.com/nanotech/feynman.html.
- Eric Drexler. Machine phase nanotechnology. *Scientific American*, September 2001, p. 66.
- Richard Smalley. Of chemistry, love and nanobots. *Scientific American*, September 2001, p. 68.
- *C&EN* cover story debate on molecular assemblers. *C&EN* **81**, 37–42 (2003).

Provide a brief synopsis on what this debate is about and list/explain any/all of the physical and chemical concepts Smalley uses to counter Drexler's claims. Do outside reading if necessary. The intent here is to try and get you thinking about the physical laws that govern biology, chemistry and physics.

1.2 Terminology

Provide a brief definition/explanation for the following two terms you are likely to encounter in this text as well as in the nanoscience literature:

- "top down"
- "bottom up"

Give an example of manufacturing through either approach.

1.3 Public perception

Provide an example of nanoscience and/or nanotechnology in popular culture. This could be a reference in a movie, a book, or a commercial. It could also be an actual product.

1.4 Media

Public perception of nanoscience and nanotechnology is, to a large extent, shaped by the media. Find examples of recent news articles covering nanotechnology. Assess whether nanoscience/nanotechnology is portrayed in a positive or negative light. Is the article biased towards one view? Has the reporter done sufficient background work to provide critical coverage of the issue?

1.5 Industry

Provide examples of some companies that are currently producing nano-related products.

1.6 A comparison of the field

Compare the scale of nano research in different countries. One can make a comparison based on published government funding estimates as well as private sector funding estimates. Alternatively, one can compare the number of scientific papers containing the keyword "nano" published each year by different countries.

1.7 Nanoscience education

Before 1998, there were very few university programs dedicated to nanoscience. Today, there are a number of institutions that have nanoscience or nanotechnology degree programs. Find a few of these programs and discuss what subject matter they believe is relevant towards achieving a nanoscience and/or nanotechnology degree.

1.8 Running the numbers

Assume Drexler is right and one "assembler" can assemble a million atoms a second. Roughly how long would it take for one of these assemblers to put together a golf ball. Make assumptions where necessary. State your assumptions explicitly. The idea here is to get you used to "running the numbers."

1.9 Scientific fraud

Just like any "hot" field, there is always the risk of scientific fraud. As an example, consider the case involving Hendrik Schön in the area of molecular electronics (note that some consider this to be part of nanoscience). Read about the Schön case in Service (2002). The post-investigation report (Beasley et al. 2002)—known as the Beasley Report—is quite illuminating. Along these lines you may also find the book *Plastic Fantastic* (Reich 2009) interesting. Discuss the likelihood of scientific fraud in the nanoscience/nanotechnology field and the mechanisms in place to prevent this. How adequate do you think they are?

1.10 Health risks

One of the major concerns these days with nanoscience and nanotechnology has to do with health risks. There are those who think that all nano research should be stopped until its health risks have been fully evaluated. Provide a brief summary and overview of the debate and weigh the pros and cons of each side. Provide all suitable literature references and examples.

1.11 Ethical implications

What are some of the broader moral and ethical implications of nanoscience and nanotechnology?

1.12 Nanoscience future trends

Nanoscience, like every other field, evolves. This is especially noticeable when it comes to research funding. Explore where the field is going by comparing the themes that were popular ten years ago with what is popular today and presumably tomorrow.

1.5 REFERENCES

The following are general references on nanoscience and nanotechnology that the reader may consult in addition to this text:

Beasley MR (Chair), Datta S, Kogelnik H, Kroemer H, Monroe D (2002) *Report of the Investigation Committee on the Possibility of Scientific Misconduct in the Work of Hendrik Schön and Coauthors.* Lucent Technologies Report (distributed by the American Physical Society). Available at publish.aps.org/reports/lucentrep.pdf.

Lindsay SM (2010) *Introduction to Nanoscience.* Oxford University Press, Oxford, UK.

Ozin GA, Arsenault AC (2008) *Nanochemistry*, 2nd edn. RSC Publishing, Cambridge, UK.

Poole CP, Owens FJ (2003) *Introduction to Nanotechnology.* Wiley, Hoboken, NJ.

Ratner M, Ratner D (2003) *Nanotechnology. A Gentle Introduction to the Next Big Idea.* Prentice Hall, Upper Saddle River, NJ.

Reich ES (2009) *Plastic Fantastic: How the Biggest Fraud in Physics Shook the Scientific World.* Palgrave Macmillan, New York.

Scientific American (2002) *Understanding Nanotechnology.* Warner Books, New York.

Service RF (2002) Bell Labs fires star physicist found guilty of forging data. *Science* **298**, 30.

Structure

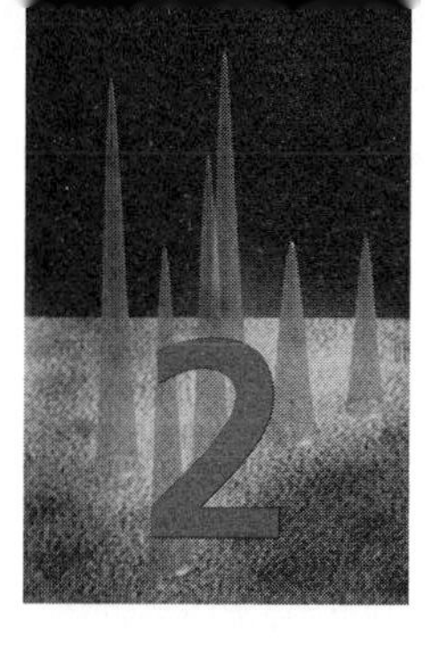

2.1 INTRODUCTION

Solids generally appear in three forms: amorphous, polycrystalline, and crystalline. Briefly, amorphous solids such as glass or plastics lack long-range order in the arrangement of their constituent atoms. Conversely, single crystals such as quartz consist of atoms arranged in a regular, periodic manner and hence exhibit long-range order. In between are polycrystalline solids that consist of multiple regions, called grains, where the atoms within a grain are ordered but where the grains themselves exhibit disorder. Since the nanostructures we will be concerned with in this text are crystalline metal or semiconductor nanocrystals, nanowires, and quantum wells, having a basic perspective for how elements arrange themselves within these low-dimensional systems is important. In this regard, the crystal structure is responsible for many properties of a material, ranging from its density to its electronic spectra. We therefore provide a brief overview about the fundamental properties of crystals. Note that our discussion is not meant to be comprehensive, and for more details the interested reader is encouraged to consult the books by Ashcroft and Mermin (1976), de Graef and McHenry (2007), and Kittel (2005).

2.2 BASIC PROPERTIES

Atoms in a crystal are generally pictured as being arranged on an imaginary lattice in space. Crystallography has established seven different crystal systems: cubic, tetragonal, rhombohedral, hexagonal, monoclinic, triclinic, and orthorhombic. **Table 2.1** lists their defining angles and cell length parameters, which, among other things, enable us to distinguish one system from another. The meaning of these unit-cell lengths as well as angles will become clearer in what follows when we define a unit-cell coordinate system. For now, just note the differences. Note also that more specialized structures exist within each of these seven crystal systems, resulting in 14 three-dimensional Bravais lattices. They are illustrated in **Figure 2.1**, with **Table 2.2** providing their corresponding volumes.

The consecutive repetition of these unit cells into larger three-dimensional assemblies then enables one to begin constructing a crystal. However, this alone does not generate a crystal, since an atomic basis is required. To construct an actual crystal, then, individual atoms (or groups of atoms) must be hung off the unit cell lattice points much like Christmas tree ornaments. These individual atoms (or groups of atoms) are called the basis, and the endless repetition of these basis atom(s), tied to a unit cell, allows one to construct a full crystal of interest.

Table 2.1 Crystal system parameters. Angles and lengths are defined in Section 2.2.1

Crystal system	Unit cell lengths	Angles
Triclinic	$a \neq b \neq c$	$\alpha \neq \beta \neq \gamma$
Monoclinic	$a \neq b \neq c$	$\alpha = \gamma = 90°, \beta \neq 90°$
Orthorhombic	$a \neq b \neq c$	$\alpha = \beta = \gamma = 90°$
Tetragonal	$a = b \neq c$	$\alpha = \beta = \gamma = 90°$
Rhombohedral	$a = b = c$	$\alpha = \beta = \gamma \neq 90°$
Cubic	$a = b = c$	$\alpha = \beta = \gamma = 90°$
Hexagonal	$a = b \neq c$	$\alpha = \beta = 90°, \gamma = 120°$

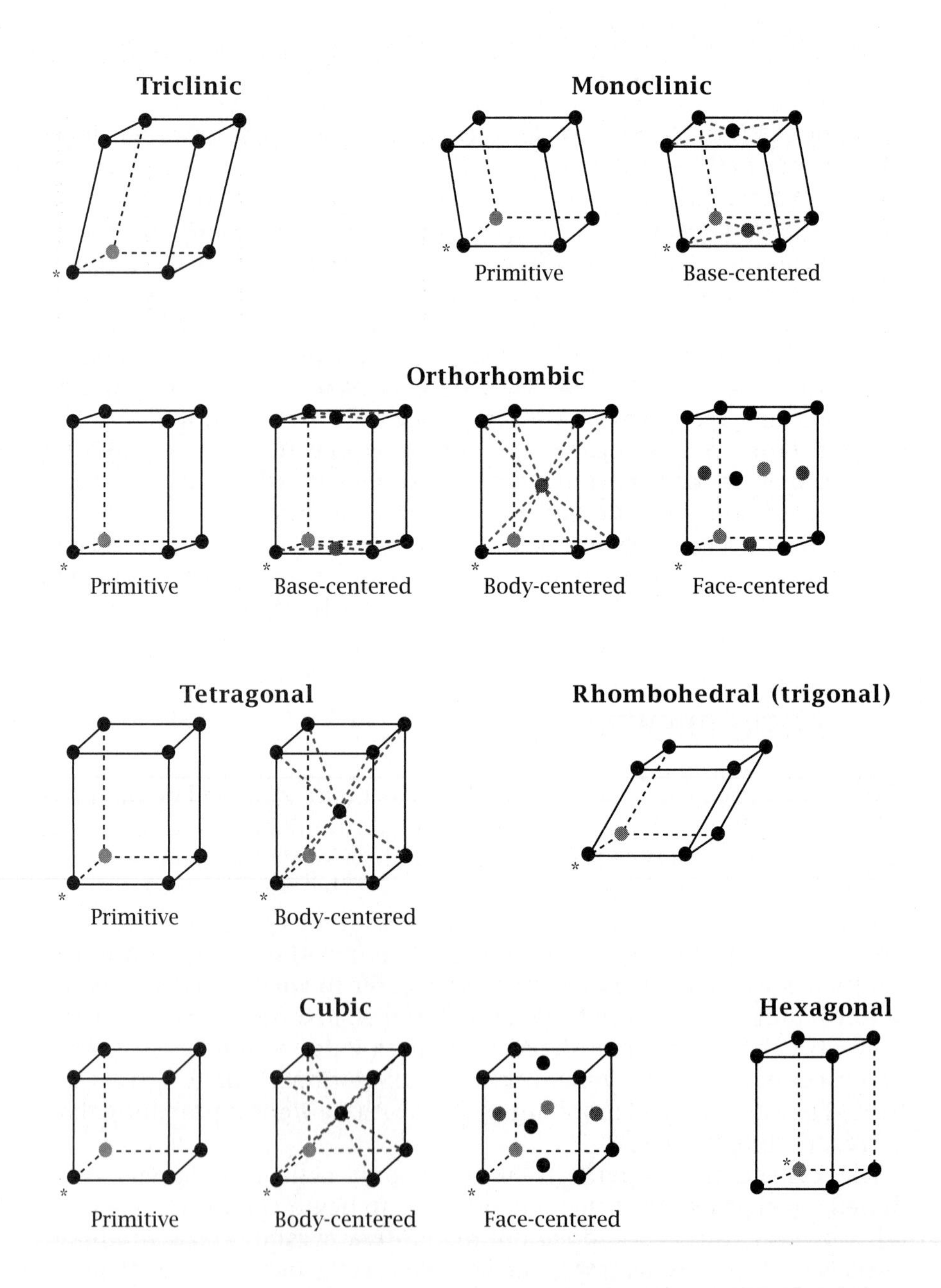

Figure 2.1 Illustration of the 14 Bravais lattices. Asterisks denote the origin of each unit cell.

In the simplest case, the basis consists of a given atom with each positioned directly over a lattice point (top, **Figure 2.2**). However, it is also common to have bases consisting of multiple atoms and even of different elements, especially when dealing with binary, ternary, or

Table 2.2 Unit-cell volumes. Angles and lengths are defined in Section 2.2.1

Crystal system	Unit-cell volume
Triclinic	$abc\sqrt{1-\cos^2\alpha-\cos^2\beta-\cos^2\gamma+2\cos\alpha\cos\beta\cos\gamma}$
Monoclinic	$abc\sin\beta$
Orthorhombic	abc
Tetragonal	a^2c
Rhombohedral	$a^3\sqrt{1-3\cos^2\alpha+2\cos^3\alpha}$
Cubic	a^3
Hexagonal	$\frac{1}{2}\sqrt{3}a^2c$

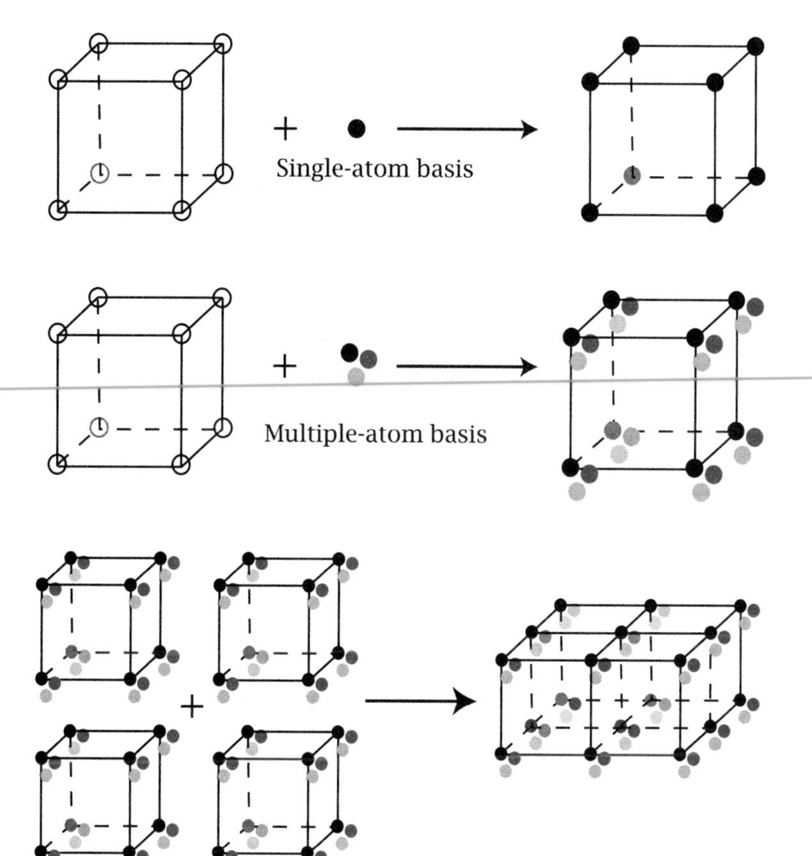

Figure 2.2 Illustration of a single-atom and a multiple-atom basis. The bottom illustration shows that the repeated arrangement of unit cells with a basis leads to a crystal of interest.

mixed-composition materials. This will be important when discussing semiconductors, since they often consist of different elements. Component atoms of the basis therefore do not necessarily sit atop a lattice point. Rather, the center, connecting the basis to the unit cell, could be an imaginary point in between constituent atoms. This concept is illustrated in **Figure 2.2** (middle and bottom) using a simple cubic unit cell with a multiple-atom basis.

2.2.1 Coordinate System

When referring to the 14 unit cells shown in **Figure 2.1**, as well as to their defining parameters in **Tables 2.1** and **2.2**, there exists an implicit coordinate system. We illustrate this in **Figure 2.3** (top) for the special case of a cubic lattice where the unit-cell lengths a, b, and c reside along the x, y, and z directions of a Cartesian coordinate system. Notice that a right-handed coordinate system is used. Furthermore, note that α, β,

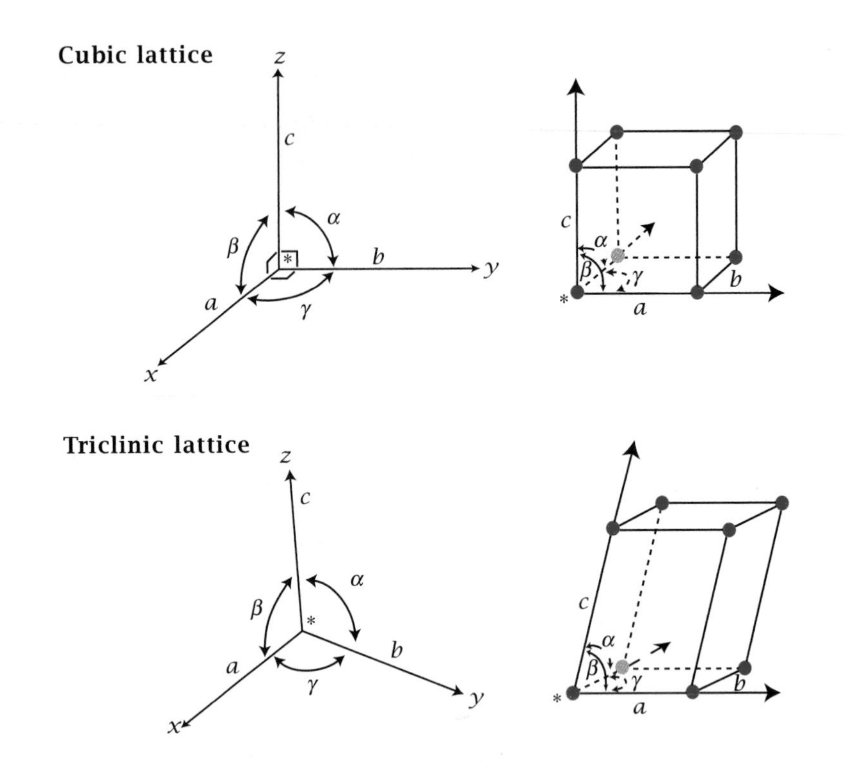

Figure 2.3 Illustration of the unit-cell coordinate system for a cubic and a triclinic lattice. Asterisks denote the unit-cell origin in each case.

and γ denote the angles between the axes defining the planes *normal* to a, b, and c, respectively.

More generally, unit cells exhibit nonorthogonal axes (see **Table 2.1**). Under these circumstances, the coordinate system is made *relative* to the actual unit-cell axes depicted in **Figure 2.1**. An example is shown in **Figure 2.3** (bottom) for a triclinic lattice, where a, b, and c refer to the lengths of the relative unit-cell "x," "y," and "z" axes. Similarly, α, β, and γ refer to the angles between (b and c), (a and c), and (a and b), respectively.

With this in mind, it should be possible to examine the 14 structures shown in **Figure 2.1** and verify that each indeed belongs to a given crystallographic family. The reader may do this as an exercise.

2.2.2 Atoms per Unit Cell

Now, given a classification system for unit cells, their coordinate system, and their defining angles and cell lengths, it is often convenient to know the number of atoms contained in a given unit cell. This will be especially useful when we estimate the total number of atoms in a nanostructure or even the total number of surface atoms it possesses. To illustrate, when referring to a complex, multielement structure, we might claim that it possesses a face-centered cubic unit cell with a two-atom basis. Subsequently, we might use this information to estimate the total number of atoms in the structure, the number of surface atoms it possesses, and even its surface-to-volume ratio.

The following rules therefore illustrate a general counting scheme for determining the number of atoms in a unit cell. They are applicable to all Bravais lattices shown in **Figure 2.1**. In this regard, as depicted in **Figure 2.4**,

- Atoms contained entirely within the volume of the unit cell are counted fully (i.e., as 1).
- Atoms localized on unit-cell corners are counted as $\frac{1}{8}$ of a full atom.

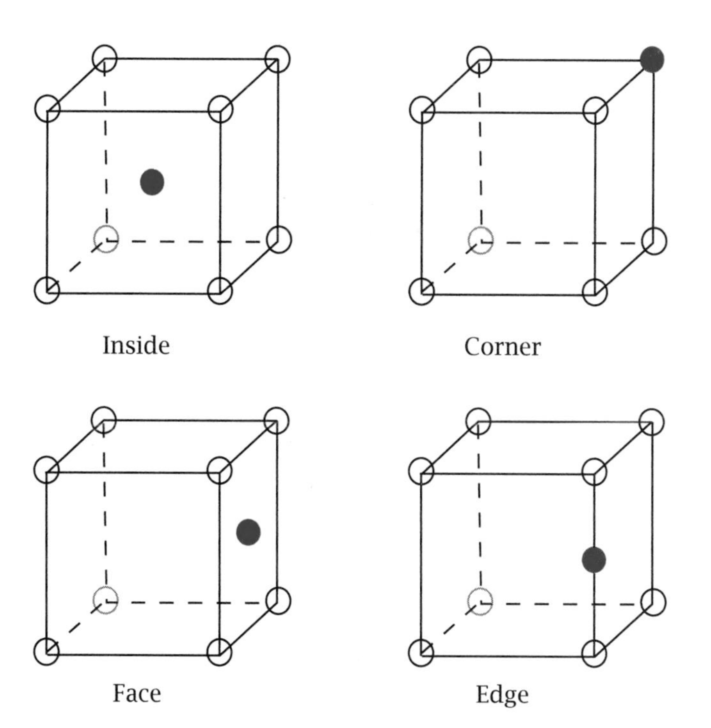

Figure 2.4 Illustration of atoms localized on the interior, corner, face, and edge of a unit cell, using the simple cubic lattice as an example. The solid sphere represents the atom of interest in each case.

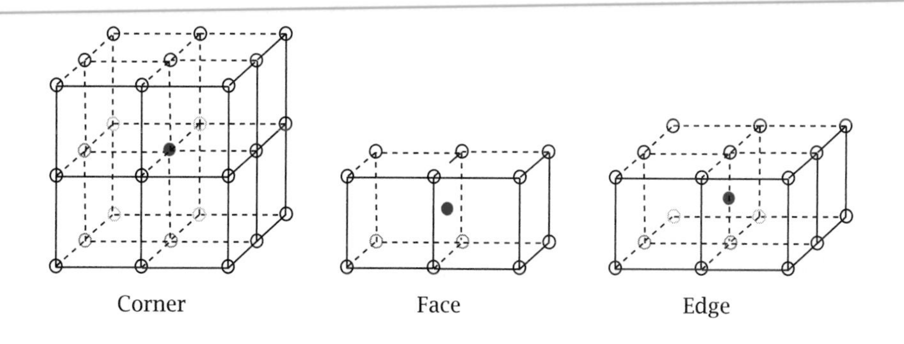

Figure 2.5 Illustration of atom sharing when localized on the corner, face, and edge of a unit cell. The solid sphere represents the atom of interest in each case.

- Atoms localized on unit-cell faces are counted as $\frac{1}{2}$ of a full atom.
- Atoms localized on unit-cell edges are counted as $\frac{1}{4}$ of a full atom.

This weighting scheme follows because when unit cells are repeated along all directions of an extended lattice, atoms on corners, faces, and edges are shared by other unit cells. Corner atoms are therefore common to eight unit cells. Face atoms are shared by two neighboring unit cells, while edge atoms are shared by four. Only internal atoms remain exclusive to a particular cell. These concepts are illustrated in **Figure 2.5**, where we demonstrate atom sharing along a unit cell corner, face, and edge.

2.3 EXAMPLES OF CRYSTAL STRUCTURES

We now list a few examples of crystal structures of interest to us. We begin with a discussion of single-element crystals and then move to more complicated multielement systems.

2.3.1 Single-Element Crystals

Single-element crystals represent the easiest examples of actual crystals, since the resulting lattice plus basis often resembles one of the 14 Bravais lattices shown in **Figure 2.1**. In the case of metals, cubic lattices are important, particularly the face-centered cubic (FCC) and body-centered cubic (BCC) structures.

Simple Cubic (SC)

The simple cubic structure is not common in the systems we will generally encounter in this text. This has to do, in part, with its low overall packing density and its low atomic coordination number, making this atomic arrangement less favorable than others from an energetic standpoint. However, it has been adopted by some systems, especially when a multiple-atom basis is involved. For the current case, the number of atoms per unit cell is 1; we can see this by noting that the simple, single-element, cubic structure has 8 corner atoms.

In summary, the SC structure has the following important parameters:

- Lattice: SC
- Atoms/unit cell: 1

Face-Centered Cubic (FCC)

A number of elements crystallize with the FCC structure. Many of them are routinely encountered in the literature, including copper, silver, gold, aluminum, palladium, and platinum. (Note that an alternative name for the face-centered cubic structure is cubic close-packed (CCP).) This is therefore an important structure, since many metallic nanoparticles and nanowires possess this lattice.

The number of atoms per unit cell is 4, which we can verify using our counting scheme. Specifically, an examination of the FCC structure shows that it has 8 corner atoms and 6 face atoms. The total number of atoms per unit cell is therefore $8\left(\frac{1}{8}\right) + 6\left(\frac{1}{2}\right) = 4$.

In summary, the FCC structure has the following important parameters:

- Lattice: FCC
- Atoms/unit cell: 4

Note that alternatively we can view the FCC lattice as SC with a 4-atom basis as follows:

- Lattice: SC
- Basis, first-atom coordinate $(0, 0, 0)$
- Basis, second-atom coordinate $\left(\frac{1}{2}a, \frac{1}{2}a, 0\right)$
- Basis, third-atom coordinate $\left(\frac{1}{2}a, 0, \frac{1}{2}a\right)$
- Basis, fourth-atom coordinate $\left(0, \frac{1}{2}a, \frac{1}{2}a\right)$

In all cases, a represents the unit-cell lattice constant.

Body-Centered Cubic (BCC)

This lattice is not as commonly seen as the FCC structure. However, important elements such as iron crystallize with it. The number of atoms per BCC unit cell is 2. This can be verified using our counting scheme. Namely, an examination of the BCC structure shows that it possesses 1 interior atom and 8 corner atoms. The number of atoms per unit cell is therefore $1(1) + 8\left(\frac{1}{8}\right) = 2$.

In summary, the BCC structure has the following important parameters:

- Lattice: BCC
- Atoms/unit cell: 2

Note that alternatively we can view the BCC lattice as SC with a 2-atom basis as follows:

- Lattice: SC
- Basis, first-atom coordinate $(0, 0, 0)$
- Basis, second-atom coordinate $\left(\frac{1}{2}a, \frac{1}{2}a, \frac{1}{2}a\right)$

with a the lattice constant of the unit cell.

The Diamond Structure, FCC with a multiple-atom basis

Another lattice commonly encountered in the literature is the diamond structure. This arrangement is generally adopted by elements that form strong covalent bonds, resulting in a tetrahedral bonding geometry. Common elements that crystallize with this structure include carbon, silicon, and germanium. All are technologically important materials.

Figure 2.6 provides an illustration of the diamond lattice conventional unit cell. The structure differs from its FCC and BCC counterparts primarily because it has a multiple-atom basis, even though all of its atoms consist of the same element. Therefore, it does not immediately resemble any of the 14 Bravais lattices seen in **Figure 2.1**. However, the diamond structure has an underlying FCC lattice with a two-atom basis. Note that another way to visualize this structure is by imagining two interpenetrating FCC lattices, offset from one other by $\left(\frac{1}{4}a, \frac{1}{4}a, \frac{1}{4}a\right)$ in the cubic coordinate system shown in **Figure 2.3**.

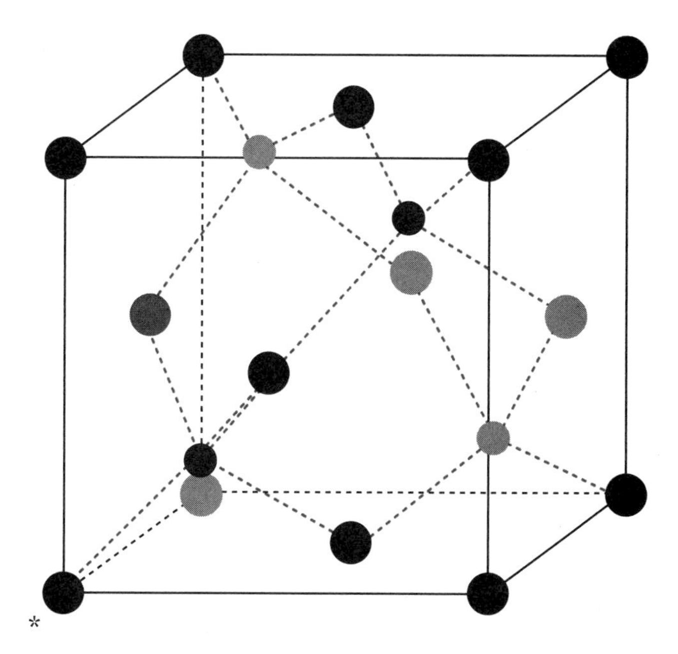

Figure 2.6 Illustration of the diamond structure. The asterisk denotes the unit-cell origin and the degree of shading of the atoms represents their depth within the page. Lighter-colored atoms are further away from the reader.

The number of atoms per unit cell is 8, since an examination shows that the diamond structure has 8 corner atoms, 4 interior atoms, and 6 face atoms.

In summary, the diamond structure has the following important parameters:

- Lattice: FCC
- Atoms/unit cell: 8
- Basis, first-atom coordinate $(0, 0, 0)$
- Basis, second-atom coordinate $\left(\frac{1}{4}a, \frac{1}{4}a, \frac{1}{4}a\right)$

2.3.2 Compound, Multiple-Element, Crystals

In the case of binary compounds, such as the III–V and II–VI semiconductors, things become a little more complicated. One does not have the benefit of a conventional unit cell plus a single-atom basis, leading to structures resembling any of the 14 standard Bravais lattices shown in **Figure 2.1**. Instead, these structures have multiple-atom/multiple-element bases with historical names such as the NaCl structure, the ZnS structure, and the CsCl structure. **Figure 2.7** illustrates what they look like.

ZnS Lattice

The ZnS lattice is also referred to as the zinc blende (ZB) or sphalerite structure. Binary compounds that crystallize with this atomic arrangement include ZnS, GaAs, ZnTe, and CdTe. It is identical to the diamond structure seen earlier in the case of single-element crystals in that it has a FCC conventional unit cell. However, it possesses a

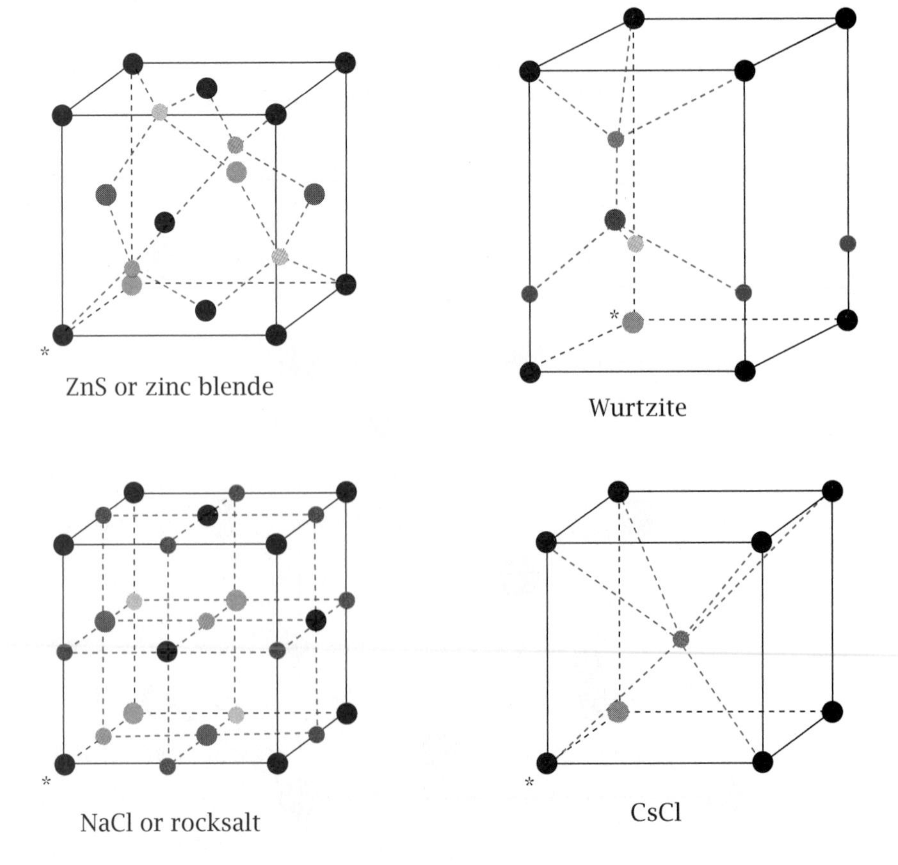

Figure 2.7 Illustration of the ZnS, wurtzite, NaCl, and CsCl unit cells. Asterisks denote the unit-cell origin in each case. Furthermore, atoms of different elements are depicted with different-sized spheres. Shading is used to illustrate distance from the reader, with lighter-colored atoms further into the page.

two-atom/two-element basis with atoms located at $(0, 0, 0)$ (first element) and $\left(\frac{1}{4}a, \frac{1}{4}a, \frac{1}{4}a\right)$ (second element). Note that, just as with the diamond unit cell, it is possible to visualize the ZnS structure as two interpenetrating FCC lattices offset by $\left(\frac{1}{4}a, \frac{1}{4}a, \frac{1}{4}a\right)$.

The ZnS structure contains 8 atoms per unit cell, of which 4 belong to the first element and 4 to the second. This can be seen in **Figure 2.7**, where 4 atoms of one element are located completely inside the unit cell. The remaining 4 atoms reside as 8 corner atoms and 6 face atoms.

In summary, the ZnS structure has the following important parameters:

- Lattice: FCC
- Atoms/unit cell: 8
- Basis, first-element coordinate $(0, 0, 0)$
- Basis, second-element coordinate $\left(\frac{1}{4}a, \frac{1}{4}a, \frac{1}{4}a\right)$

Wurtzite Lattice

The wurtzite structure is the compound material version of the single-element hexagonal close-packed structure. Compounds that crystallize with this arrangement include CdS and CdSe. The underlying lattice is hexagonal with a four-atom/two-element basis. These atoms are located at $(0, 0, 0)$ and $\left(\frac{2}{3}a, \frac{1}{3}a, \frac{1}{2}c\right)$ for the first element and $(0, 0, \delta)$ and $\left(\frac{2}{3}a, \frac{1}{3}a, \frac{1}{2}c + \delta\right)$ for the second element, where δ is just an offset along the z direction of the unit cell. The unit cell is shown in **Figure 2.7** and contains four atoms per unit cell. They exist as 8 corner atoms and 1 interior atom of the first element along with 4 edge atoms and 1 interior atom of the second.

In summary, the wurtzite crystal structure has the following important parameters:

- Lattice: hexagonal
- Atoms/unit cell: 4
- Basis, first-element coordinate $(0, 0, 0)$
- Basis, first-element coordinate $\left(\frac{2}{3}a, \frac{1}{3}a, \frac{1}{2}c\right)$
- Basis, second-element coordinate $(0, 0, \delta)$
- Basis, second-element coordinate $\left(\frac{2}{3}a, \frac{1}{3}a, \frac{1}{2}c + \delta\right)$

NaCl Lattice

The NaCl structure is also called the rocksalt structure. Examples of materials that crystallize this way include PbS, PbSe, and PbTe. These are all interesting materials from a thermoelectric or photovoltaic standpoint. In all cases, the underlying lattice is FCC with a two-atom/two-element basis. These atoms are located at $(0, 0, 0)$ for the first element and $\left(\frac{1}{2}a, \frac{1}{2}a, \frac{1}{2}a\right)$ for the second element. Note that the NaCl structure can alternatively be viewed as two interpenetrating FCC lattices offset by $\left(\frac{1}{2}a, \frac{1}{2}a, \frac{1}{2}a\right)$.

The NaCl structure contains 8 atoms per unit cell, with 4 atoms from the first element and 4 from the second. This can be seen in **Figure 2.7**, where there are 8 corner and 6 edge atoms of the first element. In tandem, there are 12 edge atoms and 1 interior atom from the second.

In summary, the NaCl structure has the following important parameters:

- Lattice: FCC
- Atoms/unit cell: 8
- Basis, first-element coordinate $(0, 0, 0)$
- Basis, second-element coordinate $\left(\frac{1}{2}a, \frac{1}{2}a, \frac{1}{2}a\right)$

CsCl Lattice

Finally, the CsCl structure is the compound material version of the single-element BCC unit cell. The underlying lattice is simple cubic with a two-atom/two-element basis. The first basis atom is located at $(0, 0, 0)$ and the second can be found at $\left(\frac{1}{2}a, \frac{1}{2}a, \frac{1}{2}a\right)$. Two atoms exist per unit cell. This can be seen in **Figure 2.7**, where it is apparent that there are 8 corner atoms of the first element and 1 interior atom of the second.

In summary, the CsCl structure has the following important parameters:

- Lattice: SC
- Atoms/unit cell: 2
- Basis, first-element coordinate $(0, 0, 0)$
- Basis, second-element coordinate $\left(\frac{1}{2}a, \frac{1}{2}a, \frac{1}{2}a\right)$

2.4 MILLER INDICES

At this point, having established basic information about different unit cells and their atomic bases, and having described how the assembly of the lattice plus basis leads to actual crystals, we now describe a naming convention for different crystallographic planes and directions. In this regard, the Miller index system was developed as a means of denoting planes, families of symmetrically equivalent planes, crystallographic directions, and families of equivalent crystallographic directions in lattices. The notation is often encountered in the literature, for example, when describing nanowire growth directions, faces for the oriented attachment of crystals, crystal faces on which scanning probe experiments are conducted, and even directions in band structure calculations. Many other examples exist and the reader will no doubt encounter them in due course. Therefore, we briefly review this notation.

2.4.1 Miller Notation

In general, in Miller notation,

- (hkl) refers to a crystalline plane of atoms.
- $\{hkl\}$ refers to a family of equivalent atomic planes.
- $[hkl]$ refers to a given crystallographic direction.
- $\langle hkl \rangle$ refers to a family of equivalent crystallographic directions.

In all cases, h, k, and l are integers with rules for obtaining them described below.

Planes

To obtain Miller indices for a plane of atoms, we do the following (consider this a "recipe"):

- Take a desired plane and determine where it intersects the x, y, and z axes of the unit cell in terms of multiples of the lattice constant.
- Find the reciprocal of each intersection point and reduce all values to their lowest integer equivalent. These numbers are then associated with the indices h, k, and l along the x, y, and z directions of the lattice, respectively.
- Miller indices for the plane of interest are then expressed using the three integers in parenthesis without commas. For example, one writes (mnp) if the integers found are m, n, and p.
- Should one of the planes not intersect with an axis, use 0 as the index. For example, if the plane does not intersect the cell's x axis write $(0kl)$.
- Furthermore, if the intercept occurs on the negative side of an axis, relative to the unit cell origin, place a bar over the number to indicate a negative direction. For example, one might write $(m\bar{n}p)$ or, more specifically, $(1\bar{2}1)$, which is pronounced "one, two bar, one".

Figure 2.8 depicts this procedure carried out for a cubic lattice. The reader may verify that the above procedure leads to the Miller indices depicted there.

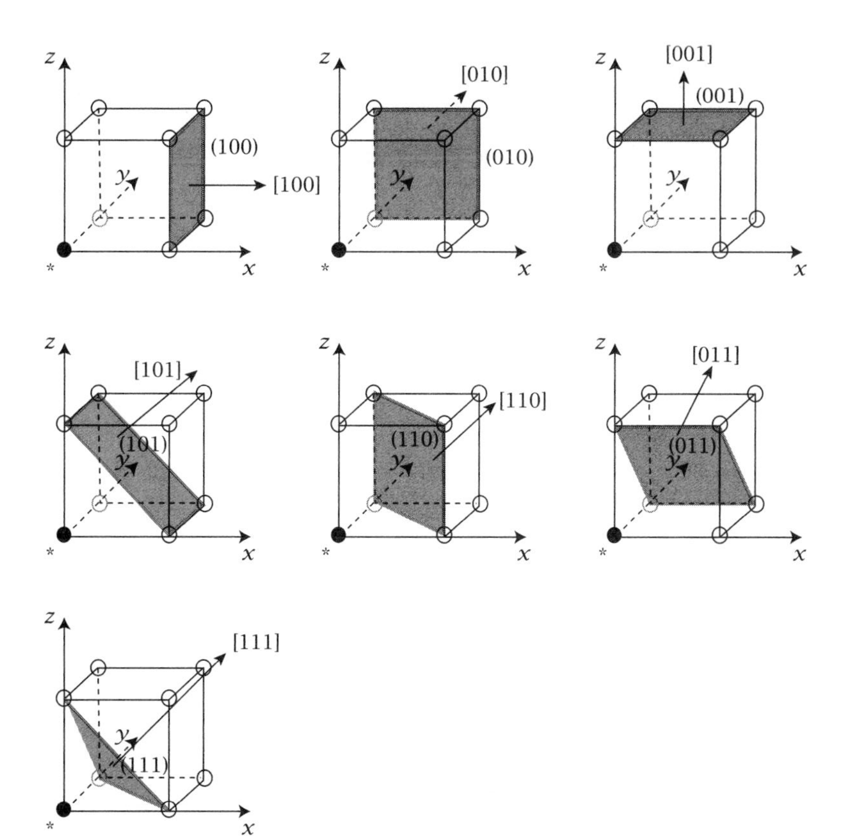

Figure 2.8 Illustration of Miller indices for the special case of a cubic lattice. Asterisks denote the unit-cell origin.

Equivalent Planes

Next, we illustrate the family of equivalent planes, $\{hkl\}$, for the case of a cubic lattice. Again, the emphasis is on cubic systems, since many nanostructures consist of elements that crystallize this way.

- For {100}, the following planes are symmetrically equivalent: (100), (010), (001), $(\bar{1}00)$, $(0\bar{1}0)$, and $(00\bar{1})$.
- Likewise, for {110}, the following planes are symmetrically equivalent: (110), (101), (011), $(\bar{1}10)$, $(\bar{1}01)$, $(0\bar{1}1)$, $(1\bar{1}0)$, $(10\bar{1})$, $(01\bar{1})$, $(\bar{1}\bar{1}0)$, $(\bar{1}0\bar{1})$, and $(0\bar{1}\bar{1})$.
- Finally, for {111}, the following planes are symmetrically equivalent: (111), $(\bar{1}11)$, $(1\bar{1}1)$, $(11\bar{1})$, $(\bar{1}\bar{1}1)$, $(1\bar{1}\bar{1})$, $(\bar{1}1\bar{1})$, and $(\bar{1}\bar{1}\bar{1})$.

Note that the number of equivalent planes in a family is called the multiplicity.

Directions

Crystallographic directions in a lattice are denoted by three integers within square brackets, $[hkl]$. To visualize this, simply construct a vector that starts from the unit-cell origin and ends at the point in space (within our unit cell coordinate system) described by the three coordinate integers. So [100] corresponds to a direction from the unit-cell origin along the x axis exclusively. Likewise, [001] corresponds to a direction aligned solely along the cell's z axis.

Note that in the special case of a cubic system, indices are shared between the vector normal to a given plane and the plane itself. For example, [100] is normal to (100). Likewise, [110] is normal to (110). However, this is not generally true, and **Figure 2.9** illustrates Miller

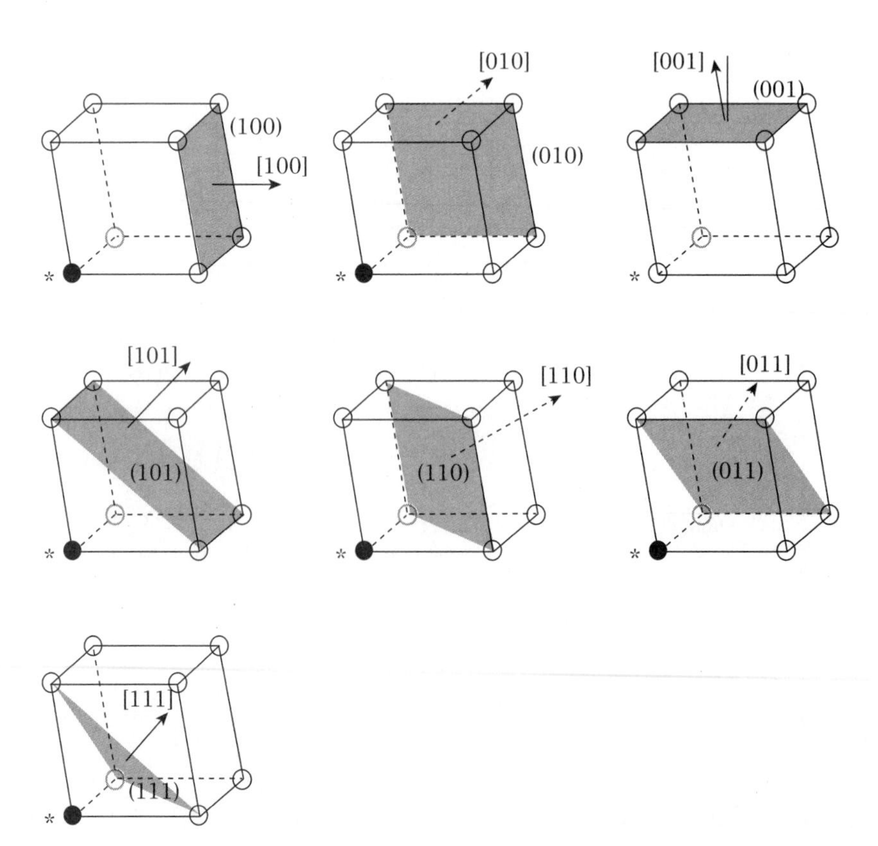

Figure 2.9 Illustration of Miller indices for the case of a monoclinic lattice. Asterisks denote the unit-cell origin.

notation for a monoclinic lattice, where close examination reveals that vectors are often not normal to their corresponding surfaces with the same *hkl* indices.

Equivalent Directions

We now illustrate the family of equivalent directions, $\langle hkl \rangle$, for a cubic lattice,

- For $\langle 100 \rangle$, the following directions are symmetrically equivalent: $[100]$, $[010]$, $[001]$, $[\bar{1}00]$, $[0\bar{1}0]$, and $[00\bar{1}]$.
- For $\langle 110 \rangle$, the following directions are symmetrically equivalent: $[110]$, $[101]$, $[011]$, $[\bar{1}10]$, $[\bar{1}01]$, $[0\bar{1}1]$, $[1\bar{1}0]$, $[10\bar{1}]$, $[01\bar{1}]$, $[\bar{1}\bar{1}0]$, $[\bar{1}0\bar{1}]$, and $[0\bar{1}\bar{1}]$.
- For $\langle 111 \rangle$, the following directions are symmetrically equivalent: $[111]$, $[\bar{1}11]$, $[1\bar{1}1]$, $[11\bar{1}]$, $[\bar{1}\bar{1}1]$, $[1\bar{1}\bar{1}]$, $[\bar{1}1\bar{1}]$, and $[\bar{1}\bar{1}\bar{1}]$.

Finally, note that a slightly different notation exists for hexagonal structures. Specifically, a four-index notation called Miller–Bravais notation is employed. It appears as $(hkil)$ or $[hkil]$, where the new integer is i. For brevity, we shall not go into this naming convention. The reader may consult the cited references for more details. Simply notice that the fourth integer is actually related to the others through $i = -(h + k)$.

2.5 SURFACE-TO-VOLUME RATIO

We now describe other aspects of nanostructures apart from their underlying lattice. Namely, a characteristic feature of nanoscale materials is their large surface-to-volume ratio R. This arises because the small size of nanostructures means that a large fraction of their component atoms reside on the surface. There are thus both good and bad consequences for the optical and electrical properties of such materials. For example, the benefits of an increased surface-to-volume ratio include the development of more efficient catalysts. At the same time, disadvantages include lower emission quantum yields, stemming from the presence of surface defects, which leads to the nonradiative recombination of carriers. The optical properties of nanostructures are described more thoroughly in later chapters.

In what follows, we demonstrate that the surface-to-volume ratio of nanostructures increases with decreasing size. Specifically, R follows an inverse power law (i.e., it varies as 1 over the defining size of the structure). This can already be seen qualitatively, since surface areas scale as L^2 while corresponding volumes scale as L^3. Their ratio then exhibits an L^{-1} dependence. Thus, it is apparent that as the size of the object decreases, R increases dramatically. Let us now consider a number of highly idealized model nanostructure geometries.

2.5.1 Plane

Our first model geometry is a flat square plane of finite thickness, which we shall use to represent a quantum well. (Quantum wells are introduced in Chapter 4. For now, we just assume that they are planar.) The total surface area of this structure is the sum of six face areas, where we assume the following dimensions: a length l along with an accompanying thickness a. If the structure is sufficiently thin, $l \gg a$, and the effective area presented to us is

$$S_{\text{plane}} \simeq 2l^2.$$

The corresponding volume V_{plane} of the structure is

$$V_{\text{plane}} = al^2.$$

We therefore obtain an effective surface-to-volume ratio of

$$R_{\text{plane}} = \frac{S_{\text{plane}}}{V_{\text{plane}}} \simeq \frac{2l^2}{al^2},$$

or

$$R_{\text{plane}} \simeq \frac{2}{a}. \tag{2.1}$$

R behaves as an inverse power law with the defining dimension of the structure, a.

2.5.2 Cylinder

Next, to represent a nanowire, we consider a cylinder as our rudimentary geometrical model. (Nanowires are introduced in Chapter 4. For now, just assume that they are cylindrical.) The cylinder's surface area is

$$S_{\text{cylinder}} = 2\pi al,$$

where a is the wire's radius and l is its length. The corresponding volume is

$$V_{\text{cylinder}} = \pi a^2 l,$$

and hence the surface-to-volume ratio is

$$R_{\text{cylinder}} = \frac{S_{\text{cylinder}}}{V_{\text{cylinder}}} = \frac{2\pi al}{\pi a^2 l} = \frac{2}{a}. \tag{2.2}$$

R is again an inverse power law with the defining dimension a of the wire.

2.5.3 Cube

In the case of a quantum dot, we employ a cube as our rudimentary geometrical model. (Quantum dots are introduced in Chapter 4. For now, assume that they possess this geometry.) The cube's surface area is

$$S_{\text{cube}} = 6a^2,$$

where a is the length of the cube along a given side. Its volume is

$$V_{\text{cube}} = a^3,$$

which leads to the following surface-to-volume ratio:

$$R_{\text{cube}} = \frac{S_{\text{cube}}}{V_{\text{cube}}} = \frac{6a^2}{a^3} = \frac{6}{a}. \tag{2.3}$$

We again obtain a power law with the critical dimension of the nanostructure.

2.5.4 Sphere

Alternatively, we can use a sphere as a more realistic geometrical representation of a quantum dot. In this case, the surface area is

$$S_{\text{sphere}} = 4\pi a^2,$$

with a corresponding volume

$$V_{\text{sphere}} = \frac{4}{3}\pi a^3.$$

From these two expressions, the surface-to-volume ratio is

$$R_{\text{sphere}} = \frac{S_{\text{sphere}}}{V_{\text{sphere}}} = \frac{4\pi a^2}{\frac{4}{3}\pi a^3} = \frac{3}{a}, \tag{2.4}$$

which exhibits power-law behavior, as expected.

In all cases, the surface-to-volume ratio of low-dimensional systems scales as 1 over the nanostructure's critical size.

2.5.5 A Comparison

Let us now illustrate the dramatic increase in a nanostructure's surface-to-volume ratio through an example. In this regard, since surfaces are useful for applications such as catalysis, there is often an incentive for making nanostructured systems. The hope is that the large increase in R will lead to marked improvements in catalytic activity for the same material, with the only difference being physical size.

In the specific case of dye-sensitized solar cells, such devices have become an important area of nanoscience since their initial discovery by O'Regan and Grätzel in 1991. More about Grätzel cells will be seen in Chapter 15 when we discuss applications of nanostructures. However, the basic idea involves creating an efficient, low-cost, solar cell using sintered titanium dioxide nanoparticles as substrates for light absorbing dye molecules. Upon excitation, these sensitizers inject electrons into the TiO_2 film, creating an electrical current that can be used to power a device. Now, part of the reason behind the device's success is that a large dye loading is possible when nanoparticle surfaces, as opposed to flat substrates, are used. In this respect, the device's light absorption efficiency increases since more dye molecules can be incorporated into the same amount of real estate.

We can model this surface area increase as follows. Let us compare the total surface area of a flat TiO_2 slab 1 cm × 1 cm × 25 nm in size with the total surface area of an equivalent number of 25 nm diameter TiO_2 nanoparticles that occupy the same volume. A simple geometrical analysis then shows that the volume of the flat surface is $V_{\text{surface}} = 25 \times 10^{-7}\ \text{cm}^3$ while the volume of an individual sphere is $V_{\text{sphere}} = 8.2 \times 10^{-18}\ \text{cm}^3$. As a consequence, $n_{\text{sphere}} = 3 \times 10^{11}$ particles occupy the same volume as the macroscopic slab.

Next, the total surface area of the flat substrate is

$$S_{\text{surface}} \simeq 2\ \text{cm}^2$$

while that of an individual sphere is

$$S_{\text{sphere}} = 1.96 \times 10^{-11}\ \text{cm}^2.$$

Although S_{sphere} is quite small, the *total* surface area of a nanoparticle ensemble is

$$S_{\text{tot}} = 5.89\,\text{cm}^2$$

(S_{sphere} is multiplied by 3×10^{11} total spheres). This therefore represents an approximate threefold increase in total surface area, achieved by simply reducing the size of the constituent material. Even more dramatic changes occur when the particle size decreases, further illustrating the potential usefulness of nanometer-sized materials.

2.5.6 An Estimate of the Number of Atoms in a Nanostructure Using a Unit-Cell Approach

We now end our discussion by demonstrating how knowing a nanostructure's crystal structure enables us to estimate the number of atoms present within it or on its surface. There are a number of approaches for carrying out the first task, and the one we will discuss involves using information about the crystal's underlying unit cell. An alternative approach uses its bulk density. However, both lead to the same answer, since they ultimately assume the same thing, namely, that there are no distortions of atoms in a nanostructure relative to its parent bulk lattice. This is sometimes not the case, but often enough the assumption holds.

Let us illustrate the approach using a spherical particle of radius r and possessing a cubic FCC unit cell with a lattice constant a. The volume of the particle is then

$$V_{\text{particle}} = \tfrac{4}{3}\pi r^3,$$

with a corresponding unit-cell volume of

$$V_{\text{unit}} = a^3.$$

The total number of unit cells (rounded to an appropriate integer) is

$$N = \frac{V_{\text{particle}}}{V_{\text{unit}}},$$

or

$$N = \frac{4}{3}\pi \left(\frac{r}{a}\right)^3. \tag{2.5}$$

Since we know the number of atoms present per unit cell (n_{unit}), the total number of atoms in the entire particle is

$$n_{\text{tot}} = n_{\text{unit}} N = n_{\text{unit}} \frac{4}{3}\pi \left(\frac{r}{a}\right)^3.$$

Thus, in the case of a 12.5 nm radius gold nanoparticle with an underlying FCC lattice ($n_{\text{unit}} = 4$) and with a lattice constant of $a = 4.08$ Å (0.408 nm), we find that the particle contains about 480 000 atoms. By comparison, a 5 nm radius gold nanoparticle has about 31 000 atoms.

2.5.7 An Estimate of the Number of Surface Atoms in a Nanostructure using a Unit-Cell Approach

It may also be of interest to determine the number of surface atoms in a nanostructure. Note again that there are no conventional approaches for this calculation, since a variety of estimates exist, each with its own advantages and disadvantages. Since we have just illustrated a unit-cell approach, let us continue with it here.

We again employ a spherical nanoparticle to illustrate the calculation. Analogous estimates can be conducted using other geometries. As before, we assume an underlying cubic unit cell with a lattice constant a. Our strategy involves calculating the volume of the particle's outer shell with a thickness a. Then, from a unit-cell perspective, let us call the outer layer of unit cells the particle's "surface."

The total volume of the particle is

$$V_1 = \tfrac{4}{3}\pi r^3,$$

while the associated volume of a slightly smaller sphere with radius $r - a$ is

$$V_2 = \tfrac{4}{3}\pi (r - a)^3.$$

The volume difference is the desired shell volume

$$\begin{aligned} V_{\text{shell}} &= V_1 - V_2 \\ &= \tfrac{4}{3}\pi [r^3 - (r - a)^3]. \end{aligned}$$

At this point, if we assume that $r \gg a$, we can approximate the term in square brackets as $r^3 - (r - a)^3 \simeq 3r^2 a$. We then have

$$V_{\text{shell}} \simeq 4\pi r^2 a.$$

The number of unit cells occupying the same volume as the outer shell is then

$$\begin{aligned} N &= \frac{V_{\text{shell}}}{V_{\text{unit}}} \\ &\simeq 4\pi \left(\frac{r}{a}\right)^2. \end{aligned}$$

Finally, the number of surface atoms is the product of N and the number of atoms per unit cell:

$$n_{\text{surface}} = n_{\text{unit}} N = n_{\text{unit}} 4\pi \left(\frac{r}{a}\right)^2.$$

In the case of a 12.5 nm radius gold nanoparticle ($a = 4.08$ Å and $n_{\text{unit}} = 4$), we find that the particle surface contains about 47 000 atoms. By comparison, a 5 nm radius gold nanoparticle has about 7500 surface atoms.

2.6 SUMMARY

With this, we end our brief introduction to crystal structures, underlying both bulk and nanoscale systems. We now discuss appropriate length scales for nano in Chapter 3, since not everything small necessarily exhibits the size-dependent optical and electrical properties that make nanostructures distinct from bulk systems. Having done this, we will introduce common low-dimensional systems (quantum wells, quantum wires, and quantum dots) in Chapter 4.

2.7 THOUGHT PROBLEMS

2.1 Number of atoms

The lattice constant of silicon is $a = 5.43$ Å. Using this value, calculate the number of silicon atoms contained in a cubic centimeter of the bulk solid.

2.2 Number of atoms

Calculate the number of gallium atoms per cubic centimeter in a bulk GaAs crystal. The lattice constant of GaAs is $a = 5.65$ Å. Do the same for the As atoms.

2.3 Number of atoms

Consider an actual silicon device on a wafer with physical dimensions $2\,\mu m \times 2\,\mu m \times 1\,\mu m$. Calculate the total number of atoms in the device.

2.4 Densities

Estimate the density of gold for yourself given a unit-cell length of $a = 4.08$ Å, $n_{unit} = 4$ atoms/unit cell, and an atomic mass of 196.97 g/mol.

2.5 Densities

The density of bulk silver is $\rho = 10.5\ g/cm^3$. From this, calculate its unit-cell length a.

2.6 Unit cells

Consider the cubic unit cell shown in **Figure 2.10**.

(a) How many atoms are there per unit cell?

(b) This unit cell is representative of what kind of Bravais lattice?

2.7 Unit cells

Consider the cubic unit cell shown in **Figure 2.11**.

(a) How many atoms are there per unit cell?

(b) This unit cell is representative of what kind of Bravais lattice?

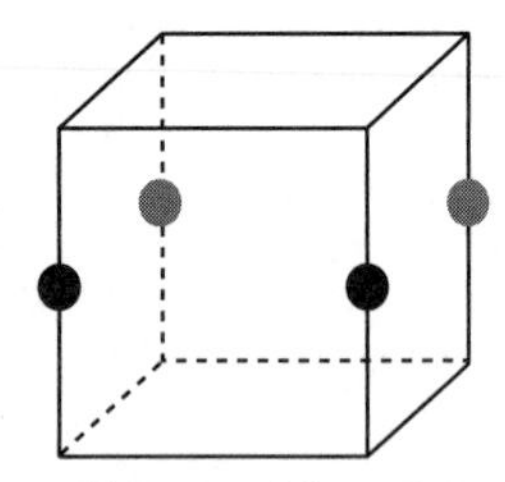

Figure 2.10 Unit cell for Problem 2.6.

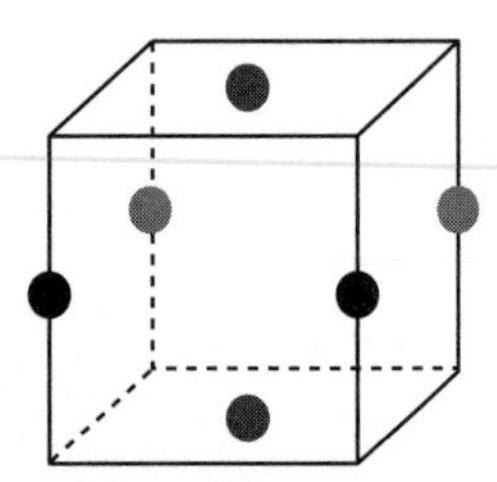

Figure 2.11 Unit cell for Problem 2.7.

2.8 Structure

In a simple cubic lattice (single-element basis) with atoms over each lattice point and with a lattice constant a, what is the density of atoms (number per unit area) on a (100) surface?

2.9 Structure

Draw the surface of a silver crystal cut along the (111) and (100) planes.

2.10 Structure

In a cubic BCC unit cell,

(a) Draw the appearance of the atoms on a (110) surface.

(b) What is the density of atoms (number per unit area) on this plane?

2.11 Back of the envelope

Calculate the number of atoms in a 1.4 nm diameter platinum nanoparticle using the total number of unit cells present. Consider a FCC unit cell with a lattice constant of $a = 3.91$ Å. Explain the interest in small platinum or palladium nanoparticles for catalysis. As an example, see Narayanan and El-Sayed (2004). Small gold nanoparticles have also been reported to possess catalytic abilities (Campbell 2004; Chen and Goodman 2004). Note, though, that these claims have been contested by some.

2.12 Back of the envelope

Calculate the number of atoms in a 1.4 nm diameter platinum nanoparticle using the bulk density of platinum. Consider $\rho = 21.5\ g/cm^3$. How does it compare with the previous value?

2.13 Back of the envelope

Cobalt usually possesses a hexagonal crystal structure. It was recently found to crystallize with a simple cubic structure, now called ϵ-cobalt. The lattice constant of ϵ-cobalt is $a = 6.097$ Å. The corresponding density is $\rho = 8.635\ g/cm^3$, with a unit cell containing 20 atoms. See Dinega and Bawendi (1999) for more information about this material. Calculate the number of atoms in a 2 nm diameter ϵ-cobalt nanocrystal using a unit-cell argument. Verify this number using a density argument.

2.14 A computer program

Write a computer program that creates a spherical 5 nm diameter cobalt nanoparticle. The program should be able to depict all atoms that are present within it. Repeat the same calculation except this time for a 5 nm diameter CdSe nanocrystal. Again, both the Cd and Se atoms should be explicitly depicted.

2.15 Surface-to-volume ratio

Estimate the number of surface atoms and percentage of surface atoms in a 1.4 nm diameter platinum nanoparticle. This can be done using a unit-cell approach. However, use whatever method suits you. Repeat the calculation for 2, 3, 5, and 10 nm diameter particles. Indicate roughly where the fraction of surface atoms falls below 50%.

2.16 Surface-to-volume ratio

Wurtzite CdSe has a hexagonal unit cell (unit cell volume $\simeq$ 112 Å^3). Its lattice constants are $a = 4.3$ Å and $c = 7$ Å. Calculate the total number of atoms in a CdSe quantum dot for 1, 2, 3, 4, 5, and 6 nm diameter particles. How many atoms of each element are there? For the same nanocrystals considered, calculate the fraction of surface atoms in each and plot this on a graph.

2.17 Literature

Find an example from the literature where the surface-to-volume ratio is invoked to explain some special property of nanostructures. Briefly summarize the arguments used and provide the surface-to-volume ratio of this material. Calculate the surface-to-volume ratio yourself using any of the approaches you have learned. How close is your number to theirs?

2.18 History

Surface-enhanced Raman scattering (SERS) spectroscopy and microscopy is a popular research area these days. The surface enhancement effect was first discovered by Fleischmann of Pons and Fleischmann cold fusion fame. (This is another interesting story akin to the Schön case. If interested, you can read an account of it in Taubes (1993).) Now, the initial explanation for the effect given by Fleischmann was essentially an argument saying that electrochemically roughened metal surfaces provided more area for molecules to physisorb onto and hence this increased the analyte's concentration, accounting for any observed Raman enhancements (Fleischmann et al. 1974). However, Van Duyne and Creighton were not convinced of this and later came up with alternative enhancement mechanisms still used today (Albrecht and Creighton 1977; Jeanmaire and Van Duyne 1977). Reproduce Fleischmann's train of thought with a back-of-the-envelope calculation for the concentration increase on a roughened silver surface when compared with a perfectly flat electrode. Make any reasonable estimates as needed. Then explain why Van Duyne and Creighton might have a point.

2.19 Additional literature

The subject of nanoparticle alignment in solution to form nanowires was initially controversial when first reported by Kotov and co-workers (Tang et al. 2002). This is because CdTe generally possesses a cubic zinc blende lattice. Thus, at first glance, one might expect only symmetric structures to form from this material. By contrast, structures with a hexagonal wurtzite lattice (i.e., having a unique c axis) might be expected to yield anisotropic structures, since they often have a permanent dipole moment associated with this direction due to the unit-cell asymmetry.

Murray and co-workers (Cho et al. 2005) reported the alignment of PbSe particles in solution to make nanowires. Interestingly, they invoked a dipole moment argument just like Kotov. Now, PbSe possesses a rocksalt lattice. Explain Murray and Talapin's argument for an intrinsic dipole in the nanoparticles. Furthermore, calculate and attempt to reproduce their numbers for the probability of encountering a dipole along different crystal directions of the lattice.

2.20 Additional literature

It is worth noting that nanoscale solids can behave in ways that are noticeably different than those of comparable bulk solids. As an example, even though we tend to think of crystals as being rigid, it has been shown by Alivisatos that cation exchange is possible in CdSe nanocrystals. Read and discuss the article by Son et al. (2004).

2.21 Additional literature

The suppression of a system's melting point is another well-known property of nanoscale materials. Goldstein et al. (1992) and Dick et al. (2002) describe this property in a semiconductor and in a metal. Think of any potential applications that exploit this property.

2.22 Additional literature

An important step forward in the synthesis of nanoscale materials came with the development of anisotropic semiconductor structures. One of the first studies in this area describes the growth of CdSe nanorods. Read Peng et al. (2000) and Scher et al. (2003) and rationalize for yourself the postulated growth mechanism. Since then, a number of nanostructures with different morphologies have been made. Find another article from the literature and see whether one can apply the concepts described in the above two articles to rationalize the observed shape.

2.23 Additional literature

Many colloidal II–VI nanostructures contain mixtures of zinc blende and wurtzite. See, for example, Grebinski et al. (2004). Explain how such a phase admixture can coexist in a crystal.

2.8 REFERENCES

Albrecht MG, Creighton JA (1977) Anomalously intense Raman spectra of pyridine at a silver electrode. *J. Am. Chem. Soc.* **99**, 5215.

Ashcroft NW, Mermin ND (1976) *Solid State Physics.* Brooks/Cole, Belmont, CA.

Campbell CT. (2004) The active site in nanoparticle gold catalysis (Perspective). *Science* **306**, 234.

Chen, MS, Goodman DW (2004) The structure of catalytically active gold on titania. *Science* **306**, 252.

Cho KS, Talapin DV, Gaschler W, Murray CB (2005) Designing PbSe nanowires and nanorings through oriented attachment of nanoparticles. *J. Am. Chem. Soc.* **127**, 7140.

de Graef M, McHenry ME (2007) *Structure of Materials: An Introduction to Crystallography, Diffraction, and Symmetry.* Cambridge University Press, Cambridge, UK.

Dick K, Dhanasekaran T, Zhang Z, Meisel D (2002) Size-dependent melting of silica-encapsulated gold nanoparticles *J. Am. Chem. Soc.* **124**, 2312.

Dinega DP, Bawendi MG (1999) A solution-phase chemical approach to a new crystal structure of cobalt. *Angew. Chem. Int. Ed.* **38**, 1788.

Fleischmann M, Hendra PJ, McQuillan AJ (1974) Raman spectra of pyridine adsorbed at a silver electrode. *Chem. Phys. Lett.* **26**, 163.

Goldstein AN, Echer CM, Alivisatos AP (1992) Melting in semiconductor nanocrystals. *Science* **256**, 1425.

Grebinski JW, Hull KL, Zhang J, et al. (2004) Solution based straight and branched CdSe nanowires. *Chem. Mater.* **16**, 5260.

Jeanmaire DL, Van Duyne RP (1977) Surface Raman spectroelectrochemistry: Part I. Heterocyclic, aromatic, and aliphatic amines adsorbed on the anodized silver electrode. *J. Electroanal. Chem.* **84**, 1.

Kittel C (2005) *Introduction to Solid State Physics*, 8th edn. Wiley, Hoboken, NJ.

Narayanan R, El-Sayed MA (2004) Shape-dependent catalytic activity of platinum nanoparticles in colloidal solution. *Nano Lett.* **4**, 1343.

Peng X, Manna L, Yang W, et al. (2000) Shape control of CdSe nanocrystals. *Nature* **404**, 59.

Scher EC, Manna L, Alivisatos AP (2003) Shape control and applications of nanocrystals. *Phil. Trans. R. Soc. Lond. A* **361**, 241.

Son DH, Hughes SM, Yin Y, Alivisatos AP (2004) Cation exchange reactions in ionic nanocrystals. *Science* **306**, 1009.

Tang Z, Kotov NA, Giersig M (2002) Spontaneous organization of single CdTe nanoparticles into luminescent nanowires. *Science* **297**, 237.

Taubes G (1993) *Bad Science: The Short Life and Weird Times of Cold Fusion.* Random House, New York.

Length Scales

3.1 INTRODUCTION

What are the relevant length scales for nano? The answer to this question depends on who you talk to. Some will call nano anything smaller than things on the micrometer level (10^{-6} m). This includes objects with sizes hundreds of nanometers in length. However, engineers have been building things with such dimensions for a while. For example, Intel's Pentium III computer chips built during the late 1990s already featured transistors with element sizes approximately 200 nm in width. More recently, Intel has introduced chips containing much smaller features with widths of 45 nm and even 32 nm.

Others may tell you that nano is anything extremely small with molecular dimensions. However, one might ask if this isn't just traditional chemistry or biochemistry. After all, chemists, biologists, and physicists have been making and dealing with small things for a long time, even before nano became a popular buzzword.

A useful perspective on an appropriate length scale for nano is therefore a regime where the optical and electrical properties of matter become size- and shape-dependent. For semiconducting materials, which we will speak of extensively, this is given by the bulk exciton Bohr radius or alternatively by the de Broglie wavelength of carriers in the material. Namely, nano occurs when the physical dimensions of the system become comparable to or smaller than the natural size of electrons and holes in it. This is because in this regime quantum mechanical confinement effects begin to occur.

3.2 de BROGLIE WAVELENGTH

Let us therefore begin our discussion about appropriate length scales for nano with the de Broglie wavelength. The historical origin of this concept stems from a debate about the nature of light, namely, whether it consisted of a stream of particles or whether it was a wave. On one side, Huygens thought light was a wave. On the other, those such as Newton thought light consisted of particles (the corpuscular theory of light). de Broglie's contribution to ending this debate was essentially to suggest that light was both. It had both particle-like and wave-like properties. The resulting hypothesis was and still is called his wave–particle duality.

However, de Broglie's suggestion had implications beyond light, because what it ultimately implied was that matter—whether you, me, a car, an atom, an electron, or a quantum dot—possessed both wave-like and particle-like properties. Now, the wave-like properties of matter are important when dealing with very small things. By contrast, even though macroscopic objects have wave-like aspects, they are better described using Newton's laws. In what follows, we shall therefore deal extensively with the wave-like properties of electrons and complementary holes, since they ultimately dictate the optical and electrical response of materials.

According to de Broglie, the wavelength associated with each object is

$$\lambda = \frac{h}{p}, \tag{3.1}$$

where $h = 6.62 \times 10^{-34}\,\mathrm{J\,s}$ is Planck's constant, $p = mv$ is the object's momentum, and m is its associated mass. To put these de Broglie wavelengths into perspective, let us calculate λ for various objects.

EXAMPLE 3.1

The de Broglie Wavelength of a Free Electron Let us first consider the case of a free electron traveling in vacuum at a nonrelativistic velocity of $v = 0.01c$ ($\frac{1}{100}$ the speed of light). The momentum of the electron is

$$\begin{aligned} p = mv &= 9.11 \times 10^{-31}\,\mathrm{kg}\,(0.01)(3 \times 10^{8}\,\mathrm{m/s}) \\ &= 2.73 \times 10^{-24}\,\mathrm{kg\,m/s} \end{aligned}$$

and, when inserted into Equation 3.1, we obtain a corresponding de Broglie wavelength of

$$\lambda = 2.4 \times 10^{-10}\,\mathrm{m} \simeq 0.24\,\mathrm{nm}.$$

This is quite small. Note that it is comparable to the interatomic spacing of atoms (10^{-10} m) in crystals, as discussed earlier in Chapter 2. Electrons can therefore be put to good use, since their wavelength can be controlled by simply varying their momentum. As a consequence, a standard nanoscience characterization tool, called the transmission electron microscope (TEM), uses electrons to image nanostructures. We will see more of this later in Chapter 14.

EXAMPLE 3.2

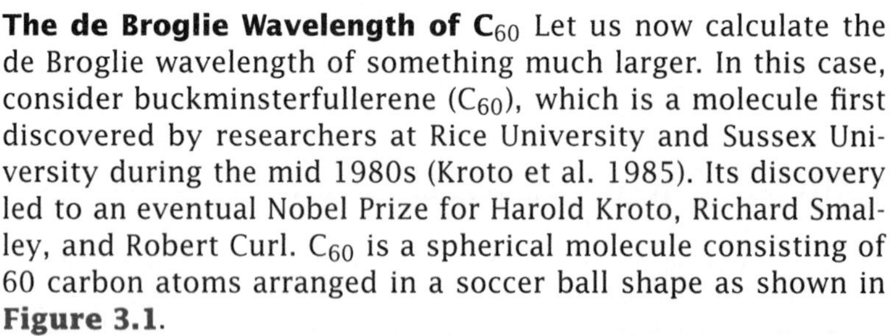

The de Broglie Wavelength of C_{60} Let us now calculate the de Broglie wavelength of something much larger. In this case, consider buckminsterfullerene (C_{60}), which is a molecule first discovered by researchers at Rice University and Sussex University during the mid 1980s (Kroto et al. 1985). Its discovery led to an eventual Nobel Prize for Harold Kroto, Richard Smalley, and Robert Curl. C_{60} is a spherical molecule consisting of 60 carbon atoms arranged in a soccer ball shape as shown in **Figure 3.1**.

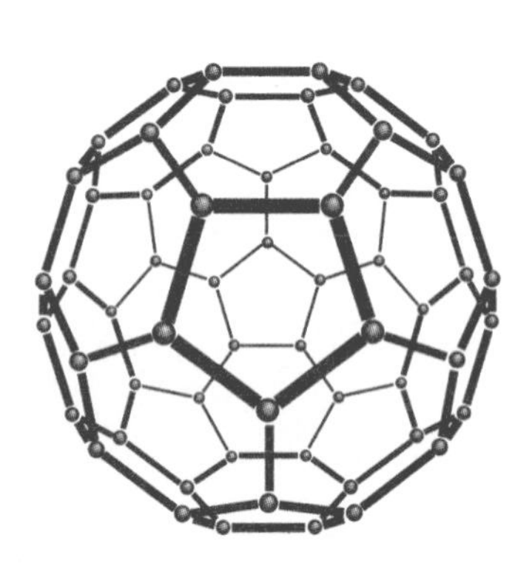

Figure 3.1 Illustration of a C_{60} molecule.

We now proceed in an identical fashion to the above free electron example by first determining the momentum of an individual C_{60} molecule. Equation 3.1 is then used to determine the associated de Broglie wavelength. Towards this end, the mass of an individual C_{60} molecule is

$$m_{C_{60}} = \frac{12.011}{N_a \times 1000}(60) = 1.2 \times 10^{-24}\,\mathrm{kg}$$

with $N_a = 6.022 \times 10^{23}$ being Avogadro's number. Assuming that the C_{60} molecule moves at a given speed, say $|v| = 220\,\mathrm{m/s}$, Equation 3.1 yields a corresponding wavelength of

$$\lambda_{C_{60}} = 2.5\,\mathrm{pm}.$$

Notice how much smaller the de Broglie wavelength of more macroscopic objects tend to be when compared with the free electron case. This is the reason why the wave-like aspects of macroscopic objects are generally not as significant as their particle-like properties. To further illustrate this, the reader may repeat the calculation for a C_{70} molecule. The de Broglie wavelength of C_{60} has recently been measured by Austrian researchers; the reader may consult Arndt et al. (1999) and Brezger et al. (2002) for more details.

EXAMPLE 3.3

The de Broglie Wavelength of an Electron in a Semiconductor Next, let us calculate the de Broglie wavelength of an electron inside a semiconductor. This turns out to be different from the free electron case, because the apparent mass of an electron in a semiconductor (m_{eff}) differs from the free electron mass. In fact, it varies from one semiconductor to another. In addition, m_{eff} differs along different crystallographic directions. All of this stems from the fact that electrons in a crystal move within a potential dictated by the regular arrangement of atoms in the solid. We will see more of this later in Chapter 10 when we talk about bands and how they form due to the periodic arrangement of atoms.

For simplicity, let us assume a generic isotropic electron mass. Specifically, $m_{\text{eff}} = 0.1 m_0$, where $m_0 = 9.11 \times 10^{-31}$ kg is the free electron mass. For the purpose of this calculation, let us also assume a nonrelativistic electron velocity of $v = 1 \times 10^5$ m/s. When these values are inserted into Equation 3.1, we obtain

$$\lambda_e = 7.3 \times 10^{-8}\,\text{m} \simeq 73\,\text{nm}.$$

This wavelength is again much larger than that of the free electron.

EXAMPLE 3.4

The de Broglie Wavelength of a Hole in a Semiconductor In addition to electrons, holes also exist in semiconductors. They are complementary particles, which are positively charged, identical in magnitude, and result from the absence of electrons. Like the electron, the hole is also mobile and can move about the semiconductor. In fact, it is the flow of electrons and holes in response to an externally applied electric field that results in an electrical current running through the material.

Although electrons and holes have identical charge magnitudes, their effective masses differ. In general, holes possess larger m_{eff} values than those of corresponding electrons. To illustrate, **Table 3.1** lists representative effective masses for both electrons and holes in various II–VI, III–V, and IV–VI semiconductors.

For the purpose of this example, let us use a hole effective mass $m_{\text{eff}} = 0.4 m_0$ (i.e., 0.4 times the free electron mass). Then, assuming a hole velocity of 10^5 m/s, Equation 3.1 yields the following de Broglie wavelength:

$$\lambda_h \simeq 18.2\,\text{nm}.$$

Table 3.1 Representative electron and hole effective masses in various II–VI, III–V, and IV–VI semiconductors

III–V	$m_e(m_0)$	$m_h(m_0)$
AlAs	0.150	0.5
GaAs	0.067	0.5
GaSb	0.041	0.28
InP	0.077	0.6
InAs	0.022	0.4
InSb	0.014	0.4
II–VI	$m_e(m_0)$	$m_h(m_0)$
CdS	0.2	0.7
CdSe	0.13	0.45
CdTe	0.11	0.35
HgTe	0.029	0.3
IV–VI	$m_e(m_0)$	$m_h(m_0)$
PbS	0.1	0.1
PbSe	0.07	0.06

Notice how much smaller λ_h is compared with the analogous electron de Broglie wavelength. This difference and others that the reader can calculate using **Table 3.1** will become important later when we refer to various confinement regimes for nanoscale materials.

3.3 THE BOHR RADIUS

Having introduced the de Broglie wavelength, let us now illustrate another length scale that defines nano in semiconductors. It is called the bulk exciton Bohr radius. The reason for this is that sometimes in the literature one sees a statement that a nanomaterial is in the quantum confinement regime because its physical size is smaller than the corresponding electron or hole de Broglie wavelength. At other times, one sees a statement that a nanomaterial is quantum-confined because its size is smaller than the corresponding bulk exciton Bohr radius. These statements are not contradictory and, in fact, are essentially the same thing, as we shall see.

3.3.1 The Textbook Bohr Radius a_0

First, let us discuss the textbook Bohr radius a_0. In physics, a_0 is the closest stable orbit to the nucleus in Bohr's planetary model of the atom. As an illustration, the first three stable orbits of the electron are shown schematically in **Figure 3.2**.

Next, the textbook Bohr radius of an electron is given by the following equation

$$a_0 = \frac{4\pi\epsilon_0\hbar^2}{m_0 q^2}, \tag{3.2}$$

where $\epsilon_0 = 8.854 \times 10^{-12}$ F/m is the permittivity of free space, $\hbar = h/2\pi = 1.054 \times 10^{-34}$ J s, $m_0 = 9.11 \times 10^{-31}$ kg (the free electron mass), and $q = 1.602 \times 10^{-19}$ C. When all of these numbers are inserted into

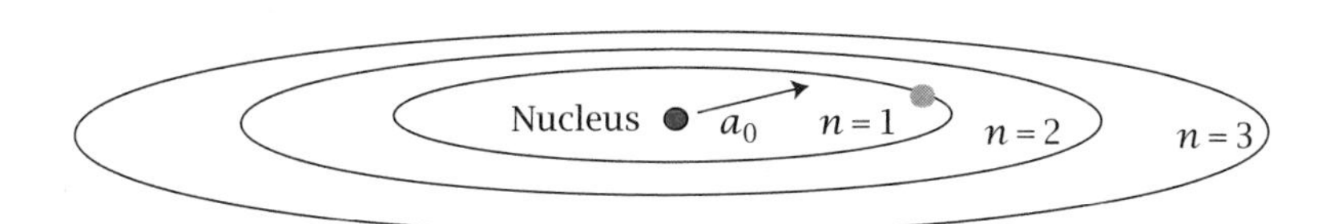

Figure 3.2 Illustration of the first three stable orbits in Bohr's planetary model of the atom.

Equation 3.2, we obtain a value of

$$a_0 = 5.28 \times 10^{-11}\ \mathrm{m} = 0.528\,\text{Å}. \tag{3.3}$$

3.3.2 Derivation of the Bohr Radius

Equation 3.2 is obtained by equating the centripetal force of an electron circling an infinitely heavy, positively charged, nucleus with their mutual Coulomb attractive force:

$$\frac{mv^2}{r} = \frac{q^2}{4\pi\epsilon_0 r^2}.$$

We also make use of a relationship between the wavelength of the particle and its associated Bohr radius:

$$n\lambda = 2\pi r. \tag{3.4}$$

In Equation 3.4, n is an integer and the equivalence suggests that an integer number of electron wavelengths must "fit" into the circumference of a classic Bohr orbit if it is to be allowed. This relationship is illustrated schematically in **Figure 3.3**.

The de Broglie relationship then enters into the picture by associating λ with the particle's momentum through $\lambda = h/p$. We obtain the following expression:

$$\lambda = \frac{h}{p} = \frac{2\pi r}{n} = \frac{h}{mv}.$$

At this point, solving for the particle's velocity and expressing it in terms of r gives

$$v = \frac{nh}{2\pi mr} = \frac{n\hbar}{mr}. \tag{3.5}$$

This quantized velocity is subsequently inserted into the original equivalence between the Coulomb and centripetal forces, yielding

$$\frac{n^2\hbar^2}{mr} = \frac{q^2}{4\pi\epsilon_0}.$$

Finally, solving for the radius r gives a generic Bohr radius

$$r = \frac{4\pi\epsilon_0 n^2\hbar^2}{mq^2},$$

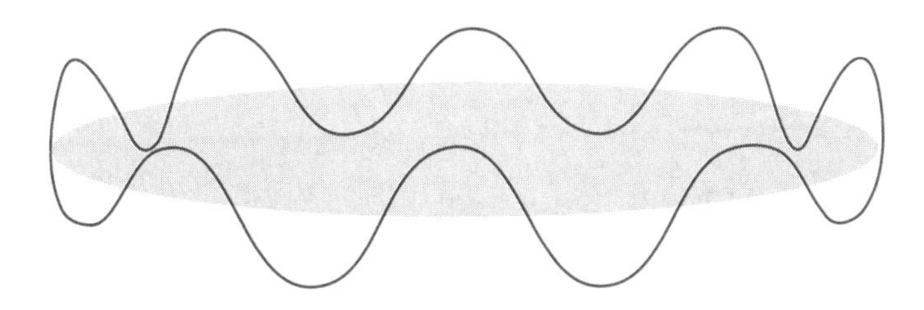

Figure 3.3 Integer wavelengths fitting into the circumference of a ring.

Table 3.2 Representative static dielectric constants for various semiconductors and organic compounds

III–V	ϵ_{static}
GaAs	12.5
GaSb	15
InP	12.1
InAs	12.5
InSb	18
II–VI	ϵ_{static}
ZnS	8.3
ZnSe	8.1
ZnTe	9.7
CdS	8.9
CdSe	10.6
CdTe	10.9
HgS	10.9
HgSe	25
IV–VI	ϵ_{static}
PbS	170
PbSe	250
Organic	ϵ_{static}
Pentane	1.818
Hexane	1.886
Benzene	2.271
Cyclohexane	2.030
Carbon disulfide	2.643
Trichloromethane	4.950
Methanol	32.66
Ethanol	25.13

where the standard expression (Equation 3.2) for a_0 is then obtained by letting $n = 1$ and assuming $m = m_0$:

$$a_0 = \frac{4\pi\epsilon_0\hbar^2}{m_0 q^2}.$$

Note that if the electron is not in vacuum, the expression for r above must be modified to take into account the dielectric constant of the medium. Basically, this means replacing ϵ_0 in the numerator of Equation 3.2 with $\epsilon\epsilon_0$, where ϵ is the relative dielectric constant of the material. This yields the following generic expression:

$$r = \frac{4\pi\epsilon\epsilon_0 n^2\hbar^2}{mq^2}. \tag{3.6}$$

The reason for this change is that if the electron is not in vacuum (e.g., if it is in a semiconductor), one must account for other free charges or dipoles in the medium, which respond to its electric field. Specifically, mobile free charges present in the medium will move in response to the charge. Likewise, any dipoles present will re-orient to oppose the carrier's electric field. The net result is that the Coulomb potential between the electron and the positively charged nucleus is diminished by the response of these free charges and dipoles. The factor by which it is suppressed is called the relative dielectric constant and is denoted by ϵ. **Table 3.2** lists representative *static* dielectric constants of various materials so that the reader can get a sense for what typical ϵ values look like. Notice how ϵ increases in polar solvents.

Finally, notice that analogous expressions to a_0 can be written for a positively (negatively) charged particle orbiting a negatively (positively) charged nucleus. This leads to the following Bohr radius expressions for the electron (e) and hole (h):

$$a_e = \frac{4\pi\epsilon\epsilon_0\hbar^2}{m_e q^2}, \tag{3.7}$$

$$a_h = \frac{4\pi\epsilon\epsilon_0\hbar^2}{m_h q^2}, \tag{3.8}$$

where m_e and m_h are the respective electron and hole effective masses in the material. Electron/hole Bohr radii are shown schematically in **Figure 3.4**, centered about a generic origin. Since holes often possess heavier effective masses than electrons ($m_h > m_e$, see **Table 3.1**), a_h is usually smaller than a_e.

3.3.3 Connecting the de Broglie Wavelength to the Bohr Radius

To close the loop, let us connect the de Broglie wavelength to the Bohr radius just found. The claim is that both are essentially the same thing. We focus on the electron for simplicity. Starting with the electron Bohr

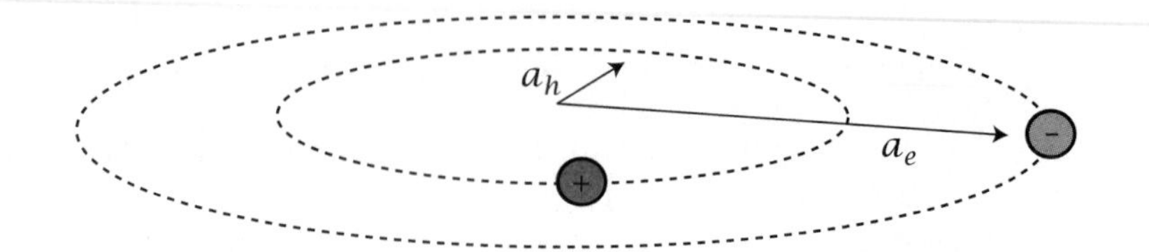

Figure 3.4 A qualitative comparison of the electron and hole Bohr radii. For illustration purposes, a generic origin is used.

radius (Equation 3.7)

$$a_e = \frac{4\pi\epsilon\epsilon_0\hbar^2}{m_e q^2},$$

we have an associated velocity (Equation 3.5)

$$\begin{aligned} v_e &= \frac{\hbar}{m_e a_e} \\ &= \frac{q^2}{4\pi\epsilon\epsilon_0\hbar}. \end{aligned}$$

The reader may calculate the numerical value of v_e in this and other orbits. It might also be interesting to compare these numbers with the speed of light.

We now use Equation 3.1 to calculate the associated electron de Broglie wavelength. This gives

$$\begin{aligned} \lambda_e &= \frac{h}{p} = \frac{h}{m_e v_e} \\ &= \frac{2\epsilon\epsilon_0 h^2}{m_e q^2}. \end{aligned}$$

At this point, Equation 3.4 requires that

$$\lambda_e = 2\pi a_e, \tag{3.9}$$

where we have explicit expressions for λ_e and a_e above, illustrating that the electron's de Broglie wavelength and its Bohr radius are essentially the same thing. A similar calculation can be done for the hole and ultimately for the exciton, a species that we will describe next.

3.4 EXCITONS

Excitons are Coulombically bound electron–hole pairs in a semiconductor. They form upon the absorption of light, which promotes an electron from the semiconductor's valence band into its conduction band. This leaves behind a positively charged hole (i.e., the absence of an electron). The formation of bands is discussed in Chapter 10. For now, the reader may simply assume that they exist.

Two general types of excitons exist: Mott–Wannier excitons and Frenkel excitons. They are distinguished by their electron–hole binding energies. Specifically, Mott–Wannier excitons have weak electron–hole interactions caused by a relatively small Coulomb attraction between component electrons and holes. Corresponding binding energies are of the order of 10 meV, and consequently the carriers remain relatively far apart.

By contrast, Frenkel excitons have much stronger Coulomb interactions. Corresponding binding energies are of the order of 100 meV. As a consequence, the electron and hole remain in close proximity. This difference between Frenkel and Mott–Wannier excitons is illustrated in **Figure 3.5**, where the latter species is delocalized over larger distances, spanning multiple lattice sites of the crystal. In both cases, since the exciton is composed of two oppositely charged species, it is charge-neutral.

The excitons that we will usually deal with in semiconductors are Mott–Wannier excitons. This is because semiconductor dielectric constants are generally large, $\epsilon \sim 10$. By contrast, Frenkel excitons are often

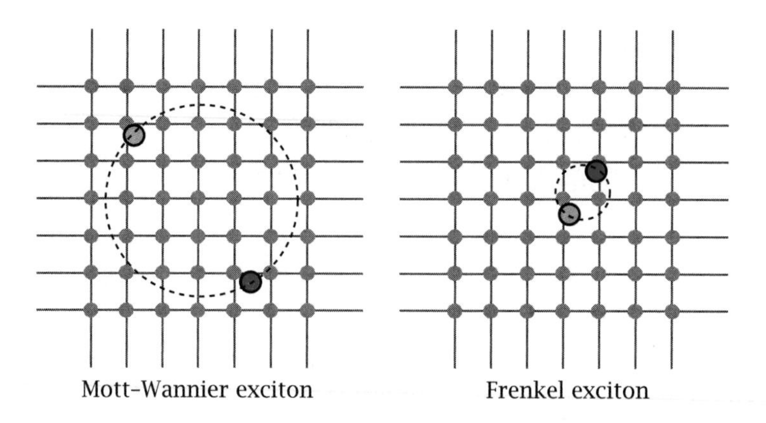

Figure 3.5 A comparison of the Mott–Wannier and Frenkel excitons. The shaded spheres in the background represent the location of atoms in the lattice.

seen in organic materials where relative ϵ values are small, $\epsilon \sim 2$. The reader may verify this by referring to **Table 3.2**, which lists static ϵ values for various organic and semiconductor compounds.

Next, excitons can sometimes be seen in the absorption spectrum of bulk semiconductors. They generally appear just below the band-edge transition because the exciton's energy is smaller than the semiconductor's band gap (analogous to the HOMO–LUMO gap in molecular systems) by an amount equal to its binding energy. In particular, the exciton formation energy is

$$E_X = E_g - E_{\text{bind}},$$

where E_g is the semiconductor band gap and E_{bind} is the exciton binding energy. Such transitions are illustrated schematically in **Figure 3.6**. Note that the absorption of light by semiconductors will be discussed phenomenologically in Chapter 5 and more fully in Chapter 12.

Since (Mott–Wannier) exciton binding energies have orders of magnitude of 10 meV, they are generally not stable at room temperature. This is because thermal energy at room temperature is $kT \simeq 25$ meV, and

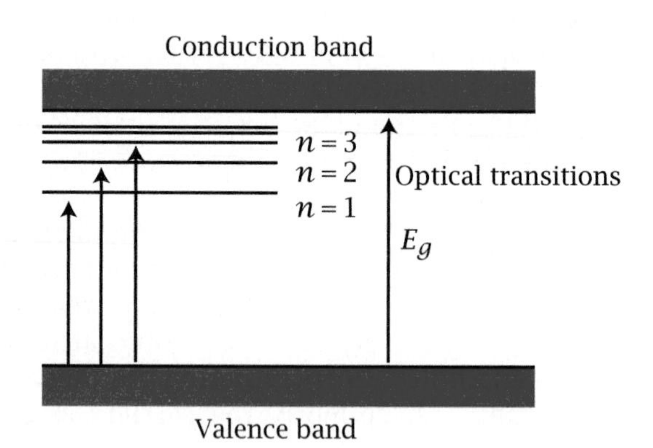

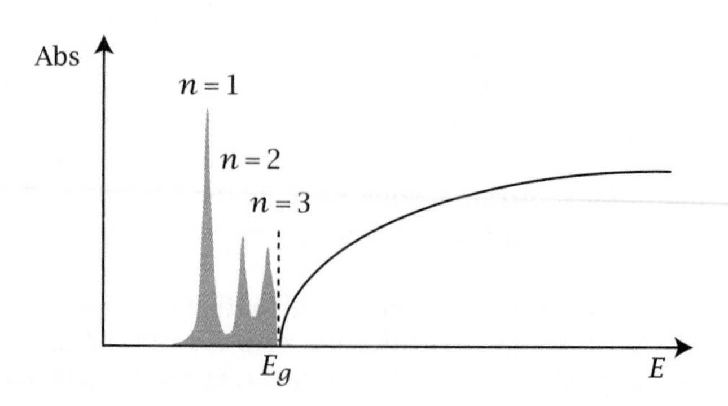

Figure 3.6 Schematic of the absorption spectrum (solid line) of a bulk semiconductor accompanied by excitonic transitions (shaded peaks) below its band edge at low temperatures.

consequently the exciton easily dissociates into its component electron and hole. Excitons are therefore apparent in the absorption spectrum of bulk semiconductors at low temperatures where $kT \simeq 1$ meV. The situation begins to change in nanostructures because component electrons and holes are forced to remain in close proximity to each other, due to physical size constraints imposed by the system.

3.4.1 The Bulk Exciton Bohr Radius

Now, associated with the exciton is a Bohr radius. We have previously seen that the electron (or hole) Bohr radius has the form (Equations 3.7 and 3.8)

$$a_{e(h)} = \frac{4\pi\epsilon\epsilon_0\hbar^2}{m_{e(h)}q^2}.$$

To find an analogous expression for the exciton, we simply alter the above expression by replacing the mass of the electron (or hole) with the reduced mass of the electron–hole pair, μ, where

$$\frac{1}{\mu} = \frac{1}{m_e} + \frac{1}{m_h}. \tag{3.10}$$

In this expression, m_e (m_h) is the electron (hole) effective mass. The resulting bulk exciton Bohr radius is then

$$a_B = \frac{4\pi\epsilon\epsilon_0\hbar^2}{\mu q^2} = a_e + a_h, \tag{3.11}$$

where one often sees a_B written in terms of a_0,

$$a_B = \frac{\epsilon m_0}{\mu} a_0. \tag{3.12}$$

Equation 3.12 is a convenient way of calculating a_B for various semiconductors, given that a_0 is constant. Note that ϵ is frequency-dependent and hence there is occasionally some ambiguity as to which value of the dielectric constant to use, whether it be the static dielectric constant (i.e., those listed in **Table 3.2**), the high-frequency dielectric constant ϵ_∞, or something in between. In general, we will try to employ the dielectric constant at optical frequencies used to create the exciton.

Finally, one can relate the above exciton Bohr radius to the corresponding de Broglie wavelength using Equation 3.4. This again highlights the fact that both are essentially the same and either is suitable for defining an appropriate length scale for nano in semiconductors.

EXAMPLE 3.5

The CdSe Bulk Exciton Bohr Radius To illustrate what representative Bohr radii look like, we will evaluate the bulk exciton Bohr radius of a common II–VI semiconductor, CdSe, using the following literature values:

$$m_e = 0.13 m_0,$$

$$m_h = 0.45 m_0,$$

$$\epsilon_{\text{optical}} \sim 9.7.$$

From m_e and m_h, we find that the CdSe exciton reduced mass is $\mu = 9.2 \times 10^{-32}$ kg. The exciton Bohr radius is then found using Equation 3.12, giving

$$
\begin{aligned}
a_B &= \frac{\epsilon m_0}{\mu} a_0 \\
&= \frac{(9.7)(9.11 \times 10^{-31})}{8.9 \times 10^{-32}}(5.28 \times 10^{-11}) \\
&= 5.07 \times 10^{-9}\,\text{m} \\
&= 5.07\,\text{nm}.
\end{aligned}
$$

EXAMPLE 3.6

The ZnO Bulk Exciton Bohr Radius For ZnO, a common II–VI semiconductor, we find

$$
\begin{aligned}
m_e &= 0.24 m_0, \\
m_h &= 0.45 m_0, \\
\epsilon_{\text{optical}} &= 3.7.
\end{aligned}
$$

The associated reduced mass is then $\mu = 1.43 \times 10^{-31}$ kg, whereupon the desired bulk exciton Bohr radius is

$$
\begin{aligned}
a_B &= \frac{\epsilon m_0}{\mu} a_0 \\
&= \frac{(3.7)(9.11 \times 10^{-31})}{1.43 \times 10^{-31}}(5.28 \times 10^{-11}) \\
&= 1.24 \times 10^{-9}\,\text{m} \\
&= 1.24\,\text{nm}.
\end{aligned}
$$

It is apparent that a_B here is significantly smaller than in the CdSe case. As a consequence, the electron and hole in ZnO are in close proximity to each other. Note that this is largely due to the small dielectric constant of ZnO. In what follows, we shall see that one implication of having a small Bohr radius is that the electron–hole binding energy will be large.

3.4.2 The Exciton Energy and Associated Electron–Hole Binding Energy

Next, we shall find an expression for the associated bulk exciton binding energy. In this regard, the exciton energy is the sum of its kinetic energy and the Coulomb (attractive) potential energy between constituent electrons and holes:

$$E_x = \frac{mv^2}{2} - \frac{q^2}{4\pi\epsilon\epsilon_0 r}.$$

In the Coulomb term, we have included ϵ to account for the fact that we are working in a semiconductor. Note that we can simplify this expression since we have previously seen from our derivation of the Bohr radius that $mv^2/r = q^2/(4\pi\epsilon\epsilon_0 r^2)$. As a consequence, $mv^2 = q^2/(4\pi\epsilon\epsilon_0 r)$, and E_x can therefore be written as

$$E_x = -\frac{1}{2}\left(\frac{q^2}{4\pi\epsilon\epsilon_0 r}\right).$$

At this point, recall that we found $r = 4\pi\epsilon\epsilon_0 n^2\hbar^2/\mu q^2$ for excitons. By introducing this into the above equation, the exciton energy can alternatively be written as

$$E_x = -\frac{1}{2}\left[\frac{\mu q^4}{(4\pi\epsilon\epsilon_0)^2\hbar^2}\right]\frac{1}{n^2}, \tag{3.13}$$

whereupon Equation 3.13 is often simplified further by writing it in terms of the so-called Rydberg

$$R_y = \frac{1}{2}\left[\frac{m_0 q^4}{(4\pi\epsilon_0)^2\hbar^2}\right]. \tag{3.14}$$

The Rydberg is the energy (13.6 eV) required to remove an electron from the first Bohr orbit of hydrogen to a location infinitely far away. As an exercise, the reader may verify that Equation 3.14 corresponds to this energy.

Returning to the problem, we then have

$$E_x = -R_y\left(\frac{\mu}{\epsilon^2 m_0}\right)\frac{1}{n^2},$$

and defining an exciton Rydberg as the product $R_x = R_y(\mu/\epsilon^2 m_0)$ gives our desired (simplified) exciton energy expression

$$E_x = -\frac{R_x}{n^2}. \tag{3.15}$$

Finally, we can find the associated exciton binding energy E_{bind}. Namely, we take the difference in energy between the electron–hole pair in a given orbit and the case where they are infinitely separated ($n = \infty$). We thus have

$$E_{\text{bind}} = E_{x,n=\infty} - E_{x,n},$$

and, if $n = 1$,

$$E_{\text{bind}} = R_y\left(\frac{\mu}{\epsilon^2 m_0}\right) = R_x. \tag{3.16}$$

More generally, for an intermediate starting n,

$$E_{\text{bind}} = \frac{R_x}{n^2}. \tag{3.17}$$

Clearly, the larger n is, the easier it is to separate the component electron and hole.

EXAMPLE 3.7

The Bulk CdSe Exciton Binding Energy Let us now calculate actual binding energies to see what these values look like. Using our previous CdSe exciton reduced mass $\mu = 9.2 \times 10^{-32}$ kg and a dielectric constant $\epsilon_{\text{optical}} = 9.7$, we find from Equation 3.16 that

$$\begin{aligned} E_{\text{bind}} &= R_y\left(\frac{\mu}{\epsilon^2 m_0}\right) \\ &= 13.6\left(\frac{9.2 \times 10^{-32}}{(9.7)^2(9.11 \times 10^{-31})}\right) \\ &= 0.015\ \text{eV}. \end{aligned}$$

The estimated CdSe exciton binding energy is therefore $E_{\text{bind}} \simeq$ 15 meV. From this, one deduces that such excitons exist only at low temperatures, since this value is smaller than kT at room temperature.

EXAMPLE 3.8

The Bulk ZnO Exciton Binding Energy By contrast, we can show that ZnO possesses a significantly larger exciton binding energy. We could have already predicted this because of its smaller dielectric constant and smaller bulk exciton Bohr radius. However, let us show this explicitly.

Using our previous ZnO reduced mass $\mu = 1.43 \times 10^{-31}$ kg, as well as $\epsilon_{\text{optical}} = 3.7$, we find that

$$\begin{aligned} a_B &= R_y\left(\frac{\mu}{\epsilon^2 m_0}\right) \\ &= 13.6\left(\frac{1.43 \times 10^{-31}}{(3.7)^2(9.11 \times 10^{-31})}\right) \\ &= 0.156\,\text{eV}. \end{aligned}$$

The estimated ZnO exciton binding energy is therefore $E_{\text{bind}} \simeq$ 156 meV and is much larger than in CdSe. Furthermore, it is greater than kT at room temperature. As a consequence, room temperature excitonic effects can be seen in bulk ZnO.

3.5 CONFINEMENT REGIMES

At the heart of the binding energy is the Coulomb attraction between oppositely charged electrons and holes. Recall that this Coulomb term is proportional to r^{-1}. In Chapters 7 and 8, we will see that the confinement energy associated with either the electron or hole is proportional to r^{-2}. Thus, this latter confinement term grows faster than r^{-1} when the size of the system becomes small. As a consequence, while the Coulomb term is important in bulk systems, in nanoscale materials of appropriate symmetry the confinement of individual electrons and holes dominates at small enough sizes. It is therefore in this regime where the optical and electrical properties of materials become size- and shape-dependent and where some of the most fascinating aspects of nano begin.

We therefore highlight the three confinement regimes that exist. They are referred to as the strong, intermediate, and weak confinement regimes and are characterized by the following criteria:

- Strong confinement: $a < a_e, a_h$
- Intermediate confinement: $a_h < a < a_e$
- Weak confinement: $a > a_e, a_h$

In all cases, a is the critical dimension of the nanostructure. Below, we describe these various confinement regimes in more detail.

3.5.1 The Weak Confinement Regime

In this scenario, $a > (a_e, a_h)$; i.e., the critical dimension of the nanostructure is larger than both the individual electron and hole Bohr radii

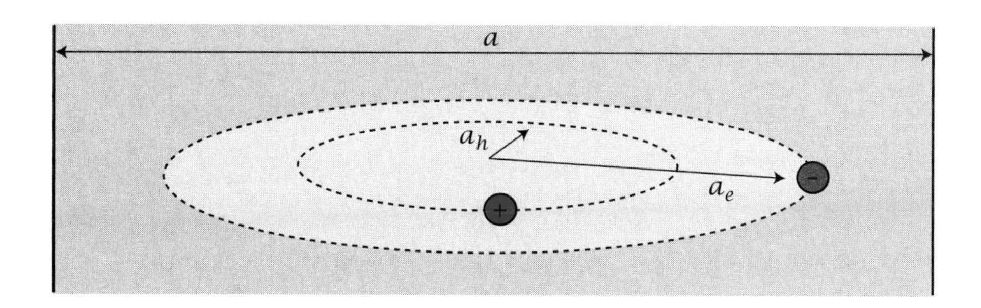

Figure 3.7 Illustration of the weak confinement regime, where $a > (a_e, a_h)$. The shaded region denotes the critical length scale a. For illustration purposes, a generic origin is used for both the electron and hole.

(**Figure 3.7**). As a consequence, the exciton binding energy is weak as in bulk systems. Furthermore, the optical and electrical properties of these nanostructures are essentially bulk-like. Note that implicit to this are some simplifying assumptions, since, in general, nanostructure Coulomb interations can be quite large. However, a more in-depth discussion of so-called dielectric contrast effects is beyond the scope of this text.

3.5.2 The Intermediate Confinement Regime

In this case, the critical dimension of the material is smaller than one carrier's Bohr radius but larger than the other's. Since m_e is generally smaller than m_h (**Table 3.1**), this criterion usually means that $a_h < a < a_e$ (**Figure 3.8**).

At this point, quantization effects should start to become apparent in the material. A good example of a system in this regime is CdSe, where structure is seen in the linear absorption of colloidal nanoparticles made of CdSe. This indicates the onset of quantization effects, as we will see in later chapters.

3.5.3 The Strong Confinement Regime

Finally, the strong confinement regime is often implicitly assumed when talking about nanoscale materials. The criterion is readily met in (very) small nanomaterials, as well as in systems where both electron and hole effective masses are small and corresponding dielectric constants are large. Examples of this latter case include the lead chalcogenides such as PbS and PbSe, which have small electron and hole effective masses and large ϵ values. See **Tables 3.1** and **3.2**. This leads to corresponding bulk exciton Bohr radii of $a_B \simeq 18$ nm (PbS) and $a_B \simeq 46$ nm (PbSe). As a consequence, it is easy to achieve conditions where $a < (a_e, a_h)$ (**Figure 3.9**). The optical properties of these materials are therefore dominated by confinement effects. Again, we will see more of this in Chapters 7 and 8 when we describe quantum mechanical models for

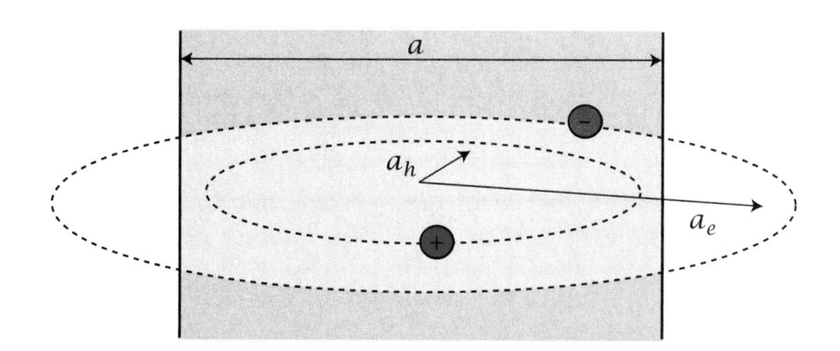

Figure 3.8 Illustration of the intermediate confinement regime, where, generally speaking, $a_h < a < a_e$. The shaded region represents the critical length scale a. For illustration purposes, a generic origin is used for both the electron and hole.

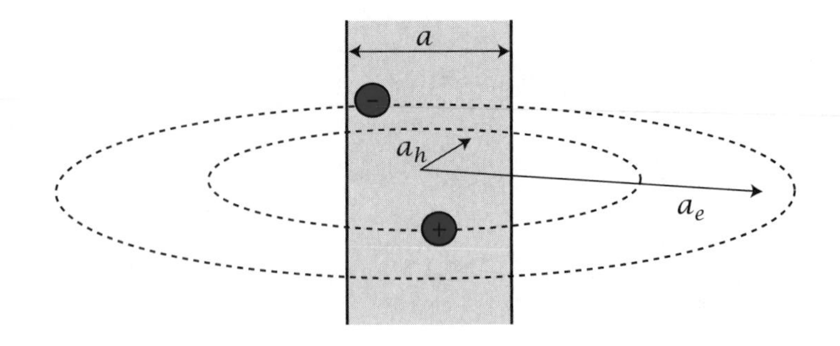

Figure 3.9 Illustration of the strong confinement regime, where $a < (a_e, a_h)$. The shaded region denotes the critical length scale a. For illustration purposes, a generic origin is used for both the electron and hole.

carrier confinement. Additional information about these three confinement regimes as well as examples of systems found within them can be found in Gaponenko (1999) and Efros and Rosen (2000).

3.6 METALS

The above discussion has focused on appropriate length scales describing nano in semiconductors. Defining an analogous length scale in metals, however, is more problematic. This is because most metal nanostructures exhibit bulk-like properties even at quite small sizes, $a \simeq 2\,\text{nm}$. Furthermore, in terms of their optical properties, much of their response can be described using classical bulk models. As a consequence, unlike in semiconductors, we cannot focus solely on one type of length scale, whether it be the de Broglie wavelength or the bulk exciton Bohr radius. Instead, it turns out that there are at least *three* different critical sizes to consider, depending on whether one is interested in size effects on the optical or electrical properties of metal nanostructures. They are outlined schematically in **Figure 3.10**.

Specifically, we have one length scale associated with carrier confinement. This is called the Fermi wavelength λ_F, and is identical to the electron de Broglie wavelength discussed above. The reader will notice that this represents the smallest length scale shown in **Figure 3.10**. Moving to progressively larger sizes, we find another critical size, d_{charging}, associated with the electrical properties of small metal nanostructures. Namely, this is a size below which the Coulomb charging energy of such structures becomes significant compared with kT. Depending on the dielectric constant of the medium, d_{charging} can be on the order of tens of nanometers. Finally, the last length scale relates again to the optical properties of metals and occurs when the nanostructure possesses a size, d_{mean}, below the mean free path of electrons within it. In this case, the metal's optical properties change due to variations of its dielectric constant. The length scale associated with d_{mean} likewise turns out to be on the order of tens of nanometers.

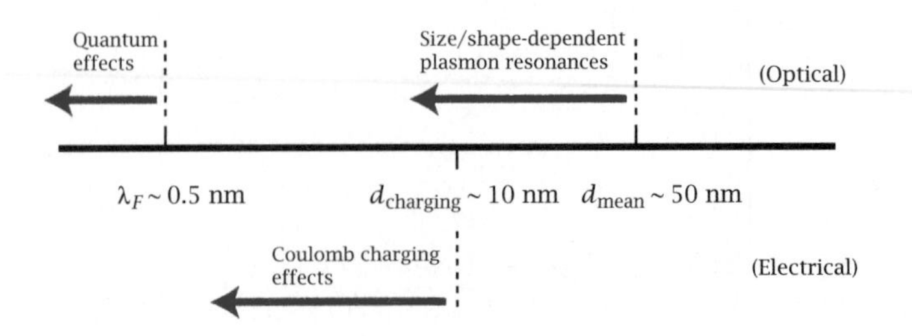

Figure 3.10 Illustration of the various critical length scales in metals.

3.7 THE FERMI ENERGY, FERMI VELOCITY, AND KUBO GAP

We begin our discussion with the smallest length scale associated with confinement effects in metal nanostructures. Specifically, at a small enough size, discreteness in their energy levels sets in. In turn, they start to take molecular-like properties and a HOMO–LUMO gap opens up. In fact, at such sizes, one can see emission from the metal itself. This is the origin of emission seen in small noble metal clusters containing several atoms. More about this can be found in Zheng et al. (2007).

As discussed earlier, such quantization effects occur when the physical size of the structure becomes comparable to the electron's de Broglie wavelength. For a metal, this occurs when its physical size becomes comparable to or smaller than what is referred to as the Fermi wavelength λ_F.

In what follows, we therefore calculate λ_F for different metals. To do this, we will first evaluate the Fermi energy E_F of the metal. We will then obtain the associated electron velocity, called the Fermi velocity v_F. Finally, employing Equation 3.1 yields the associated electron de Broglie wavelength, which is our desired Fermi wavelength. The section concludes with a calculation of the energy difference between electronic levels called the Kubo gap.

Let us begin with what is known as a density-of-states calculation. More about this will be seen later in Chapter 9, and, if desired, the interested reader may jump ahead to see more details about the calculation. However, this section is intended to be self-contained and can be followed without any additional information, provided that the reader accepts some assumptions we will make.

3.7.1 The Density of States, Carrier Concentration, and Fermi Energy

The purpose of the following calculation will be to estimate the number of states in the metal per unit volume per unit energy. However, rather than calculate these states directly in terms of energies, we will employ the wavevector k. Note that a direct relationship between k and E exists through $E = \hbar^2 k^2/2m$, with m the electron mass.

We first picture a sphere in "k-space" having a radius k. The associated volume is

$$V_k = \tfrac{4}{3}\pi k^3.$$

The sphere possesses a constant-energy surface and also contains many states with smaller energies, characterized by smaller radii. The "volume" of a given state in k-space is simply the product

$$V_{\text{state}} = k_x k_y k_z,$$

where $k_x = 2\pi/L_x$, $k_y = 2\pi/L_y$, $k_z = 2\pi/L_z$ are components of k along the various coordinates in k-space and L_x, L_y, L_z are characteristic physical lengths along the x, y, z directions of the material.

The number of energy states, N_1, contained within the sphere is then the ratio of the sphere's volume to that of a given state:

$$N_1 = \frac{\frac{4}{3}\pi k^3}{k_x k_y k_z} = \frac{k^3}{6\pi^2} L_x L_y L_z.$$

We now need to deal with the spin degeneracy of carriers. Since electrons possess two spin orientations that are equal in energy, we multiply

N_1 by two to obtain

$$N_2 = 2N_1 = \frac{k^3}{3\pi^2} L_x L_y L_z.$$

At this point, we can define a carrier density ρ as the number of carriers per unit volume, by dividing N_2 by $L_x L_y L_z$ to obtain

$$\rho = \frac{k^3}{3\pi^2}.$$

The relationship $E = \hbar^2 k^2/2m$ (i.e., $k = \sqrt{2mE/\hbar^2}$) is then invoked to obtain

$$\rho = \frac{1}{3\pi^2}\left(\frac{2mE}{\hbar^2}\right)^{3/2},$$

whereupon we can define an energy density, $\rho_{\text{energy}} = d\rho/dE$, giving

$$\rho_{\text{energy}} = \frac{1}{2\pi^2}\left(\frac{2m}{\hbar^2}\right)^{3/2}\sqrt{E}, \tag{3.18}$$

with units of number per unit volume per unit energy.

The next part of our calculation involves finding the total concentration of carriers, n, in the metal. We do this by integrating ρ_{energy} over all energies, weighted by the probability that an electron occupies a given state. The desired expression is

$$n = \int_0^\infty \rho_{\text{energy}}(E) f(E)\, dE,$$

where $f(E)$ is called the Fermi–Dirac distribution and has the form

$$f(E) = \frac{1}{1 + e^{(E-E_F)/kT}}.$$

In this expression, E_F is called the Fermi level and represents the energy where $f(E_F) = \frac{1}{2}$. At 0 K, E_F equals our desired Fermi energy. However, this is not true at higher temperatures. More about $f(E)$ and a corresponding function for holes will be seen later in Chapter 9.

At this point, we have

$$\begin{aligned} n &= \int_0^\infty \frac{1}{2\pi^2}\left(\frac{2m}{\hbar^2}\right)^{3/2}\sqrt{E}\left(\frac{1}{1+e^{(E-E_F)/kT}}\right) dE \\ &= \frac{1}{2\pi^2}\left(\frac{2m}{\hbar^2}\right)^{3/2}\int_0^\infty \sqrt{E}\left(\frac{1}{1+e^{(E-E_F)/kT}}\right) dE, \end{aligned}$$

and at $T = 0$ K the Fermi–Dirac distribution equals 1 for all energies up to E_F. Physically, this means that all electron states below E_F are occupied and all states above it are empty. Try to rationalize this for yourself. As a consequence, we have for the 0 K carrier concentration

$$\begin{aligned} n(0\,\text{K}) &= \frac{1}{2\pi^2}\left(\frac{2m}{\hbar^2}\right)^{3/2}\int_0^{E_F(0\,\text{K})} \sqrt{E}\, dE \\ &= \frac{1}{2\pi^2}\left(\frac{2m}{\hbar^2}\right)^{3/2} \frac{2}{3} E^{3/2}\Big|_0^{E_F(0\,\text{K})}. \end{aligned}$$

The resulting carrier density, with units of number per unit volume, is then

$$n(0\,\text{K}) = \frac{1}{3\pi^2}\left(\frac{2m}{\hbar^2}\right)^{3/2} E_F(0\,\text{K})^{3/2}. \tag{3.19}$$

Solving for $E_F(0\,\text{K})$ then gives

$$E_F(0\,\text{K}) = (3\pi^2)^{2/3}\left[\frac{\hbar^2}{2m}n(0\,\text{K})^{2/3}\right], \tag{3.20}$$

which is also our desired Fermi energy. The only thing left to do is to determine a numerical value for $E_F(0\,\text{K})$. This means finding an actual value for $n(0\,\text{K})$. Let us now calculate this 0 K electron density for some common metals.

EXAMPLE 3.9

The Fermi Energy of Copper As an illustration, we begin with copper. From a periodic table, we find that its atomic number is $Z = 29$. As a consequence, using the Aufbau principle, we fill up its atomic orbitals as follows:

$$1s^2 2s^2 2p^6 3s^2 3p^6 4s^1 3d^{10}.$$

The reader will recall that filling the 3d orbital occurs prior to filling the 4s orbital because of the improved stability of this electronic configuration. Every copper atom therefore contributes one valence electron (i.e., the 4s electron) to the system's overall carrier density.

The density of bulk copper is $\rho = 8.96\,\text{g/cm}^3$ and since the atomic mass of copper is 63.55 g/mol, we also have $\rho = 1.41 \times 10^5\,\text{mol/m}^3$. An equivalent atomic density is

$$\rho = 8.49 \times 10^{28}\,\text{atoms/m}^3.$$

Since each copper atom contributes 1 electron to the solid, the desired electron density is

$$n(0\,\text{K}) = 8.49 \times 10^{28}\,\text{electrons/m}^3.$$

From Equation 3.20, we then find that

$$E_F = (3\pi^2)^{2/3}\left[\frac{(1.054 \times 10^{-34})^2}{2(9.11 \times 10^{-31})}\right](8.49 \times 10^{28})^{2/3}$$
$$= 1.126 \times 10^{-18}\,\text{J}.$$

Note that we have implicitly suggested that $m \simeq m_0$ in a metal. This is generally a reasonable approximation, since we shall see later on that the electron effective mass in a metal is essentially the free electron mass, unlike in a semiconductor. The Fermi energy of copper is therefore ($1\,\text{eV} = 1.602 \times 10^{-19}\,\text{J}$)

$$E_F = 7.03\,\text{eV}.$$

EXAMPLE 3.10

The Fermi Energy of Aluminum Next, let us find the Fermi energy of aluminum. From a periodic table, its atomic number is $Z = 13$. By the Aufbau principle, we then have the following electronic configuration:

$$1s^2 2s^2 2p^6 3s^2 3p^1.$$

This tells us that there are three outermost valence electrons: two from the 3s orbital and one from the 3p orbital. Each aluminum atom therefore contributes three electrons to the overall electron density.

The bulk density of aluminum is $\rho = 2.7\,\text{g/cm}^3$ and its atomic weight is 26.98 g/mol. The corresponding atomic density is thus

$$\rho = 6.03 \times 10^{28}\,\text{atoms/m}^3$$

with a corresponding electron density three times this value,

$$n = 3 \times (6.03 \times 10^{28})\,\text{electrons/m}^3.$$

Using Equation 3.20, we then find that

$$E_F = (3\pi^2)^{2/3}\left[\frac{(1.054 \times 10^{-34})^2}{2(9.11 \times 10^{-31})}\right][3 \times (6.03 \times 10^{28})]^{2/3}$$
$$= 1.87 \times 10^{-18}\,\text{J}.$$

As before note that we have assumed the free electron mass in the calculation. The desired Fermi energy of aluminum is therefore

$$E_F = 11.65\,\text{eV}.$$

Other Metals

The reader may find the analogous Fermi energies of gold and silver as an exercise. **Table 3.3** summarizes these values as well as those found above. Note that Table 1.1 of Ashcroft and Mermin (1976) may be useful in finding the Fermi energies of other elements.

3.7.2 Fermi Velocity

Let us now use these Fermi energies to find the associated electron velocity, called the Fermi velocity. From this, we will use the de Broglie relationship to obtain the corresponding electron wavelength, called the Fermi wavelength. This will be our first critical length scale in metals.

Given the relationship between kinetic energy and velocity

$$E = \tfrac{1}{2}mv^2,$$

we substitute $E_F(0\,\text{K})$ for E and rearrange the expression to obtain the desired Fermi velocity

$$v_F = \sqrt{\frac{2E_F(0K)}{m}}. \tag{3.21}$$

Table 3.3 Representative electron densities, Fermi energies, Fermi velocities, and Fermi wavelengths of various metals

Metal	n (electrons/m^3)	Fermi energy (eV)	Fermi velocity (m/s)	Fermi wavelength (nm)
Gold	5.9×10^{28}	5.52	1.39×10^6	0.52
Silver	5.86×10^{28}	5.50	1.39×10^6	0.52
Copper	8.49×10^{28}	7.03	1.57×10^6	0.46
Aluminum	1.81×10^{29}	11.65	2.02×10^6	0.36

We therefore find for copper that

$$v_F = 1.57 \times 10^6 \, \text{m/s}.$$

The reader may likewise calculate the Fermi velocities of the other metals just considered. Resulting values can be compared with those listed in **Table 3.3**.

3.7.3 Fermi Wavelength

To evaluate the associated electron de Broglie wavelength, called the Fermi wavelength, we employ the de Broglie relationship $\lambda = h/p = h/mv$ (Equation 3.1). Namely, we find that the Fermi wavelength of an electron in copper is

$$\begin{aligned}\lambda_F &= \frac{6.62 \times 10^{-34}}{(9.11 \times 10^{-31})(1.57 \times 10^6)} \\ &= 4.63 \times 10^{-10} \, \text{m} \\ &= 0.46 \, \text{nm},\end{aligned}$$

which is extremely small. In fact, it starts to become comparable to the interatomic spacings first discussed in Chapter 2. As a consequence, quantization effects in metal nanostructures occur *only* at very small sizes. Thus, with few exceptions, the majority of metal nanostructures seen in the literature do not exhibit confinement effects. The reader may verify this as an exercise. As a further exercise, the reader may calculate the corresponding Fermi wavelengths of the other metals considered. Evaluated λ_F values are listed in **Table 3.3**.

3.7.4 The Kubo Gap

Finally, let us calculate the energy spacing between levels of small metal nanostructures as originally done by Kubo (1962). Starting with Equations 3.18 and 3.19, we have

$$\rho_{\text{energy}}(E) = \frac{1}{2\pi^2} \left(\frac{2m}{\hbar^2} \right)^{3/2} \sqrt{E},$$

as well as

$$n(0\,\text{K}) = \frac{1}{3\pi^2} \left(\frac{2m}{\hbar^2} \right)^{3/2} E_F(0\,\text{K})^{3/2}.$$

These equations express the carrier density and carrier concentration in a metal with units of number per unit volume per unit energy and number per unit volume, respectively. We first convert $\rho_{\text{energy}}(E)$ to a state density by dividing by 2 to remove the effects of spin degeneracy. This gives

$$\rho_{\text{state}}(E) = \tfrac{1}{2} \rho_{\text{energy}}(E).$$

Next, when $E = E_F(0\,\text{K})$ in ρ_{state}, we have

$$\rho_{\text{state}}(E_F(0\,\text{K})) = \frac{1}{4\pi^2} \left(\frac{2m}{\hbar^2} \right)^{3/2} \sqrt{E_F(0\,\text{K})}.$$

Then taking the ratio

$$\frac{\rho_{state}(E_F(0\,\mathrm{K}))}{n(0\,\mathrm{K})} = \frac{\frac{1}{4\pi^2}\left(\frac{2m}{\hbar^2}\right)^{3/2}\sqrt{E_F(0\,\mathrm{K})}}{\frac{1}{3\pi^2}\left(\frac{2m}{\hbar^2}\right)^{3/2} E_F(0K)^{3/2}}$$
$$= \frac{3}{4}\frac{1}{E_F(0\,\mathrm{K})}$$

leads to

$$\rho_{state}(E_F(0\,\mathrm{K})) = \frac{3}{4}n(0\,\mathrm{K})\frac{1}{E_F(0\,\mathrm{K})},$$

which is an alternative expression for our original state density. We now define a final state density $D(E_F)$ as

$$D(E_F(0\,\mathrm{K})) = \rho_{state}(E_F(0K))V_{tot}$$
$$= \frac{3}{4}[n(0\,\mathrm{K})V_{tot}]\frac{1}{E_F(0\,\mathrm{K})},$$

with units of number per unit energy and with V_{tot} a volume. The product $N(0\,\mathrm{K}) = n(0\,\mathrm{K})V_{tot}$ represents the total number of electrons in the nanostructure. Finally, the energy spacing between states is just

$$\delta\epsilon = \frac{1}{D(E_F(0\,\mathrm{K}))} = \frac{4}{3}\frac{E_F(0\,\mathrm{K})}{N(0\,\mathrm{K})} \tag{3.22}$$

and is called the Kubo gap. Equation 3.22 allows us to estimate the significance of confinement in metal nanostructures.

EXAMPLE 3.11

Numerical Examples Using Equation 3.22, let us numerically evaluate the magnitude of the Kubo gap for various metal nanostructures. For simplicity, assume spherical particles made of copper, silver, gold, or aluminum. Their Fermi energies can be found in **Table 3.3**. From Chapter 2, we also know that all of these metals possess a FCC crystal structure with respective unit cell lengths of

- Au, $a = 4.08\,\text{Å}$
- Ag, $a = 4.09\,\text{Å}$
- Cu, $a = 3.61\,\text{Å}$
- Al, $a = 4.05\,\text{Å}$

In addition, recall that in FCC lattices, there are four atoms per unit cell.

For illustration purposes, consider a 3 nm ($r = 15\,\text{Å}$) diameter particle in each case. The associated volume is

$$V_{sphere} = \tfrac{4}{3}\pi(15)^3\,\text{Å}^3,$$

where $V_{sphere} = 14\,137.2\,\text{Å}^3$. The corresponding unit-cell volumes a^3 for Au, Ag, Cu, and Al are then

- Au, $V_{unit} = 67.92\,\text{Å}^3$
- Ag, $V_{unit} = 68.42\,\text{Å}^3$
- Cu, $V_{unit} = 47.05\,\text{Å}^3$
- Al, $V_{unit} = 66.43\,\text{Å}^3$

At this point, the number of unit cells, n_{unit}, contained within each nanoparticle is the ratio V_{sphere}/V_{unit}, resulting in

- Au, $n_{unit} = 208$
- Ag, $n_{unit} = 207$
- Cu, $n_{unit} = 301$
- Al, $n_{unit} = 213$

Given that there are four atoms per unit cell, the total number of atoms within each 3 nm diameter particle is

- Au, $N = 832$
- Ag, $N = 828$
- Cu, $N = 1204$
- Al, $N = 852$

Since each Au, Ag, and Cu atom contributes one electron while each Al atom contributes three, we find

- Au, $N(0\,\text{K}) = 832$
- Ag, $N(0\,\text{K}) = 828$
- Cu, $N(0\,\text{K}) = 1204$
- Al, $N(0\,\text{K}) = 2556$

Inserting these numbers into Equation 3.22 with appropriate units then gives the corresponding Kubo gaps:

- Au, $\delta\epsilon = 8.85 \times 10^{-3}\,\text{eV} = 8.9\,\text{meV}$
- Ag, $\delta\epsilon = 8.86 \times 10^{-3}\,\text{eV} = 8.9\,\text{meV}$
- Cu, $\delta\epsilon = 7.8 \times 10^{-3}\,\text{eV} = 7.8\,\text{meV}$
- Al, $\delta\epsilon = 1.82 \times 10^{-3}\,\text{eV} = 6.1\,\text{meV}$

All of these energies are smaller than kT at room temperature. As a consequence, quantum size effects in metal nanostructures occur only in very small particles where $r < 1$ nm or when temperatures are very low. The reader can drive this point home more strongly by calculating the expected Kubo gap when the nanoparticle diameter equals the Fermi wavelength of each metal listed in **Table 3.3**.

3.8 THE MEAN FREE PATH IN METALS

We now work our way up to larger sizes by discussing the second of our three critical length scales in metals. This is the mean free path of electrons in metals and turns out to be of the order of 10 nm. Recall that the mean free path represents the average distance a particle (an electron in this case) can travel in a system without suffering a collision. In metals, these collisions can be electron–electron or electron–phonon in nature. Note, however, that we will not discuss phonons here, and the reader may consult more advanced texts for additional information about this subject.

The claim is that when metal nanostructures have sizes comparable to or smaller than the electron mean free path d_{mean}, changes occur to their optical properties. One can qualitatively see this because when the size of the nanostructure becomes small compared with d_{mean}, there will be a surface into which the carrier eventually runs into and scatters from. We therefore show that this ultimately leads to size-dependent changes in its dielectric constant, which, in turn, affects the metal's absorption spectrum. In what follows, we will therefore derive an expression for the dielectric constant of a metal using the so-called Drude–Lorentz model. We will then relate this to the absorption coefficient of the system, which dictates the appearance of the metal's absorption spectrum. Absorption coefficients are more fully described in Chapter 5.

Let us first evaluate the mean free path of an electron in a metal. We start with the definition of the current density J:

$$J = nqv_d = \frac{I}{A}, \tag{3.23}$$

where n is the electron concentration with units of number per unit volume, q is the elementary unit of charge, v_d is the so-called electron drift velocity, I represents the current, and A is a cross-sectional area. From Ohm's law, we also have

$$V = IR, \tag{3.24}$$

where R is a resistance with units of Ω. Alternatively, R can be defined in terms of a resistivity ρ with units of Ω m (not to be confused with the densities ρ seen earlier) and a length L:

$$R = \frac{\rho L}{A}. \tag{3.25}$$

Likewise, a conductivity σ can be expressed as

$$\sigma = \frac{1}{\rho}, \tag{3.26}$$

with units of S/m. When Equation 3.23 for the current density is rewritten, we find

$$\begin{aligned} nqv_d &= \frac{I}{A} \\ &= \frac{1}{A}\left(\frac{V}{R}\right) = \frac{1}{A}\left(\frac{VA}{\rho L}\right) = \frac{1}{\rho}\frac{V}{L} \\ &= \sigma \frac{V}{L}. \end{aligned}$$

Since V/L is equivalent to an electric field E, we alternatively write

$$nqv_d = \sigma E. \tag{3.27}$$

We now need an expression for the drift velocity v_d in terms of E (note that we are working with the magnitude of the field). Given Newton's second law

$$F = ma,$$

Table 3.4 Conductivity, electron mean free path, and effective electron mass of various metals

Metal	Conductivity (S/m)	Electron mass (m_0)	Mean free path (nm)
Gold	4.52×10^7	1.01	38.1
Silver	6.30×10^7	1.03	54.6
Copper	5.90×10^7	1.47	56.9
Aluminum	3.54×10^7	1.16	16.3

where the force on a charge due to the presence of a field is written as $F = qE$, we find the following equivalence:

$$ma = qE,$$

$$m\frac{dv}{dt} = qE.$$

We now approximate $dv/dt \sim v_d/\tau$, where $\tau = d_{\text{mean}}/v_F$ is a characteristic time between collisions, given a mean free path d_{mean} and an electron Fermi velocity v_F. We therefore have

$$m\frac{v_d}{\tau} = qE,$$

whereupon solving for the drift velocity gives $v_d = (qE/m)\tau$, or

$$v_d = \frac{qE}{m}\left(\frac{d_{\text{mean}}}{v_F}\right). \tag{3.28}$$

Finally, when replaced back into Equation 3.27, we find that

$$\frac{nq^2E}{m}\left(\frac{d_{\text{mean}}}{v_F}\right) = \sigma E,$$

from which

$$d_{\text{mean}} = \frac{mv_F\sigma}{nq^2}. \tag{3.29}$$

At this point, we can use experimental values of the conductivity σ to find representative d_{mean} values. Parameters for several common metals are listed in **Table 3.4**.

EXAMPLE 3.12

The Electron Mean Free Path in Copper Let us begin by calculating d_{mean} for copper. Using Equation 3.29 and values from **Table 3.3** as well as **Table 3.4**, we find

$$\begin{aligned} d_{\text{mean}} &= \frac{mv_F\sigma}{nq^2} \\ &= \frac{[1.47(9.11 \times 10^{-31})](1.57 \times 10^6)(5.9 \times 10^7)}{(8.49 \times 10^{28})(1.602 \times 10^{-19})^2}\ \text{m} \\ &= 5.69 \times 10^{-8}\ \text{m}. \end{aligned}$$

The resulting mean free path of the electron in copper is therefore

$$d_{\text{mean}} = 56.9\ \text{nm}.$$

As an exercise, the reader may calculate the mean free paths of electrons in silver, gold, and aluminum to verify the values listed in **Table 3.4**.

Thus, when copper nanostructures become smaller than d_{mean}, their optical properties change due to the additional scattering of electrons from the surface. We will see this explicitly in what follows when we use the so-called Drude–Lorentz model to obtain an expression for the real and imaginary parts of a metal's dielectric constant. The latter term relates in turn to the metal's absorption spectrum.

3.8.1 The Drude–Lorentz Model

To complete our discussion, we introduce the Drude–Lorentz model, which describes the interaction between light and a free electron in a metal. Results from this model can subsequently be used to describe the system's intraband transitions. However, our intent here is to find an expression for the metal's dielectric constant, since we have just asserted that it will change below a critical size. The reader can skip ahead to the results, Equations 3.41 and 3.42, if desired.

Let us begin with the following second-order differential equation:

$$\frac{d^2x(t)}{dt^2} + \frac{\Gamma}{m}\frac{dx(t)}{dt} = -\frac{q}{m}E(t), \tag{3.30}$$

which describes the time-dependent spatial displacement $x(t)$ of an electron in a metal interacting with light. The equation can be constructed using Hooke's law as follows. Specifically, recall that the force on a spring is given by

$$F_{\text{Hooke}} = -k_{\text{Hooke}}x(t) = -m\omega_0^2 x(t),$$

where k_{Hooke} is the spring constant related to the natural resonant frequency of the system, $\omega_0 = \sqrt{k_{\text{Hooke}}/m}$. The physical picture here is of an electron loosely bound to its parent atom.

Next, the restoring force on the charge is counteracted by the force exerted on it by the electric field of the incident light. This force is simply

$$F = -qE(t),$$

where $E(t)$ is the time-dependent electric field. See **Figure 3.11**. The net force on the electron is therefore the sum of two terms:

$$F_{\text{net}} = -m\omega_0^2 x(t) - qE(t).$$

Upon invoking Newton's second law ($F = ma$), this equation can be written as

$$F_{\text{net}} = m\frac{d^2x(t)}{dt^2} = -m\omega_0^2 x(t) - qE(t).$$

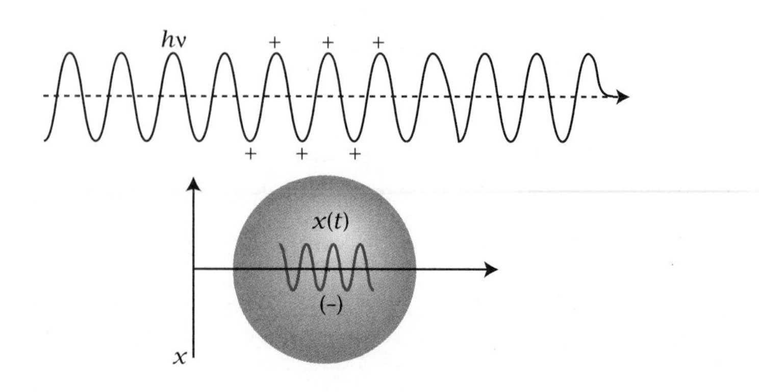

Figure 3.11 Schematic of the time-dependent electron displacement in a metal under illumination.

Finally, introducing a phenomenological damping force $F_{damp} = -\Gamma dx(t)/dt$ accounts for "friction" in the metal, which suppresses the motion of the oscillating charge. The net force on the electron is therefore

$$F_{net} = m\frac{d^2x(t)}{dt^2} = -m\omega_0^2 x(t) - qE(t) - \Gamma\frac{dx(t)}{dt}.$$

Rearranging terms gives

$$m\frac{d^2x(t)}{dt^2} + \Gamma\frac{dx(t)}{dt} + m\omega_0^2 x(t) = -qE(t), \tag{3.31}$$

whereupon letting $\omega_0 = 0$ (a completely free electron with no restoring force) and dividing by m yields Equation 3.30.

If we describe the incident light as a traveling wave (traveling waves are discussed more fully in Chapter 5), we can write its time-dependent electric field as

$$\begin{aligned} E(x,t) &= E_0 e^{i(kx-\omega t)} \\ &= E_0 e^{ikx} e^{-i\omega t}, \end{aligned}$$

where E_0 is the amplitude of the electric field, $k = 2\pi/\lambda$ is the wavevector of light (of wavelength λ), and $\omega = 2\pi\nu$ is its angular frequency. We have implicitly assumed that the light propagates along the x direction of a Cartesian coordinate system.

We can simplify the expression further if $kx \ll 1$, since, in this case,

$$e^{ikx} \simeq 1.$$

Physically, what this says is that if the size of the nanostructure is much smaller than the wavelength of light, spatial variations of the electric field can be ignored. Under these conditions, we can therefore write

$$E(t) = E_0 e^{-i\omega t}, \tag{3.32}$$

which is sometimes called the quasistatic approximation.

Next, to solve Equation 3.30, we assume a solution of the form

$$x(t) = x_0 e^{-i\omega t}, \tag{3.33}$$

where x_0 is a prefactor with units of length. Qualitatively, this makes sense since the position of the electron in the metal oscillates under the influence of the light's electric field as shown in **Figure 3.11**. When this solution is inserted into Equation 3.30, we obtain

$$\begin{aligned} -\omega^2 x_0 e^{-i\omega t} + \frac{\Gamma}{m}\left(-i\omega x_0 e^{-i\omega t}\right) &= -\frac{q}{m}E_0 e^{-i\omega t} \\ -\omega^2 x_0 + \frac{\Gamma}{m}(-i\omega x_0) &= -\frac{qE_0}{m}, \end{aligned}$$

whereupon solving for x_0 gives

$$x_0 = \frac{qE_0/m}{\omega^2 + i\Gamma\omega/m}. \tag{3.34}$$

The following expression therefore describes our time-dependent electron displacement

$$x(t) = \frac{q/m}{\omega^2 + i\gamma\omega}E_0 e^{-i\omega t}, \tag{3.35}$$

with $\gamma = \Gamma/m$.

To evaluate an expression for the metal's dielectric constant, we first find the *microscopic* dipole moment (also called the microscopic polarization) μ from

$$\mu(t) = -qx(t). \tag{3.36}$$

Note that the Greek letter μ is also often used to represent a reduced mass; however, the reader should be able to discern which one is being referred to from the context of the discussion.

We thus have

$$\mu(t) = -\frac{q^2/m}{\omega^2 + i\gamma\omega} E_0 e^{-i\omega t}, \tag{3.37}$$

whereupon the *macroscopic* polarization P can be written as $P = n\mu$, with n the number of charges present per unit volume in the metal. As a consequence,

$$P = -\frac{nq^2/m}{\omega^2 + i\gamma\omega} E_0 e^{-i\omega t}. \tag{3.38}$$

In turn, we find an expression for the dielectric constant of the metal through the displacement D:

$$D = \epsilon\epsilon_0 E = \epsilon_0 E + P. \tag{3.39}$$

From this equivalence, we find

$$\epsilon\epsilon_0 E_0 e^{-i\omega t} = \epsilon_0 E_0 e^{-i\omega t} - \frac{nq^2/m}{\omega^2 + i\gamma\omega} E_0 e^{-i\omega t},$$

which ultimately leads to the following expression for the metal's dielectric constant:

$$\epsilon = 1 - \frac{nq^2/\epsilon_0 m}{\omega^2 + i\gamma\omega}.$$

Multiplying the numerator and denominator of the last term by $\omega^2 - i\gamma\omega$ yields

$$\epsilon = \left[1 - \frac{\frac{nq^2}{\epsilon_0 m}}{\omega^2 + \gamma^2}\right] + i\left[\frac{\left(\frac{\gamma}{\omega}\right)\frac{nq^2}{\epsilon_0 m}}{\omega^2 + \gamma^2}\right],$$

where we have split ϵ into a real part and an imaginary part. The reader may verify this result.

To further simplify the expression, let us define a plasma frequency

$$\omega_p^2 = \frac{nq^2}{\epsilon_0 m} \tag{3.40}$$

such that

$$\epsilon(\omega) = \left[1 - \frac{\omega_p^2}{\omega^2 + \gamma^2}\right] + i\left[\frac{\left(\frac{\gamma}{\omega}\right)\omega_p^2}{\omega^2 + \gamma^2}\right].$$

By inspection, the real and imaginary parts of the dielectric constant are then

$$\epsilon_1 = 1 - \frac{\omega_p^2}{\omega^2 + \gamma^2} \tag{3.41}$$

and

$$\epsilon_2 = \frac{\left(\frac{\gamma}{\omega}\right)\omega_p^2}{\omega^2 + \gamma^2}. \tag{3.42}$$

At this point, note that the imaginary part of the dielectric constant (Equation 3.42) underlies the absorption of a material because ϵ_2 directly relates to the system's absorption coefficient α. We will see more of this in Chapter 5. However, for now, just note that the relationship between the two is given by

$$\alpha = \frac{\omega\epsilon_2}{cn}. \tag{3.43}$$

Additional information about these expressions can be found in Fox (2010).

Now, where the size of the metal nanostructure begins to matter is in γ, within the denominator of both Equations 3.41 and 3.42. This term in effect reflects a mean electron collision time in the metal, $\gamma = \tau^{-1}$, where, in bulk materials, τ is simply the mean free path divided by the Fermi velocity, $\tau = d_{\text{mean}}/v_F$. In nanostructures, smaller than d_{mean}, γ changes due to surface scattering. An expression used in the literature to reflect this is

$$\gamma = \gamma_0 + g\frac{v_F}{D},$$

where D is the particle diameter and g is a unitless number close to 1 that depends on the system. As a consequence, γ has a size dependence, causing ϵ_2 and, in turn, the absorption, to change. The electron mean free path is therefore one more critical length scale in metal nanostructures.

3.9 CHARGING ENERGY

Finally, we discuss the last of our three critical length scales that is related to the electrical properties of metal nanostructures. This is a size, d_{charging}, at which the Coulomb charging energy in the material becomes comparable to kT. The charging energy is the potential that must be overcome in order to add an additional electron to the system. Thus, systems with sizes below d_{charging} show discreteness in their electrical response at room temperature, in contrast to the behavior of their bulk counterparts.

We begin with an expression for the capacitance of a small metal sphere (a representative nanoparticle) with a radius a, surrounded by a medium with a dielectric constant ϵ:

$$C_{\text{sphere}} = 4\pi\epsilon\epsilon_0 a.$$

The total energy stored in this system is then

$$U \simeq \frac{q^2}{2C_{\text{sphere}}},$$

where q is the charge on the sphere. This yields our desired expression for the charging energy:

$$U = \frac{q^2}{8\pi\epsilon\epsilon_0 a}, \tag{3.44}$$

which becomes large compared with kT when a is small.

EXAMPLE 3.13

The Charging Energy of a 3 nm Diameter Metal Sphere To illustrate Equation 3.44, let us consider a 3 nm diameter metal sphere embedded in a dielectric medium ($\epsilon = 2$). We can then find the particle's associated charging energy as follows:

$$\begin{aligned} U &= \frac{q^2}{8\pi\epsilon\epsilon_0 a} \\ &= \frac{(1.602 \times 10^{-19})^2}{8\pi(2)(8.854 \times 10^{-12})(1.5 \times 10^{-9})} \\ &= 3.84 \times 10^{-20}\,\text{J}. \end{aligned}$$

This is equivalent to 0.24 eV and is large compared with kT at room temperature ($\simeq$25 meV). As a consequence, it is readily apparent that the charging energy becomes significant when the size of the object becomes small.

We can also see where U becomes comparable to kT. Specifically, at room temperature, $kT \simeq 0.025$ eV or 4×10^{-21} J. So, working backwards, we have

$$4 \times 10^{-21} = \frac{(1.602 \times 10^{-19})^2}{8\pi(2)(8.854 \times 10^{-12})(a)},$$

whereupon solving for a gives a value of 14.4 nm. Obviously, the radius where the particle's charging energy becomes significant depends on the dielectric constant of the local environment as well as on the actual temperature of interest.

Due to these significant charging energies, discreteness will appear in electrical measurements conducted on such metal nanostructures. Namely, steps occur in the current during current/voltage measurements. This has, in turn, led to the development of single-electron transistors, which we will describe briefly in Chapter 15. A literature example of this phenomenon can be found in Chen et al. (1998).

3.10 SUMMARY

With this, we conclude our discussion of critical length scales defining nano in semiconductors and metals. We have previously seen that size-dependent properties in semiconductors occur only when the nanostructure's size becomes comparable to or smaller than the de Broglie wavelength or associated Bohr radius of carriers in the material. In metals, this means that only molecular-like species exhibit confinement effects. However, other characteristic length scales exist in metals where their optical and electrical properties become size-dependent. In Chapter 4, we introduce quantum wells, nanowires, and quantum dots as representative low-dimensional systems and go on to provide phenomenological descriptions of their optical properties in Chapter 5.

3.11 THOUGHT PROBLEMS

3.1 Conversions

What is the wavelength in nanometers of

(a) a 1 eV photon

(b) a 3 eV photon?

Derive a simple rule-of-thumb expression that allows you to convert between the two. It should be

$$\lambda(\mathrm{nm}) \simeq \frac{1240}{E(\mathrm{eV})}.$$

3.2 Thermal energy

Provide a value for the thermal energy kT in units of meV at the following temperatures: 300 K, 77 K, 10 K, and 4 K. Low-temperature experiments are commonly conducted at 4 K and 77 K. Suggest why.

3.3 de Broglie wavelength

What is your de Broglie wavelength when moving at 10 m/s? What is the de Broglie wavelength of a baseball moving at 50 m/s?

3.4 Bohr model

Using the Bohr model, show that the radius of the first Bohr orbit in hydrogen is $a_0 = 0.528\,\text{Å}$. Find the associated speed of the electron in this lowest orbit and its ratio with the speed of light. This ratio is called the fine structure constant α and has a value of $\alpha \simeq \frac{1}{137}$.

3.5 Bohr model

Estimate the magnitude of the electric field experienced by an electron in the first Bohr orbit.

3.6 Bohr radii

For InP, which has the parameters $m_e = 0.077m_0$, $m_h = 0.6m_0$, and $\epsilon = 12.4$, find the individual electron and hole Bohr radii. Provide physical size ranges for nanostructures made of this material to be in the strong, intermediate, and weak confinement regimes.

3.7 Bulk exciton Bohr radius

Calculate the bulk exciton Bohr radius of the following materials: PbSe, PbS, PbTe, CdTe, CdSe, CdS, InAs, and Si. Use effective electron and hole masses, obtained from a reference such as Landolt–Börnstein. If needed, use $m_{\text{eff}} = (m_t^2 m_l)^{1/3}$, as well as the "heavy hole" mass. Note that the effective mass of a carrier, in general, depends on direction.

3.8 Bulk exciton Bohr radius

InP has a zinc blende crystal structure with a lattice constant $a = 5.9\,\text{Å}$. It has a dielectric constant $\epsilon = 12.4$ and an electron (hole) effective mass $m_e = 0.077m_0$ ($m_h = 0.60m_0$). Provide the following:

(a) The bulk exciton Bohr radius

(b) The number of unit cells contained within the lowest $n = 1$ exciton orbit

(c) The highest temperature at which stable excitons can be found

3.9 Confinement

Explain what the size of the exciton Bohr radius means for achieving quantum confinement. What systems are easiest for achieving this effect?

3.10 Bulk exciton Bohr radius

Qualitatively explain the role of the dielectric constant in the expression for the bulk exciton Bohr radius. Consider the case of a nanowire where now instead of a homogeneous bulk environment there exists a dramatic difference in dielectric constant with the surrounding environment. What happens to the bulk exciton Bohr radius and exciton binding energy? See, for example, Muljarov et al. (2000).

3.11 Literature

Locate an article in the literature that states something along the lines of "the size of the material is below the corresponding bulk exciton Bohr radius." Alternatively, locate an instance where an author mentions structures being smaller than the de Broglie wavelength of carriers in the material.

3.12 Literature

Find two examples in the current literature where materials are said to be in any one of the various confinement regimes described in the main text.

3.13 Excitons

Calculate the binding energy and radius of the $n = 1$ and $n = 2$ Mott–Wannier excitons in CdS, where $m_e = 0.2m_0$, $m_h = 0.8m_0$, and $\epsilon = 8.9$. Would you expect these excitons to be stable at room temperature?

3.14 Excitons

Calculate the wavelength difference of the $n = 1$ and $n = 2$ excitons in bulk CdTe, which has a band gap $E_g \simeq 1.5\,\text{eV}$, an electron (hole) effective mass $m_e = 0.11m_0$ ($m_h = 0.35m_0$), and $\epsilon = 10.9$.

3.15 Metals

Calculate E_F, v_F, and λ_F for Zn and Mn. Assume two valence electrons for each. To help you, see Table 1.1 of Ashcroft and Mermin (1976).

3.16 Drude–Lorentz model

An expression that relates the susceptibility of a metal to its dielectric constant is

$$\epsilon = 1 + \chi.$$

Find expressions for the real (χ_1) and imaginary (χ_2) parts of the susceptibility.

3.17 Drude–Lorentz model

It is often useful to look at plots of ϵ_1, ϵ_2, n, k, and α for metals. Plot all of the above functions for the following conditions in relative frequency units.

(a) $\omega_p = 10$ and $\gamma = 0.0005$

(b) $\omega_p = 10$ and $\gamma = 0.5$

3.18 Drude–Lorentz model

Derive the following expression for the absorption coefficient α of a material in terms of its susceptibility:

$$\alpha = \frac{\omega}{cn}\chi_2,$$

where χ_2 is the imaginary part of the complex susceptibility χ. Next show that if the real part of the susceptibility (χ_1) is much larger than χ_2 (i.e., $\chi_1 \gg \chi_2$), the expression can be rewritten as

$$\alpha = \frac{\omega}{c}\frac{\chi_2}{\sqrt{1+\chi_1}}.$$

Hint: consider how the complex refractive index relates to the susceptibility.

3.19 Lorentz model

In addition to the Drude–Lorentz model, there is a complementary Lorentz model. It is a classic oscillator model that describes the interaction of a bound electron in an atom with light using the same ball-and-spring description. The second-order differential equation describing the displacement of the bound electron is

$$\frac{d^2x(t)}{dt^2} + \frac{\Gamma}{m}\frac{dx(t)}{dt} + \omega_0^2 x(t) = -\frac{eE(t)}{m},$$

where $x(t)$ is the electron displacement, Γ is a phenomenological oscillation damping factor, m is the electron mass, ω_0 is the spring's resonant frequency, and $E(t)$ is the electric field of the incident light. Notice that the main difference between the Drude–Lorentz model and the Lorentz model is the absence of the $\omega_0^2 x(t)$ term in the former since electrons in metals are more or less free.

Assume the following general solution to the displacement above:

$$x(t) = x_0 e^{-i\omega t},$$

with $E(t) = E_0 e^{-i\omega t}$. Derive the following expressions:

(a) The time-dependent displacement

$$x(t) = \frac{-eE_0/m}{(\omega_0^2 - \omega^2) - i\gamma\omega} e^{-i\omega t},$$

where $\gamma = \Gamma/m$ is a phenomenological damping frequency.

(b) The microscopic dipole moment

$$\mu(t) = \frac{e^2E_0/m}{(\omega_0^2 - \omega^2) - i\gamma\omega} e^{-i\omega t}$$

(c) The polarizability (with $\mu = \alpha E(t)$)

$$\alpha(\omega) = \frac{e^2/m}{(\omega_0^2 - \omega^2) - i\gamma\omega}$$

(d) The real and imaginary parts of the polarizability (α_1 and α_2)

(e) The macroscopic polarization ($P = n\mu$)

$$P = \frac{ne^2E_0/m}{(\omega_0^2 - \omega^2) - i\gamma\omega} e^{-i\omega t},$$

(f) The complex dielectric constant through the equivalence $D = \epsilon\epsilon_0 E(t) = \epsilon_0 E + P$

(g) The real and imaginary parts of the dielectric constant (ϵ_1, ϵ_2)

(h) The real and imaginary parts of the refractive index (n, k)

3.20 Lorentz model

The ϵ_1 and ϵ_2 expressions from the Lorentz model above can sometimes be written as

$$\epsilon_1(\Delta\omega) = \epsilon_\infty + [\epsilon(0) - \epsilon_\infty]\frac{2\omega_0\Delta\omega}{4(\Delta\omega)^2 + \gamma^2},$$

$$\epsilon_2(\Delta\omega) = [\epsilon(0) - \epsilon_\infty]\frac{\gamma\omega_0}{4\Delta\omega^2 + \gamma^2},$$

where $\Delta\omega = \omega_0 - \omega$ is the "detuning" off resonance. Use the parameters $\omega_0 = 10^{14}$ rad/s, $\gamma = 5 \times 10^{12}$ rad/s, $\epsilon(0) = 12.1$, and $\epsilon_\infty = 10$ to plot the two functions. The second ϵ_2 plot yields a Lorentzian function.

3.21 Lorentz model

Armed with expressions for ϵ_1, ϵ_2, n, k, ω_p, α, and R, it is often illustrative to plot n, k, α, and R for different values of ω_0, ω_p, and γ. Plot these functions using relative units for both the frequency and linewidth. Consider the following cases:

(a) $\omega_0 = 4$, $\omega_p = 8$, and $\gamma = 1$

(b) $\omega_0 = 4$, $\omega_p = 8$, and $\gamma = 0.3$

3.22 Lorentz model

The polarization of a solid is related to its susceptibility χ by $P = \epsilon_0\chi E$. Derive an expression for the susceptibility and express it as a sum of a real part χ_1 and an imaginary part χ_2. Then provide approximations for both when near resonance. The imaginary part should end up as a Lorentzian. To simplify things, let $Ne^2/\epsilon_0 m_0 = \omega_p^2$ (the plasma frequency), where N is the number of bound charges per unit volume.

3.23 Lorentz model

For a hypothetical solid with three well-separated resonances, plot the refractive index and absorption coefficient, both as functions of frequency, given

- $\omega_1 = 4 \times 10^{13}$ rad/s and $\gamma_1 = 4 \times 10^{12}$ rad/s
- $\omega_2 = 4 \times 10^{15}$ rad/s and $\gamma_2 = 5 \times 10^{14}$ rad/s
- $\omega_3 = 1 \times 10^{17}$ rad/s and $\gamma_3 = 3 \times 10^{13}$ rad/s

The mass associated with the first transition is 10 000 times the free electron mass. The mass for both the second and third transitions is the free electron mass. Also, the number of bound charges per unit volume is $N = 10^{22}$ m^{-3}. Explain what these three transitions are and why the first mass is so much larger than the rest.

3.24 Lorentz model

Show that on resonance the absorption coefficient of a Lorentz oscillator has no ω_0 dependence.

3.12 REFERENCES

Arndt M, Nairz O, Vos-Andreae J, et al. (1999) Wave–particle duality of C_{60} molecules. *Nature* **401**, 680.

Ashcroft NW, Mermin ND (1976) *Solid State Physics*. Brooks/Cole, Belmont, CA.

Brezger B, Hackermüller L, Uttenthaler S, et al. (2002) Matter-wave interferometer for large molecules. *Phys. Rev. Lett.* **88**, 100404-1.

Chen S, Ingram RS, Hostetler MJ, et al. (1998) Gold nanoelectrodes of varied size: Transition to molecule-like charging. *Science* **280**, 2098.

Efros AlL, Rosen M (2000) The electronic structure of semiconductor nanocrystals. *Annu. Rev. Mater. Sci.* **30**, 475.

Fox M (2010) *Optical Properties of Solids*, 2nd edn. Oxford University Press, Oxford, UK.

Gaponenko SV (1999) *Optical Properties of Semiconductor Nanocrystals*. Cambridge University Press, Cambridge, UK.

Kroto HW, Heath JR, O'Brien SC, et al. (1985) C_{60}: buckminsterfullerene. *Nature* **318**, 162.

Kubo R (1962) Electronic properties of metallic fine particles. I. *J. Phys. Soc. Jpn* **17**, 975.

Muljarov EA, Zhukov EA, Dneprovskii VS, Masumoto Y (2000) Dielectrically enhanced excitons in semiconductor–insulator quantum wires: theory and experiment. *Phys. Rev. B* **62**, 7420.

Zheng J, Nicovich PR, Dickson RM (2007) Highly fluorescent noble-metal quantum dots. *Annu. Rev. Phys. Chem.* **58**, 409.

Types of Nanostructures

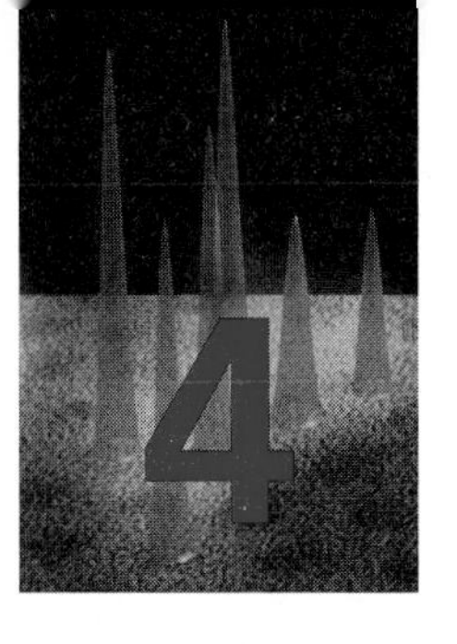

4.1 INTRODUCTION

In this chapter, we catalog the various types of nanostructures that we will encounter in this text. Salient terminology and defining features are highlighted. Furthermore, images of actual nanostructures are provided to enable the reader to see what these materials look like in practice. The chapter ends by describing two general strategies for making nanostructures, namely, bottom-up synthesis and top-down manufacturing. However, specific details about actual syntheses are left for Chapter 13.

4.1.1 Semiconductors

Among semiconductor systems, there are three generic types of nanostructures important to us. They are referred to as

- Quantum wells
- Quantum wires (also called nanowires)
- Quantum dots (also called nanocrystals, colloidal quantum dots, or self-assembled quantum dots, depending on how they were made).

These structures are alternatively referred to as "low-dimensional" systems and possess a given dimensionality. Specifically, they are called

- two-dimensional (2D) systems (quantum wells)
- one-dimensional (1D) systems (quantum wires)
- zero-dimensional (0D) systems (quantum dots)

where the stated dimensionality reflects the degrees of freedom available for carriers (i.e., electrons and holes) in the material. These three types of semiconductor nanostructures are illustrated schematically in **Figure 4.1**.

From **Figure 4.1(a)**, one sees that a quantum well is essentially a thin slab of semiconductor sandwiched by an insulator on either side. When its physical thickness becomes narrow such that a, depicted along the z direction of a Cartesian coordinate system, is smaller than the characteristic length of electrons and holes in the material, carrier confinement occurs. Recall our discussion about appropriate length scales for nano in Chapter 3, where we introduced both the electron and hole de Broglie wavelength as well as the bulk exciton Bohr radius a_B.

At the same time, since the dimensions of the material along x and y are much larger than either of these critical lengths, carriers experience no additional confinement effects. They therefore possess two degrees of freedom. In summary, electrons and holes in quantum wells have one degree of confinement and two degrees of freedom.

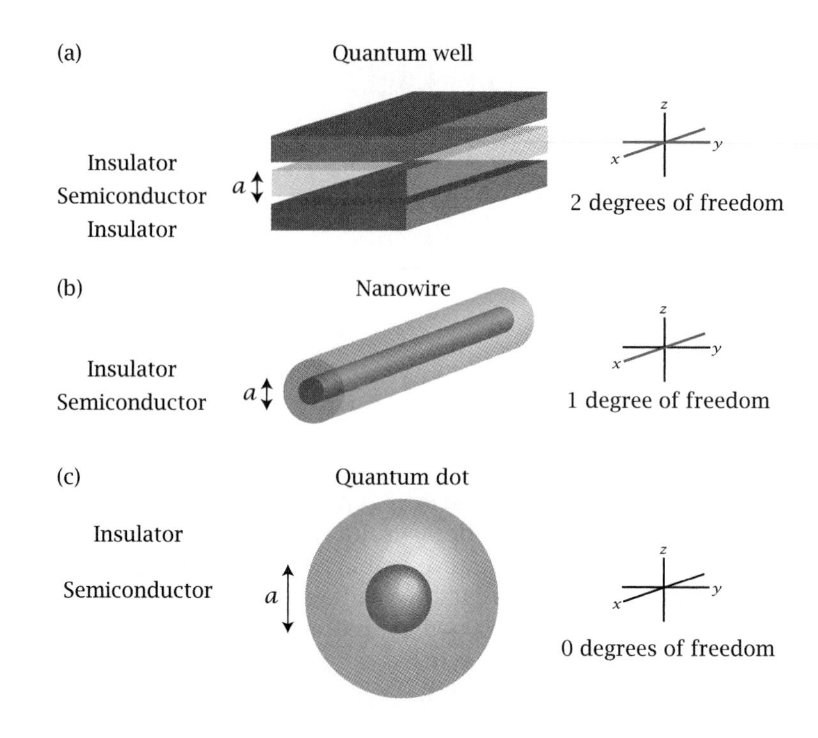

Figure 4.1 Schematic illustration of (a) a quantum well (2D system), (b) a quantum wire (1D system), and (c) a quantum dot (0D system). In all cases, the inner dark red region represents the semiconductor nanostructure of interest and the surrounding light red region denotes an insulator.

Next, a nanowire is essentially a semiconductor elongated along a single direction with two associated (orthogonal) dimensions that are much narrower. **Figure 4.1(b)** shows a generic nanowire where the x direction of a Cartesian coordinate system depicts its growth axis. The two complementary (narrow) dimensions along y and z then result in the confinement of carriers, since their physical dimensions are smaller than the characteristic lengths of electrons and holes in the material. Carriers in nanowires therefore possess one degree of freedom with two degrees of confinement. Note that the actual nanowire geometry differs depending on the synthesis. For convenience, we have simply illustrated a cylindrical wire in **Figure 4.1(b)**. However, rectangular geometries or other shapes are possible. In all cases, an insulator surrounds the structure along all three directions.

Quantum dots are the logical (and limiting) extension to the above carrier confinement trend. In this case, the structure is narrow along all three directions of a Cartesian coordinate system. A generic quantum dot is illustrated in **Figure 4.1(c)**. It is depicted as a small spherical semiconductor particle with a radius a smaller than the characteristic de Broglie wavelength or Bohr radius of carriers in the material. Note that, as with nanowires, other quantum dot geometries are possible, depending on how they were made. In all cases, confinement occurs along all three directions of a Cartesian coordinate system. As such, carriers possess no degrees of freedom. **Table 4.1** summarizes the degrees of confinement as well as the degrees of freedom for carriers in quantum wells, wires, and dots.

4.1.2 Examples

We now illustrate actual examples of nanowires and quantum dots. For convenience, let us skip quantum wells, since this is a well-established area. The interested reader may refer to Sridhara Rao et al. (1999) and references therein for example micrographs that depict these 2D structures. A brief history of their development can also be found in Zory (1993).

Table 4.1 Degrees of confinement and degrees of freedom for low-dimensional materials

Structure	Degrees of confinement	Degrees of freedom	Nomenclature
Bulk	0	3	Bulk material
Quantum well	1	2	2D system
Nanowire	2	1	1D system
Quantum dot	3	0	0D system

Nanowires

Figure 4.2 illustrates a low-resolution transmission electron micrograph (TEM) of a small ensemble of solution-synthesized CdSe nanowires. Depending on how they were made, the wires have mean diameters between 5 and 12 nm and overall size (i.e., diameter) distributions between 15% and 25%. Lengths exceed 1 μm. One sees that, based on our discussion of appropriate length scales for nano (Chapter 3), such wires are within the intermediate to weak confinement regimes of CdSe. Hence, carrier confinement effects may appear in their optical properties.

Figure 4.3 is an accompanying high-resolution micrograph of these nanowires. In the image, one sees fringes from the underlying crystal lattice, stemming from the regular arrangement of atoms in each wire. In fact, the fringes represent the planes of atoms in the underlying lattice and highlight the high degree of crystallinity of the wires.

From the appearance of the fringes, it is also possible to determine the identity of the underlying crystal structure. We have already discussed in Chapter 2 the various types of crystal lattices that exist. In the case of CdSe, there are two low-energy crystallographic phases: zinc blende (ZB) and wurtzite (W). The former has a cubic unit cell and the latter a hexagonal unit cell. These two crystal structures therefore exhibit different lattice patterns when viewed down certain crystallographic directions (e.g., $\langle 110 \rangle$) due to their different atomic arrangements.

This difference is illustrated in **Figure 4.4(a)**, which is a photograph of a nanowire model viewed down the $\langle 110 \rangle$ zone, normal to its $\langle 111 \rangle / \langle 0001 \rangle$ growth axis. The model is constructed so as to show two sections of zinc blende separated by a so-called twin boundary (T). A third section on the far right is exclusively wurtzite. In the model, one sees characteristic "zig-zag" lattice fringes from either ZB or W sections,

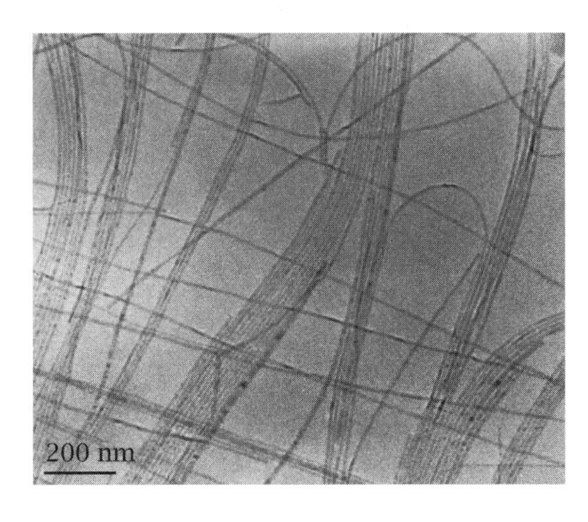

Figure 4.2 TEM micrograph of a small collection of solution-synthesized CdSe nanowires.

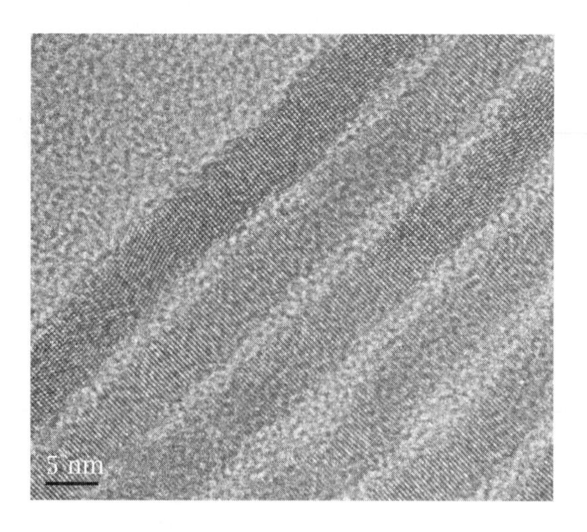

Figure 4.3 High-resolution TEM micrograph of closely packed CdSe nanowires.

as suggested by the dashed lines. This allows one to identify the wire's underlying crystal structure by visual inspection of its lattice fringes.

Now, when the same model is rotated and viewed down a non-⟨110⟩ direction, only parallel lattice fringes are seen. This is true for both ZB and W sections. **Figure 4.4(b)** illustrates an image of a CdSe nanowire viewed down a non-⟨110⟩ zone, showing the parallel fringes that result. As a consequence, the visual appearance of the lattice only allows one to readily identify the underlying crystal lattice in certain orientations.

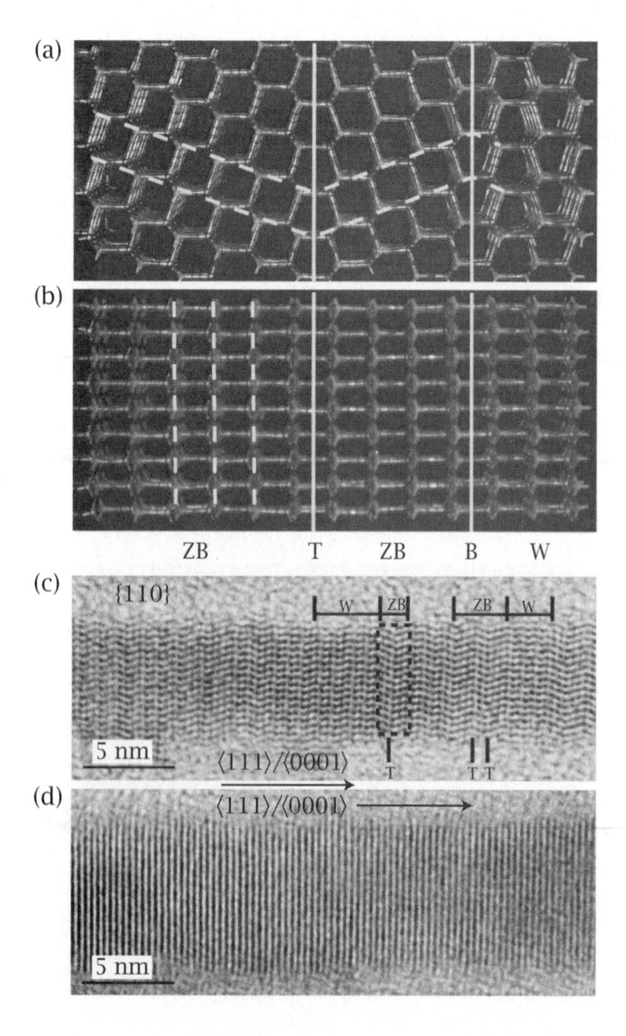

Figure 4.4 (a) Model showing the appearance of zinc blende (ZB) and wurtzite (W) sections in a CdSe nanowire when viewed along the ⟨110⟩ zone, normal to the ⟨111⟩/⟨0001⟩ growth axis. A twin plane between ZB sections is denoted by T. The boundary between ZB and W is denoted by B. (b) Appearance of the same model when rotated and viewed along a different (non-⟨110⟩) crystallographic direction. (c) and (d) Actual high-resolution TEM images, illustrating (a) and (b) in CdSe nanowires. Figure used with permission.

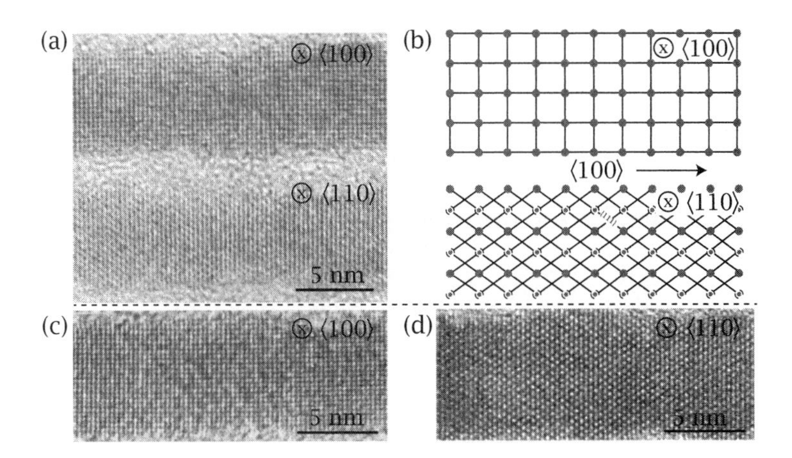

Figure 4.5 High-resolution TEM micrographs of PbSe nanowires with accompanying model illustrations showing their appearance along different crystallographic directions normal to the ⟨100⟩ growth axis. Figure used with permission.

Next, apart from showing that the wires are crystalline, the high-resolution micrograph in **Figure 4.3** shows that the wires are uniform (i.e., their diameters do not vary much along their length). This corroborates the initial impression given by the low-resolution image in **Figure 4.2**. In fact, experimentally determined intrawire diameter distributions are of the order of 5%.

Note that the choice of CdSe is not unique. Nanowires have been made from other semiconductors, including Si, Ge, and most of the binary II–VI, III–V, and IV–VI semiconductors. More information about nanowire synthesis can be found in Chapter 13. However, to illustrate an example of these other systems, **Figure 4.5(a)**, **(c)**, and **(d)** show high-resolution TEM micrographs of solution-synthesized PbSe nanowires. The accompanying model drawings in **Figure 4.5(b)** illustrate a PbSe wire oriented along different crystallographic directions. It can be seen that the underlying rocksalt lattice of PbSe implies the appearance of a characteristic checkerboard pattern in lattice-resolved ⟨100⟩-oriented wires. This is apparent in the high-resolution TEM micrographs shown in **Figure 4.5(a, top)** and **(c)**.

By contrast, when the same wire is oriented along a non-⟨100⟩ direction, only parallel lattice fringes are seen. This is illustrated in **Figure 4.4(a, bottom)** and **(d)**. Thus, as with CdSe, the lattice resolved images reveal that the wires are highly crystalline. Furthermore, under certain circumstances, it is possible to identify the underlying crystal structure from the image alone.

Quantum Dots

Next, we illustrate several examples of colloidal quantum dots. **Figure 4.6** shows a TEM image of a close-packed array of CdSe quantum dots made using solution chemistry. More about their synthesis can be found in Chapter 13.

In the micrograph, each of the spherical features is an individual nanocrystal. One sees that all of the dots have roughly the same size and shape. In fact, current syntheses routinely yield nanocrystals with ensemble size distributions of the order of 5–8%. As a consequence, when solutions of such uniform particles are deposited onto substrates, they tend to arrange in a billiard-ball-like, close-packed, fashion.

The reader may have noticed that the spacing between quantum dots is fairly uniform. This is also true of the solution-synthesized CdSe nanowires seen earlier in **Figure 4.3**. The reason for this stems from the presence of organic ligands on the quantum dot or nanowire surface. These ligands are used in solution-phase nanostructure syntheses to

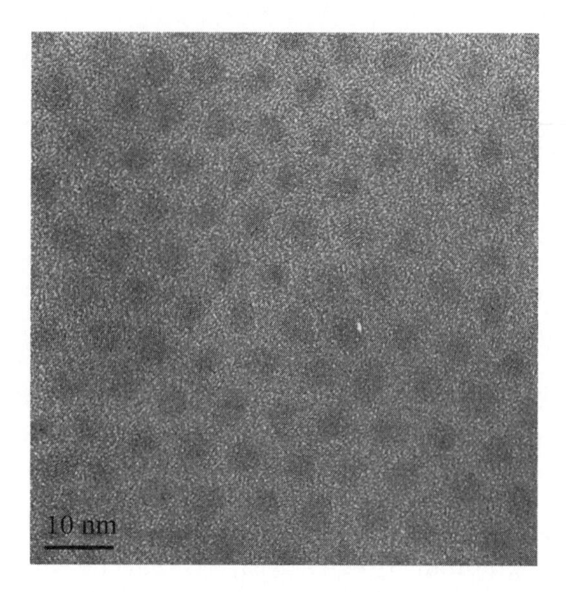

Figure 4.6 TEM micrograph of a close-packed array of colloidal CdSe quantum dots.

help control growth kinetics. In tandem, they passivate dangling bonds on the nanostructure surface resulting from the abrupt termination of the underlying crystal lattice. These ligands also sterically stabilize resulting nanostructures and prevent them from touching/aggregating. This results in uniformly spaced quantum dot and nanowire arrays.

Figure 4.7 shows a collage of high-resolution TEM micrographs of the same nanocrystals. From these images, lattice fringes are apparent, revealing the planes of atoms that make up each particle. As with the CdSe nanowires seen earlier (**Figures 4.3** and **4.4**), both zinc blende and wurtzite phases can coexist within the dots. Hence, one occasionally sees the characteristic lattice fringes of a given phase in ⟨110⟩-oriented nanocrystals. However, when such particles are aligned in more common non-⟨110⟩ orientations, such as those shown in **Figure 4.7**, only parallel lattice fringes are seen. Any additional structural information is therefore lost. There are, however, other ways

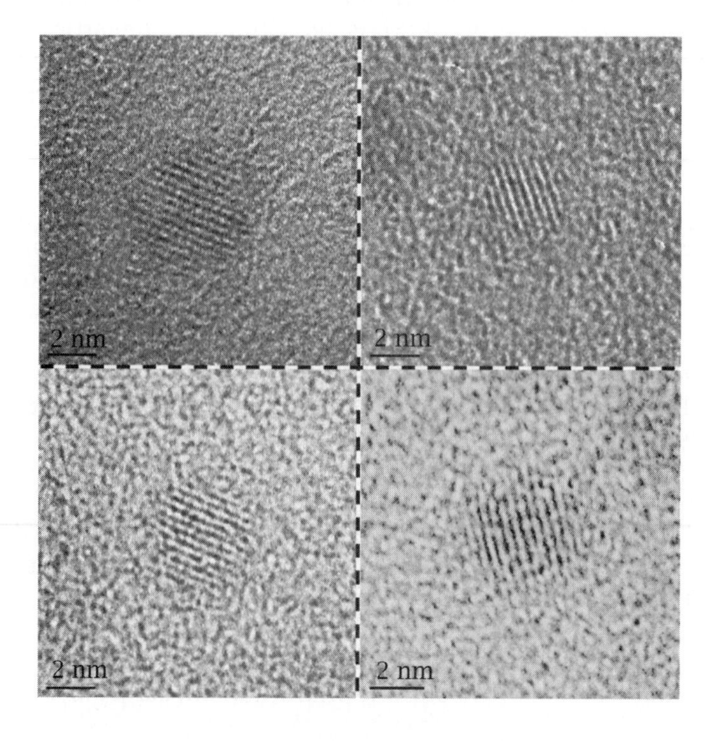

Figure 4.7 High-resolution TEM micrographs of individual colloidal CdSe quantum dots.

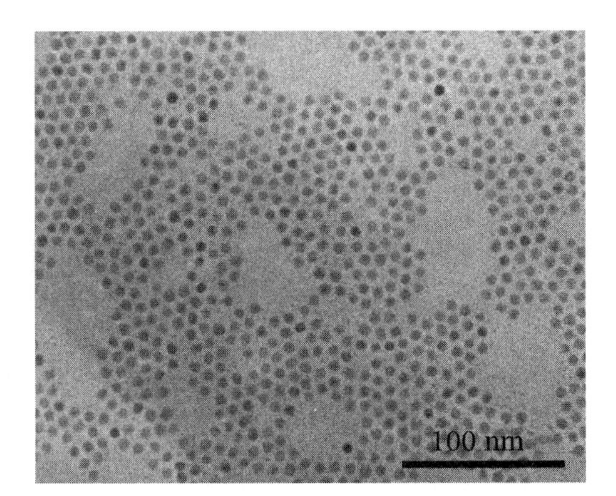

Figure 4.8 TEM micrograph of close-packed arrays of colloidal PbSe quantum dots. Figure courtesy of I. Lightcap.

to determine the crystallographic phase of a nanostructure; these approaches will be discussed in Chapter 14 when we review various nanostructure characterization techniques.

Note that while the above TEM images show that the quantum dots are highly crystalline, a close examination of **Figure 4.7** reveals that their abrupt termination leads to facets, irregular steps, and edges. Thus, the actual structure of a quantum dot is complicated. However, to a good approximation, its optical and electrical properties can be modeled by assuming a spherical geometry.

Note that since mean nanoparticle diameters range from 2 to 6 nm, the above CdSe dots are in the intermediate confinement regime. Recall that the CdSe bulk exciton Bohr radius is $a_B \simeq$ 5–6 nm. As a consequence, confinement effects can be seen in their optical properties. This will be the subject of Chapters 7 and 8.

Finally, the choice of nanocrystal material is not unique, and quantum dots have been made out of most common semiconductors, including Si, Ge, and many of the binary compounds in the II–VI, III–V, and IV–VI families. In what follows, we illustrate examples of colloidal PbSe quantum dots.

Figure 4.8 shows a close-packed array of PbSe quantum dots made using solution chemistry. As with the earlier CdSe nanocrystals, the synthesis yields particles with near-identical sizes and shapes. Size distributions are small and are on the order of 5%. **Figure 4.9** is an accompanying collage of high-resolution TEM images from the same ensemble. One readily sees lattice fringes in each particle resulting from its underlying crystal structure. Furthermore, one sees in properly oriented dots such as those in **Figure 4.9(a)** and **(b)** the characteristic checkerboard pattern of the rocksalt lattice. However, non-$\langle 100 \rangle$-oriented dots in **Figure 4.9(c)** and **(d)** exhibit only parallel lattice fringes.

4.1.3 Metallic Systems

There are equivalent metal nanostructures to the above low-dimensional semiconductors. The primary difference between these structures and their semiconductor counterparts, however, is that most of the optical and electrical properties of metallic systems can be described in a bulk-like fashion. We saw this earlier in Chapter 3. Only in the very smallest metal nanostructures do carrier confinement effects occur. As a consequence, while size-dependent changes take place in these materials, they are somewhat different in spirit than those occurring in low-dimensional semiconductors.

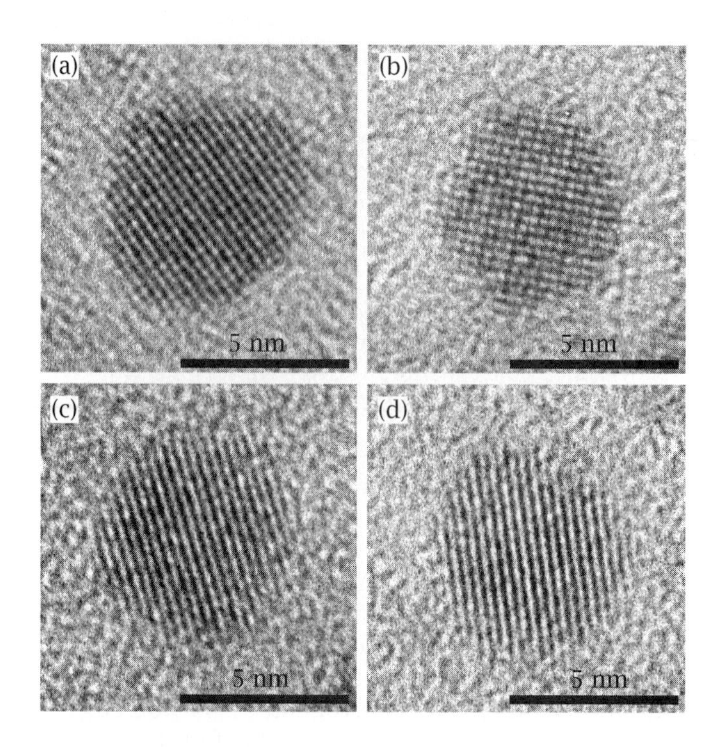

Figure 4.9 High-resolution TEM micrographs of (a,b) ⟨100⟩-oriented and (c,d) non ⟨100⟩-oriented colloidal PbSe quantum dots. Figure courtesy of I Lightcap.

Nanowires

Our first metallic nanostructure involves metal nanowires. A variety of such systems have been made in recent years using chemical or template-based approaches. As a specific example, **Figure 4.10** shows a TEM image of a small collection of silver nanowires with a mean diameter of 62 nm (±12 nm). Lengths range from 2 to 20 μm. Note that the clear separation between wires in **Figure 4.10** stems from the presence of organic ligands on the wire surface. They passivate the abrupt termination of each crystal and provide steric stabilization, preventing them from touching/aggregating. High-resolution images of the same wires (not shown) reveal that they exhibit the FCC crystal lattice.

Nanoparticles

Complementary colloidal metal nanoparticles have a long history. As an interesting aside, they have been used (unknowingly) for decorative purposes for at least 2000 years. An example is the Roman Lycurgus cup from the 4th century AD, which contains gold nanoparticles. More about it can be found in Freestone et al. (2007), as well as in Chapter 13.

Figure 4.10 Low-resolution TEM micrograph of solution-synthesized silver nanowires. Figure used with permission.

Figure 4.11 shows a TEM micrograph of a collection of small colloidal gold nanoparticles. They are analogous to the semiconductor quantum dots shown above. The particles all have roughly the same size and have an average diameter of 14 nm, with a corresponding size (i.e., diameter) distribution of 7%. Note that one difference between this material and the other semiconductor or metal nanostructures shown above is that the particles have little or no interparticle spacing. This is because the gold nanocrystals here have been passivated electrostatically using small citrate molecules on their surface. As a consequence, when the gold sol is rendered charge-neutral, component particles touch and aggregate.

Figure 4.12 is a high-resolution TEM micrograph of these gold nanoparticles. One sees crystalline lattice fringes in many of them. Furthermore, depending on the particle orientation, the observed lattice pattern helps us identify its underlying crystal structure. Recall from Chapter 2 that gold generally takes the FCC lattice.

4.1.4 Carbon Nanostructures

Finally, we briefly mention three general classes of carbon nanostructures, since this is an entire field in itself. The interested reader may

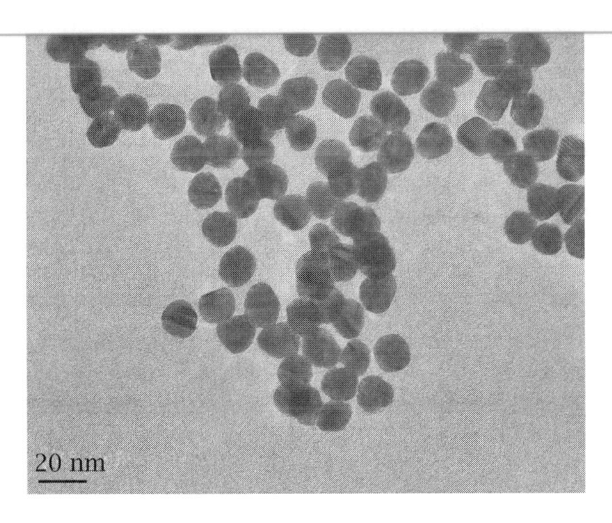

Figure 4.11 TEM micrograph of a small assembly of colloidal gold nanoparticles.

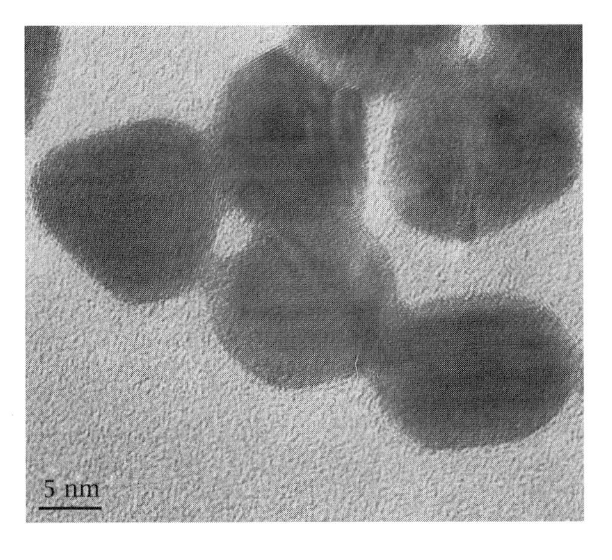

Figure 4.12 High-resolution TEM micrograph of colloidal gold nanoparticles.

consult Dresselhaus et al. (1996) for a more thorough overview of carbon nanoscience. In our case, the systems of interest are buckyballs, carbon nanotubes and graphene. We saw an example of a buckyball earlier in Chapter 3 when we introduced C_{60}. This is a 60-carbon molecule that structurally resembles a soccer ball (**Figure 3.1**). It was discovered in 1985 by Harold Kroto, Richard Smalley, and Robert Curl, for which they were eventually awarded a Nobel Prize in Chemistry (Kroto et al. 1985). Since then, other variants of C_{60} have been found, with different numbers of carbon atoms. Examples include C_{70} and C_{84}. These "buckyballs" are similar to the 0D semiconductor quantum dots seen earlier. However, they are more on the molecular side of things due to their small (~1 nm) diameters.

Single-wall carbon nanotubes (SWCNTs) are essentially sheets of graphite rolled into tubes with diameters as small as 0.4 nm (**Figure 4.13**) (Ijima and Ichihashi 1993; Bethune et al. 1993), while multiwall carbon nanotubes (MWCNTs) are nested variants of these SWCNTs. The latter were discovered by Ijima in 1991 (Ijima 1991). See also Monthioux et al. (2006). In general, nanotubes are complementary 1D analogues to the nanowires seen earlier. However, they possess an interesting property in that they can be either semiconducting or metallic, depending on their structure. Unfortunately, while this property is potentially very useful, current syntheses do not produce exclusively one form or the other. Instead, mixtures of semiconducting and metallic tubes are obtained. Many efforts have therefore aimed at isolating these different semiconductor/metal variants, with varying degrees of success. More information about nanotube purification can be found in Hersam et al. (2008).

Finally, graphene is an isolated (single) sheet of graphite. An example is shown in **Figure 4.14**. This system has recently been studied extensively for its unique electronic properties, especially after the discovery in 2004 that it could be isolated using Scotch tape. This discovery led to the 2010 Nobel Prize in Physics for Andre Geim and Konstantin Novoselov. As with buckyballs and nanotubes, graphene can be compared with counterpart low-dimensional systems discussed earlier. However, its properties are quite unique, preventing direct comparisons. More about graphene, its history, its properties, and its applications can be found in Geim et al. (2007).

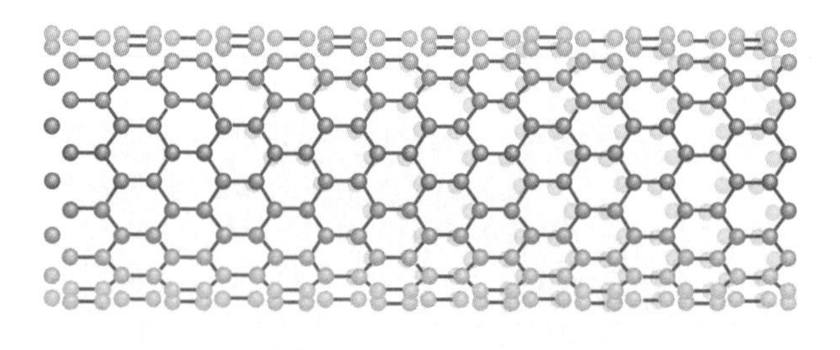

Figure 4.13 Schematic illustrating a single-wall carbon nanotube.

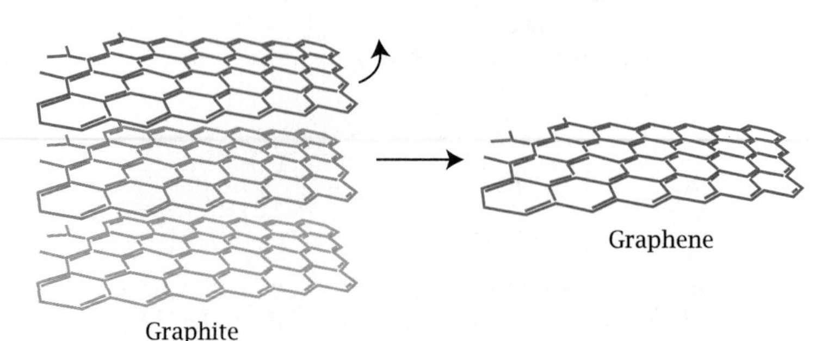

Figure 4.14 Illustration of an exfoliated graphene sheet.

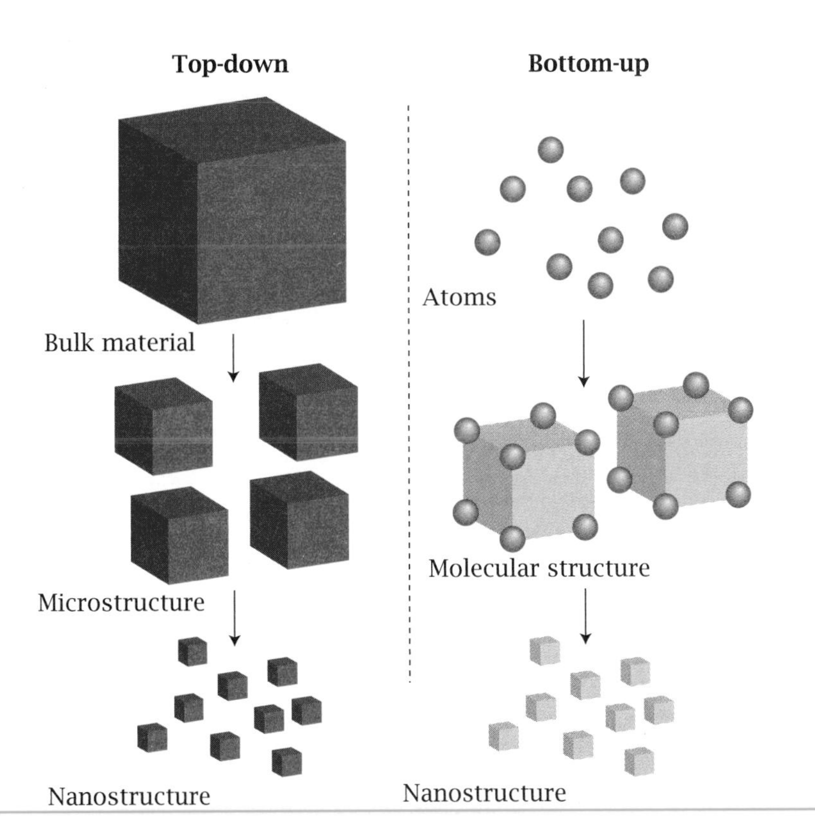

Figure 4.15 Illustration outlining the top-down and bottom-up manufacturing of nanostructures.

4.2 BOTTOM-UP OR TOP-DOWN

Finally, to conclude, we outline two general strategies for synthesizing the abovementioned low-dimensional materials. The first is bottom-up synthesis while the other is called top-down manufacturing. "Bottom-up" refers to making nanostructures starting from individual atoms and working up to larger structures. This approach is generally practiced by chemists, physicists, and others pursing low-tech chemical syntheses as well as alternative high-tech approaches such as molecular-beam epitaxy. More details about chemical means of making nanostructures will be discussed in Chapter 13.

Top-down manufacturing, on the other hand, begins with large macroscopic semiconductor fragments. Optical lithography is then used to work down to nanostructures. This approach is, in fact, what engineers have been using for many years to create semiconductor computer chips with progressively smaller transistor sizes. A drawback of the method, however, is that since radiation is used to define features, the optical diffraction limit represents a lower limit to the length scales that can be obtained. Generally speaking, this is half the wavelength of light. As a consequence, to move to progressively smaller sizes, radiation with shorter wavelengths must be used. This means migrating from visible-light, to ultraviolet-light, and eventually to electron-beam lithographies. Recall from Chapter 3 that the wavelength of electrons can be controlled by altering their momentum. The general contrast between top-down and bottom-up manufacturing is outlined in **Figure 4.15**.

4.3 SUMMARY

Having introduced the low-dimensional systems of interest to us, we now move on to describing their size-dependent optical properties.

While this journey begins in Chapter 5, we really need a better handle on the underlying quantum mechanics behind absorption and emission processes in these systems. As a consequence, starting with Chapter 6, we review basic quantum mechanics and, in turn, begin a more involved path towards Chapter 12, where the absorption of these materials is described in detail.

4.4 THOUGHT PROBLEMS

4.1 Top down

Investigate a common top-down manufacturing approach called electron-beam lithography. Explain what this technique entails in practice.

4.2 Literature

Find examples in the literature of the following nanostructures:

(a) A "self-assembled" quantum dot

(b) A colloidal quantum dot.

Explain whether the approaches used to make these quantum dots are bottom-up or top-down.

4.3 Literature

Sometimes, self-assembled quantum dots possess a pyramidal shape, as opposed to the spherical morphologies shown in the main text. Find an example in the literature and explain the reason for this shape.

4.4 Literature

Find examples of the following nanostructures in the literature:

(a) Nanowires made using a technique referred to as "vapor–liquid–solid" growth

(b) Nanowires made using a technique referred to as "solution–liquid–solid" growth

Explain whether the approaches used to make these wires are bottom-up or top-down.

4.5 Literature

Sometimes, in the older literature, one sees reference to so-called "v-groove" quantum wires. Find an example of such a 1D system in the literature and explain what the name implies.

4.6 Literature

In addition to the quantum dots and nanowires seen in this chapter, there are other nanostructures that bridge these materials. Nanorods are one such example. See if you can find a definition in the literature that distinguishes nanorods from nanowires.

4.7 Literature

Find an example of a nanostructure in the literature that has a more complex morphology than what we have seen and that does not lend itself to easy characterization as a 0D, 1D, or 2D nanostructure.

4.8 Literature

There is considerable interest in the community in going beyond the generic classes of nanostructures discussed in this chapter. For example, there is interest in combining individual quantum dots to make so-called quantum dot "molecules." Alternatively, one can make more extended structures, leading to artificial "solids" sometimes called superlattices. Find examples of each in the literature.

4.9 Literature

Find examples in the literature of solution-synthesized nanostructures that have a sheet-like morphology. These are chemical analogues of more traditional quantum wells.

4.5 REFERENCES

Bethune DS, Klang CH, de Vries MS, et al. (1993) Cobalt-catalysed growth of carbon nanotubes with single-atomic-layer walls. *Nature* **363**, 605.

Dresselhaus MS, Dresselhaus G, Eklund PC (1996) *Science of Fullerenes and Carbon Nanotubes: Their Properties and Applications*. Academic Press, San Diego, CA.

Freestone I, Meeks N, Sax M, Higgitt C (2007) The Lycurgus cup—a Roman nanotechnology. *Gold. Bull.* **40/4**, 270.

Geim AK, Novoselov KS (2007) The rise of graphene. *Nature Mater.* **6**, 183.

Hersam MC (2008) Progress towards monodisperse single-walled carbon nanotubes. *Nature Nanotechnol.* **3**, 387.

Iijima S (1991) Helical microtubules of graphitic carbon. *Nature* **354**, 56.

Iijima S, Ichihashi T (1993) Single-shell carbon nanotubes of 1-nm diameter. *Nature* **363**, 603.

Kroto HW, Heath JR, O'Brien SC, et al. (1985) C_{60}: buckminsterfullerene. *Nature* **318**, 162.

Monthioux M, Kuznetsov VL (2006) Who should be given the credit for the discovery of carbon nanotubes? *Carbon* **44**, 1621.

Sridhara Rao DV, Muraleedharan K, Dey GK, et al. (1999) Transmission electron microscopy and X-ray diffraction studies of quantum wells. *Bull. Mater. Sci.* **22**, 947.

Zory PS Jr (ed.) (1993) *Quantum Well Lasers*. Academic Press, San Diego, CA.

Absorption and Emission Basics

5.1 INTRODUCTION

In this chapter, we introduce basic concepts related to the phenomenological absorption of light by nanostructures. Of particular interest is the behavior of the quantum wells, wires, and dots just introduced in Chapter 4. The principles and relationships discussed here can then be used to quantify how efficiently these structures absorb (or emit) light and can also be used to determine the number of photogenerated carriers within them. These are all useful things from a practical perspective, since they enable us to quantify the efficiency of photodetectors, solar cells and other devices made of nanostructured materials. Our ultimate goal, though, is to develop a more quantitative understanding of their size-dependent optical and electrical properties. This requires us to revisit their underlying quantum mechanics in Chapters 6–11, with our ultimate destination being Chapter 12. In this chapter, we begin with simple bulk expressions related to the absorption of light. We then focus on corresponding expressions for quantum wells, wires, and dots and subsequently move to describing the emission process. We summarize by showing how one can connect two important parameters: the absorption cross section and the emission lifetime.

5.2 EXPONENTIAL ATTENUATION LAW

Empirically, it has been found that in bulk solids, an exponential attenuation relationship exists, describing the intensity of light as it passes through a material. This relationship is called the exponential attenuation law and has the general form

$$I = I_0 e^{-\alpha l}. \tag{5.1}$$

In Equation 5.1, I_0 is the incident light intensity and l is the thickness traversed by the light. The parameter α is called the absorption coefficient and has units of inverse length. Unfortunately, the Greek letter α is also used to denote a material's polarizability, as first seen in Chapter 3. The absorption coefficient is a frequency-dependent value and is usually expressed in units of cm^{-1}. Many bulk semiconductors have values ranging from 10^3 to 10^5cm^{-1} and **Table 5.1** lists α values for some common materials. **Figure 5.1** shows how α changes with frequency for three bulk systems: silicon, CdSe, and GaAs.

5.2.1 A Geometric Derivation of the Exponential Attenuation Law

The above exponential attenuation relationship can be derived using a simple geometrical model. To illustrate, picture a slab of a homogeneous material with an overall thickness l (**Figure 5.2a**). Incident

Table 5.1 Common bulk semiconductor band-edge absorption coefficients. Values from Landolt–Börnstein (1982)

Semiconductor	Room-temperature band gap (eV)	Associated band-edge absorption coefficient (cm^{-1})
Si	1.10	1×10^5
ZnS	3.70	5×10^2
CdS	2.50	5.5×10^4
CdSe	1.74	8×10^3
GaAs	1.43	8.2×10^3

light with an intensity I_0 impinges on it along the surface normal. Any transmitted light exits the other side and continues propagating in the same direction. Imagine now that the bulk slab is cut up into many thinner slabs, each with a thickness Δl, all stacked back to back. Thus, for a given slab with a thickness l, there exist $m = l/\Delta l$ sub-slabs (**Figure 5.2b**).

Assume that each sub-slab absorbs a finite fraction f of the incident light. The remaining portion, $1 - f$, is transmitted. As a consequence, the amount of light after the first sub-slab is

$$I_1 = (1 - f)I_0,$$

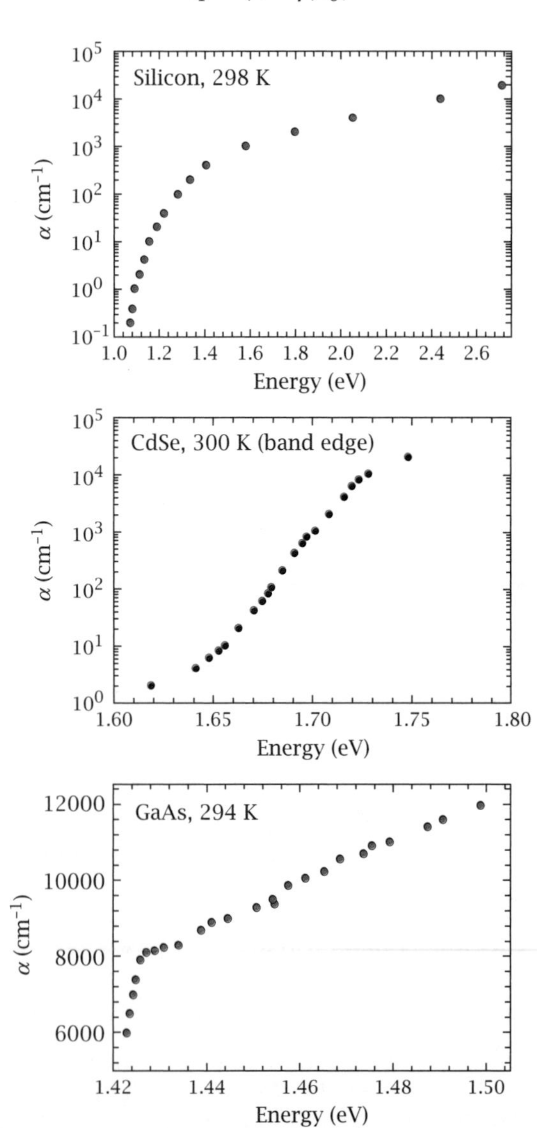

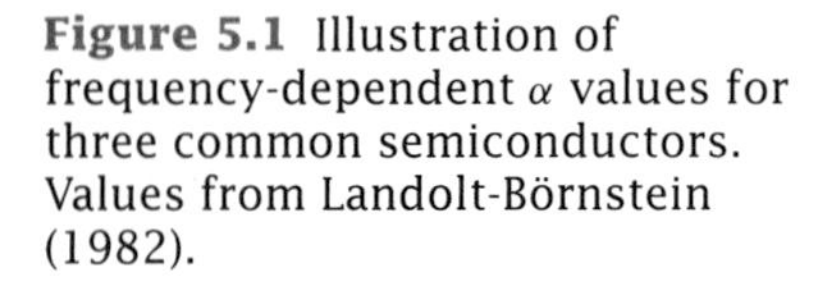

Figure 5.1 Illustration of frequency-dependent α values for three common semiconductors. Values from Landolt-Börnstein (1982).

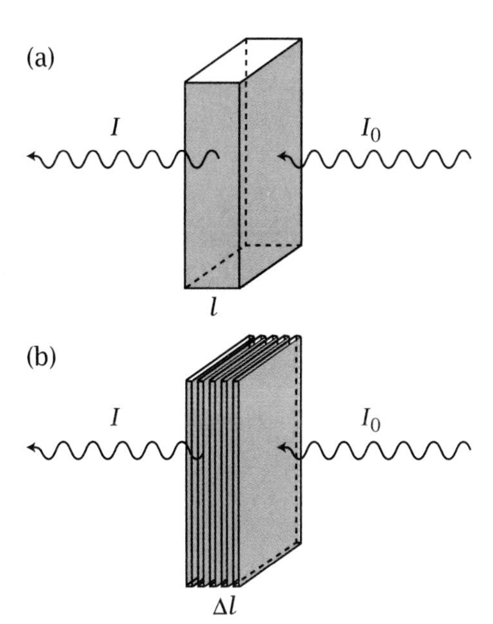

Figure 5.2 Geometric derivation of the exponential attenuation law.

while the amount of transmitted light after the second sub-slab is

$$\begin{aligned} I_2 &= (1 - f)I_1 \\ &= (1 - f)^2 I_0. \end{aligned}$$

We can continue in this manner to evaluate the intensity after m sub-slabs.

The general relationship obtained for the transmitted light intensity is

$$I = (1 - f)^m I_0,$$

which one can alternatively write as

$$\begin{aligned} I &= (1 - f)^{l/\Delta l} I_0, \\ \frac{I}{I_0} &= (1 - f)^{l/\Delta l}. \end{aligned}$$

To simplify the expression, make a slight change of notation and express the transmitted light fraction as $z = 1 - f$. This yields

$$\begin{aligned} \frac{I}{I_0} &= z^{l/\Delta l} \\ &= (z^l)^{1/\Delta l} \\ &= \left[\left(\frac{1}{z}\right)^{-l}\right]^{1/\Delta l} \\ &= \left[\left(\frac{1}{z}\right)^{1/\Delta l}\right]^{-l}, \end{aligned}$$

where both z and Δl are constants. The term in square brackets can therefore be called by another name, say y. This gives

$$\frac{I}{I_0} = y^{-l}.$$

We can alternatively express y as $y = e^{\ln y}$ to obtain

$$\frac{I}{I_0} = e^{-l \ln y}$$

with $\ln y$ simply another constant. For convenience, call it α, the absorption coefficient. The resulting expression is our desired exponential attenuation relationship (Equation 5.1):

$$\frac{I}{I_0} = e^{-\alpha l}.$$

5.3 OTHER CONVENTIONS

When dealing with the absorption of light, there exist a number of conventions used by chemists, physicists, and engineers. In bulk solids, it is common to see the exponential attenuation relationship written in terms of the absorption coefficient

$$I = I_0 e^{-\alpha l}.$$

By contrast, in molecular systems, one writes

$$I = I_0 e^{-\sigma [n] l}, \quad (5.2)$$

where σ is called the absorption cross section (with units of cm^2). This is an area effectively presented by the absorber to incident photons, which, in turn, relates to the likelihood of it absorbing light. The coefficient $[n]$ is the concentration of absorbers (with units of number/cm^3), while l is the sample's overall thickness (with units of cm).

Typical absorption cross sections for molecules are of the order of 10^{-16} cm^2. Cross sections for colloidal quantum dots, however, are generally of the order of 10^{-15} cm^2. As such, they are more efficient absorbers than organic systems, making them better suited for some applications. This will be seen in Chapter 15. **Table 5.2** lists various molecular and nanostructure cross sections compiled from the literature. It is by no means complete and simply illustrates the sorts of numbers that one routinely encounters. Associated literature can be found in the Further Reading.

5.3.1 Beer's Law

When dealing with molecular systems (let us include nanowires and quantum dots among these materials), there is an alternative expression for the exponential attenuation relationship. The expression is called Beer's law or the Beer–Lambert law:

$$A = \epsilon_{\text{molar}} [c] l, \quad (5.3)$$

Table 5.2 Compilation of various absorption cross sections from the literature

System	Wavelength (nm)	Associated absorption cross section (cm^2)
Coumarin-120	365	3.9×10^{-17}
Coumarin-307	397	5.4×10^{-17}
2,5-Diphenylfuran (PPF)	325	1.3×10^{-16}
C_{60}	315	1.6×10^{-16}
C_{70}	389	9.0×10^{-17}
CdSe QDs ($r = 3$ nm)	350	1.5×10^{-14}
InAs QDs ($r = 3$ nm)	450	1.7×10^{-14}
Carbon nanotubes (6,5)	~1000	1×10^{-13}
CdS NWs ($r = 7$ nm, $l = 1$ μm)	405	3.3×10^{-11}
CdSe NWs ($r = 11$ nm, $l = 1$ μm)	387	6.2×10^{-11}
CdTe NWs ($r = 5$ nm, $l = 1$ μm)	387	2.0×10^{-11}

QD, quantum dot; NW, nanowire.

Table 5.3 Sample molar extinction coefficients.

System	Wavelength (nm)	Associated molar extinction coefficient (M^{-1} cm^{-1})
Coumarin 343	445	4.4×10^4
Rhodamine 6G	530	1.2×10^5
Eosin Y	525	1.1×10^5
Perylene	434	3.2×10^4
Anthracene	357	7.9×10^3
2-Aminopurine	305	6.3×10^3
CdSe QDs ($r = 3$ nm)	350	3.9×10^6
InAs QDs ($r = 3$ nm)	450	4.4×10^6

QD, quantum dot.

where A is the absorbance of the material (unitless, also called the optical density), $[c]$ is its molar concentration (with units of mol/L, or M), l is the sample's pathlength or thickness (with units of cm), and ϵ_{molar} is a material-dependent parameter called the molar extinction coefficient (with units of M^{-1} cm^{-1}). Like σ, ϵ_{molar} describes how efficiently a material absorbs light and **Table 5.3** lists representative ϵ_{molar} values for molecules and nanostructures compiled from the literature. Associated citations can be found in the Further Reading.

In Equation 5.3, a large value of A means that the sample absorbs strongly. For example, $A = 0.3$ means that 50% of the incident light is absorbed. Conversely, an absorbance of $A = 1.0$ means that 90% of the light is absorbed. This nonlinear scaling occurs because absorbances are expressed on a logarithmic (base 10) scale. To see this, note the following relationship between absorbance and transmittance:

$$A = \log\left(\frac{1}{T}\right), \tag{5.4}$$

where

$$T = \frac{I}{I_0}. \tag{5.5}$$

Equation 5.4 will be useful when we relate ϵ_{molar} to σ in what follows.

5.4 RELATING ϵ_{molar} TO σ

At this point, having both ϵ_{molar} and σ, it is often useful to relate the two. This is because ϵ_{molar} is a molar quantity and represents the collective behavior of Avogadro's number of absorbers. However, recent advances in single-molecule and single-particle microscopy, as well as the ability to make devices employing single nanowires or quantum dots, means that one can interrogate individual nanostructures both optically and electrically. As a consequence, to make sense of these latter studies, we need σ rather than ϵ_{molar}.

In this section, we therefore derive a common relationship that connects these two quantities. Specifically, given Beer's law and $A = \log(1/T)$,

$$\log\left(\frac{1}{T}\right) = \log\left(\frac{I_0}{I}\right) = \epsilon_{molar}[c]l.$$

Next, from the exponential attenuation law, expressed in terms of σ, $I/I_0 = e^{-\sigma[n]l}$,

$$\log e^{\sigma[n]l} = \epsilon_{\text{molar}}[c]l,$$
$$\sigma[n]l \log e = \epsilon_{\text{molar}}[c]l,$$
$$\sigma[n] \log e = \epsilon_{\text{molar}}[c].$$

Note that $[n]$ has units of number/cm^3 while $[c]$ is a molar concentration with units of mol/L (M). To compare the two, their units must be made consistent. We therefore modify $[c]$ so that it takes the same units as $[n]$:

$$[c] \rightarrow \frac{[c]N_a}{\text{dm}^3} \rightarrow \frac{[c]N_a}{1000\,\text{cm}^3} = [n].$$

In the above transformation, N_a is Avogadro's number ($N_a = 6.022 \times 10^{23}$/mol). One therefore finds that $[c] = [n](1000)/N_a$, resulting in a familiar expression, linking σ to ϵ_{molar}:

$$\sigma(\text{cm}^2) = \frac{(2.303)(1000)\epsilon_{\text{molar}}(\text{M}^{-1}\,\text{cm}^{-1})}{N_a}. \tag{5.6}$$

In Equation 5.6, σ and ϵ_{molar} have units of cm^2 and M^{-1} cm^{-1}, respectively. The 2.303 in the numerator originates from $1/\log e$.

Let us now see whether common molar extinction coefficients of $\epsilon_{\text{molar}} \simeq 10^5\,\text{M}^{-1}\,\text{cm}^{-1}$ correspond to commonly seen ballpark absorption cross sections of $10^{-16}\,\text{cm}^2$. Inserting ϵ_{molar} into Equation 5.6 yields $\sigma \simeq 4 \times 10^{-16}\,\text{cm}^2$ and shows that the correspondence does indeed make sense. The reader may verify this.

Calculating σ at Other Frequencies

Knowing σ at a given frequency enables one to calculate its value at another using experimental data. To illustrate, assume that we know σ at one wavelength. From Equation 5.6, we also know that

$$\sigma = \frac{(2.303)(1000)}{N_a}\epsilon_{\text{molar}}.$$

Then, from Beer's law, ϵ_{molar} is proportional to the sample's absorbance through

$$\epsilon_{\text{molar}} = \frac{A}{[c]l}.$$

When these two expressions are combined,

$$\sigma = \frac{(2.303)(1000)}{N_a}\frac{A}{[c]l}$$

such that a ratio of cross sections at two different frequencies yields

$$\frac{\sigma_2}{\sigma_1} = \frac{A_2}{A_1}.$$

The subscripts 1 and 2 simply distinguish σ and A at different wavelengths. This means that we can find the cross section of an absorber at any wavelength provided that we know A and σ at a given frequency along with A at the desired wavelength. Namely,

$$\sigma_2 = \sigma_1\frac{A_2}{A_1}. \tag{5.7}$$

5.5 ESTIMATING α AND σ

Cross sections and molar extinction coefficients are important values for any quantitative analysis of molecular or nanostructured systems. However, such parameters are generally not known beforehand. They must be measured experimentally. For molecular systems, this task is trivial since molecules generally have well-defined molecular weights. Exceptions include polymeric materials with a weight distribution about a mean value. In the majority of cases, then, the concentration of a molecular species in solution can easily be obtained by knowing the precise mass of a sample, its molecular weight, and the volume of solvent used. Subsequent use of Beer's law (Equation 5.3) along with Equation 5.6 enables one to determine ϵ_{molar} and σ.

A problem arises with nanostructures. Namely, whether colloidal quantum dots, nanowires or some other species, all possess a distribution of sizes and possibly lengths (i.e., no synthesis today produces an ensemble of dots or wires where all constituent species have precisely the same dimensions; this will be discussed in Chapter 13). Furthermore, many colloidal nanostructures have organic ligands on their surfaces, which provide either steric or electrostatic stabilization. As a consequence, the molecular weight of chemically synthesized nanostructures is generally ill defined.

It is possible, however, to make a priori estimates of the absorption coefficient and cross section of low-dimensional materials by assuming that they are bulk-like. What this means is that if we assume they possess a bulk-like density of states in certain wavelength regions (the concept of a density of states will be discussed in Chapter 9), it is possible to estimate α or σ.

5.5.1 Traveling Waves

Before starting our back-of-the-envelope calculation for nanostructure α or σ values, we need some background information about traveling waves. This is because α will eventually be determined by looking at the intensity of the light as it passes through a material. Readers unfamiliar with waves are encouraged to consult an introductory physics text to put things into perspective.

Let us therefore consider light as a traveling wave, moving in a given direction. For the purpose of this discussion, assume that it propagates along the x direction of a Cartesian coordinate system. The associated electric field of the light is then

$$E(x,t) = E_0 e^{i(k_{\text{light}}x - \omega t)}, \tag{5.8}$$

where E_0 is the amplitude of the electric field (with units of V/m), ω is its angular frequency (with units of rad/s), and $k_{\text{light}} = 2\pi/\lambda$ (with units of m^{-1}) is called the wavevector with λ the wavelength of the light. Note that Equation 5.8 is just the complex version of the electric field of the light. One can alternatively describe $E(x,t)$ in terms of cosines or sines. The complex form, however, happens to be more convenient in what follows.

Next, given two possible propagation directions (i.e., towards increasing or decreasing x values), it is possible to tell the direction of motion of the wave by keeping its phase (i.e., $k_{\text{light}}x - \omega t$ in Equation 5.8) constant and letting time increase. Thus if $k_{\text{light}}x - \omega t = \text{const}$ and time increases, x must also increase. As a consequence, Equation 5.8

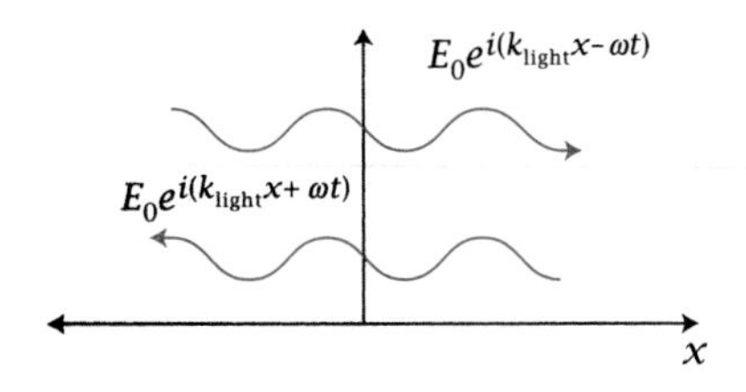

Figure 5.3 Schematic of right- and left-going traveling waves.

describes a wave moving towards the right. Conversely, one writes

$$E(x,t) = E_0 e^{i(k_{light}x+\omega t)}$$

to describe a left-going wave. The reader may verify this. Right- and left-going traveling waves are highlighted in **Figure 5.3**.

In either case, $k_{light} = 2\pi/\lambda$ is the wavevector and λ is the wavelength of light. Both λ and c (the speed of light) remain unchanged in vacuum. On moving through a nonabsorbing medium with a refractive index n_m (e.g., glass), both c and λ change.

Before going on, note that the refractive index is a complex number with a real and an imaginary part:

$$n_m = n + ik. \tag{5.9}$$

In Equation 5.9, n is the real part of the refractive index while k is its complementary, imaginary part called the extinction coefficient (this is not to be confused with the wavevector!). The name is suggestive and k is indeed related to the absorption coefficient of the material, as we shall see.

Returning to our discussion, in a medium with refractive index n_m, both the speed of light and the wavelength change, from c and λ to c/n_m and λ/n_m, respectively. As a consequence, $c = \lambda\nu$ becomes $c/n_m = (\lambda/n_m)\nu$. In this way, the wavevector changes to

$$k_{light} = \frac{2\pi}{\lambda} \rightarrow \frac{2\pi}{\lambda/n_m} = \frac{2\pi\nu}{c/n_m} = \frac{n_m\omega}{c},$$

yielding

$$k_{light} = \frac{n_m\omega}{c}. \tag{5.10}$$

Since n_m is complex (Equation 5.9), we have $k_{light} = (n+ik)\omega/c$, which we can replace back into Equation 5.8 to obtain

$$E(x,t) = E_0 \exp\left\{i[(n+ik)\frac{\omega}{c}x - \omega t]\right\}.$$

Consolidating the real and imaginary parts gives

$$E(x,t) = E_0 e^{-k\omega x/c} \exp\left[i\left(\frac{n\omega x}{c} - \omega t\right)\right].$$

Finally, we square the electric field to find its associated intensity I (the asterisk denotes the complex conjugate of the function; in practice, all this means is replacing i with $-i$):

$$I = |E(x,t)|^2 = E^*(x,t)E(x,t) = |E_0|^2 e^{-2k\omega x/c},$$

or

$$I = I_0 e^{-2k\omega x/c} \tag{5.11}$$

with $I_0 = |E_0|^2$. The intensity therefore decays exponentially with distance, just like in Equation 5.1. A comparison between the two suggests that

$$\alpha = \frac{2\omega k}{c}, \tag{5.12}$$

which is our desired relationship between the absorption coefficient and the extinction coefficient.

We can stop here. However, note that we can also modify the expression to yield another common form of α by recalling relationships between the real and imaginary parts of the refractive index as well as the real and imaginary parts of the dielectric constant ϵ.

In general, $n_m = \sqrt{\epsilon}$. Like the refractive index, the dielectric constant ϵ is a complex number:

$$\epsilon = \epsilon_1 + i\epsilon_2, \tag{5.13}$$

where $\epsilon_1(\epsilon_2)$ is its real (imaginary) part. Note that ϵ_2 is related to the absorption of light, just like k. Towards seeing this, the reader may show as an exercise that

$$\epsilon_1 = n^2 - k^2 \tag{5.14}$$

and

$$\epsilon_2 = 2nk. \tag{5.15}$$

Thus, when ϵ_2 is used in Equation 5.12, one has the following, alternative, expression for the absorption coefficient:

$$\alpha = \frac{\omega\epsilon_2}{nc}. \tag{5.16}$$

Figure 5.4 illustrates n, k, ϵ_1, and ϵ_2 for bulk CdSe to show what these values look like. Again, all are frequency-dependent quantities. More information about these optical constants can be found in Fox (2010).

We now provide back-of-the-envelope expressions for the absorption coefficients and associated absorption cross sections of bulk semiconductors, quantum wells, nanowires, and quantum dots. As suggested earlier, these expressions are approximations that formally apply only in certain frequency regions, which turn out to be at frequencies far to the blue of each system's absorption edge.

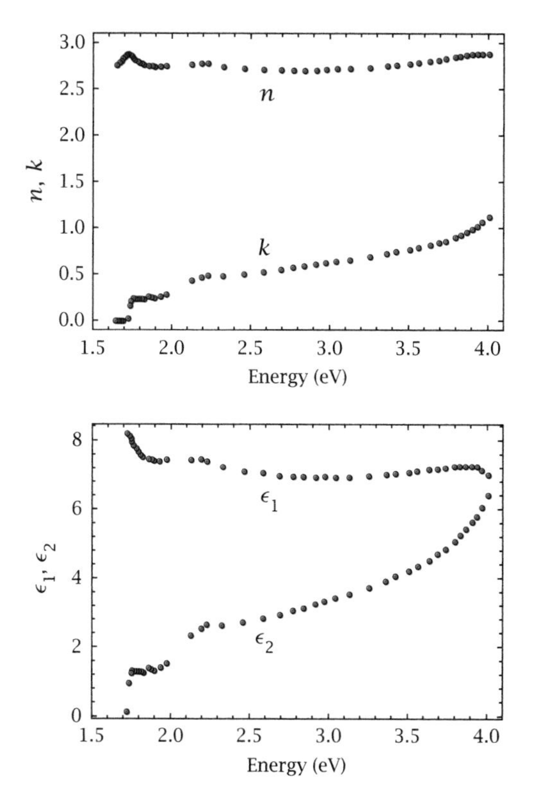

Figure 5.4 Illustration of frequency-dependent n, k, ϵ_1, and ϵ_2 values for CdSe. Values from Palik (1998).

5.5.2 Bulk Absorption Coefficient

We have just shown that for a homogeneous bulk system,

$$\alpha_{\text{bulk}} = \frac{\omega\epsilon_2}{nc}. \tag{5.17}$$

EXAMPLE 5.1

Estimate of the Bulk, Band-Edge, Absorption Coefficient for CdSe Let us use this expression to crudely estimate α at the band edge for a model semiconductor such as CdSe. We can then compare our approximation with the literature value for CdSe in **Table 5.1**. The only parameters needed are ϵ_2 and n at 1.74 eV. By consulting a table of bulk refractive indices and dielectric constants (Palik 1998, also illustrated in **Figure 5.4**), we find

$$n = 2.856,$$
$$k = 0.162,$$
$$\epsilon_1 = 8.13,$$
$$\epsilon_2 = 0.925.$$

Inserting these values in Equation 5.17 and ensuring proper units then gives $\alpha = 2.9 \times 10^4\,\text{cm}^{-1}$ at 1.74 eV, in reasonable agreement with the literature value of $\alpha = 8.0 \times 10^3\,\text{cm}^{-1}$. **Figure 5.5** shows calculated α values at other frequencies.

5.5.3 Nanostructure Absorption Coefficient

The case for nanostructures is slightly different. Namely, the quantum wells, wires, and dots we will refer to are conceptually embedded in a homogeneous, nonabsorbing dielectric medium. As a consequence, we will formally have a volume fraction P to consider. Furthermore, electromagnetic boundary conditions mean that a local field factor $f(\omega)$, related to the ratio of the electric field magnitude inside the nanostructure to that outside, will have to be taken into account. As a consequence, a generic expression for the absorption coefficient of the system, consisting of a nonabsorbing host with nanostructure inclusions, is

$$\alpha(\omega) = \frac{\omega}{n_m c} P|f(\omega)|^2\epsilon_2. \tag{5.18}$$

Although qualitatively similar to Equation 5.17, the reader will notice some subtle differences. Apart from P and $f(\omega)$, n_m in the denominator

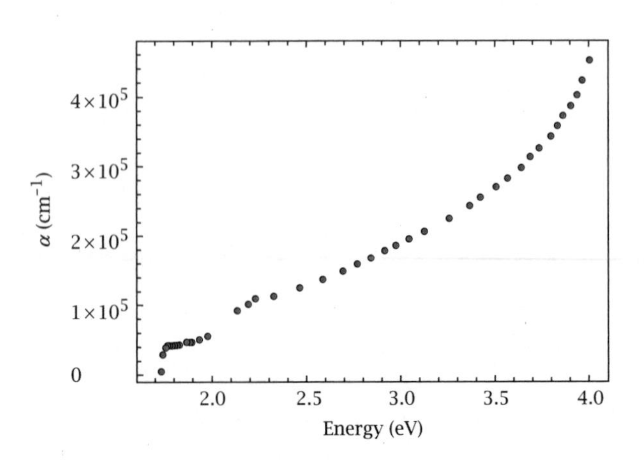

Figure 5.5 Calculated α values for bulk CdSe using numbers from Palik (1998).

refers to the refractive index of the nanostructure's surrounding medium, not of the semiconductor itself. More information about Equation 5.18 can be found in Giblin and Kuno (2010).

5.5.4 Quantum Well Absorption Coefficient

In the case of a quantum well, the absorption coefficient can be approximated by

$$\alpha_{QW}(\omega) = \frac{\omega}{nc}\epsilon_2, \quad (5.19)$$

where we have implicitly considered the limit $P \to 1$, which, in turn, means replacing n_m by n given that we are just interested in the well. Furthermore, for light incident normal to the quantum well plane, $f(\omega) = 1$.

5.5.5 Nanowire Absorption Cross Section

The absorption cross section of a nanowire has additional complications because the electric field of the incident light is not as simple as in the bulk semiconductor or quantum well cases above. This has to do with electromagnetic boundary conditions imposed by the wire's cylindrical shape. As a consequence, the local field factor must be explicitly considered in our expression. Recall that this local field factor is a unitless ratio of the magnitude of the electric field inside the wire to that outside. More about it can be found in Landau et al. (1984).

In the nanowire case, we find, after some angle averaging to account for the wire's asymmetric shape (this assumes normal incidence of the light on the nanowire growth axis), that

$$f(\omega) = 2\left(\frac{2\epsilon_m}{\epsilon_m + \epsilon_s(\omega)}\right). \quad (5.20)$$

In the expression, ϵ_m is the dielectric constant of the wire's surrounding medium and ϵ_s is the dielectric constant of the semiconductor. Inserting $f(\omega)$ into Equation 5.18 then yields

$$\alpha(\omega) = \frac{\omega}{n_m c} P |f(\omega)|^2 \epsilon_{2,s},$$

where P is again a volume fraction for the nanowire in the host, and $\epsilon_{2,s}$ is the imaginary part of ϵ_s. We can also find the wire's associated absorption cross section σ, since $\alpha = R\sigma$, where R represents the number of wires per unit volume in the host (i.e., $R = N/V_{tot}$). This gives $\sigma_{NW}(\omega) = \alpha/R = \alpha V_{tot}/N$ and yields

$$\sigma_{NW}(\omega) = \frac{\omega}{n_m c} P \frac{V_{tot}}{N} |f(\omega)|^2 \epsilon_{2,s}.$$

If $P = NV_{NW}/V_{tot}$ is the corresponding volume fraction of wires in the host, we find

$$\sigma_{NW}(\omega) = \frac{\omega}{n_m c} \frac{NV_{NW}}{V_{tot}} \frac{V_{tot}}{N} |f(\omega)|^2 \epsilon_{2,s},$$

or

$$\sigma_{NW}(\omega) = \frac{\omega}{n_m c} V_{NW} |f(\omega)|^2 \epsilon_{2,s},$$

with $V_{NW} = \pi r^2 l$. The end result is

$$\sigma_{NW}(\omega) = \frac{\omega}{n_m c} (\pi r^2 l) |f(\omega)|^2 \epsilon_{2,s}. \quad (5.21)$$

Alternatively, since $\epsilon_{2,s} = 2n_s k_s$ (Equation 5.15), an equivalent expression sometimes seen in the literature is

$$\sigma_{NW}(\omega) = \frac{\omega}{n_m c}(\pi r^2 l)|f(\omega)|^2 2n_s k_s. \tag{5.22}$$

EXAMPLE 5.2

CdS Nanowire Absorption Cross Section Estimate Let us now utilize Equation 5.22 to calculate σ for a CdS nanowire. Specifically, assume the following literature parameters to calculate the cross section of a 14 nm diameter, 1 μm long wire at 405 nm:

$\epsilon_s = 6.72 + i2.31$,
$n_s = 2.63$ at 3.0 eV,
$k_s = 0.44$ at 3.0 eV,
$\epsilon_m = 2.2$, assuming a surrounding toluene matrix,
$n_m = 1.5$, assuming a surrounding toluene matrix,
$\omega = 4.65 \times 10^{15}\ \text{s}^{-1}$ ($\lambda = 405$ nm),
radius $r = 7$ nm,
length $l = 1\ \mu$m.

When these parameters are inserted into Equation 5.22, we find

$$\sigma_{\text{CdS NW,405 nm}}(\omega) = 4.06 \times 10^{-11}\ \text{cm}^2,$$

which is consistent with the cross-sections provided in **Table 5.2**. Note that bulk values of n_s, k_s, ϵ_1, and ϵ_2 are used in the absence of actual nanowire optical constants. As an exercise, the reader may carry out the same calculation for a CdSe nanowire excited as the same frequency.

5.5.6 Quantum Dot Absorption Cross Section

The final case of a quantum dot is straightforward and follows the nanowire example above. The only difference is a change in the local field factor due to the dot's spherical shape and associated electromagnetic boundary conditions. Namely,

$$f(\omega) = \frac{3\epsilon_m}{2\epsilon_m + \epsilon_s(\omega)} \tag{5.23}$$

(Ricard et al. 1994; Klimov 2000). Inserting this into Equation 5.18 gives

$$\alpha(\omega) = \frac{\omega}{n_m c} P|f(\omega)|^2 \epsilon_{2,s}.$$

An associated single quantum dot cross section is found using $\alpha = R\sigma$, where R is the number of dots per unit volume in the host (i.e., $R = N/V_{\text{tot}}$). We obtain $\sigma = \alpha/R = \alpha V_{\text{tot}}/N$. This then gives

$$\sigma(\omega) = \frac{\omega}{n_m c} P \frac{V_{\text{tot}}}{N} |f(\omega)|^2 \epsilon_{2,s},$$

where $P = NV_{\text{QD}}/V_{\text{tot}}$ is the volume fraction of nanocrystals. We thus obtain

$$\sigma_{\text{QD}}(\omega) = \frac{\omega}{n_m c} \frac{NV_{\text{QD}}}{V_{\text{tot}}} \frac{V_{\text{tot}}}{N} |f(\omega)|^2 \epsilon_{2,s},$$

or

$$\sigma_{\text{QD}}(\omega) = \frac{\omega}{n_m c} V_{\text{QD}} |f(\omega)|^2 \epsilon_{2,s},$$

with $V_{QD} = \frac{4}{3}\pi r^3$. This yields

$$\sigma_{QD}(\omega) = \frac{\omega}{n_m c}\left(\frac{4}{3}\pi r^3\right)|f(\omega)|^2 \epsilon_{2,s}. \tag{5.24}$$

As before, an alternative literature expression can be found by recalling that $\epsilon_{2,s} = 2n_s k_s$:

$$\sigma_{QD}(\omega) = \frac{\omega}{n_m c}\left(\frac{4}{3}\pi r^3\right)|f(\omega)|^2 2n_s k_s. \tag{5.25}$$

More about these sorts of approximations to nanostructure absorption coefficients and cross sections can be found in the cited references.

EXAMPLE 5.3

CdSe Quantum Dot Absorption Cross-Section Estimate Let us now illustrate Equation 5.25 by finding σ for a 6 nm diameter CdSe quantum dot using the following literature values:

$\epsilon_s = 7.08 + i4.28$,
$n_s = 2.77$ at 3.54 eV,
$k_s = 0.772$ at 3.54 eV,
$\epsilon_m = 2.2$, assuming a surrounding toluene matrix,
$n_m = 1.5$, assuming a surrounding toluene matrix,
$\omega = 5.39 \times 10^{15}\ \text{s}^{-1}$ ($\lambda = 350$ nm),
radius $r = 3$ nm.

When these parameters are inserted into Equation 5.25, we obtain

$$\sigma_{\text{CdSe QD},350\,\text{nm}}(\omega) = 1.7 \times 10^{-14}\ \text{cm}^2,$$

which is in good agreement with the quantum dot cross sections listed in **Table 5.2**. Note that bulk n_s, k_s, ϵ_1, and ϵ_2 values are used in the absence of actual quantum dot optical constants.

5.6 USING THE ABSORPTION CROSS SECTION

At this point, we can demonstrate the use of σ to estimate the concentration of carriers in a nanostructure upon photoexcitation. Namely, upon absorbing light, carriers are promoted from a system's valence band to its conduction band. These carriers then relax, occasionally resulting in the emission of light. Alternatively, in photoconductivity measurements, they are extracted to obtain a photocurrent. The excited-state carrier concentrations we will evaluate can then be used to model the system's photocurrent or emission decay kinetics more quantitatively and will also be useful in determining its photoluminescence efficiency.

We begin with an intensity (more formally speaking, the irradiance) of the incident light, I (with units of W/cm^2). The rate of photons absorbed by a system is then

$$R = \frac{I\sigma}{h\nu}. \tag{5.26}$$

In Equation 5.26, $h\nu$ is the energy per photon (with units of J) and h is Planck's constant ($h = 6.62 \times 10^{-34}$ J s). The energy per photon at a given wavelength is obtained using

$$E(\text{eV}) \simeq \frac{1240}{\lambda(\text{nm})}, \tag{5.27}$$

Excited state
|2⟩
Absorption
Stimulated emission
Spontaneous emission
Nonradiative relaxation
|1⟩
Ground state

Figure 5.6 Diagram illustrating the absorption process and various relaxation pathways from the excited state of a two-level system. Dirac bra–ket notation is used to denote the ground (|1⟩) and excited (|2⟩) states. More about this notation will be seen in Chapter 6.

where E is the energy of the light (with units of eV) and λ is its associated wavelength (with units of nm). The conversion between λ and E in electronvolts is easy to derive and is left for the reader (see Problem 3.1 in Chapter 3). To obtain the photon energy in joules, E is multiplied by 1.602×10^{-19} J/eV.

Accompanying the absorption of light are complementary relaxation processes. These include stimulated emission, nonradiative relaxation, and photoluminescence. All are illustrated in **Figure 5.6** in the context of a simple two-level system.

If we lump these relaxation processes together, there exists one (overall) excited-state relaxation rate. Physically, the net photogenerated carrier density decreases via *all* of the abovementioned decay processes, occurring simultaneously. We now model the time-dependent ground and excited state populations through a simple kinetic analysis as follows.

First, picture a two-level system (**Figure 5.7**) with effective upward and downward rates γ_{12} and γ_{21}. The differential equations that describe the rate of population and depopulation of each state are

$$\frac{dn_1}{dt} = -\gamma_{12} n_1 + \gamma_{21} n_2,$$

$$\frac{dn_2}{dt} = \gamma_{12} n_1 - \gamma_{21} n_2,$$

where n_1 and n_2 are the ground-state and excited-state populations.

At equilibrium, the forward and reverse rates are equal, so that

$$\gamma_{12} n_1 = \gamma_{21} n_2.$$

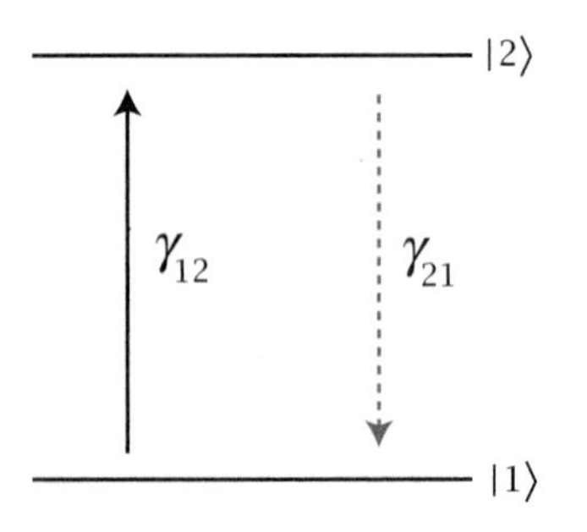

Figure 5.7 Diagram of a two-level system with effective upward and downward rates γ_{12} and γ_{21}. Dirac bra–ket notation is used to denote the ground (|1⟩) and excited (|2⟩) states. More about this notation will be seen in Chapter 6.

In parallel, the absorption rate can be described in terms of the absorption cross section σ and the incident light intensity I through (Equation 5.26), resulting in

$$\frac{I\sigma}{h\nu} = \gamma_{21} n_2.$$

We therefore find from this equivalence that

$$n_2 = \frac{I\sigma}{h\nu}\tau_{21}, \tag{5.28}$$

where $\tau_{21} = 1/\gamma_{21}$. Equation 5.28 describes the equilibrium number of carriers in the system, accounting for all relaxation processes. Note that if a pulsed laser with a temporal pulse width of τ_{pulse} is used to excite the sample and $\tau_{\text{pulse}} \ll \tau_{21}$ (i.e., a very short pulse), one can derive the following analogous expression for the steady-state excited-state population

$$n_2 = \frac{I\sigma}{h\nu}\tau_{\text{pulse}}. \tag{5.29}$$

EXAMPLE 5.4

Estimate of the Average Number of Carriers in a Quantum Dot under Continuous Excitation To illustrate Equation 5.28, let us calculate the average number of carriers in a quantum dot under continuous illumination. Assume the following parameters: a 3 nm radius quantum dot, a corresponding cross section $\sigma = 1.5 \times 10^{-14}$ cm^2 at 350 nm (**Table 5.2**), an excited-state lifetime of 20 ns, and an incident excitation intensity $I = 100$ W/cm^2. When inserted into Equation 5.28, we find

$$\langle n_2 \rangle = \left(\frac{(100\,\mathrm{W/cm^2})(1.5 \times 10^{-14}\,\mathrm{cm^2})}{5.7 \times 10^{-19}\,\mathrm{J}} \right) (20 \times 10^{-9}\,\mathrm{s})$$
$$= 0.05,$$

showing that, on average, $\langle n_2 \rangle \simeq 0.05$ electrons (or holes) exist per dot under steady-state conditions.

5.6.1 A More Detailed Look at the Excited-State Population

Equations 5.28 and 5.29 are limiting forms of a more general expression for the time-dependent excited-state population (n_2). From our preceding rate expressions, we know that

$$\frac{dn_2}{dt} = \gamma_{12} n_1 - \gamma_{21} n_2.$$

We will now solve this equation to obtain a general time-dependent expression for $n_2(t)$. Recall that $\gamma_{12} n_1 = I\sigma / h\nu$. As a consequence,

$$\frac{dn_2}{dt} = \frac{I\sigma}{h\nu} - \gamma_{21} n_2.$$

To simplify the notation, let

$$f = \frac{I\sigma}{h\nu} - \gamma_{21} n_2$$

to obtain

$$\frac{dn_2}{dt} = f.$$

From the definition of f, $df/dt = -\gamma_{21}\, dn_2/dt$, or $dn_2/dt = -(1/\gamma_{21})\, df/dt$, and so

$$\frac{df}{dt} = -\gamma_{21} f.$$

On integrating this, we find

$$\frac{df}{f} = -\gamma_{21}\, dt$$

$$\int_{f(t=0)}^{f(t=\tau_{\mathrm{pulse}})} \frac{df}{f} = -\gamma_{21} \int_{t=0}^{t=\tau_{\mathrm{pulse}}} dt,$$

where τ_{pulse} is the laser pulse width. The solution is then

$$f = Ce^{-\gamma_{21}\tau_{\mathrm{pulse}}},$$

where C is a constant. Equating this to $f = I\sigma/h\nu - \gamma_{21} n_2$ and solving for $n_2(\tau_{\mathrm{pulse}})$ yields

$$n_2(\tau_{\mathrm{pulse}}) = \frac{1}{\gamma_{21}} \left(\frac{I\sigma}{h\nu} \right) - \frac{C}{\gamma_{21}} e^{-\gamma_{21}\tau_{\mathrm{pulse}}}.$$

Finally, to determine what C is, we apply the problem's initial conditions. Namely, at $t = 0$, $n_2 = 0$ and $\tau_{\mathrm{pulse}} = 0$ since we assume no excited-state population prior to exciting the system. We therefore find that $C = I\sigma/h\nu$, and inserting this into our solution gives

$$n_2(\tau_{\mathrm{pulse}}) = \left(\frac{I\sigma}{h\nu} \right) \frac{1}{\gamma_{21}} (1 - e^{-\gamma_{21}\tau_{\mathrm{pulse}}}).$$

Since $1/\gamma_{21} = \tau_{21}$, our final expression for the time-dependent excited-state population is

$$n_2(\tau_{\text{pulse}}) = \left[\left(\frac{I\sigma}{h\nu}\right)\tau_{21}\right]\left(1 - e^{-\tau_{\text{pulse}}/\tau_{21}}\right). \tag{5.30}$$

At this point, we can readily evaluate its limiting forms in order to obtain Equations 5.28 and 5.29.

First Limit: A Short Laser Pulse

If $\tau_{\text{pulse}} \ll \tau_{21}$ (i.e., an ultrafast laser), we can apply a Taylor series expansion to the exponential in Equation 5.30 (i.e., $e^x \simeq 1 - x + \cdots$). This leads to

$$n_2(t) \simeq \left[\left(\frac{I\sigma}{h\nu}\right)\tau_{21}\right]\left[1 - \left(1 - \frac{\tau_{\text{pulse}}}{\tau_{21}}\right)\right].$$

Simplifying the result yields our desired final expression:

$$n_2(t) \simeq \left(\frac{I\sigma}{h\nu}\right)\tau_{\text{pulse}},$$

which matches Equation 5.29 shown earlier.

Second Limit: A Long Laser Pulse

Alternatively, if $\tau_{\text{pulse}} \gg \tau_{21}$ (i.e., a long pulse that approaches a continuous beam), the exponential term is small and close to zero. We can therefore ignore it. As a consequence,

$$n_2(t) \simeq \left(\frac{I\sigma}{h\nu}\right)\tau_{21}.$$

This matches Equation 5.28 shown earlier.

5.7 EMISSION PROCESSES

A major way for the system to relax after being photoexcited is by spontaneous emission (i.e., photoluminescence). In this section, we therefore model the process using the same two-level system shown earlier in **Figure 5.7**. This time, we shall be a little more explicit, noting that $\gamma_{21} = \gamma_{\text{spont}} + \gamma_{\text{nonrad}}$. We shall also ignore stimulated emission, since it is only present during the excitation.

In this picture, then, the excited state, $|2\rangle$ (Note that Dirac bra–ket notation will be introduced in Chapter 6. For now just take this to be a single label), is populated *instantaneously* using a very fast laser pulse. The resulting carrier population subsequently decays with the rate

$$\frac{dn_2(t)}{dt} = -\gamma_{21} n_2(t).$$

Solving for $n_2(t)$ yields

$$n_2(t) = Ce^{-\gamma_{21} t},$$

where C is a constant and can be found using the problem's initial conditions. Specifically, at time $t = 0$, the excited-state population is essentially the initial carrier population n_0, generated by the instantaneous laser pulse. An approximation for n_0 can be made using

Equation 5.29. The desired expression for the excited-state population is therefore

$$n_2(t) = n_0 e^{-\gamma_{21} t} \tag{5.31}$$

and decays exponentially with time. Thus, in the very simplest of scenarios, excited-state populations decay as described in Equation 5.31. The time it takes for n_2 to fall to $1/e$ ($\sim 37\%$) of its initial value is called the $1/e$ time or the lifetime. Thus, for the process at hand,

$$t_{1/e} = \frac{1}{\gamma_{21}} = \tau_{21}. \tag{5.32}$$

Typical molecular/nanostructure lifetimes range from 1 to 10 ns and such ballpark values can be estimated, a priori, if the system's absorption cross section is known. This is something we aim to demonstrate and entails using what are called Einstein A and B coefficients, which we will see shortly.

Summary

At this point, the take-home message of what we have just discussed is that semiconductor absorption coefficients are on the order of $\alpha \sim 10^4\ \text{cm}^{-1}$ (**Table 5.1**). Associated absorption cross sections are $\sigma \sim 10^{-15}$–$10^{-16}\ \text{cm}^2$ (**Table 5.2**) and corresponding molar extinction coefficients are $\epsilon \sim 10^5\ \text{M}^{-1}\ \text{cm}^{-1}$. Typical excited-state lifetimes accompanying these absorption values should then range from 1 to 10 ns.

5.7.1 Quantum Yields

To conclude our brief discussion of nanostructure emission, we introduce the emission quantum yield (QY). This is a unitless number between 0 and 1, which describes the ratio of emitted to absorbed photons in a system. It can be expressed as the fraction of the excited-state decay that occurs radiatively as opposed to nonradiatively. To illustrate, the total decay rate out of the excited state in **Figure 5.6** (excluding stimulted emission) is

$$\gamma_{\text{eff}} = \gamma_{\text{rad}} + \gamma_{\text{nonrad}}.$$

The fraction of the total excitation resulting in radiative emission is then the ratio

$$\frac{\gamma_{\text{rad}}}{\gamma_{\text{eff}}} = \text{QY} = \frac{\gamma_{\text{rad}}}{\gamma_{\text{rad}} + \gamma_{\text{nonrad}}},$$

yielding

$$\text{QY} = \frac{\gamma_{\text{rad}}}{\gamma_{\text{rad}} + \gamma_{\text{nonrad}}} = \gamma_{\text{rad}} \tau_{\text{eff}}. \tag{5.33}$$

Note that on the far right side of Equation 5.33, QY has been expressed in terms of the $1/e$ excited state decay time ($\tau_{\text{eff}} = (\gamma_{\text{rad}} + \gamma_{\text{nonrad}})^{-1}$).

QYs of many highly fluorescent organic molecules take values near unity. However, for nanostructures such as colloidal quantum dots, QYs on the order of 0.1 are common. Larger values, ranging from 0.3 to 0.6, can be obtained after "overcoating" these materials with a thin semiconductor shell, which passivates surface defects, contributing to the nonradiative decay of the excited state. This is discussed more fully in Chapter 13. For solution-grown nanowires, values as low as 0.1%

Table 5.4 Sample emission quantum yields

System	Associated emission quantum yield
Rhodamine 6G	0.95
CdSe QDs	0.1
InP QDs	0.3
CdSe/ZnS core/shell QDs	0.3–0.5
CdSe/CdS core/shell QDs	0.5
Carbon nanotubes	0.03
CdS nanowires	0.0004–0.01
CdSe nanowires	0.001
C_{60}	$\sim 2 \times 10^{-4}$
C_{70}	$\sim 9 \times 10^{-4}$

QD, quantum dot.

have been seen and likely stem from the existence of analogous surface defects. **Table 5.4** lists representative QYs for various organic and nanostructured systems. Associated literature can be found in the Further Reading.

Finally, it should be mentioned that having both the QY and the effective excited state decay time enables one to extract γ_{rad} and γ_{nonrad}. To illustrate, given a QY as well as a lifetime, one obtains γ_{rad} from Equation 5.33. Further use of Equation 5.33 yields γ_{nonrad}. This is demonstrated below.

EXAMPLE 5.5

Estimates of the Radiative and Nonradiative Recombination Rates from the QY and the Effective Excited-State Decay Time Assume an overcoated CdSe/ZnS quantum dot that has QY = 0.3 (**Table 5.4**) and an effective excited-state lifetime $\tau_{21} = \tau_{eff} = 25$ ns. We first find γ_{rad} using Equation 5.33:

$$QY = \gamma_{rad}\tau_{21}$$
$$0.3 = \gamma_{rad}(25 \times 10^{-9}\ \text{s}).$$

This yields $\gamma_{rad} = 1.2 \times 10^7\ \text{s}^{-1}$, or $\tau_{rad} = 83$ ns. Using Equation 5.33 again,

$$\gamma_{rad} + \gamma_{nonrad} = \frac{1}{\tau_{21}}$$
$$1.2 \times 10^7 + \gamma_{nonrad} = \frac{1}{25 \times 10^{-9}}$$

then gives $\gamma_{nonrad} = 2.8 \times 10^7\ \text{s}^{-1}$.

5.8 EINSTEIN *A* AND *B* COEFFICIENTS

Finally, what we want to do is illustrate how knowing a material's absorption cross section enables one to estimate its excited-state lifetime. We expect that if we were to do such a back-of-the-envelope calculation, we would find that molar extinction coefficients $\epsilon \sim 10^5\ \text{M}^{-1}\ \text{cm}^{-1}$ (or absorption cross sections $\sigma \sim 10^{-16}\ \text{cm}^2$) are associated with excited-state lifetimes of the order of a few nanoseconds (1–10 ns). In this section, we therefore derive the connection between the two via what are referred to as Einstein *A* and *B* coefficients.

Let us begin by considering the two-level system shown in **Figure 5.8**. It has an upward rate due to the absorption of light. It also has a downward rate due to stimulated as well as spontaneous emission. For convenience, we ignore all other nonradiative processes. Next, we define the following upward rate, which involves the first of two Einstein (*B*) coefficients (McHale 1999; Bernath 2005):

$$\gamma_{abs} = B_{12}\rho(\nu)N_1.$$

In this expression, B_{12} is called an Einstein *B* coefficient (with units of $m^3\,J^{-1}\,s^{-2}$), $\rho(\nu)$ is an energy density, which could be thermal or light-related, especially if the sample is illuminated with a laser, and N_1 is the population of the lower energy level, called the ground state. Similarly, the stimulated emission rate is defined as

$$\gamma_{stim} = B_{21}\rho(\nu)N_2,$$

where B_{21} is the second Einstein *B* coefficient (again with units of $m^3\,J^{-1}\,s^{-2}$) and N_2 is an excited-state population. Finally, we introduce a phenomenological spontaneous emission rate

$$\gamma_{emm} = AN_2,$$

where *A* is called the Einstein *A* coefficient (with units of s^{-1}).

We now write two rate equations that describe changes in the ground- and excited-state populations through their various population/depopulation rates:

$$\frac{dN_1}{dt} = -B_{12}\rho(\nu)N_1 + B_{21}\rho(\nu)N_2 + AN_2,$$

$$\frac{dN_2}{dt} = B_{12}\rho(\nu)N_1 - B_{21}\rho(\nu)N_2 - AN_2.$$

At equilibrium, both forward and reverse rates are equal, so that

$$B_{12}\rho(\nu)N_1 = B_{21}\rho(\nu)N_2 + AN_2.$$

Solving for $\rho(\nu)$ then yields

$$\rho(\nu) = \frac{A}{B_{12}N_1/N_2 - B_{21}}$$

and describes the energy density of the incident light illuminating the system. Note that the ratio N_1/N_2 in the denominator is related to the Boltzmann distribution through $N_2/N_1 = e^{-\Delta E/kT}$, or

$$\frac{N_1}{N_2} = e^{\Delta E/kT}, \qquad (5.34)$$

where $\Delta E = h\nu$ is the transition energy between the two states. A plot of the Boltzmann distribution is shown in **Figure 5.9**. As an aside, it is one of three statistical distribution functions that we will encounter throughout this text. The other two are called the Fermi–Dirac and Bose–Einstein distributions. For comparison purposes, the Bose–Einstein distribution is plotted as a function of ΔE in **Figure 5.9**. The Fermi–Dirac distribution was first seen in Chapter 3 and will be used more extensively in Chapter 9.

When Equation 5.34 is inserted into $\rho(\nu)$, the following energy density results:

$$\rho(\nu) = \frac{A}{B_{12}e^{\Delta E/kT} - B_{21}}. \qquad (5.35)$$

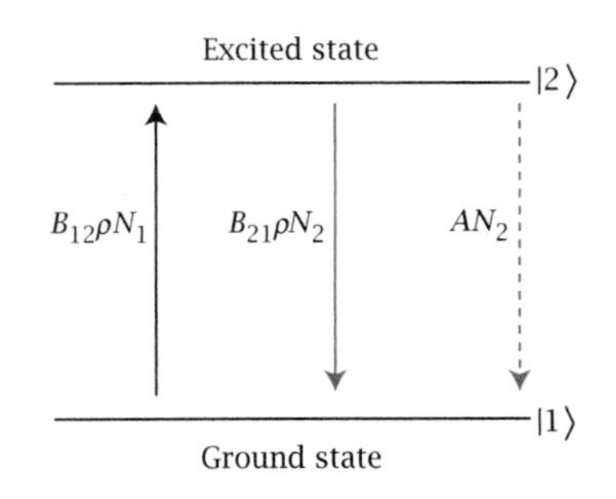

Figure 5.8 Diagram of a two-level system with associated Einstein upward and downward rates. Dirac bra–ket notation is used to denote the ground (|1⟩) and excited (|2⟩) states. More about this notation will be seen in Chapter 6.

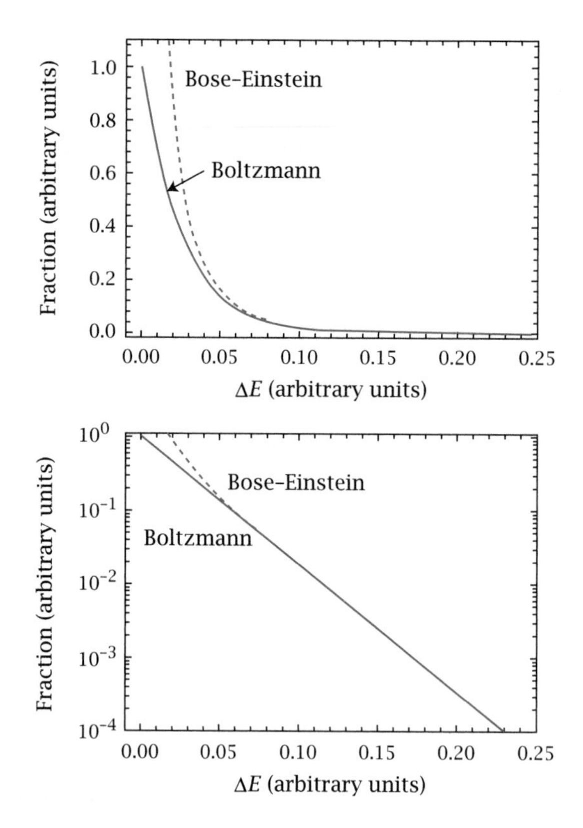

Figure 5.9 Comparison of the Boltzmann and Bose–Einstein distributions.

Explicit expressions for A and B_{12} (B_{21}) are then obtained by relating Equation 5.35 to the Planck blackbody radiation distribution

$$\rho(\nu) = \frac{8\pi h\nu^3}{c^3}\left(\frac{1}{e^{h\nu/kT} - 1}\right), \tag{5.36}$$

with units of energy per unit volume per unit frequency. As an aside, note that the term in parenthesis is the Bose–Einstein distribution, which applies to photons. When $h\nu$ is large compared with kT, this distribution converges to the Boltzmann distribution, as illustrated in **Figure 5.9**. Note further that the Planck blackbody distribution can alternatively be written in terms of λ:

$$\rho(\lambda) = \frac{8\pi hc}{\lambda^5}\left(\frac{1}{e^{hc/\lambda kT} - 1}\right), \tag{5.37}$$

with units of energy per unit volume per unit wavelength. We shall use $\rho(\nu)$ in our analysis with the conversion between $\rho(\nu)$ and $\rho(\lambda)$ being left as an exercise for the reader.

We now equate these two versions of the energy density, the first by Einstein and the second by Planck:

$$\frac{A}{B_{12}e^{h\nu/kT} - B_{21}} = \frac{8\pi h\nu^3}{c^3}\left(\frac{1}{e^{h\nu/kT} - 1}\right).$$

When $B_{12} = B_{21}$,

$$\frac{A}{B_{12}\left(e^{h\nu/kT} - 1\right)} = \frac{8\pi h\nu^3}{c^3}\left(\frac{1}{e^{h\nu/kT} - 1}\right).$$

Then, by comparison,

$$\frac{A}{B_{12}} = \frac{8\pi h\nu^3}{c^3},$$

leading to the classic Einstein A and B relationships

$$A = \frac{8\pi h\nu^3}{c^3} B_{12}, \tag{5.38}$$

$$B_{12} = B_{21}. \tag{5.39}$$

Note that texts will sometimes consider the degeneracy of the ground and excited states. This value reflects the number of levels having the same energy. We shall see this shortly when we discuss quantum mechanical models for carrier confinement (Chapter 7). For now, however, assume that both the ground and excited states in our picture have identical degeneracies, causing any dependency to disappear. Note also that when not in vacuum, c in the denominator of Equation 5.38 is replaced with c/n_m, where n_m is the refractive index of the surrounding medium.

5.8.1 A Word of Caution

Finally, a word of caution. The Planck blackbody radiation distribution can be written in a number of ways. As a consequence, different texts have what appear to be completely different Einstein A and B relationships. These differences arise because of different definitions of the Planck density used by authors. Previously, we saw two expressions. The first, in terms of frequency (Equation 5.36), had units of energy per unit volume per unit frequency. The second, in terms of wavelength (Equation 5.37), had units of energy per unit volume per unit wavelength. All of this becomes confusing when comparing derived Einstein A and B relationships. One should therefore look very carefully at what is meant by "density." Other confusing aspects about Einstein A and B coefficients exist, and the reader is encouraged to consult Hilborn (1982) for more details.

5.9 RELATING ABSORPTION CROSS SECTIONS TO EXCITED-STATE LIFETIMES

We are now at the point where we can relate B_{12} to the absorption cross section discussed earlier. This will allow us to estimate, a priori, excited-state lifetimes for a two-level system. Namely, provided $\sigma \sim 10^{-15}$–$10^{-16}\,\mathrm{cm}^2$, we should be able to find associated excited state lifetimes of the order of 1–10 ns. Alternatively, given a lifetime, we should be able to work backwards to find σ.

Although the previous relationships between the Einstein A and B coefficients are valid, it is not straightforward to relate them to σ, because we have assumed the Planck blackbody energy density when deriving B_{12}. As a consequence, we need to consider the more realistic case of a laser exciting our sample. The underlying energy density therefore differs and the analysis changes as follows. This derivation follows the work of Hilborn (1982).

We start with an expression for the differential absorption rate by a sample, $w_{12}(\nu)\,d\nu$ (with units of number/s), expressed as

$$[w_{12}(\nu)]\,d\nu = [b_{12}(\nu)\rho(\nu)N_1]\,d\nu$$

with $b_{12} = B_{12}f(\nu)$, B_{12} the Einstein B coefficient (with units of $\mathrm{m^3\,J^{-1}\,s^{-2}}$), $\rho(\nu)$ the laser's energy density (with units of $\mathrm{J\,m^{-3}\,Hz^{-1}}$) and $f(\nu)$ a transition lineshape (with units of inverse frequency, $\mathrm{Hz^{-1}}$).

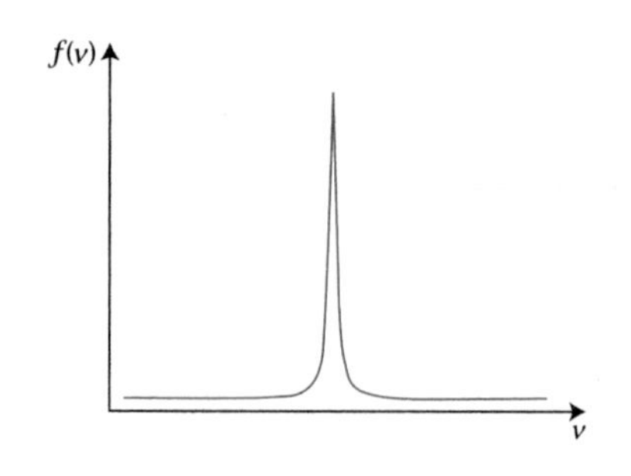

Figure 5.10 Sharply peaked Lorentzian lineshape.

A schematic illustration of $f(\nu)$ is shown in **Figure 5.10**. For simplicity, a Lorentzian lineshape is drawn, although other lineshapes are possible, as will be seen shortly. N_1 is the ground-state population.

Next, we define the laser intensity $I(\nu)$ (with units of W/m^2) as

$$I(\nu) = c\rho(\nu)\,d\nu = i(\nu)\,d\nu, \tag{5.40}$$

where $i(\nu)$ is called the spectral irradiance (with units of W m^{-2} Hz^{-1}). This can be seen through a unit analysis. We then have the following relationship:

$$\rho(\nu) = \frac{i(\nu)}{c}, \tag{5.41}$$

which relates the energy density to the spectral irradiance and which we will use shortly.

At this point, the differential absorption rate is multiplied by $h\nu$ to obtain the differential power change on absorbing light:

$$-dP = (h\nu)[w_{12}(\nu)]\,d\nu = (h\nu)[b_{12}(\nu)\rho(\nu)N_1]\,d\nu.$$

The negative sign indicates a loss of power due to absorption. On a more macroscopic scale,

$$-\Delta P = (h\nu)[b_{12}(\nu)\rho(\nu)(n_1 A\,\Delta x)]\,d\nu,$$

where N_1 has been replaced by $n_1 A\,\Delta x$, with A the physical cross-sectional area of the beam (with units m^2), Δx is its pathlength (with units of m) and n_1 is the number density of absorbers (with units of number/m^3).

Since the spectral irradiance represents the change in power per unit area per unit frequency, we independently write

$$\begin{aligned}\frac{di(\nu)}{dx} &= \frac{\left(\dfrac{\text{change in power}}{\text{unit area}\cdot\text{unit frequency}}\right)}{(\text{unit length})}\\ &= \frac{\left(\dfrac{\Delta P}{A\,d\nu}\right)}{(\Delta x)} = \frac{\Delta P}{A\,\Delta x\,d\nu}\\ &= -(h\nu)\left[b_{12}(\nu)\rho(\nu)n_1\right].\end{aligned}$$

Then, given $\rho(\nu) = i(\nu)/c$ (Equation 5.41),

$$\frac{di(\nu)}{dx} = -(h\nu)\left[b_{12}(\nu)n_1\frac{i(\nu)}{c}\right]$$

$$\frac{1}{i(\nu)}\frac{di(\nu)}{dx} = -\sigma(\nu)n_1,$$

where

$$\sigma(\nu) = \frac{h\nu b_{12}(\nu)}{c}.$$

The result is connected to the Einstein B coefficient by recalling that $b_{12} = B_{12}f(\nu)$:

$$\sigma(\nu) = \frac{h\nu B_{12}}{c}f(\nu). \tag{5.42}$$

Alternatively, since $B_{12} = (c^3/8\pi h\nu^3)A$ (Equation 5.38),

$$\sigma(\nu) = \frac{c^2 A}{8\pi\nu^2}f(\nu). \tag{5.43}$$

Equation 5.43 can now be integrated to obtain a more convenient relationship between an *integrated* absorption cross section σ_{tot} (with units of m^2/s) and A:

$$\sigma_{\text{tot}} = \int_{-\infty}^{\infty} \sigma(\nu)\, d\nu = \int_{-\infty}^{\infty} \frac{c^2 A}{8\pi\nu^2} f(\nu)\, d\nu.$$

This will in turn enable us to obtain our desired relationship between $\sigma(\nu)$ and A as seen below. At this point, if we assume a narrow linewidth such that $\nu \simeq \nu_{12}$, then

$$\sigma_{\text{tot}} \simeq \frac{c^2 A}{8\pi\nu_{12}^2} \int_{-\infty}^{\infty} f(\nu)\, d\nu.$$

Next, provided that the linewidth is normalized (i.e., $\int_{-\infty}^{\infty} f(\nu)\, d\nu = 1$), we obtain

$$\sigma_{\text{tot}} = \int_{-\infty}^{\infty} \sigma(\nu)\, d\nu \simeq \frac{c^2}{8\pi\nu_{12}^2} A. \tag{5.44}$$

Finally, even though the integral of $\sigma(\nu)$ in Equation 5.44 should be carried out explicitly, it can be shown that $\int_{-\infty}^{\infty} \sigma(\nu)\, d\nu \sim \sigma(\nu_{12})\nu_{12}$, where $\sigma(\nu_{12})$ is the peak absorption cross section of the sharp transition. We therefore find the relationship

$$\sigma(\nu_{12})\nu_{12} \simeq \frac{c^2}{8\pi\nu_{12}^2} A,$$

which leads to our desired approximation

$$\sigma(\nu_{12}) \simeq \frac{c^2}{8\pi\nu_{12}^3} A \tag{5.45}$$

that can be used to obtain ballpark lifetimes, provided $\sigma(\nu_{12})$.

EXAMPLE 5.6

Estimate of the Excited-State Lifetime from the Absorption Cross Section To demonstrate Equation 5.45, assume a quantum dot that has a band-edge transition at 532 nm and a peak absorption cross section of $\sigma(\nu_{12}) = 1 \times 10^{-16}\,\text{cm}^2$. Using Equation 5.45, we have

$$\nu_{12} = 5.64 \times 10^{14}\,\text{Hz}\ (\lambda = 532\,\text{nm}),$$
$$\sigma(\nu_{12}) = 1 \times 10^{-16}\,\text{cm}^2 = 1 \times 10^{-20}\,\text{m}^2,$$

resulting in $A = 5 \times 10^8\,\text{s}^{-1}$. Since $A = 1/\tau_{\text{rad}}$, $\tau_{\text{rad}} \sim 2\,\text{ns}$. This illustrates that typical cross sections between 10^{-15} and $10^{-16}\,\text{cm}^2$ have associated excited-state lifetimes between 1 and 10 ns.

5.9.1 Lineshapes

To conclude, note that there are many possible lineshape functions. For example, we just saw $f(\nu)$ as a Lorentzian (**Figure 5.11a**):

$$f(\nu) = \frac{1}{\pi} \frac{\left(\frac{1}{2}A\right)}{(\nu - \nu_{12})^2 + \left(\frac{1}{2}A\right)^2}, \tag{5.46}$$

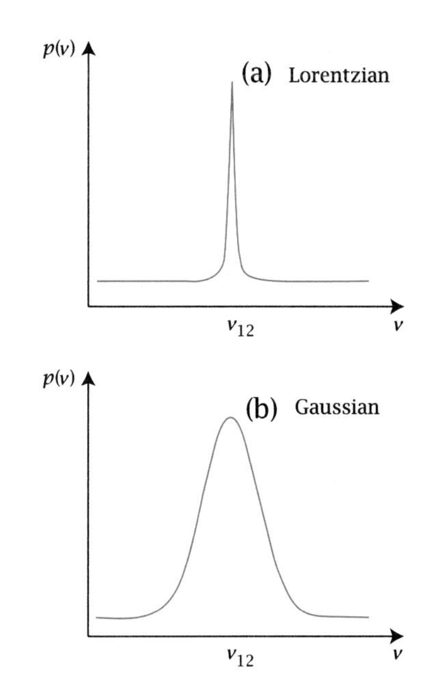

Figure 5.11 Illustration of (a) the Lorentzian and (b) the Gaussian lineshape functions.

with A the spontaneous emission rate related to the lineshape's width. The function is also normalized

$$\int_{-\infty}^{\infty} f(\nu)\, d\nu = 1$$

and has a peak value of

$$f(\nu_{12}) = \frac{2}{\pi A}.$$

The reader may verify both as an exercise.

Alternatively, the lineshape can have a Gaussian profile (**Figure 5.11b**):

$$f(\nu) = \frac{1}{\sqrt{2\pi} A} e^{-(\nu-\nu_{12})^2/2A^2}. \quad (5.47)$$

Equation 5.47 is normalized and the reader may likewise show that its peak value is

$$f(\nu_{12}) = \frac{1}{\sqrt{2\pi} A}.$$

In either case, the connection between A and the transition's spectral width arises from the Heisenberg uncertainty principle (Chapter 6), which states that there exists an inverse relationship between the transition linewidth and the system's excited-state lifetime.

5.10 SUMMARY

This completes our initial foray into the optical properties of low-dimensional materials. In the next chapter, we will revisit basic quantum mechanics to begin making these descriptions of their phenomenological optical constants more quantitative. This also requires us to introduce new concepts such as carrier confinement (Chapters 7 and 8), the density of states of nanostructures (Chapter 9), the concept of bands (Chapter 10), and time-dependent perturbation theory (Chapter 11). Using this information, we will then develop a more quantitative description of the size-dependent interband transitions seen in nanostructures, which will be the topic of Chapter 12.

5.11 THOUGHT PROBLEMS

5.1 Traveling wave

Consider a plane wave in SI units described by

$$E = 5\cos\left[3\pi \times 10^{14}\left(t - \frac{x}{c}\right) + \frac{\pi}{2}\right].$$

What is the wave's

(a) Frequency?

(b) Wavelength?

(c) Amplitude?

(d) Direction of propagation?

5.2 Traveling wave

While traveling through a material, the electric field of z-polarized light is given by

$$E_z = E_0\cos\left[2\pi \times 10^{15}\left(t - \frac{x}{0.5c}\right)\right].$$

Find

(a) The light's frequency

(b) Its wavelength

(c) The refractive index of the medium the light travels through.

5.3 Wave equation

Maxwell's equations under conditions where no free charges exist lead to a wave equation for both the electric and magnetic fields. Namely, starting with

$$\nabla \times \mathbf{E} = -\frac{\partial \mathbf{B}}{\partial t}$$

one can derive

$$\nabla^2 \mathbf{E} = \mu\mu_0\epsilon\epsilon_0 \frac{\partial^2 \mathbf{E}}{\partial t^2},$$

which is an expression for the electric field in the form of a wave equation. Likewise, starting from

$$\nabla \times \mathbf{B} = \mu\mu_0\epsilon\epsilon_0 \frac{\partial \mathbf{E}}{\partial t}$$

one can derive an expression for the magnetic field also in the form of a wave equation:

$$\nabla^2 \mathbf{H} = \mu\mu_0\epsilon\epsilon_0 \frac{\partial^2 \mathbf{H}}{\partial t^2}.$$

General solutions to either **E** or **H** take the form

$$f(x,t) = C\cos[k_{\text{light}}(x - vt) + \delta],$$

where v is the velocity of the wave, x is the propagation direction, k_{light} is the wavevector ($k_{\text{light}} = 2\pi/\lambda$), δ is a phase delay, C is an amplitude, and $k_{\text{light}}(x - vt) + \delta$ is the phase. Show that one can use a similar expression to solve these equations except with sine instead of cosine. The only difference is a phase shift, which you will determine.

5.4 Refractive index

CdS has a refractive index $n \simeq 2.5$ and a band-edge absorption coefficient $\alpha = 5.5 \times 10^4\ \text{cm}^{-1}$. Calculate the associated optical density of a 1 μm thick piece of CdS.

5.5 Back of the envelope

Many nanostructured Schottky junction solar cells today employ thin films of quantum dots as absorbing layers (see, e.g., Sargent 2009). A key requirement is the device's ability to absorb as much incident sunlight as possible using the thinnest nanocrystal layer achievable. This is because increasing the dot film thickness beyond a critical value increases the pathlength carriers must travel to their respective electrodes. As a consequence, unwanted recombination losses might occur. Furthermore, carriers might undergo diffusion over drift since they do not experience built-in fields close to the Schottky junction. Estimate the thickness of a CdSe quantum dot film needed to absorb 90% of the incident light. Make simplifying assumptions where needed.

5.6 Colors

Consider the well-known progression of colors possessed by ensembles of different-sized colloidal CdSe quantum dots. The smaller particles appear yellow. Larger ones appear red. As an illustration, see Boatman et al. (2005). Explain why small-diameter ensembles appear yellow while larger-diameter ensembles appear red. Similarly, it is well know that gold nanoparticles generally have a distinct ruby red color. By contrast, aggregated gold ensembles appear blue to the eye. Explain these colors as well.

5.7 Absorption coefficient

The complex dielectric constant of bulk CdTe is $\epsilon = 8.99 + i2.29$ at a given frequency. Calculate the associated absorption coefficient α.

5.8 Absorption coefficient

Plot the frequency-dependent absorption coefficient of bulk GaAs. Use values from tables of bulk real and imaginary refractive indices or dielectric constants. This might require a trip to the library.

5.9 Absorption cross sections

Colloidal quantum dots are popular 0D materials today. They are now commercially available from a number of sources. An important step in their development was the characterization of their absorption cross sections. See Yu et al. (2003) and the accompanying erratum. From the data provided there, find the range of band-edge absorption cross sections in cm^2 for CdSe quantum dots with sizes between 2 and 6 nm. Then express all results in Å^2. Compare these values with the physical dimensions of the nanocrystals responsible for the absorption.

5.10 Absorption cross sections

Estimate the absorption cross section of a 4 nm diameter PbSe quantum dot at 400 nm using expressions provided in the main text. Compare the obtained value with results from the literature (see, e.g., Moreels et al. 2007).

5.11 Absorption cross sections

Estimate the absorption cross section of a 10 nm diameter, 1 μm long CdTe nanowire at 488 nm, using expressions provided in the main text. Compare the obtained value with results from the literature (see, e.g., Protasenko et al. 2006).

5.12 Back of the envelope

Photothermal heterodyne imaging is a way to monitor the absorption of individual nanostructures (see, e.g., Cognet et al. 2008). Assume that one is working with a single gold nanoparticle surrounded by water. If the particle is excited on resonance with its main plasmon resonance at approximately 520 nm, estimate how much the temperature of the local environment changes. Assume that the particle is nonemissive and that all excess energy is released as heat.

5.13 Back of the envelope

Overcoated colloidal CdSe/ZnS core/shell quantum dots are popular fluorophores today in many single-nanostructure optical experiments. This is because they exhibit a phenomenon referred to as fluorescence intermittency or "blinking." Namely, under continuous excitation, the dots emit light in an intermittent "on/off" manner. For more details, see Nirmal et al. (1996). The same behavior is true of many organic dyes. In all cases, if one wants to image an individual fluorophore, one needs to attain sufficiently low coverages on a surface to ensure that only a single dot or dye is being studied. Estimate the concentration of a dot or dye solution needed to achieve coverages of the order of 10 fluorophores per square micrometer. Make simplifying assumptions as needed.

5.14 Back of the envelope

There is interest these days in conducting direct absorption experiments on nanostructures. This often entails using modulation approaches. See, e.g., Arbouet et al. (2004). Assume that one is doing a direct absorption experiment on a single semiconductor nanowire at 488 nm. For simplicity, assume that the beam is collimated and has a width equivalent to the diffraction-limited spot size if the beam were focused down. The intensity of the light across the collimated width is also assumed to be uniform. Estimate the fraction of light absorbed and transmitted.

5.15 Absorption saturation

Optical experiments on nanostructures are often carried out under conditions where the material's response scales linearly with excitation intensity. However, any system with discrete energy levels will exhibit absorption saturation at large enough intensities. Show for a two-level system that the probability P_2 of being in the excited state scales with excitation intensity I as

$$P_2 = \frac{1}{1 + I_{sat}/I},$$

where $I_{sat} = \gamma_{21} h\nu/\sigma$ is the system's saturation intensity, γ_{21} is the first-order decay rate constant from the excited state, and σ is the system's absorption cross section. Assume equilibrium conditions and the following rate equations that describe the population/depopulation of the ground and excited states:

$$\frac{dP_1}{dt} = -\gamma_{12} P_1 + \gamma_{21} P_2,$$

$$\frac{dP_2}{dt} = \gamma_{12} P_1 - \gamma_{21} P_2.$$

Recall that for the probabilities P_1 and P_2, $P_1 + P_2 = 1$. Show also that the associated emission rate is

$$R = \gamma_{21} \left(\frac{1}{1 + I_{sat}/I} \right).$$

5.16 Polarization anisotropy

In a number of experiments conducted on single quantum dots and nanowires, the system's polarization anisotropy is measured. Namely, in cases where there exists a unique orientation of the transition dipole moment, a polarization sensitivity can be seen in both the absorption and emission. Often this polarization sensitivity is quantified by the parameter

$$r = \frac{I_{\parallel} - I_{\perp}}{I_{\parallel} + 2I_{\perp}}$$

where $I_{\parallel}$ ($I_{\perp}$) is the absorbed/emitted intensity parallel (perpendicular) to the polarization of the incident light. Show that r can be expressed as

$$r = \tfrac{1}{2}(3\cos^2\theta - 1),$$

where θ is the angle between the incident light polarization vector and the system's transition moment. Assume vertically polarized light along the z direction of a Cartesian coordinate system and that the intensity of light absorbed or radiated in a given orientation is proportional to the square of the projection of the induced dipole's electric field onto a given axis. To obtain the final expression, make sure to angle-average the expression within the (x, y) plane. Show that when $\theta = 54.7°$, $r = 0$.

5.17 Emission decays

Derive a relationship between the half-life of a first-order decay and its corresponding $1/e$ lifetime.

5.18 Emission decays

Show mathematically how one extracts the average decay time of an exponential process. Compare this with the $1/e$ lifetime.

5.19 Dark exciton

The linear absorption of high-quality colloidal CdSe quantum dots was first understood in the early 1990s. However, the origin of their near-band-edge emission remained controversial. This was because it exhibited unusual features such as long microsecond lifetimes at low temperatures (by contrast, bulk recombination times are on the nanosecond time scale) and nanosecond lifetimes at room temperature. Furthermore, size-dependent "resonant" and "nonresonant" Stokes shifts were seen in corresponding emission spectra. These unusual properties are described in Nirmal et al. (1995). As a consequence, the emission was often thought to be surface-defect-related. In the mid 1990s, a revised effective mass model appeared that could readily explain all of these unusual features in a concerted manner (Efros et al. 1996). Key to the model was the existence of a dark, triplet-like, exciton state with an energetically close, optically active, bright state.

Show that the temperature dependence of CdSe quantum dot lifetimes can readily be rationalized by assuming a low-energy dark exciton state along with a slightly higher-energy bright exciton state. The energy difference between states is ΔE. The first-order rate constants for emission out of each state are γ_D and γ_B, where the subscripts refer to dark and bright. To simplify things, assume an instantaneous population of the excited states after pulsed excitation and that the population of these states reaches thermal equilibrium. At low temperatures, the decay should thus occur with a rate constant γ_D, whereas at higher temperatures, the decay occurs as the sum of rates, $\gamma_D + \gamma_B$.

5.20 Auger kinetics

There are many quantum dot and nanorod studies that show evidence of Auger two-carrier and three-carrier decays at high carrier densities (see, e.g., Htoon et al. 2003). Explain the Auger two- and three-carrier processes in quantum dots. Then solve the following differential equation describing the excited-state carrier density decay with time via Auger processes and find the general time-dependent carrier concentration $n(t)$:

$$\frac{dn}{dt} = -C_D n^D.$$

Finally, consider the specific cases $D = 2$ and $D = 3$ and show that plotting $n(0)/n(t) - 1$ $(D = 2)$ and $[n(0)/n(t)]^2 - 1$ $(D = 3)$ versus t yields straight lines.

5.21 Emission quantum yields

Explain how one measures the emission quantum yield of a nanostructure in real life. Find an example of someone doing this in the literature.

5.22 Emission quantum yields

Come up with an alternative means of measuring the quantum yield using a system's emission decay profile. Assume first-order kinetics for simplicity.

5.23 Emission decays and Lorentzian lineshape

Consider the two-level system described in the main text. Assume that it undergoes spontaneous emission when excited.

We have seen that the emitted light intensity decays exponentially with time. If we model the electric field of the emitted light as a damped oscillating function

$$E(t) = \begin{cases} 0 & (t < 0), \\ E_0 \cos(\omega_0 t)\, e^{-\gamma t} & (t \geq 0), \end{cases}$$

show through a Fourier transform that the spectrum of the emitted light exhibits a Lorentzian lineshape. Recall that

$$E(\omega) \propto \int_{-\infty}^{\infty} E(t) e^{i\omega t}\, dt$$

where $I(\omega) = |E(\omega)|^2$. Assume $\omega_0 \gg \gamma$ and conditions near resonance (i.e., $\omega \sim \omega_0$).

5.24 Equilibrium

A continuous laser beam is incident on a nanostructure that has an absorption coefficient α at the laser frequency ν.

(a) Show that electron–hole pairs in the material are generated at a rate equal to $I\alpha/h\nu$ per unit volume per unit time, where I is the laser's intensity.

(b) By considering the balance between the above carrier generation rate and the recombination rate under equilibrium, show that the carrier density n within the illuminated volume is equal to $I\alpha\tau/h\nu$, where τ is the electron–hole recombination lifetime.

5.25 Equilibrium

Imagine studying the optical properties of individual nanowires. For simplicity, model the absorption and emission of this system using a two-level system. Assume a typical absorption cross section for a 1 μm long nanowire of $\sigma = 10^{-11}$ cm^2. The excitation wavelength is 488 nm and the emission lifetime, assuming first-order kinetics, is 700 ps (unity quantum yield). Estimate the laser intensity resulting in the excited state being populated 50% of the time. Ignore stimulated emission and assume equilibrium conditions.

5.12 REFERENCES

Arbouet A, Christofilos D, Del Fatti N, et al. (2004) Direct measurement of the single-metal-cluster optical absorption. *Phys. Rev. Lett.* **93**, 127401-1.

Bernath PF (1995) *Spectra of Atoms and Molecules*, 2nd edn. Oxford University Press, New York.

Boatman EM, Liesensky GC, Nordell KJ (2005) A safer, easier, faster synthesis for CdSe quantum dot nanocrystals. *J. Chem. Educ.* **82**, 1697.

Cognet L, Berciaud S, Lasne D, Lounis B (2008) Photothermal methods for single nonluminescent nano-objects. *Anal. Chem.* **80**, 2288.

Efros AlL, Rosen M, Kuno M, et al. (1996) Band-edge exciton in quantum dots of semiconductors with a degenerate valence band: dark and bright exciton states. *Phys. Rev. B* **54**, 4843.

Fox M (2010) *Optical Properties of Solids*, 2nd edn. Oxford University Press, Oxford, UK.

Giblin J, Kuno M (2010) Nanostructure absorption: a comparison of nanowire and colloidal quantum dot cross sections. *J. Phys. Chem. Lett.* **1**, 3340.

Hilborn RC (1982) Einstein coefficients, cross sections, *f* values, dipole moments, and all that. *Am. J. Phys.* **50**, 982.

Htoon H, Hollingsworth JA, Dickerson R, Klimov VI (2003) Effect of zero- to one-dimensional transformation on multiparticle Auger recombination in semiconductor quantum rods. *Phys. Rev. Lett.* **91**, 227401.

Klimov VI (2000) Optical nonlinearities and ultrafast carrier dynamics in semiconductor nanocrystals. *J. Phys. Chem. B* **104**, 6112.

Landau LD, Lifshitz EM, Pitaevskii LP (1984) *Electrodynamics of Continuous Media*, 2nd edn. Pergamon, Oxford, UK.

Landolt–Börnstein (1982) *Numerical Data and Functional Relationships in Science and Technology. New series* (ed. KH Hellwege). Springer-Verlag, Berlin.

McHale JL (1999) *Molecular Spectroscopy*. Prentice Hall, Upper Saddle River, NJ.

Moreels I, Lambert K, De Muynck D, et al. (2007) Composition and size-dependent extinction coefficient of colloidal PbSe quantum dots. *Chem. Mater.* **19**, 6101.

Nirmal M, Dabbousi BO, Bawendi MG, et al. (1996) Fluorescence intermittency in single cadmium selenide nanocrystals. *Nature* **383**, 802.

Nirmal M, Norris DJ, Kuno M, Bawendi MG (1995) Observation of the "dark exciton" in CdSe quantum dots. *Phys. Rev. Lett.* **75**, 3728.

Palik ED (ed.) (1998) *Handbook of Optical Constants of Solids II*. Academic Press, San Diego, CA.

Protasenko V, Bacinello D, Kuno M (2006) Experimental determination of the absorption cross-section and molar extinction coefficient of CdSe and CdTe nanowires. *J. Phys. Chem. B* **110**, 25322.

Ricard D, Ghanassi M, Schanne-Klein MC (1994) Dielectric confinement and linear and nonlinear optical properties of semiconductor doped glasses. *Opt. Commun.* **108**, 311.

Sargent EH (2009) Infrared photovoltaics made by solution processing. *Nature Photonics* **3**, 325.

Yu WW, Qu L, Guo W, Peng X (2003) Experimental determination of the extinction coefficient of CdTe, CdSe, and CdS nanocrystals. *Chem. Mater.* **15**, 2854.

5.13 FURTHER READING

Absorption Cross Sections

Carlson LJ, Maccagnano SE, Zheng M et al. (2007) Fluorescence efficiency of individual carbon nanotubes. *Nano Lett.* **7**, 3698.

Coheur PF, Carleer M, Colin R (1996) The absorption cross sections of C_{60} and C_{70} in the visible–UV region. *J. Phys. B At. Mol. Opt. Phys.* **29**, 4987.

Eggeling C, Brand L, Seidel CAM (1997) Laser-induced fluorescence of coumarin derivatives in aqueous solution: Photochemical aspects for single molecule detection. *Bioimaging* **5**, 105.

Leatherdale CA, Woo WK, Mikulec FV, Bawendi MG (2002) On the absorption cross section of CdSe nanocrystal quantum dots. *J. Phys. Chem. B* **106**, 7619.

Protasenko V, Bacinello D, Kuno M (2006) Experimental determination of the absorption cross-section and molar extinction coefficient of CdSe and CdTe nanowires. *J. Phys. Chem. B* **110**, 25322.

Puthussery J, Lan A, Kosel TH, Kuno M (2008) Band-filling of solution-synthesized CdS nanowires. *ACS Nano* **2**, 357.

Schmidt J, Penzkofer A (1989) Absorption cross sections, saturated vapor pressures, sublimation energies, and evaporation energies of some organic laser dye vapors. *J. Chem. Phys.* **91**, 1403.

Yu P, Beard MC, Ellingson RJ, et al. (2005) Absorption cross-section and related optical properties of colloidal InAs quantum dots. *J. Phys. Chem. B* **109**, 7084.

Molar Extinction Coefficient

Birge RR (1987) *Kodak Laser Dyes.* Kodak Publication JJ-169, Eastman Kodak Co., Rochester, NY.

Hirayama K (1967) *Handbook of Ultraviolet and Visible Absorption Spectra of Organic Compounds.* Plenum Press, New York.

Leatherdale CA, Woo WK, Mikulec FV, Bawendi MG (2002) On the absorption cross section of CdSe nanocrystal quantum dots. *J. Phys. Chem. B* **106**, 7619.

Reynolds GA, Drexhage KH (1975) New coumarin dyes with rigidized structure for flashlamp-pumped dye lasers. *Opt. Commun.* **13**, 222.

Seybold PG, Gouterman M, Callis J (1969) Calorimetric, photometric and lifetime determinations of fluorescence yields of fluorescein dyes. *Photochem. Photobiol.* **9**, 229.

Yu P, Beard MC, Ellingson RJ, et al. (2005) Absorption cross-section and related optical properties of colloidal InAs quantum dots. *J. Phys. Chem. B* **109**, 7084.

Yu WW, Qu L, Guo W, Peng X (2003) Experimental determination of the extinction coefficient of CdTe, CdSe, and CdS nanocrystals. *Chem. Mater.* **15**, 2854.

Emission Quantum Yield

Arbogast JW, Foote CS (1991) Photophysical properties of C_{70}. *J. Am. Chem. Soc.* **113**, 8886.

Carlson LJ, Maccagnano SE, Zheng M, et al. (2007) Fluorescence efficiency of individual carbon nanotubes. *Nano Lett.* **7**, 3698.

Dabbousi BO, Viejo JR, Mikulec FV, et al. (1997) (CdSe)ZnS core-shell quantum dots: synthesis and characterization of a size series of highly luminescent nanocrystallites. *J. Phys. Chem. B* **101**, 9463.

Goebl JA, Black RW, Puthussery J, et al. (2008) Solution-based II–VI core/shell nanowire heterostructures. *J. Am. Chem. Soc.* **130**, 14822.

Hines MA, Guyot-Sionnest P (1996) Synthesis and characterization of strongly luminescing ZnS-capped CdSe nanocrystals. *J. Phys. Chem.* **100**, 468.

Kubin RF, Fletcher AN (1982) Fluorescence quantum yields of some rhodamine dyes. *J. Lumin.* **27**, 455.

Micic OI, Sprague J, Lu Z, Nozik AJ (1996) Highly efficient band-edge emission from InP quantum dots. *Appl. Phys. Lett.* **68**, 3150.

Murray CB, Norris DJ, Bawendi MG (1993) Synthesis and characterization of nearly monodisperse CdE (E = S, Se, Te) semiconductor nanocrystallites. *J. Am. Chem. Soc.* **115**, 8706.

Peng X, Schlamp MC, Kadavanich AV, Alivisatos AP (1997) Epitaxial growth of highly luminescent CdSe/CdS core/shell nanocrystals with photostability and electronic accessibility. *J. Am. Chem. Soc.* **119**, 7019.

Protasenko VV, Hull KL, Kuno M (2005) Disorder-induced optical heterogeneity in single CdSe nanowires. *Adv. Mater.* **17**, 2942.

Puthussery J, Lan A, Kosel TH, Kuno M (2008) Band-filling of solution-synthesized CdS nanowires. *ACS Nano* **2**, 357.

Sun YP, Wang P, Hamilton NB (1993) Fluorescence spectra and quantum yields of buckminsterfullerene (C_{60}) in room-temperature solutions. No excitation wavelength dependence. *J. Am. Chem. Soc.* **115**, 6378.

A Quantum Mechanics Review

6.1 INTRODUCTION

Matter has both wave-like and particle-like properties. It is this realization that underlies the quantum mechanics important to us in nanoscience. This is because when dealing with the optical and electrical properties of nanostructures, we will describe carriers (electrons and holes) in terms of their wave-like properties.

In what follows, we briefly review some key concepts of quantum mechanics. Note that this chapter is not meant to be comprehensive. It is simply meant to provide the reader an appropriate foundation for the sections to follow. The interested reader is therefore encouraged to consult the introductory texts listed in the Further Reading for more details.

6.2 WAVEFUNCTIONS

Given the de Broglie wave–particle duality, which states that matter has both wave-like and particle-like properties, it turns out that one can mathematically describe particles as waves using so-called wavefunctions. Alternatively, a more suggestive name is matter waves. In either case, wavefunctions (or matter waves) are usually denoted by the symbol ψ.

Now, for macroscopic objects, such as a football, a car, or even a person, their wave-like properties are not very significant. These objects, for all practical purposes, are adequately described by classical mechanics. What is different in nanoscience, however, is that when the size of matter becomes small, its quantum mechanical properties become significant. In fact, it was this realization at the turn of the 20th century that revolutionized physics. Thus, in our studies, we will be interested in the wave-like properties of carriers such as electrons and holes, since they ultimately underlie the optical and electrical properties of nanostructures.

The wavefunction of a "free" particle, an electron for example, has the form

$$\psi = e^{ikx}, \tag{6.1}$$

where k is called the wavevector and is related to the particle's wavelength, λ, through $k = 2\pi/\lambda$. **Figure 6.1** depicts the free particle wavefunction drawn along the x axis. In general, wavefunctions also exhibit time dependences. Therefore, if one needed to, one could multiply Equation 6.1 by $e^{-i\omega t}$ to obtain a traveling wave, which moves towards the right. (Traveling waves were discussed in Chapter 5.)

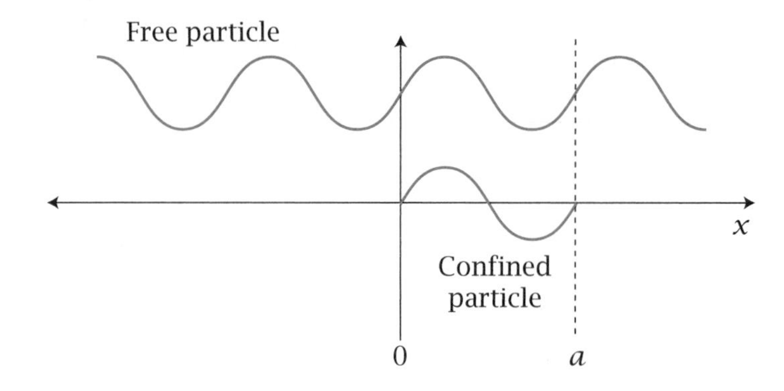

Figure 6.1 Illustration of a free particle and a spatially confined particle wavefunction along the x axis.

Alternatively, for an electron confined (i.e., trapped) along one dimension, one could describe its wavefunction using the expression

$$\psi = \sqrt{\frac{2}{a}} \sin \frac{n\pi x}{a}, \tag{6.2}$$

with a the length of the region to which the particle is confined and $n = 1, 2, 3, \ldots$ an integer. Note that Equation 6.2 is not general and was derived under a set of circumstances limiting it to the region between $x = 0$ and $x = a$. In fact, we will see various forms for carrier wavefunctions in Chapter 7 when we model carriers confined to quantum wells, wires, and dots. For now, though, let us simply use Equation 6.2 as a representative function that describes confined carriers. **Figure 6.1** illustrates ψ drawn on the x axis.

In general, the wavefunction replaces the classical Newtonian concept of a trajectory. It contains all the dynamical information about a system one can know. As a consequence, much of the work we will do from here on (i.e., Chapters 7, 8, and 10) consists in finding out what the wavefunction of a particle looks like given certain constraints on the system, called boundary conditions. This is what led earlier to our particle being confined to the interval $0 \leq x \leq a$.

6.2.1 Probabilistic Interpretation of Wavefunctions

There exists a probabilistic interpretation of wavefunctions called the Born interpretation. In it, the square of the wavefunction, written as

$$|\psi|^2 = \psi^* \psi, \tag{6.3}$$

represents what is called the probability density (i.e., the probability per unit length in one dimension or the probability per unit volume in three dimensions) of a particle. The asterisk in Equation 6.3 represents the complex conjugate of the wavefunction and implies that it can, in general, be complex-valued. Next,

$$|\psi|^2 \, dx = \psi^* \psi \, dx \tag{6.4}$$

represents an actual probability in one dimension. Similarly, in three dimensions, one writes $|\psi|^2 \, d^3r = \psi^* \psi \, d^3r$.

Through these quantities, one can find the probability that a particle will be found somewhere in the region where its wavefunction is defined. From a physical perspective, only $|\psi|^2$ has physical significance, because $|\psi|^2$ is real and positive, as illustrated in **Figure 6.2**. By contrast, ψ itself can be real or complex.

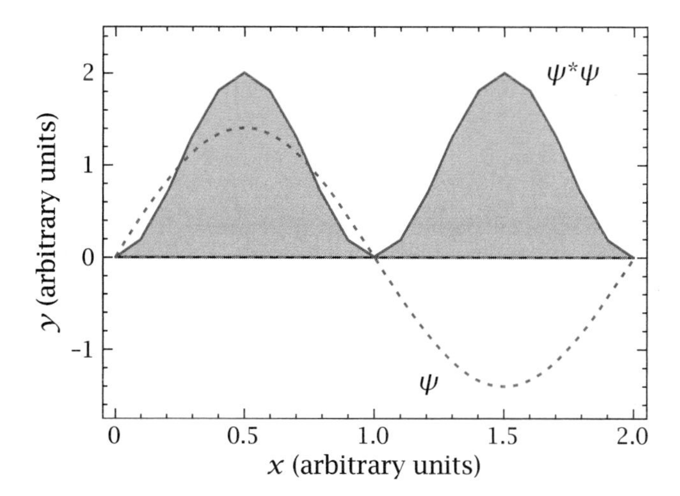

Figure 6.2 Illustration of a wavefunction ψ and its associated probability density $\psi^*\psi$.

Because of this probabilistic interpretation, the wavefunction must be normalized:

$$\int |\psi|^2\, dx = 1.$$

This means that the probability of finding the particle somewhere within the region within which it is defined must be unity. A larger value would be unphysical.

We illustrate this concept as follows. Assume that a confined particle can be described by the following wavefunction defined along the x coordinate between $x = 0$ and $x = a$:

$$\psi = \sqrt{\frac{2}{a}} \sin\frac{\pi x}{a}.$$

To verify normalization, we evaluate the integral

$$\int_0^a \psi^*\psi\, dx = \int_0^a \left(\sqrt{\frac{2}{a}} \sin\frac{\pi x}{a}\right)^* \left(\sqrt{\frac{2}{a}} \sin\frac{\pi x}{a}\right) dx,$$

which is simply

$$= \frac{2}{a}\int_0^a \sin^2\frac{\pi x}{a}\, dx$$

since sine is a real function and taking its complex conjugate does not alter anything. Next, by employing the trigonometric identity $\sin^2 m = \frac{1}{2}(1 - \cos 2m)$, we find

$$\int_0^a \psi^*\psi\, dx = \frac{1}{a}\int_0^a 1 - \cos\frac{2\pi x}{a}\, dx.$$

This integrates to

$$= \frac{1}{a}\left(x|_0^a - \frac{a}{2\pi} \sin\frac{2\pi x}{a}\Big|_0^a \right),$$
$$= 1,$$

showing that ψ is normalized.

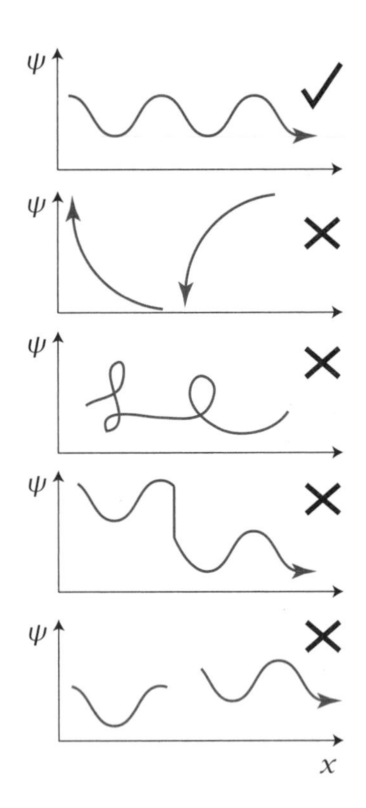

Figure 6.3 Illustration of valid and invalid particle wavefunctions.

6.2.2 Other Properties to Remember

Additional mathematical properties or constraints on the wavefunction are as follows:

- ψ must be continuous and must have a first and a second derivative.
- It must be single-valued.
- It must be normalizable and, as a consequence, must not diverge or show singularities.

Schematic examples of these constraints are shown in **Figure 6.3**.

- Furthermore, normalization means that $\int |\psi|^2\, d^3r = 1$ (in three dimensions). Namely, integration of the function over all space where it is defined yields a value of one.
- Wavefunctions can also be orthogonal. This will often be the case when they are eigenfunctions of the Hamiltonian operator, as we will see shortly. But, for now, what we mean to say is that for two independent wavefunctions ψ_1 and ψ_2, it is possible to have $\int \psi_1^* \psi_2\, d^3r = 0$, which says physically that the two wavefunctions have no net mutual spatial overlap.

  We illustrate orthogonality using the following two confined wavefunctions:

$$\psi_1 = \sqrt{\frac{2}{a}} \sin \frac{\pi x}{a},$$

$$\psi_2 = \sqrt{\frac{2}{a}} \sin \frac{2\pi x}{a},$$

based on Equation 6.2. Both functions are defined over the interval $x = 0$ to $x = a$ (**Figure 6.4**). We now show that these two wavefunctions are orthogonal by evaluating the following overlap integral:

$$\int_0^a \psi_1^* \psi_2\, dx = \frac{2}{a} \int_0^a \sin \frac{\pi x}{a} \sin \frac{2\pi x}{a}\, dx.$$

To simplify the right-hand side, we invoke the trigonometric identity $\sin m \sin n = \frac{1}{2}[\cos(m-n) - \cos(m+n)]$, which results in

$$\frac{1}{a} \int_0^a \left[\cos\left(-\frac{\pi x}{a}\right) - \cos \frac{3\pi x}{a}\right] dx.$$

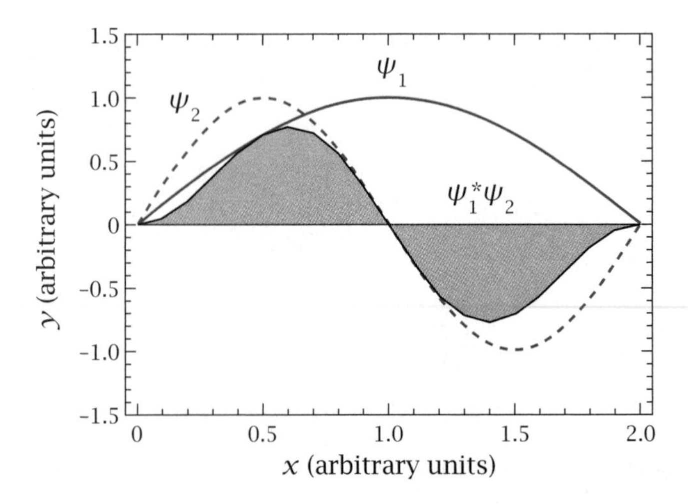

Figure 6.4 Illustration of orthogonality between two wavefunctions ψ_1 and ψ_2.

This equals

$$\frac{1}{a}\int_0^a \left(\cos\frac{\pi x}{a} - \cos\frac{3\pi x}{a}\right) dx,$$

since $\cos(-m) = \cos(m)$. Carrying out the integration, we obtain

$$\int_0^a \psi_1^* \psi_2 \, dx = \frac{1}{a}\left(\frac{a}{\pi}\sin\frac{\pi x}{a}\bigg|_0^a - \frac{a}{3\pi}\sin\frac{3\pi x}{a}\bigg|_0^a\right)$$

with the quantity in parentheses equaling zero since $\sin\pi = \sin 3\pi = \sin 0 = 0$. As a consequence, the two wavefunctions are said to be orthogonal:

$$\int \psi_1^* \psi_2 \, dx = 0.$$

Figure 6.4 illustrates this visually. One sees that the product $\psi_1^*\psi_2$ is symmetric about the y axis. As a consequence, its integrated area will be zero.

An additional point on terminology: if both wavefunctions ψ_1 and ψ_2 are simultaneously normalized (i.e., $\int \psi_1^*\psi_1 \, dx = 1$ and $\int \psi_2^*\psi_2 \, dx = 1$) and are also orthogonal, then they are said to be *orthonormal*.

- Finally, if the wavefunction can be solved piecewise for different regions in space, then at their interfaces, the wavefunctions in these regions must satisfy so-called matching conditions. For example, in the case where the wavefunctions in regions 1 and 2 meet at a given value x (**Figure 6.5**), we simultaneously require that

$$\psi_1(x) = \psi_2(x) \tag{6.5}$$

$$\psi_1'(x) = \psi_2'(x). \tag{6.6}$$

Namely, the wavefunctions must be equal where they meet and so too must their first derivatives. We will see in Chapter 10 that these matching conditions are extremely important when developing the concept of bands.

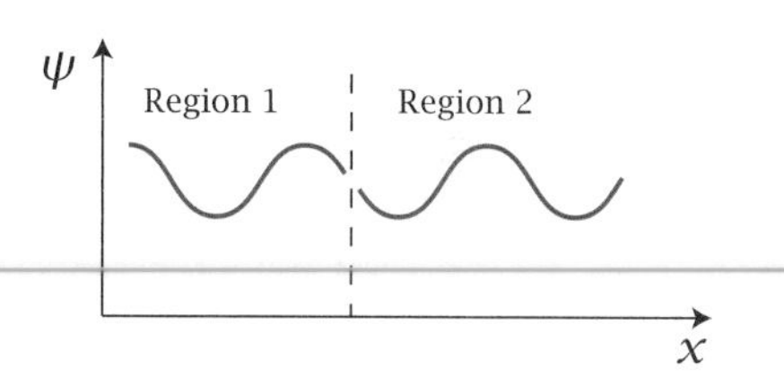

Figure 6.5 Illustration of wavefunction matching conditions.

6.3 OBSERVABLES AND THE CORRESPONDENCE PRINCIPLE

In quantum mechanics, all physical observables have corresponding operators. In this regard, just as particles are represented mathematically by wavefunctions, observables such as

- position
- momentum
- kinetic energy
- angular momentum

are represented mathematically by *operators*. Specifically, an operator (indicated by a "hat" ^) acts on a wavefunction to transform it into another function. Examples include

- $\hat{x}$ (multiply by x)
- $\partial/\partial x$ (read as "take the partial derivative of the function with respect to x")
- $\partial/\partial t$ (read as "take the partial derivative of the function with respect to time")

Table 6.1 Common operators in quantum mechanics

Operator name	Associated symbol	Operation
Position	$\hat{x}$	Multiply by x
Momentum, x component	$\hat{p}_x$	$-i\hbar\frac{\partial}{\partial x}$
Momentum, y component	$\hat{p}_y$	$-i\hbar\frac{\partial}{\partial y}$
Momentum, z component	$\hat{p}_z$	$-i\hbar\frac{\partial}{\partial z}$
Momentum, general	$\hat{p}$	$-i\hbar\left(\frac{\partial}{\partial x}+\frac{\partial}{\partial y}+\frac{\partial}{\partial z}\right)$
Kinetic energy	$\hat{T}$	$-\frac{\hbar^2}{2m}\left(\frac{\partial^2}{\partial x^2}+\frac{\partial^2}{\partial y^2}+\frac{\partial^2}{\partial z^2}\right)$
Potential energy	$\hat{V}(\mathbf{r})$	Multiply by $V(\mathbf{r})$
Total energy (Hamiltonian)	$\hat{H}$	$-\frac{\hbar^2}{2m}\left(\frac{\partial^2}{\partial x^2}+\frac{\partial^2}{\partial y^2}+\frac{\partial^2}{\partial z^2}\right)+V(r)$
Angular momentum, x component	$\hat{l}_x$	$-i\hbar\left(y\frac{\partial}{\partial z}-z\frac{\partial}{\partial y}\right)$
Angular momentum, y component	$\hat{l}_y$	$-i\hbar\left(z\frac{\partial}{\partial x}-x\frac{\partial}{\partial z}\right)$
Angular momentum, z component	$\hat{l}_z$	$-i\hbar\left(x\frac{\partial}{\partial y}-y\frac{\partial}{\partial x}\right)$
Angular momentum squared	$\hat{L}^2$	$-\hbar^2\left[\frac{1}{\sin\theta}\frac{\partial}{\partial\theta}\left(\sin\theta\frac{\partial}{\partial\theta}\right)-\frac{1}{\sin^2\theta}\frac{\partial^2}{\partial\phi^2}\right]$

and so forth. **Table 6.1** lists some common operators encountered in quantum mechanics.

Two fundamental operators are the position operator $\hat{x}$ (we will assume the x direction for now, but the concept is general) and the momentum operator $\hat{p}$. The position operator is simply

$$\hat{x} = x,$$

which we read as "multiply the function by x." The momentum operator along the x direction is likewise

$$\hat{p}_x = -i\hbar\nabla = -i\hbar\frac{\partial}{\partial x},$$

which we read as "take the derivative of the function with respect to x and then multiply by $-i\hbar$."

Note that many other operators can be constructed from combinations of these two fundamental operators. For example, the kinetic energy operator is illustrated below. For simplicity, assume that we work in one dimension, namely along the x direction of a Cartesian coordinate system.

Kinetic Energy Operator $\hat{T}$

The kinetic energy operator is defined as

$$\hat{T} = \frac{\hat{p}^2}{2m},$$

where m is the particle's mass. In one dimension, since $\hat{p}_x = -i\hbar\, d/dx$, this becomes

$$= \frac{1}{2m}\left(-i\hbar\frac{d}{dx}\right)\left(-i\hbar\frac{d}{dx}\right)$$

$$= \frac{1}{2m}\left(-\hbar^2\frac{d^2}{dx^2}\right),$$

leading to an expression for the one-dimensional kinetic energy operator:

$$\hat{T} = -\frac{\hbar^2}{2m}\frac{d^2}{dx^2}. \tag{6.7}$$

We understand this operator as simply taking the second derivative of the wavefunction and multiplying the result by $-\hbar^2/2m$. In this manner, one obtains information about the kinetic energy of the particle. We will see more about this when we talk about eigenfunctions and eigenvalues below.

Potential Energy Operator $\hat{V}(\mathbf{r})$

An associated potential energy operator exists and is defined as

$$\hat{V}(\mathbf{r}) = V(\mathbf{r}).$$

We interpret this as multiplying any wavefunction by $V(\mathbf{r})$. For the potential energy operator with only an x dependence, we have

$$\hat{V}(x) = V(x). \tag{6.8}$$

Total Energy Operator $\hat{H}$

The total energy operator contains both the kinetic and potential energy operators. It is called the Hamiltonian operator, or Hamiltonian for short:

$$\hat{H} = \hat{T} + \hat{V}.$$

If we consider only the x direction, we have

$$\hat{H} = -\frac{\hbar^2}{2m}\frac{d^2}{dx^2} + \hat{V}(x). \tag{6.9}$$

We interpret this as operating on a wavefunction with the kinetic and potential energy operators together. What is returned is the total energy of the particle.

Although the above examples focus only on operators expressed in one dimension, they can readily be extended to multiple dimensions. For example, in three dimensions, the Hamiltonian can be written as

$$\hat{H} = \left(\frac{\hat{p}_x^2}{2m} + \frac{\hat{p}_y^2}{2m} + \frac{\hat{p}_z^2}{2m}\right) + \hat{V}(x, y, z),$$

where the momentum operators along the three Cartesian coordinates are

$$\hat{p}_x = -i\hbar\frac{\partial}{\partial x}, \quad \hat{p}_y = -i\hbar\frac{\partial}{\partial y}, \quad \hat{p}_z = -i\hbar\frac{\partial}{\partial z}.$$

This results in

$$\hat{H} = -\frac{\hbar^2}{2m}\left(\frac{\partial^2}{\partial x^2} + \frac{\partial^2}{\partial y^2} + \frac{\partial^2}{\partial z^2}\right) + \hat{V}(x, y, z). \tag{6.10}$$

Alternatively, one can write

$$\hat{H} = -\frac{\hbar^2}{2m}\nabla^2 + \hat{V}(x, y, z), \tag{6.11}$$

where $\nabla^2 = \nabla \cdot \nabla$ is called the Laplacian operator, or Laplacian for short:

$$\nabla^2 = \frac{\partial^2}{\partial x^2} + \frac{\partial^2}{\partial y^2} + \frac{\partial^2}{\partial z^2}, \tag{6.12}$$

with

$$\nabla = \left(\frac{\partial}{\partial x}, \frac{\partial}{\partial y}, \frac{\partial}{\partial z}\right). \tag{6.13}$$

Note that in some texts, ∇^2 is sometimes written Δ. This is just a notational change. Finally, the Laplacian will look different in other coordinate systems. For example, in polar coordinates, it has the form

$$\nabla^2 = \frac{1}{r}\frac{\partial}{\partial r}\left(r\frac{\partial}{\partial r}\right) + \frac{1}{r^2}\frac{\partial^2}{\partial \theta^2}, \tag{6.14}$$

while in spherical coordinates it is

$$\nabla^2 = \frac{1}{r^2}\frac{\partial}{\partial r}\left(r^2\frac{\partial}{\partial r}\right) + \frac{1}{r^2 \sin\theta}\frac{\partial}{\partial \theta}\left(\sin\theta\frac{\partial}{\partial \theta}\right) + \frac{1}{r^2 \sin^2\theta}\frac{\partial^2}{\partial \phi^2}. \tag{6.15}$$

We will see these other versions of the Laplacian in Chapters 7 and 8 when we discuss model problems for the quantum well, quantum wire, quantum dot, and other systems of interest.

To summarize, there are many important operators built from the fundamental operators $\hat{x}$ and $\hat{p}_x$ (more generally, $\hat{\mathbf{r}}$ and $\hat{\mathbf{p}}$). In what follows, we will primarily be interested in the total energy (Hamiltonian) operator. This is because, provided an appropriate particle wavefunction, one can find its total energy by simply applying the Hamiltonian to it.

6.4 EIGENVALUES AND EIGENFUNCTIONS

There exist functions associated with an operator that, when operated on, yield the same function multiplied by a constant. Such functions are called eigenfunctions of the operator, with the constant being called an eigenvalue of the operator. We are interested in such functions because wavefunctions describing particles are often eigenfunctions of the Hamiltonian operator.

The general mathematical form of an eigenvalue/eigenfunction relationship is

$$\hat{A}f = af, \tag{6.16}$$

where $\hat{A}$ is an arbitrary operator, f is an eigenfunction of the operator, and a is the associated eigenvalue.

More specifically, suppose that we have an operator $\hat{A} = d/dx$ and a function $f(x) = e^{\alpha x}$. We can show that $f(x)$ is an eigenfunction of $\hat{A}$ by operating on it as follows:

$$\frac{d}{dx}\left(e^{\alpha x}\right) = \alpha\left(e^{\alpha x}\right).$$

One sees that the result is a constant, α (the associated eigenvalue), times the original function itself.

Let us illustrate the eigenvalue/eigenfunction concept again with the kinetic energy operator, since we will use it extensively in the following chapters. Assuming a given free particle wavefunction (see Equation 6.1)

$$\psi = Ae^{ikx},$$

where A is a constant, and the kinetic energy operator

$$\hat{T} = -\frac{\hbar^2}{2m}\frac{d^2}{dx^2},$$

we have

$$\begin{aligned}\hat{T}\psi &= -\frac{\hbar^2}{2m}\frac{d^2}{dx^2}(Ae^{ikx}) \\ &= -\frac{\hbar^2}{2m}(ik)\frac{d}{dx}(Ae^{ikx}) \\ &= -\frac{\hbar^2}{2m}(ik)^2(Ae^{ikx}) \\ &= \left(\frac{\hbar^2 k^2}{2m}\right)\psi.\end{aligned}$$

ψ is therefore an eigenfunction of the kinetic energy operator, with $\hbar^2 k^2/2m$ being its associated eigenvalue. In fact, the latter is the kinetic energy associated with the free particle.

We can also illustrate the concept by operating on a confined particle wavefunction. Given Equation 6.2 where $n = 1$,

$$\psi = \sqrt{\frac{2}{a}}\sin\frac{\pi x}{a},$$

we find that

$$\begin{aligned}\hat{T}\psi &= -\frac{\hbar^2\nabla^2}{2m}\left(\sqrt{\frac{2}{a}}\sin\frac{\pi x}{a}\right) \\ &= -\frac{\hbar^2}{2m}\frac{d}{dx}\left(\frac{d}{dx}\right)\left(\sqrt{\frac{2}{a}}\sin\frac{\pi x}{a}\right) \\ &= -\frac{\hbar^2}{2m}\left(\frac{\pi}{a}\right)\frac{d}{dx}\left(\sqrt{\frac{2}{a}}\cos\frac{\pi x}{a}\right) \\ &= \frac{\hbar^2}{2m}\left(\frac{\pi}{a}\right)^2\left(\sqrt{\frac{2}{a}}\sin\frac{\pi x}{a}\right).\end{aligned}$$

This shows that ψ is an eigenfunction of $\hat{T}$ with a corresponding eigenvalue

$$E = \frac{\hbar^2}{2m}\left(\frac{\pi}{a}\right)^2 = \frac{h^2}{8ma^2}.$$

6.5 WAVEPACKETS

Many of the wavefunctions we have seen so far (or will see) are eigenfunctions of the Hamiltonian operator. However, a more general way to describe particles is to use linear combinations of these functions. So, for example, assuming Equation 6.2 for a confined particle between $x = 0$ and $x = a$, one could, in principle, write

$$\psi = A\left(\sqrt{\frac{2}{a}}\sin\frac{\pi x}{a}\right) + B\left(\sqrt{\frac{2}{a}}\sin\frac{2\pi x}{a}\right) + C\left(\sqrt{\frac{2}{a}}\sin\frac{3\pi x}{a}\right) + \cdots,$$

where A, B, C, etc. are real or imaginary weights. Now, it is possible that all but one of these weights is zero. In this case, the particle is purely a given eigenfunction of the Hamiltonian. More generally, though, many of the weights will be nonzero and, as a consequence, eigenstates contribute to the particle's wavefunction with an amount proportional to their weight squared.

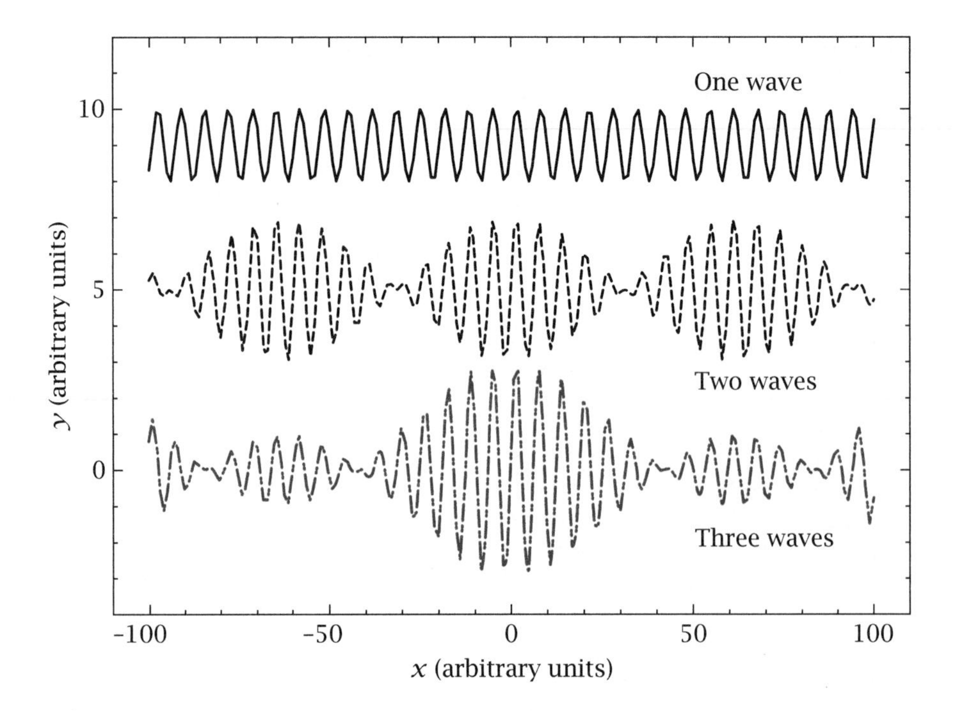

Figure 6.6 Illustration of the wavepacket principle.

Physically, what these linear combinations begin to do is localize the particle. In this regard, you might have asked yourself earlier, on examining **Figure 6.1**, why the particle, even though it was confined, was still spread out (i.e., delocalized) over the entire width of the box. This is contrary to your physical view of particles.

To begin achieving any sort of particle-like localization (mathematically) requires a linear combination of eigenfunctions. In principle, such a linear combination is infinite. However, in practice, one occasionally sees that only a few states contribute strongly to the particle's overall wavefunction. Such linear combinations are called "wavepackets" to imply the wave's localization.

Figure 6.6 shows the appearance of a wavefunction consisting of a single wave, two waves, then three waves added together. The frequency of each wave is slightly different. In the figure, one sees that the single wave is just a delocalized plane wave. When an additional plane wave is added, nodes develop and the effective wave breaks up into three smaller packets over the region shown. On adding a third wave, even more localization occurs and the spacing between packets increases. Although we will not show it, if one continues in this manner, by adding waves together and weighting their contributions, it is possible to obtain what is essentially a single wavepacket along the x axis. In this way, we can demonstrate particle localization using a simple linear combination of plane waves.

6.6 EXPECTATION VALUES

Now that wavefunctions, operators, and eigenfunctions, as well as eigenvalues, have been introduced, we show that the mean value of an observable acting on a wavefunction is by definition

$$\langle A \rangle = \int \psi^* \hat{A} \psi \, d^3 r, \tag{6.17}$$

where $\hat{A}$ is an arbitrary operator and the integral runs over all space. In fact, this is one of the postulates of quantum mechanics, as we will see later on.

In the special case where ψ is an eigenfunction of the operator (i.e., $\hat{A}\psi = a\psi$, with a the eigenvalue), one has

$$\begin{aligned}\langle A \rangle &= \int \psi^* \hat{A}\psi \, d^3r \\ &= \int \psi^* a\psi \, d^3r \\ &= a\int \psi^* \psi \, d^3r \\ &= a.\end{aligned}$$

This is because the integral equals 1 if the wavefunction is normalized. As a consequence, we find that

$$\langle A \rangle = a$$

for the special case of the expectation value of an operator's eigenfunction.

Note that even if the wavefunction is not an eigenfunction of an operator, the concept is still useful since, generally speaking, the wavefunction can be expressed as a linear combination of the operator's native eigenfunctions. We saw this earlier when we talked about wavepackets. In this sense, such eigenfunctions are said to form a "basis" (note that this is not the same basis as in crystallography, Chapter 2) from which the particle's overall wavefunction is constructed. The associated expectation value is then the weighted sum of eigenvalues from each eigenfunction.

EXAMPLE 6.1

Position Operator Expectation Value for a Confined Particle

We illustrate finding the position operator's expectation value for the following one-dimensional wavefunction that describes a confined particle between $x = 0$ and $x = a$:

$$\psi = \sqrt{\frac{2}{a}} \sin \frac{\pi x}{a}.$$

The expectation value for its position is obtained by evaluating

$$\begin{aligned}\langle x \rangle &= \int_0^a \psi^* x \psi \, dx \\ &= \frac{2}{a} \int_0^a x \sin^2 \frac{\pi x}{a} \, dx,\end{aligned}$$

where we employ the trigonometric identity

$$\sin^2 \frac{\pi x}{a} = \frac{1}{2}\left(1 - \cos \frac{2\pi x}{a}\right).$$

We then have

$$\langle x \rangle = \frac{1}{a} \int_0^a x \left(1 - \cos \frac{2\pi x}{a}\right) dx,$$

which integrates to

$$\begin{aligned}&= \frac{1}{a}\left(\frac{x^2}{2}\bigg|_0^a - \int_0^a x \cos \frac{2\pi x}{a} dx\right) \\ &= \frac{a}{2} - \frac{1}{a} \int_0^a x \cos \frac{2\pi x}{a} dx.\end{aligned}$$

The remaining integral is evaluated by parts. Let

$$u = x, \quad du = dx, \quad dv = \cos\frac{2\pi x}{a}dx, \quad v = \frac{a}{2\pi}\sin\frac{2\pi x}{a}$$

to get

$$= \frac{a}{2} - \frac{1}{a}\left(\frac{a}{2\pi}x\sin\frac{2\pi x}{a}\bigg|_0^a - \frac{a}{2\pi}\int_0^a \sin\frac{2\pi x}{a}dx\right).$$

The first term in parentheses is zero. Next, when the second term is integrated, we obtain

$$= \frac{a}{2} - \frac{1}{2\pi}\frac{a}{2\pi}\cos\frac{2\pi x}{a}\bigg|_0^a,$$

which equals zero since $\cos 2\pi = \cos 0$. The result is

$$\langle x \rangle = \frac{a}{2}.$$

The particle's mean position is therefore halfway between 0 and a, the interval over which its wavefunction is defined. Your intuition might have already told you this.

6.7 DIRAC BRA–KET NOTATION

Here we briefly touch on notation. Up to now, we have seen wavefunctions and operators expressed in their full functional forms. This involves variables, integrals, and derivatives (i.e., calculus). Dirac, however, suggested a shorthand notation for writing and dealing with wavefunctions as well as operators within the context of an alternative matrix view of quantum mechanics. Today, this notation is called Dirac *bra–ket* notation and is often used in quantum mechanics because it leads to more compact expressions.

In bra–ket notation,

- Wavefunctions ψ are denoted by $|\psi\rangle$ and are called "kets." For example, we could have a sinusoidal wavefunction confined along the x axis between 0 and a of the form

$$\psi = \sqrt{\frac{2}{a}}\sin\frac{n\pi x}{a} \equiv |\psi\rangle,$$

where a is a constant and $n = 1, 2, 3, \ldots$ is an integer. An illustration of these functions is shown in **Figure 6.7** where $n = 1, \ldots, 5$ and $a = 5$. The traces have been offset for clarity.

Note that wavefunctions are often characterized by a quantum number n. These integers arise naturally from solving what we will see later is called the Schrödinger equation. As a consequence, when describing wavefunctions in bra–ket notation, one can further simplify the notation by referring directly to n. For example, one could write

$$\psi = \sqrt{\frac{2}{a}}\sin\frac{n\pi x}{a} \equiv |n\rangle$$

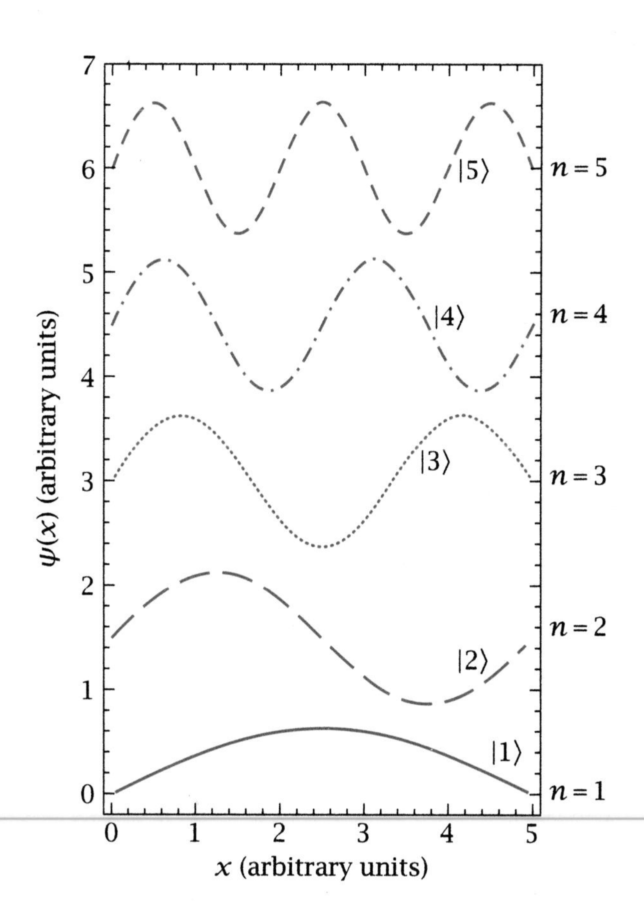

Figure 6.7 Illustration of five sinusoidal wavefunctions valid between 0 and a. The traces have been constructed assuming $a = 5$. They have also been offset for clarity.

and more specifically for three representative wavefunctions where $n = 1, 2$, and 3,

$$|\psi_1\rangle = \sqrt{\frac{2}{a}} \sin \frac{\pi x}{a} \equiv |1\rangle,$$

$$|\psi_2\rangle = \sqrt{\frac{2}{a}} \sin \frac{2\pi x}{a} \equiv |2\rangle,$$

$$|\psi_3\rangle = \sqrt{\frac{2}{a}} \sin \frac{3\pi x}{a} \equiv |3\rangle.$$

- Complex conjugates of wavefunctions, ψ^*, are denoted by $\langle\psi|$ and are called "bras." Using the above wavefunctions, we then write

$$\psi^* = \left(\sqrt{\frac{2}{a}} \sin \frac{n\pi x}{a}\right)^* = \sqrt{\frac{2}{a}} \sin \frac{n\pi x}{a} \equiv \langle\psi|,$$

since the function is real. Alternatively, if the wavefunction were complex, for example $\psi = e^{ikx}$, we would write

$$\psi^* = (e^{ikx})^* = e^{-ikx} = \langle\psi|.$$

In practice, as we have seen earlier, taking the complex conjugate of a function means replacing every i in it with $-i$.

Both bras and kets follow the rules of linear algebra:

- $|a\psi\rangle = a|\psi\rangle$,

where a is a constant that could be complex;

- $\langle \psi a| = a^* \langle \psi|$,
  where a is again a constant that could be complex;
- $\langle \psi_1|\psi_2\rangle = \langle \psi_2|\psi_1\rangle^*$,
  where, as usual, the asterisk denotes the complex conjugate and the meaning of the bra and ket together will be described below;
- $\langle \psi_1|a\psi_2 + b\psi_3\rangle = a\langle \psi_1|\psi_2\rangle + b\langle \psi_1|\psi_3\rangle$,
  where a and b are constants and again the meaning of the bra and ket together will be discussed next.

Integrals in bra–ket notation are represented by

$$\langle \psi_n|\psi_m\rangle \equiv \int \psi_n^* \psi_m \, d^3r.$$

In addition, for our previous confined wavefunctions ψ_1 and ψ_2, we have

$$\begin{aligned}\langle \psi_1|\psi_1\rangle &\equiv \int \psi_1^* \psi_1 \, d^3r \\ &= 1, \quad \text{by normalization}\end{aligned}$$

and

$$\begin{aligned}\langle \psi_2|\psi_1\rangle &\equiv \int \psi_2^* \psi_1 \, d^3r \\ &= 0, \quad \text{by orthogonality.}\end{aligned}$$

For an arbitrary operator $\hat{A}$,

$$\langle \psi_1|\hat{A}|\psi_1\rangle \equiv \int \psi_1^* \hat{A} \psi_1 \, d^3r$$

and

$$\langle \psi_2|\hat{A}|\psi_1\rangle \equiv \int \psi_2^* \hat{A} \psi_1 \, d^3r.$$

As a consequence, in the above example, we interpret $\langle \psi_1|a\psi_2 + b\psi_3\rangle$ by the following integrals

$$\begin{aligned}\langle \psi_1|a\psi_2 + b\psi_3\rangle &= \int \psi_1^*(a\psi_2 + b\psi_3) \, d^3r \\ &= a\int \psi_1^* \psi_2 \, d^3r + b\int \psi_1^* \psi_3 \, d^3r.\end{aligned}$$

6.8 OPERATOR MATH

Just as wavefunctions obey certain constraints in order to represent physical particles, operators for observable quantities also have important restrictions. These constraints are called linearity and Hermiticity. We illustrate them below.

Linearity

An operator $\hat{A}$ is said to be *linear* for constants a and b and wavefunctions $|\psi_1\rangle$ and $|\psi_2\rangle$ if

$$\hat{A}(a|\psi_1\rangle + b|\psi_2\rangle) = a\hat{A}|\psi_1\rangle + b\hat{A}|\psi_2\rangle.$$

Basically, what this says is that the operator acting on a total wavefunction, consisting of a linear combination of wavefunctions, is the same as the operator acting individually on these component wavefunctions.

Hermiticity

An operator $\hat{A}$ is said to be *Hermitian* if its integrals obey the following relationship for all valid ψ_1 and ψ_2:

$$\int \psi_1^* \hat{A} \psi_2 \, d^3r = \int \psi_2^* \hat{A} \psi_1 \, d^3r.$$

In bra–ket notation, this is equivalent to saying

$$\langle \psi_1 | \hat{A} | \psi_2 \rangle^* = \langle \psi_2 | \hat{A} | \psi_1 \rangle.$$

The restriction ensures that all eigenvalues of $\hat{A}$ are real as opposed to potentially complex. This is important when dealing with physical observables, as physical quantities should be real-valued.

6.9 MORE ON OPERATORS

Other properties and relationships of operators include the following:

$$\hat{A}|\psi\rangle = |\hat{A}\psi\rangle,$$
$$\langle \psi \hat{A}^\dagger| = \langle \psi | \hat{A}^\dagger,$$

where $\hat{A}^\dagger$ is called the *adjoint* of $\hat{A}$. Note that if the operator is Hermitian then $\hat{A} = \hat{A}^\dagger$.

In the case where ψ is an eigenfunction of $\hat{A}$ with an associated eigenvalue a, we also find that

$$\hat{A}|\psi\rangle = |\hat{A}\psi\rangle = a|\psi\rangle$$
$$\langle \psi | \hat{A}^\dagger = \langle \psi \hat{A}^\dagger | = a^* \langle \psi |.$$

Again the asterisk denotes the complex conjugate of the eigenvalue.

Regarding integrals written in bra–ket notation, one has

$$\langle \psi_1 | \hat{A} | \psi_2 \rangle = \langle \psi_1 | A \psi_2 \rangle,$$
$$\langle \psi_1 | \hat{A} | \psi_2 \rangle = \langle \psi_1 \hat{A}^\dagger | \psi_2 \rangle,$$
$$\langle \psi_1 | \hat{A} | \psi_2 \rangle = \langle \psi_2 | \hat{A}^\dagger | \psi_1 \rangle^*,$$

where again $\hat{A} = \hat{A}^\dagger$ if the operator is Hermitian. The first expression represents $\hat{A}$ operating on the ket. The second represents $\hat{A}$ operating on the bra.

6.10 COMMUTATORS

A wavefunction ψ with a characteristic well-defined value of some observable is an eigenfunction of the corresponding operator. However, ψ does not have to be an eigenfunction of another operator associated with a different observable. For a wavefunction to be an eigenfunction of two observables simultaneously, the corresponding operators must "commute."

6.10.1 Definition of "Commute"

The operators $\hat{A}$ and $\hat{B}$ are said to *commute* when the result of their actions taken in succession on any wavefunction is identical to their result when taken in reverse order. So, if we operate on a wavefunction with $\hat{B}$ first and then $\hat{A}$, the result should be the same if we had operated with $\hat{A}$ first and then $\hat{B}$. We therefore have

$$\hat{A}\hat{B}|\psi\rangle = \hat{B}\hat{A}|\psi\rangle.$$

Equivalently, the "commutator" of the two operators is said to equal zero:

$$[\hat{A}, \hat{B}] = \hat{A}\hat{B} - \hat{B}\hat{A} = 0.$$

This can be seen from our earlier expression by simply consolidating terms. Specifically, if

$$\hat{A}\hat{B}|\psi\rangle = \hat{B}\hat{A}|\psi\rangle,$$

then

$$\hat{A}\hat{B}|\psi\rangle - \hat{B}\hat{A}|\psi\rangle = 0.$$

This can subsequently be rearranged to obtain

$$[\hat{A}, \hat{B}]|\psi\rangle = 0|\psi\rangle,$$

from where one sees that the commutator is zero since the wavefunction itself is not zero. So, for two operators to commute, we have the requirement that

$$[\hat{A}, \hat{B}] = 0. \tag{6.18}$$

EXAMPLE 6.2

Illustration of Two Operators Commuting for a Free Particle To illustrate the commutator concept more concretely, let us demonstrate the special case of commuting operators for the free particle. Starting with the free particle wavefunction $|\psi\rangle = Ae^{ikx}$, we can show that $\hat{p}$ and $\hat{H}$ are commuting operators. Before continuing, let us first find out what $\hat{p}|\psi\rangle$ and $\hat{H}|\psi\rangle$ look like. First,

$$\begin{aligned}\hat{p}|\psi\rangle &= -i\hbar\frac{d}{dx}(Ae^{ikx})\\ &= -(i\hbar)(ik)|\psi\rangle\\ &= \hbar k|\psi\rangle.\end{aligned}$$

Next, for the case where $V(x) = 0$,

$$\begin{aligned}\hat{H}|\psi\rangle &= -\frac{\hbar^2}{2m}\frac{d^2}{dx^2}(Ae^{ikx})\\ &= -\frac{\hbar^2}{2m}(ik)^2|\psi\rangle\\ &= \frac{\hbar^2k^2}{2m}|\psi\rangle.\end{aligned}$$

Then,

$$\begin{aligned}[\hat{p}, \hat{H}]|\psi\rangle &= (\hat{p}\hat{H} - \hat{H}\hat{p})|\psi\rangle\\ &= \hat{p}\left(\frac{\hbar^2k^2}{2m}\right)|\psi\rangle - \hat{H}(\hbar k)|\psi\rangle\end{aligned}$$

$$= (\hbar k)\frac{\hbar^2 k^2}{2m_0}|\psi\rangle - (\hbar k)\frac{\hbar^2 k^2}{2m_0}|\psi\rangle$$
$$= 0.$$

So for the special case of a free particle, $\hat{p}$ and $\hat{H}$ commute. Note that this is not generally true of other wavefunctions. The reader may verify this using the confined particle wavefunctions shown earlier.

Conversely, if $\hat{A}$ and $\hat{B}$ do not commute (i.e., $[\hat{A}, \hat{B}] \neq 0$), one cannot specify the eigenvalues of both simultaneously, given a wavefunction. This is the origin of what is called the Uncertainty Principle and will be discussed shortly.

Noncommuting operators can be illustrated using $\hat{x}$ and $\hat{p}$. Note that when working with commutators it is often convenient to employ a generic wavefunction. We will use $|\psi\rangle$ in what follows.

EXAMPLE 6.3

Illustration of Two Noncommuting Operators, $\hat{x}$ and $\hat{p}_x$, for a Generic Particle For an arbitrary wavefunction $|\psi\rangle$, we have

$$[\hat{x}, \hat{p}_x]|\psi\rangle = (\hat{x}\hat{p}_x - \hat{p}_x\hat{x})|\psi\rangle$$
$$= \hat{x}\hat{p}_x|\psi\rangle - \hat{p}_x\hat{x}|\psi\rangle$$
$$= -i\hbar\hat{x}\frac{d|\psi\rangle}{dx} + i\hbar\frac{d}{dx}(\hat{x}|\psi\rangle)$$
$$= i\hbar\left[-x\frac{d|\psi\rangle}{dx} + x\frac{d|\psi\rangle}{dx} + |\psi\rangle\right]$$
$$= i\hbar|\psi\rangle.$$

Thus,

$$[\hat{x}, \hat{p}_x]|\psi\rangle = i\hbar|\psi\rangle,$$

or, alternatively,

$$[\hat{x}, \hat{p}_x] = i\hbar. \tag{6.19}$$

The commutator is therefore nonzero and reveals that $\hat{x}$ and $\hat{p}$ do not commute.

6.10.2 Additional Examples of Commutator Math

We will illustrate additional examples of commutator math by evaluating $[\hat{p}_x, \hat{x}^2]$ and $[\hat{p}_x^2, \hat{x}]$. As before, we employ a placeholder wavefunction $|\psi\rangle$ to simplify things.

EXAMPLE 6.4

Illustration of Commutator Math using $[\hat{p}_x, \hat{x}^2]$ on a Generic Particle

$$[\hat{p}_x, \hat{x}^2]|\psi\rangle = (\hat{p}_x\hat{x}^2 - \hat{x}^2\hat{p}_x)|\psi\rangle$$
$$= -i\hbar\frac{d}{dx}(x^2|\psi\rangle) + i\hbar x^2\frac{d|\psi\rangle}{dx}$$
$$= i\hbar\left(-x^2\frac{d|\psi\rangle}{dx} - 2x|\psi\rangle + x^2\frac{d|\psi\rangle}{dx}\right)$$
$$= -2i\hbar x|\psi\rangle.$$

The commutator is therefore

$$[\hat{p}_x, \hat{x}^2] = -2i\hbar\hat{x}. \qquad (6.20)$$

EXAMPLE 6.5

Illustration of Commutator Math using $[\hat{p}_x^2, \hat{x}]$ on a Generic Particle

$$\begin{aligned}
[\hat{p}_x^2, \hat{x}]|\psi\rangle &= (\hat{p}_x^2\hat{x} - \hat{x}\hat{p}_x^2)|\psi\rangle \\
&= -i\hbar\frac{d}{dx}\left[-i\hbar\frac{d(x|\psi\rangle)}{dx}\right] - x(-i\hbar)^2\frac{d^2|\psi\rangle}{dx^2} \\
&= -\hbar^2\frac{d}{dx}\left(x\frac{d|\psi\rangle}{dx} + |\psi\rangle\right) + \hbar^2 x\frac{d^2|\psi\rangle}{dx^2} \\
&= -\hbar^2\left(x\frac{d^2|\psi\rangle}{dx^2} + 2\frac{d|\psi\rangle}{dx} - x\frac{d^2|\psi\rangle}{dx^2}\right) \\
&= -2\hbar^2\frac{d|\psi\rangle}{dx}.
\end{aligned}$$

Although this looks messy, one can show that the above expression rearranges to

$$\begin{aligned}
\left(-2\hbar^2\frac{d}{dx}\right)|\psi\rangle &= \left[\frac{2\hbar^2}{i\hbar}(-i\hbar)\frac{d}{dx}\right]|\psi\rangle \\
&= -2i\hbar\hat{p}_x|\psi\rangle,
\end{aligned}$$

resulting in

$$[\hat{p}_x^2, \hat{x}] = -2i\hbar\hat{p}_x. \qquad (6.21)$$

6.11 MORE COMMUTATOR RELATIONSHIPS

Finally, some important commutator relationships are summarized in **Table 6.2** for arbitrary operators $\hat{A}$, $\hat{B}$, $\hat{C}$, and $\hat{D}$, and a constant b.

Table 6.2 Common commutator relationships

$[\hat{A}, \hat{A}] = 0$
$[\hat{A}, \hat{B}] = -[\hat{B}, \hat{A}]$
$[\hat{A}, \hat{B}\hat{C}] = [\hat{A}, \hat{B}]\hat{C} + \hat{B}[\hat{A}, \hat{C}]$
$[\hat{A}\hat{B}, \hat{C}] = \hat{A}[\hat{B}, \hat{C}] + [\hat{A}, \hat{C}]\hat{B}$
$[\hat{A}, \hat{B} + \hat{C}] = [\hat{A}, \hat{B}] + [\hat{A}, \hat{C}]$
$[\hat{A}, b\hat{B}] = b[\hat{A}, \hat{B}]$
$\left[\hat{A}, [\hat{B}, \hat{C}]\right] = [\hat{A}, \hat{B}\hat{C}] - [\hat{A}, \hat{C}\hat{B}]$
$[\hat{A} + \hat{B}, \hat{C} + \hat{D}] = [\hat{A}, \hat{C}] + [\hat{A}, \hat{D}] + [\hat{B}, \hat{C}] + [\hat{B}, \hat{D}]$

6.12 THE UNCERTAINTY PRINCIPLE

The Heisenberg Uncertainty Principle follows from a nonzero commutator between two operators. Formally, it is written as

$$\Delta A\,\Delta B \geq \tfrac{1}{2}|\langle i[\hat{A}, \hat{B}]\rangle|, \tag{6.22}$$

where $\hat{A}$ and $\hat{B}$ are two operators.

The most common example one sees is the uncertainty relationship between a particle's position and its momentum. While we will not prove it here, Equation 6.22 shows that a particle's position/momentum uncertainty relation is

$$\Delta x\,\Delta p_x \geq \frac{\hbar}{2}, \tag{6.23}$$

where $\Delta x\,(\Delta p_x)$ is the uncertainty in the particle's position (momentum). Note that

$$\Delta x = \sqrt{\langle x^2\rangle - \langle x\rangle^2}, \tag{6.24}$$

$$\Delta p_x = \sqrt{\langle p_x^2\rangle - \langle p_x\rangle^2}, \tag{6.25}$$

where, referring back to what we learned about expectation values,

$$\langle x^2\rangle = \int \psi^* x^2 \psi\, dx = \langle\psi|x^2|\psi\rangle,$$

$$\langle x\rangle = \int \psi^* x \psi\, dx = \langle\psi|x|\psi\rangle,$$

$$\langle p_x^2\rangle = \int \psi^* \left(-i\hbar\frac{d}{dx}\right)^2 \psi\, dx = \langle\psi|p_x^2|\psi\rangle,$$

$$\langle p_x\rangle = \int \psi^* \left(-i\hbar\frac{d}{dx}\right) \psi\, dx = \langle\psi|p_x|\psi\rangle.$$

Thus Equation 6.23 says that it is impossible to precisely state a particle's position and its momentum simultaneously. By knowing more about one property, one knows less about the other. This is in complete contrast to classical Newtonian mechanics, where it is possible to simultaneously state a particle's position and momentum to as much precision as needed. Herein lies one of the key fundamental differences between quantum mechanics and classical physics.

Analogous uncertainty relationships can also be shown for other pairs of noncommuting operators. Although they will not be derived here, uncertainty relationships between time and frequency and between time and energy are

$$\Delta\omega\,\Delta t \geq \frac{1}{2}, \tag{6.26}$$

$$\Delta E\,\Delta t \geq \frac{\hbar}{2}. \tag{6.27}$$

In fact, this is the reason why short laser pulses are said to contain many colors and conversely why long pulse duration or continuous wave light sources can be nearly monochromatic.

6.13 THE SCHRÖDINGER EQUATION

Having said what we have, where do wavefunctions actually come from? So far, we have simply asserted that wavefunctions look a certain way.

For example, we claimed that a confined particle in one dimension could have a wavefunction that looked like Equation 6.2. How did we obtain these expressions? Well, it turns out that one solves what is called the Schrödinger equation to obtain a particle's wavefunction as well as its energy. This equation was developed by Erwin Schrödinger during his 1924 Christmas break in the Alps. As an aside, Schrödinger was apparently a pretty interesting character—a very readable account of his life can be found in Moore (1989).

While we cannot "prove" the Schrödinger equation and, in fact, it is unclear how Schrödinger himself came up with it, it works. As such, it is *the* equation in quantum mechanics and has the general form

$$i\hbar\frac{\partial}{\partial t}\psi(\mathbf{r},t) = \hat{H}\psi(\mathbf{r},t), \tag{6.28}$$

where $\hat{H}$ is the Hamiltonian. Alternatively, one writes

$$i\hbar\frac{\partial}{\partial t}\psi(\mathbf{r},t) = \left[-\frac{\hbar^2}{2m}\nabla^2 + V(\mathbf{r},t)\right]\psi(\mathbf{r},t), \tag{6.29}$$

where $V(\mathbf{r},t)$ is the potential energy experienced by the particle and m is its mass. We will assume here and in the next chapter that V is time-independent. The time-dependent case will be discussed in Chapter 11.

6.13.1 The Time-Independent Schrödinger Equation

Now although Equation 6.29 is time-dependent, it can be shown that there exists a time-independent form that we will deal with more commonly. To obtain it, we will assume that the potential energy operator in the Hamiltonian is time-independent $V(\mathbf{r},t) \to V(\mathbf{r})$. Thus, assuming $V(\mathbf{r})$, Equation 6.29 becomes

$$i\hbar\frac{\partial}{\partial t}\psi(\mathbf{r},t) = \left[-\frac{\hbar^2\nabla^2}{2m} + V(\mathbf{r})\right]\psi(\mathbf{r},t).$$

Note that the left hand side of this equation possesses a time dependence while the right hand side does not. To expedite the process of obtaining the time-independent Schrödinger equation, let us look for solutions of the form

$$\psi(\mathbf{r},t) = \psi(\mathbf{r})f(t) \equiv \psi f,$$

which is the product of an exclusively time-dependent function $f(t)$ and an exclusively space-dependent function $\psi(\mathbf{r})$.

Inserting this into the Schrödinger equation gives

$$i\hbar\frac{\partial(\psi f)}{\partial t} = -\frac{\hbar^2\nabla^2(\psi f)}{2m} + V(\mathbf{r})(\psi f),$$

which rearranges to

$$\frac{i\hbar}{f}\frac{df}{dt} = \frac{1}{\psi}\left[-\frac{\hbar^2\nabla^2}{2m}\psi + V(\mathbf{r})\psi\right].$$

At this point, note that the left-hand side is independent of position while the right-hand side is independent of time. Both sides therefore equal a constant, since the equality is true for any position or any time. We call this constant E, since it will turn out to be the particle's energy.

Let us now solve the left- and right-hand sides of the equation separately.

Left-Hand Side

We have

$$\frac{i\hbar}{f}\frac{df}{dt} = E,$$

which rearranges to

$$\frac{1}{f}\frac{df}{dt} = -\frac{iE}{\hbar}.$$

This integrates to yield

$$\ln f = -\frac{iEt}{\hbar},$$

if we ignore the constant of integration. Solving for $f(t)$ then gives

$$f(t) = e^{-iEt/\hbar}, \tag{6.30}$$

where one can alternatively express this using a frequency $\omega = E/\hbar$.

Right-Hand Side

On the right-hand side, we have

$$\frac{1}{\psi}\left[-\frac{\hbar^2\nabla^2}{2m}\psi + V(\mathbf{r})\psi\right] = E,$$

or

$$\left[-\frac{\hbar^2\nabla^2}{2m} + V(\mathbf{r})\right]\psi = E\psi. \tag{6.31}$$

This can be written more concisely as

$$\hat{H}\psi = E\psi, \tag{6.32}$$

where again $\hat{H}$ is the Hamiltonian operator. The above expression is what is commonly referred to as the time-independent Schrödinger equation. It is basically an eigenvalue/eigenfunction equation (cf. Equation 6.16) and will be the equation that we will solve repeatedly in Chapter 7 when modeling carriers in quantum wells, wires, and dots. Note that the time-independent Schrödinger equation describes the total energy Hamiltonian operator acting on a wavefunction. When ψ happens to be an eigenfunction of the Hamiltonian, the eigenvalue E associated with the wavefunction is returned. Furthermore, even if the wavefunction is not immediately an eigenfunction of the Hamiltonian, it can generally be expressed as a linear combination of Hamiltonian eigenfunctions, as we saw earlier when discussing wavepackets. As a consequence, the energy returned will be a weighted sum of eigenvalues.

Summary

At this point, assuming that we have solved the time-independent Schrödinger equation, we have $\psi(\mathbf{r})$ and E. Together with $f(t)$, the total wavefunction of the particle is the product of these time-dependent and time-independent terms:

$$\psi(\mathbf{r}, t) = \psi(\mathbf{r})f(t).$$

The above solutions $\psi(\mathbf{r}, t)$ derived with a time-independent potential are called stationary solutions. This because the probability density of the particle, $|\psi|^2 = \psi^*\psi$, is time-independent. To illustrate this,

$$\begin{aligned}|\psi|^2 &= \psi^*\psi \\ &= [\psi(\mathbf{r})f(t)]^*[\psi(\mathbf{r})f(t)] \\ &= \left[\psi^*(\mathbf{r})e^{iEt/\hbar}\right]\left[\psi(\mathbf{r})e^{-iEt/\hbar}\right] \\ &= \psi^*(\mathbf{r})\psi(\mathbf{r}),\end{aligned}$$

since $e^{iEt/\hbar}e^{-iEt/\hbar} = 1$. The result is therefore time-independent.

Finally, note that constraints on $\psi(\mathbf{r})$ include the following:

$\psi(\mathbf{r})$ is finite,
$\psi(\mathbf{r})$ is continuous,
$\psi'(\mathbf{r})$ is continuous,

as noted earlier.

In addition, boundary conditions on the wavefunction will induce quantization on its associated energies as well as on the actual mathematical form of the wavefunction. Chapter 7 therefore discusses a number of model systems describing carriers confined to quantum wells, wires, and dots. They have names such as

- the "particle in a box" (which can be used to describe quantum wells)
- the "particle in a cylinder" (which can be used to describe nanowires)
- the "particle in a sphere" (which can be used to describe quantum dots).

6.14 THE POSTULATES OF QUANTUM MECHANICS

Finally, we summarize some of the main concepts discussed so far. These points are often referred to as the Postulates of Quantum Mechanics.

6.14.1 Postulate 1: Wavefunctions

The state of a system is specified at time t by $|\psi(\mathbf{r}, t)\rangle$. This wavefunction contains all of the information one needs to know about the particle.

6.14.2 Postulate 2: Operators

For every physically measurable quantity A, there exists a linear Hermitian operator $\hat{A}$ whose eigenfunctions form a complete basis (i.e., any particle's wavefunction can be described as an appropriate weighted linear combination of these functions).

6.14.3 Postulate 3: Measurements

In any measurement of the observable associated with an operator $\hat{A}$, the only values ever observed are the eigenvalues of $\hat{A}$, which satisfy the eigenvalue/eigenfunction equation

$$\hat{A}\psi = a\psi.$$

After the measurement, the state of the system is the eigenfunction associated with the measured eigenvalue (note that this is sometimes stated as a separate postulate).

6.14.4 Postulate 4: Probabilistic Outcomes

If the wavefunction of a system is described by a properly weighted linear combination of eigenfunctions for an operator $\hat{A}$, then the probability of measuring the eigenvalue associated with a given eigenfunction is simply the latter's weight squared.

So, if we have

$$|\psi\rangle = A|\psi_1\rangle + B|\psi_2\rangle + C|\psi_3\rangle,$$

where $|A|^2 + |B|^2 + |C|^2 = 1$ and E_1, E_2, E_3 are the energy eigenvalues associated with $|\psi_1\rangle, |\psi_2\rangle, |\psi_3\rangle$, then the probabilities P of measuring each energy are

$$P(E_1) = |A|^2,$$
$$P(E_2) = |B|^2,$$
$$P(E_3) = |C|^2.$$

6.14.5 Postulate 5: Expectation Values

For a system described by a normalized wavefunction $|\psi\rangle$, the expectation value of any observable A is found by "sandwiching" the operator. Thus, in one dimension,

$$\langle A\rangle = \langle\psi|\hat{A}|\psi\rangle = \int \psi^* \hat{A}\psi \, dx.$$

In three dimensions, one writes

$$\langle A\rangle = \int \psi^* \hat{A}\psi \, d^3 r.$$

6.14.6 Postulate 6: Schrödinger Equation

The time dependence of a wavefunction $|\psi\rangle$ is governed by the time-dependent Schrödinger equation (written here in one dimension)

$$i\hbar \frac{\partial}{\partial t}\psi(x,t) = \hat{H}\psi(x,t),$$

where $\hat{H}$ is the Hamiltonian operator. So, what this says in the case where the potential energy is time-independent is that the total wavefunction is the product of a time-independent part and a time-dependent part, $\psi_{\text{tot}} = \psi(\mathbf{r})f(t)$, where $f(t) = e^{-iEt/\hbar}$. As a consequence, taking our confined particle wavefunction (Equation 6.2) as an example, the particle's total wavefunction would be written as

$$\psi_{\text{tot}}(x,t) = \left(\sqrt{\frac{2}{a}} \sin \frac{n\pi x}{a}\right) e^{-iE_n t/\hbar},$$

where $n = 1, 2, 3, \ldots$.

6.15 TIME-INDEPENDENT, NONDEGENERATE PERTURBATION THEORY

Real life imposes complications in evaluating a particle's wavefunctions and energies. In this regard, one could have additional potential energy contributions to the Schrödinger equation, which make it intractable to solve analytically. It turns out that there exists a way around this problem. Namely, one can approximate the solution of the complicated Schrödinger equation, by assuming that these additional time-independent potential energy terms are small in magnitude. The time-dependent case will be discussed in Chapter 11.

The basic idea is that the true Hamiltonian $\hat{H}$ closely resembles an unperturbed Hamiltonian $\hat{H}^{(0)}$ whose eigenfunctions and eigenvalues we already know. The superscript (0) here denotes that the Hamiltonian is unperturbed. The additional term that makes it different, $\hat{H}^{(1)}$, is called a perturbation. The superscript (1) implies that this contribution to the Hamiltonian is a "first"-order correction. By the same token, higher-order corrections such as $\hat{H}^{(2)}, \hat{H}^{(3)}, \ldots$ can be included if desired. However, one generally stops after the first- or second-order term, since they become increasingly small in magnitude.

In what follows, we introduce a dimensionless parameter λ to keep track of the order (alternatively, the order of refinement) of the solutions. So, first-order corrections due to $\hat{H}^{(1)}$ have a λ in front of them. Second-order corrections due to $\hat{H}^{(2)}$ have λ^2 in front of them, and so on. Note that λ does not represent any real physical quantity. It simply helps us keep track of the order of approximation. If needed, we can set $\lambda = 1$ at the end.

Our corrected Hamiltonian is therefore

$$\hat{H} = \hat{H}^{(0)} + \lambda\hat{H}^{(1)} + \lambda^2\hat{H}^{(2)} + \cdots,$$

where again $\hat{H}^{(0)}$ is the unperturbed Hamiltonian whose eigenvalues and eigenvectors we know a priori. The corresponding wavefunctions and energies are then

$$\psi = \psi^{(0)} + \lambda\psi^{(1)} + \lambda^2\psi^{(2)} + \cdots,$$
$$E = E^{(0)} + \lambda E^{(1)} + \lambda^2 E^{(2)} + \cdots,$$

which are nothing more than the original wavefunctions and energies with some extra corrections attached to them. Our goal is now to find explicit expressions for $\psi^{(1)}$, $\psi^{(2)}$, $E^{(1)}$, $E^{(2)}$, etc. in a tractable manner. Note again that corrections above second order are generally not needed since their magnitude will be small.

The results of perturbation theory will be found by taking the exact Schrödinger equation $\hat{H}\psi = E\psi$ and substituting in it the above expressions for ψ and E. To simplify things, let us group terms by their power in λ. We therefore start with

$$(\hat{H}^{(0)} + \lambda\hat{H}^{(1)} + \lambda^2\hat{H}^{(2)})(\psi^{(0)} + \lambda\psi^{(1)} + \lambda^2\psi^{(2)})$$
$$= (E^{(0)} + \lambda E^{(1)} + \lambda^2 E^{(2)})(\psi^{(0)} + \lambda\psi^{(1)} + \lambda^2\psi^{(2)})$$

and separately evaluate the left- and right-hand sides. Furthermore, let us consider terms only up to second order (i.e., we drop terms with λ^3

and higher). The left-hand side is

$$\hat{H}^{(0)}\psi^{(0)} + \lambda\hat{H}^{(0)}\psi^{(1)} + \lambda^2\hat{H}^{(0)}\psi^{(2)} + \lambda^3\hat{H}^{(1)}\psi^{(2)} + \lambda^4\hat{H}^{(2)}\psi^{(2)}$$
$$+\lambda\hat{H}^{(1)}\psi^{(0)} + \lambda^2\hat{H}^{(1)}\psi^{(1)} + \lambda^3\hat{H}^{(2)}\psi^{(1)}$$
$$+\lambda^2\hat{H}^{(2)}\psi^{(0)}.$$

When we consider terms only up to second order, we find

$$\hat{H}^{(0)}\psi^{(0)} + \lambda(\hat{H}^{(0)}\psi^{(1)} + \hat{H}^{(1)}\psi^{(0)}) + \lambda^2(\hat{H}^{(0)}\psi^{(2)} + \hat{H}^{(1)}\psi^{(1)} + \hat{H}^{(2)}\psi^{(0)}).$$

Next, the right-hand side is

$$E^{(0)}\psi^{(0)} + \lambda E^{(0)}\psi^{(1)} + \lambda^2 E^{(0)}\psi^{(2)} + \lambda^3 E^{(1)}\psi^{(2)} + \lambda^4 E^{(2)}\psi^{(2)}$$
$$+\lambda E^{(1)}\psi^{(0)} + \lambda^2 E^{(1)}\psi^{(1)} + \lambda^3 E^{(2)}\psi^{(1)}$$
$$+\lambda^2 E^{(2)}\psi^{(0)}.$$

When we consider terms only up to second order, we have

$$E^{(0)}\psi^{(0)} + \lambda(E^{(0)}\psi^{(1)} + E^{(1)}\psi^{(0)}) + \lambda^2(E^{(0)}\psi^{(2)} + E^{(1)}\psi^{(1)} + E^{(2)}\psi^{(0)}).$$

Now, because the solution must be true for all values of λ, we equate all coefficients of a given power in λ:

$$\hat{H}^{(0)}\psi^{(0)} = E^{(0)}\psi^{(0)},$$
$$\hat{H}^{(0)}\psi^{(1)} + \hat{H}^{(1)}\psi^{(0)} = E^{(0)}\psi^{(1)} + E^{(1)}\psi^{(0)},$$
$$\hat{H}^{(0)}\psi^{(2)} + \hat{H}^{(1)}\psi^{(1)} + \hat{H}^{(2)}\psi^{(0)} = E^{(0)}\psi^{(2)} + E^{(1)}\psi^{(1)} + E^{(2)}\psi^{(0)}.$$

Rearranging these expressions then gives

$$(\hat{H}^{(0)} - E^{(0)})\psi^{(0)} = 0, \tag{6.33}$$
$$(\hat{H}^{(0)} - E^{(0)})\psi^{(1)} + (\hat{H}^{(1)} - E^{(1)})\psi^{(0)} = 0, \tag{6.34}$$
$$(\hat{H}^{(0)} - E^{(0)})\psi^{(2)} + (\hat{H}^{(1)} - E^{(1)})\psi^{(1)} + (\hat{H}^{(2)} - E^{(2)})\psi^{(0)} = 0. \tag{6.35}$$

Note that the first of these three equations is our unperturbed case. It provides our zeroth-order solutions, which we already have. There is nothing special here.

We now multiply all expressions by $\psi^{*(0)}$ and integrate over all space. To avoid writing out the integrals, we shall switch to bra–ket notation. We then have

$$\langle\psi^{(0)}|\hat{H}^{(0)} - E^{(0)}|\psi^{(0)}\rangle = 0,$$
$$\langle\psi^{(0)}|\hat{H}^{(0)} - E^{(0)}|\psi^{(1)}\rangle + \langle\psi^{(0)}|\hat{H}^{(1)} - E^{(1)}|\psi^{(0)}\rangle = 0,$$
$$\langle\psi^{(0)}|\hat{H}^{(0)} - E^{(0)}|\psi^{(2)}\rangle + \langle\psi^{(0)}|\hat{H}^{(1)} - E^{(1)}|\psi^{(1)}\rangle + \langle\psi^{(0)}|\hat{H}^{(2)} - E^{(2)}|\psi^{(0)}\rangle = 0.$$

In the case of the first- and second-order expressions, the first term in each is zero; i.e.,

$$\langle\psi^{(0)}|\hat{H}^{(0)} - E^{(0)}|\psi^{(1)}\rangle = 0,$$
$$\langle\psi^{(0)}|\hat{H}^{(0)} - E^{(0)}|\psi^{(2)}\rangle = 0.$$

This is because $\hat{H}^{(0)}$ is Hermitian ($\hat{H}^{(0)} = \hat{H}^{(0)\dagger}$) and one can operate with $\hat{H}^{(0)}$ on the bra to see that the expression disappears. This leaves

$$\langle\psi^{(0)}|\hat{H}^{(1)} - E^{(1)}|\psi^{(0)}\rangle = 0,$$
$$\langle\psi^{(0)}|\hat{H}^{(1)} - E^{(1)}|\psi^{(1)}\rangle + \langle\psi^{(0)}|\hat{H}^{(2)} - E^{(2)}|\psi^{(0)}\rangle = 0.$$

Next, let us break up each expression and drop any terms that are zero. We find

$$\langle\psi^{(0)}|\hat{H}^{(1)}|\psi^{(0)}\rangle - E^{(1)}\langle\psi^{(0)}|\psi^{(0)}\rangle = 0,$$
$$\langle\psi^{(0)}|\hat{H}^{(1)}|\psi^{(1)}\rangle - E^{(1)}\langle\psi^{(0)}|\psi^{(1)}\rangle + \langle\psi^{(0)}|\hat{H}^{(2)}|\psi^{(0)}\rangle - E^{(2)}\langle\psi^{(0)}|\psi^{(0)}\rangle = 0.$$

This reduces to

$$\langle\psi^{(0)}|\hat{H}^{(1)}|\psi^{(0)}\rangle - E^{(1)} = 0,$$
$$\langle\psi^{(0)}|\hat{H}^{(1)}|\psi^{(1)}\rangle + \langle\psi^{(0)}|\hat{H}^{(2)}|\psi^{(0)}\rangle - E^{(2)} = 0.$$

(Note that we have invoked $\langle\psi^{(0)}|\psi^{(1)}\rangle = 0$. We will see this in more detail shortly when we talk about corrections to the wavefunction. For now, just assume this to be true.) Finally, we have

$$\langle\psi^{(0)}|\hat{H}^{(1)}|\psi^{(0)}\rangle = E^{(1)},$$
$$\langle\psi^{(0)}|\hat{H}^{(1)}|\psi^{(1)}\rangle + \langle\psi^{(0)}|\hat{H}^{(2)}|\psi^{(0)}\rangle = E^{(2)}.$$

Thus, for the energies, we have the following first- and second-order corrections to the zeroth-order solutions:

$$E^{(1)} = \langle\psi^{(0)}|\hat{H}^{(1)}|\psi^{(0)}\rangle, \tag{6.36}$$
$$E^{(2)} = \langle\psi^{(0)}|\hat{H}^{(1)}|\psi^{(1)}\rangle + \langle\psi^{(0)}|\hat{H}^{(2)}|\psi^{(0)}\rangle. \tag{6.37}$$

Note that to explicitly obtain a second-order energy correction, we need to know what $|\psi^{(1)}\rangle$ looks like. The same applies to higher-order corrections. So, right now, our expression for $E^{(2)}$ is not necessarily useful. We will revisit it later after finding out what $|\psi^{(1)}\rangle$ looks like.

EXAMPLE 6.6

Nondegenerate Perturbation Theory Applied to a Particle in a Slanted Box Let us now demonstrate the usefulness of the time-independent perturbation theory just worked out. Assume that we have a particle in a one-dimensional box of width a. (This model will be fully worked out in Chapter 7, but for now simply assume that we know what the wavefunctions and energies of this problem look like. The reader may also return to this and other examples after having read Chapter 7.) Next, assume that the system is perturbed by the introduction of a slant to the bottom of the box. This could occur due to the application of an electric field, for example. The newly introduced perturbation to the total Hamiltonian is then $H^{(1)} = Vx/a$, where V is the magnitude of the applied potential. The particle in a slanted box potential is illustrated in **Figure 6.8**.

We are given zeroth-order wavefunctions

$$\psi^{(0)} = \sqrt{\frac{2}{a}} \sin\frac{n\pi x}{a}$$

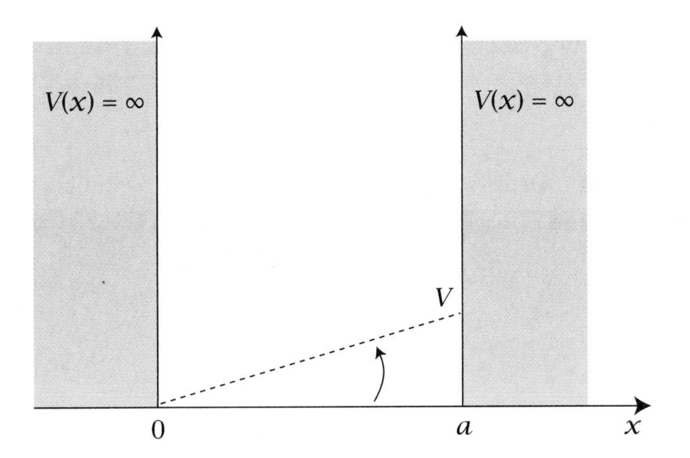

Figure 6.8 Illustration of the potential for a particle in a slanted box.

and zeroth-order energies

$$E^{(0)} = \frac{n^2 h^2}{8ma^2}.$$

What is the first-order correction to $E^{(0)}$? From nondegenerate perturbation theory, the first-order correction is simply (Equation 6.36)

$$\begin{aligned}
E^{(1)} &= \langle \psi^{(0)} | \hat{H}^{(1)} | \psi^{(0)} \rangle \\
&= \int_0^a \left(\sqrt{\frac{2}{a}} \sin \frac{n\pi x}{a} \right) \left(\frac{V}{a} x \right) \left(\sqrt{\frac{2}{a}} \sin \frac{n\pi x}{a} \right) dx \\
&= \frac{2}{a} \left(\frac{V}{a} \right) \int_0^a x \sin^2 \frac{n\pi x}{a}\, dx.
\end{aligned}$$

To simplify the integral, let $b = n\pi/a$. This gives

$$E^{(1)} = \frac{2V}{a^2} \int_0^a x \sin^2 bx\, dx,$$

where $\sin^2 bx = \frac{1}{2}(1 - \cos 2bx)$. The resulting expression is easy to evaluate, giving

$$\begin{aligned}
E^{(1)} &= \frac{2V}{a^2} \left[\frac{1}{2} \int_0^a x(1 - \cos 2bx)\, dx \right] \\
&= \frac{V}{a^2} \left(\frac{x^2}{2} \bigg|_0^a - \int_0^a x \cos 2bx\, dx \right) \\
&= \frac{V}{a^2} \left(\frac{a^2}{2} - \int_0^a x \cos 2bx\, dx \right).
\end{aligned}$$

We solve the remaining integral by parts. Let $u = x$, $du = dx$, $dv = \cos 2bx\, dx$, and $v = (1/2b) \sin 2bx$, to get

$$\begin{aligned}
E^{(1)} &= \frac{V}{2} - \frac{V}{a^2} \left(\frac{x}{2b} \sin 2bx \bigg|_0^a - \int_0^a \frac{1}{2b} \sin 2bx\, dx \right) \\
&= \frac{V}{2} - \frac{V}{a^2} \left[\frac{x}{2b} \sin 2bx \bigg|_0^a + \frac{1}{2b} \left(\frac{1}{2b} \right) \cos 2bx \bigg|_0^a \right] \\
&= \frac{V}{2} - \frac{V}{a^2} \left(\frac{1}{4b^2} \cos 2bx \bigg|_0^a \right).
\end{aligned}$$

At this point, the last term in parentheses is zero, giving

$$E^{(1)} = \frac{V}{2}$$

as the desired first-order correction to $E^{(0)}$.

The total energy of the particle in a slanted box, to first order, is then

$$E = \frac{n^2h^2}{8ma^2} + \frac{V}{2},$$

where $n = 1, 2, 3, \ldots$.

EXAMPLE 6.7

Nondegenerate Perturbation Theory Applied to an Anharmonic Oscillator Next, to continue illustrating the use of time-independent perturbation theory, consider an anharmonic oscillator whose potential is $V(x) = \frac{1}{2}kx^2 + \frac{1}{6}\gamma x^3$, where k and γ are constants. We illustrate this potential and that of the unperturbed harmonic oscillator in **Figure 6.9**. Note that this latter (harmonic oscillator) problem is worked out in Chapter 8. For now, we just assume the resulting wavefunctions and energies. Namely,

$$E^{(0)} = (n + \tfrac{1}{2})h\nu,$$
$$\psi^{(0)} = N_n H_n(\alpha^{1/2}x)e^{-\alpha x^2/2},$$

where $\alpha = \sqrt{k\mu/\hbar^2}$, $H_n(\alpha^{1/2}x)$ are Hermite polynomials (Chapter 8), and the normalization constant is

$$N_n = \frac{1}{\sqrt{2^n n!}}\left(\frac{\alpha}{\pi}\right)^{1/4}.$$

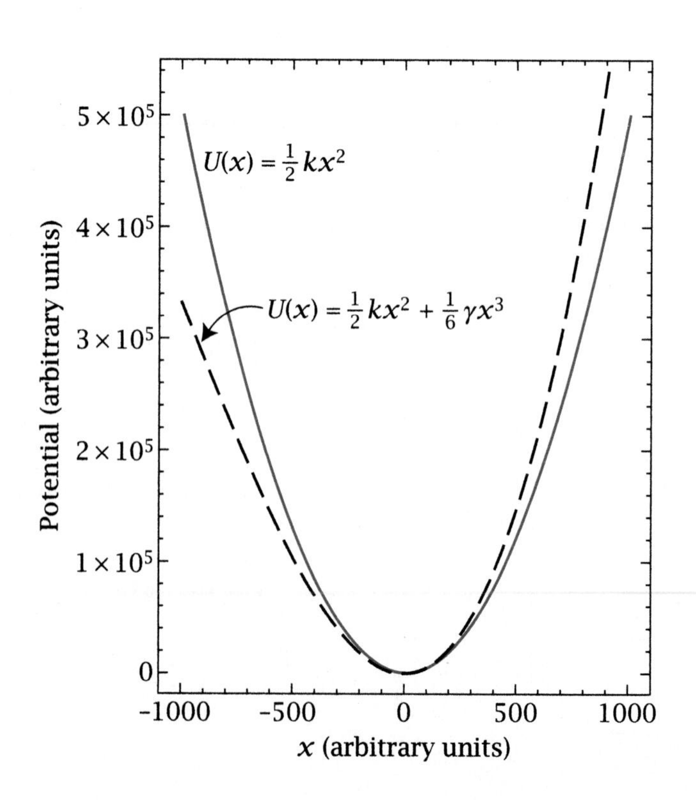

Figure 6.9 Illustration of an anharmonic oscillator potential. The unperturbed potential is shown by the solid red line while the perturbed anharmonic potential is shown by the dashed black line. The traces were constructed assuming $k = 1$ and $\gamma = 0.001$.

The first-order correction to the energy is then (Equation 6.36)

$$E^{(1)} = \langle \psi^{(0)} | \hat{H}^{(1)} | \psi^{(0)} \rangle,$$

where $\hat{H}^{(1)} = \frac{1}{6}\gamma x^3$. We have

$$E^{(1)} = \frac{\gamma}{6} \int_{-\infty}^{\infty} [N_n H_n(\alpha^{1/2}x) e^{-\alpha x^2/2}]^* (x^3) [N_n H_n(\alpha^{1/2}x) e^{-\alpha x^2/2}] \, dx$$

$$= \frac{\gamma}{6} N_n^2 \int_{-\infty}^{\infty} [H_n(\alpha^{1/2}x)](x^3)[H_n(\alpha^{1/2}x)] e^{-\alpha x^2} \, dx.$$

Now, if we consider the specific case where $n = 0$ (the ground state), $H_0(\alpha^{1/2}x) = 1$, and $N_0 = (\alpha/\pi)^{1/4}$, resulting in

$$= \frac{\gamma}{6} N_0^2 \int_{-\infty}^{\infty} x^3 e^{-\alpha x^2} \, dx.$$

The first factor (i.e., x^3) in the integral is odd. The second is even. By symmetry, their product is an overall odd function and its integral over a symmetric interval is zero. Therefore,

$$E^{(1)} = 0.$$

This is our desired first-order correction to the perturbed harmonic oscillator's ground-state energy (i.e., the lowest-energy state). As a consequence, the total ground-state energy, corrected to first order, is

$$E = \tfrac{1}{2} h\nu.$$

One sees that there is apparently no overall change in energy that arises from the perturbation. We must go to higher orders to see the effects of the perturbation.

EXAMPLE 6.8

Nondegenerate Perturbation Theory Applied to Another Anharmonic Oscillator Let us do one more example to drive home the point. We will calculate the first-order correction to the ground-state energy of another anharmonic oscillator whose potential energy is

$$V(x) = \tfrac{1}{2}kx^2 + \tfrac{1}{6}\gamma x^3 + \tfrac{1}{24}bx^4,$$

with k, γ, and b all constant. The first-order correction to the unperturbed Hamiltonian is therefore

$$\hat{H}^{(1)} = \tfrac{1}{6}\gamma x^3 + \tfrac{1}{24}bx^4.$$

Zeroth-order energies and wavefunctions of the particle are again (this will be shown in Chapter 8)

$$E = (n + \tfrac{1}{2})h\nu$$

$$\psi = N_n H_n(\alpha^{1/2}x) e^{-\alpha x^2/2},$$

where $\alpha = \sqrt{k\mu/\hbar^2}$, $H_n(\alpha^{1/2}x)$ are Hermite polynomials (Chapter 8), and the normalization constant is

$$N = \frac{1}{\sqrt{2^n n!}} \left(\frac{\alpha}{\pi}\right)^{1/4},$$

with $N_0 = (\alpha/\pi)^{1/4}$ if $n = 0$ (the ground state).

The first-order correction to the ground-state energy is then (Equation 6.36)

$$
\begin{aligned}
E^{(1)} &= \langle\psi^{(0)}|\hat{H}^{(1)}|\psi^{(0)}\rangle \\
&= \int_{-\infty}^{\infty} \left[N_0 H_0(\alpha^{1/2}x)e^{-\alpha x^2/2}\right]^* \left(\tfrac{1}{6}\gamma x^3 + \tfrac{1}{24}bx^4\right) \\
&\quad \times \left[N_0 H_0(\alpha^{1/2}x)e^{-\alpha x^2/2}\right] dx.
\end{aligned}
$$

We have already evaluated the first term. By symmetry, it drops out, leaving

$$
E^{(1)} = \frac{b}{24} N_0^2 \int_{-\infty}^{\infty} [H_0(\alpha^{1/2}x)](x^4)[H_0(\alpha^{1/2}x)]e^{-\alpha x^2}\, dx,
$$

with $H_0(\alpha^{1/2}x) = 1$. At this point, we have

$$
E^{(1)} = \frac{b}{24} N_0^2 \int_{-\infty}^{\infty} x^4 e^{-\alpha x^2}\, dx,
$$

which in this case does not disappear by symmetry. Using a table of integrals, where

$$
\int_0^{\infty} x^{2c} e^{-\alpha x^2}\, dx = \frac{1\cdot 3\cdot 5\cdot \ldots \cdot (2c-1)}{2^{c+1}\alpha^c}\sqrt{\frac{\pi}{\alpha}}, \tag{6.38}
$$

then gives

$$
\begin{aligned}
E^{(1)} &= \frac{bN_0^2}{12}\int_0^{\infty} x^4 e^{-\alpha x^2}\, dx \\
&= \frac{bN_0^2}{12}\left(\frac{3}{8\alpha^2}\right)\sqrt{\frac{\pi}{\alpha}} \\
&= \frac{bN_0^2}{32\alpha^2}\sqrt{\frac{\pi}{\alpha}}.
\end{aligned}
$$

Recall that $N_0 = (\alpha/\pi)^{1/4}$ and $\alpha = \sqrt{k\mu/\hbar^2}$, to obtain

$$
E^{(1)} = \frac{b\hbar^2}{32k\mu}
$$

as the final result for the first-order energy correction to $E^{(0)}$. The total energy of the anharmonic oscillator in its ground state ($n = 0$), corrected to first order, is therefore

$$
E = \frac{h\nu}{2} + \frac{b\hbar^2}{32k\mu}.
$$

6.15.1 Wavefunctions

Now that we have corrected the energies, what about the wavefunctions? Let us therefore find an expression for the particle's wavefunctions. We know that it has corrections of the form

$$
\psi = \psi_i^{(0)} + \lambda\psi_i^{(1)} + \lambda^2\psi_i^{(2)} + \cdots,
$$

where the index i just denotes a given basis state. Next a convenient way to express the first-order correction $\psi_i^{(1)}$ is as a linear expansion of unperturbed wavefunctions (our basis), which we denote using a different index, j, namely, $\psi_j^{(0)}$ or $|j\rangle$. We then have

$$\psi_i^{(1)} = \sum_{j \neq i} c_{ij} |j\rangle,$$

where c_{ij} are weights, essentially telling us how much of each unperturbed wavefunction is present in the expansion.

If we go back to our expression from which we derived first-order corrections to the energy, we find (Equation 6.34)

$$(\hat{H}^{(0)} - E_i^{(0)})\psi_i^{(1)} + (\hat{H}^{(1)} - E_i^{(1)})\psi_i^{(0)} = 0.$$

This rearranges to

$$\hat{H}^{(0)}\psi_i^{(1)} + \hat{H}^{(1)}\psi_i^{(0)} = E_i^{(0)}\psi_i^{(1)} + E_i^{(1)}\psi_i^{(0)}.$$

Now recall that $\psi_i^{(1)} = \sum_{j \neq i} c_{ij}|j\rangle$ and replace this into the above expression. We get

$$\hat{H}^{(0)} \sum_{j \neq i} c_{ij}|j\rangle + \hat{H}^{(1)}|i\rangle = E_i^{(0)} \sum_{j \neq i} c_{ij}|j\rangle + E_i^{(1)}|i\rangle,$$

$$\sum_{j \neq i} c_{ij}\hat{H}^{(0)}|j\rangle + \hat{H}^{(1)}|i\rangle = E_i^{(0)} \sum_{j \neq i} c_{ij}|j\rangle + E_i^{(1)}|i\rangle$$

where $|i\rangle = \psi_i^{(0)}$ and the kets $|j\rangle$ are simply eigenfunctions of $\hat{H}^{(0)}$, given that we have used an expansion in the unperturbed basis. As a consequence,

$$\sum_{j \neq i} c_{ij}E_j^{(0)}|j\rangle + \hat{H}^{(1)}|i\rangle = E_i^{(0)} \sum_{j \neq i} c_{ij}|j\rangle + E_i^{(1)}|i\rangle.$$

Next, to find the desired coefficients c_{ij}, we multiply both sides by $\langle k|$, a member of the unperturbed basis. Note that $k \neq i$. This gives

$$\sum_{j \neq i} E_j^{(0)} c_{ij}\langle k|j\rangle + \langle k|\hat{H}^{(1)}|i\rangle = E_i^{(0)} \sum_{j \neq i} c_{ij}\langle k|j\rangle + E_i^{(1)}\langle k|i\rangle,$$

$$\sum_{j \neq i} E_j^{(0)} c_{ij}\langle k|j\rangle + \langle k|\hat{H}^{(1)}|i\rangle = E_i^{(0)} \sum_{j \neq i} c_{ij}\langle k|j\rangle.$$

Now, due to the orthogonality between states composing a basis and, in particular, between $|k\rangle$ and $|j\rangle$, the only terms that survive are those where $k = j$. We then have

$$E_k^{(0)} c_{ik} + \langle k|\hat{H}^{(1)}|i\rangle = E_i^{(0)} c_{ik}.$$

Consolidating our desired c_{ik} terms gives

$$c_{ik}(E_k^{(0)} - E_i^{(0)}) = -\langle k|\hat{H}^{(1)}|i\rangle.$$

As a consequence,

$$c_{ik} = \frac{\langle k|\hat{H}^{(1)}|i\rangle}{E_i^{(0)} - E_k^{(0)}}. \tag{6.39}$$

This is our desired coefficient associated with a given state, k. Since $\psi_i^{(1)} = \sum_{j \neq i} c_{ij} |j\rangle$, the (total) corrected wavefunction of the particle to first order is

$$|\psi\rangle = |i\rangle + \sum_{k \neq i} \frac{\langle k|\hat{H}^{(1)}|i\rangle}{E_i^{(0)} - E_k^{(0)}} |k\rangle, \tag{6.40}$$

with $\psi_i^{(0)} = |i\rangle$ and

$$\psi_i^{(1)} = \sum_{k \neq i} \frac{\langle k|\hat{H}^{(1)}|i\rangle}{E_i^{(0)} - E_k^{(0)}} |k\rangle. \tag{6.41}$$

Given the above first-order correction to the wavefunction, we can now go back to Equation 6.37 for the second-order correction to the particle's energy and express it in a more user-friendly fashion. Recall that

$$E^{(2)} = \langle \psi_i^{(0)} | \hat{H}^{(1)} | \psi_i^{(1)} \rangle + \langle \psi_i^{(0)} | \hat{H}^{(2)} | \psi_i^{(0)} \rangle.$$

We substitute Equation 6.41 into this. To make life even easier, assume that the total perturbation only goes up to first order, and thus $\hat{H}^{(2)} = 0$. We get

$$\begin{aligned} E^{(2)} &= \langle \psi_i^{(0)} | \hat{H}^{(1)} | \psi_i^{(1)} \rangle \\ &= \sum_{k \neq i} \frac{\langle k|\hat{H}^{(1)}|i\rangle \langle i|\hat{H}^{(1)}|k\rangle}{E_i^{(0)} - E_k^{(0)}}. \end{aligned}$$

The final expression for a more usable second-order energy correction is therefore

$$E^{(2)} = \sum_{k \neq i} \frac{\langle k|\hat{H}^{(1)}|i\rangle \langle i|\hat{H}^{(1)}|k\rangle}{E_i^{(0)} - E_k^{(0)}}. \tag{6.42}$$

With this, we finish our brief introduction to time-independent perturbation theory. Note that in the above expression, the denominator has a difference of energies. Obviously, we cannot have the case where this difference equals zero. This means that the *nondegenerate* perturbation theory we just worked out applies only to systems where there are no degeneracies present—hence the name. There is a different approach for dealing with degeneracies (called degenerate perturbation theory), but for brevity we will not go into it. The interested reader may consult one of the texts in the Further Reading for more details about the topic.

Summary of Nondegenerate Perturbation Theory

To summarize, we have the following formulas that enable us to correct both the energies and wavefunctions of a system under the influence of a time-independent perturbation:

$$E^{(1)} = \langle i|\hat{H}^{(1)}|i\rangle, \tag{6.43}$$

$$E^{(2)} = \sum_{k \neq i} \frac{\langle k|\hat{H}^{(1)}|i\rangle \langle i|\hat{H}^{(1)}|k\rangle}{E_i^{(0)} - E_k^{(0)}} = \sum_{k \neq i} \frac{|\langle k|\hat{H}^{(1)}|i\rangle|^2}{E_i^{(0)} - E_k^{(0)}}, \tag{6.44}$$

$$|\psi_i\rangle = |i\rangle + \sum_{k \neq i} \frac{\langle k|\hat{H}^{(1)}|i\rangle}{E_i^{(0)} - E_k^{(0)}} |k\rangle. \tag{6.45}$$

Note that we will typically correct only up to second order in the energy and first order in the wavefunction.

6.16 SUMMARY

We have very briefly reviewed some of the main concepts in quantum mechanics. This review is, however, by no means complete. For example, we have left out any discussion about the matrix representation of wavefunctions and operators. Likewise, we have not discussed issues related to tunneling. We have also only touched on the important topic of time-independent perturbation theory. While we have illustrated some of its results, we have not covered degenerate perturbation theory nor have we seen the WKB approximation. These are all important topics, since they allow us to deal with the day-to-day complexities of real life. As before, the reader is encouraged to consult the Further Reading for more details.

6.17 THOUGHT PROBLEMS

6.1 Wavefunctions

Assume that we are dealing with an electron in a hydrogen atom 1s orbital. Note that the hydrogen atom problem was not discussed in our brief review of quantum mechanics. However, it is commonly found in many texts and the reader can look up the problem if needed. In our case, the electron is characterized by the wavefunction

$$\psi_{1s} = \left(\frac{1}{\pi a_0^3}\right)^{1/2} e^{-r/a_0},$$

where a_0 is the Bohr radius.

(a) Plot the wavefunction.

(b) Is there anything potentially confusing about your sketch? Is it OK for the wavefunction to have a nonzero value at the origin? Presumably there should be a zero probability for the electron to be in the nucleus. Explain how this conceptual dilemma is resolved.

6.2 Orthonormality

The hydrogen 1s and 2s electrons have the wavefunctions

$$\psi_{1s}(r) = \frac{1}{\sqrt{\pi}}\left(\frac{1}{a_o}\right)^{3/2} e^{-r/a_0},$$

$$\psi_{2s}(r) = \frac{1}{\sqrt{32\pi}}\left(\frac{1}{a_o}\right)^{3/2}\left(2-\frac{r}{a_0}\right) e^{-r/2a_0}.$$

Show that both are normalized and that they are mutually orthogonal.

6.3 Probability density

The 2s orbital of an electron in hydrogen has a radial node. Locate its position. Similarly, locate the two nodes of the hydrogen 3s orbital. Their wavefunctions are

$$\psi_{2s}(r) = \frac{1}{\sqrt{32\pi}}\left(\frac{1}{a_0}\right)^{3/2}\left(2-\frac{r}{a_0}\right) e^{-r/2a_0},$$

$$\psi_{3s}(r) = \frac{1}{81\sqrt{3\pi}}\left(\frac{1}{a_0}\right)^{3/2}\left(27-\frac{18r}{a_0}+\frac{2r^2}{a_0^2}\right) e^{-r/3a_0}.$$

6.4 Probability density

Although we did not show it in our brief review of quantum mechanics, the energy levels of the electron in a hydrogen atom are given by

$$E_n = -\frac{me^4}{8\epsilon_0^2 h^2 n^2}.$$

Notice that these energies are independent of the orbital angular momentum quantum number l (see Further Reading). Yet the Aufbau principle implies that the energies of the s, p, d, f, etc. orbitals are different. Explain why this is using the information provided in **Figure 6.10**, which illustrates the radial probability densities of the hydrogen 3s, 3p, and 3d orbitals.

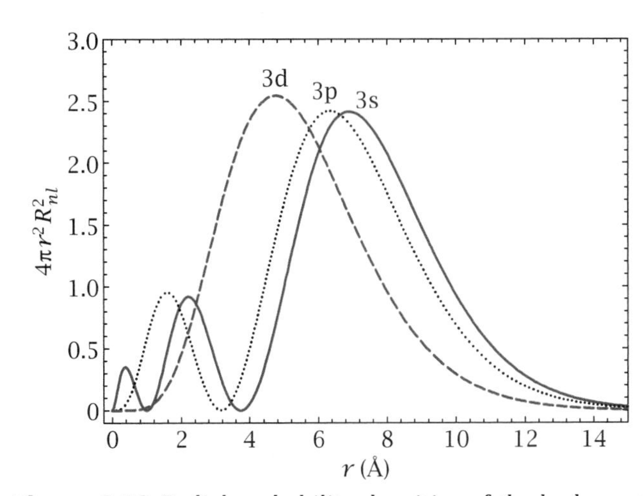

Figure 6.10 Radial probability densities of the hydrogen 3s, 3p, and 3d orbitals.

6.5 Probability density

Consider a particle localized to the region $0 \leq x \leq a$ by infinite barriers. This is the problem of a particle in a one-dimensional box, which we will see more of in Chapter 7.

Assume that the particle is defined by the wavefunction

$$\psi_3 = \sqrt{\frac{2}{a}} \sin \frac{3\pi x}{a}.$$

(a) Calculate the probability of finding the particle in the region $\frac{1}{4}a \le x \le \frac{3}{4}a$.

(b) Now indicate exactly where the probability density peaks inside the box (i.e., at what values of x will the particle most likely be found?). Conversely, where are the nodes?

6.6 Probability

The hydrogen 1s electron has the wavefunction

$$\psi_{1s}(r) = \frac{1}{\sqrt{\pi}} \left(\frac{1}{a_0}\right)^{3/2} e^{-r/a_0}.$$

Calculate the likelihood that it will be found within a distance a_0 of the nucleus. Comment on your result within the context of the Bohr model. Then calculate the probability of finding the electron within $2a_0$ of the nucleus. Recall that we are working in spherical coordinates.

6.7 Probability

Assume that a particle, confined to the region between $x = 0$ and $x = a$ by infinite potentials, has the wavefunction

$$|\psi\rangle = \frac{i}{\sqrt{2}}|\psi_1\rangle e^{-iE_1 t/\hbar} + \frac{1}{\sqrt{2}}|\psi_2\rangle e^{-iE_2 t/\hbar},$$

where E_1 is the energy of the first state and E_2 the energy of the second state and where

$$|\psi_1(x)\rangle = \sqrt{\frac{2}{a}} \sin \frac{\pi x}{a},$$

$$|\psi_2(x)\rangle = \sqrt{\frac{2}{a}} \sin \frac{2\pi x}{a}.$$

What is the probability of finding the particle between $x = \frac{1}{2}a$ and $x = a$?

6.8 Probability

For the following wavepacket that describes a particle,

$$\psi(x, t = 0) = \left(\frac{2}{\pi a^2}\right)^{1/4} \exp\left(-\frac{x^2}{a^2} + ik_0 x\right),$$

calculate the probability of finding it in the region $-\frac{1}{2}a \le x \le \frac{1}{2}a$. You can estimate the final result numerically.

6.9 Expectation values

For the hydrogen 1s and 2s electrons in Problem 6.2, find the average radial position of the electron, $\langle r\rangle$, in each case.

6.10 Expectation values

The normalized wavefunctions for a particle confined to move along the perimeter of a circle (the "particle-on-a-ring" problem) are

$$\psi(\theta) = \frac{1}{\sqrt{2\pi}} e^{-im\theta},$$

where $m = 0, \pm 1, \pm 2, \ldots$ and $0 \le \theta \le 2\pi$. Determine the expectation value $\langle\theta\rangle$.

6.11 Expectation values

The state of an electron confined to the region between $x = 0$ and $x = a$ is described by the wavefunction

$$\psi(x) = \frac{1}{2}\psi_1(x) + \frac{1}{2i}\psi_2(x) - \frac{1}{\sqrt{2}}\psi_4(x),$$

where

$$\psi_1(x) = \sqrt{\frac{2}{a}} \sin \frac{\pi x}{a},$$

$$\psi_2(x) = \sqrt{\frac{2}{a}} \sin \frac{2\pi x}{a},$$

$$\psi_4(x) = \sqrt{\frac{2}{a}} \sin \frac{4\pi x}{a}.$$

(a) When the energy of the electron is measured, what value or values are measured? Explain. Note that the Hamiltonian operator in this case has the form

$$\hat{H} = -\frac{\hbar^2}{2m}\frac{d^2}{dx^2}.$$

(b) What is the expectation value of the particle's energy? Explain.

6.12 Expectation values

Assume that the wavefunction of a particle confined to the region between $x = 0$ and $x = a$ by infinite potentials is described by the wavefunction

$$|\psi\rangle = \frac{i}{\sqrt{2}}|\psi_1(x)\rangle + \frac{1}{\sqrt{2}}|\psi_2(x)\rangle,$$

where

$$|\psi_1(x)\rangle = \sqrt{\frac{2}{a}} \sin \frac{\pi x}{a},$$

$$|\psi_2(x)\rangle = \sqrt{\frac{2}{a}} \sin \frac{2\pi x}{a}.$$

(a) If the particle's position is measured, what values will one measure and with what probability?

(b) What is the average position, $\langle x\rangle$, of the particle?

6.13 Wavefunctions and expectation values

Assume that an electron confined to the region between $x = 0$ and $x = a$ by infinite potentials has the wavefunction

$$|\psi\rangle = -\frac{2}{\sqrt{a}} \sin\left(\frac{3\pi x}{2a}\right) \cos\left(\frac{\pi x}{2a}\right).$$

The basis of the space is described completely by $|\psi_1\rangle$ and $|\psi_2\rangle$, where

$$|\psi_1(x)\rangle = \sqrt{\frac{2}{a}} \sin \frac{\pi x}{a},$$

$$|\psi_2(x)\rangle = \sqrt{\frac{2}{a}} \sin \frac{2\pi x}{a}.$$

(a) If the energy of the electron is measured, what value or values are found and with what probability? Note that the Hamiltonian operator here is

$$\hat{H} = -\frac{\hbar^2}{2m}\frac{d^2}{dx^2}.$$

(b) What is the expectation value of the energy, $\langle E\rangle$?

Hint:

$$\sin a \sin b = \tfrac{1}{2}\left[\cos(a-b) - \cos(a+b)\right]$$
$$\cos a \cos b = \tfrac{1}{2}\left[\cos(a-b) + \cos(a+b)\right].$$

6.14 Wavefunctions and expectation values

A given wavefunction in spherical coordinates is expressed by

$$|\psi\rangle = N \sin\theta \cos\phi,$$

where N is a normalization constant.

(a) Write the wavefunction in terms of spherical harmonics. (Look them up if needed. Note that they depend on two quantum numbers, l and m.)

(b) Find the required normalization constant N.

(c) What is the mean value of $\hat{L}^2$ for this state, where $\hat{L}^2|\psi\rangle = \hbar^2 l(l+1)|\psi\rangle$ (l is the state's angular momentum quantum number)?

(d) If $\hat{L}_z$ is measured, where $\hat{L}_z|\psi\rangle = m\hbar|\psi\rangle$ (m is the state's magnetic quantum number), what are the possible results of this measurement and what are their respective probabilities?

6.15 Commutators

Assume that a system is initially in a state $|\phi_3\rangle$, where ϕ_n are eigenstates of the Hamiltonian and $\hat{H}|\phi_n\rangle = n^2\epsilon_0|\phi_n\rangle$. $\hat{A}$ is an operator that behaves in the following manner:

$$\hat{A}|\phi_n\rangle = na_0|\phi_{n+1}\rangle.$$

(a) What value of the energy will be obtained if we measure $\hat{H}$?

(b) What value of the energy will be obtained if we apply $\hat{A}$ to the initial state first and then measure $\hat{H}$?

(c) Calculate $[\hat{A}, \hat{H}]|\phi_3\rangle$. Do $\hat{A}$ and $\hat{H}$ commute?

6.16 Expectation values and the Uncertainty Principle

Consider a particle confined to the region between $x = 0$ and $x = a$. The potential within the region is zero whereas for $x < 0$ and $x > a$ it is infinite.

(a) Calculate $\langle x\rangle$, $\langle p\rangle$, $\langle x^2\rangle$, and $\langle p^2\rangle$.

(b) Evaluate the uncertainty in x (i.e., Δx) and in p (i.e., Δp). Recall that for some operator $\hat{A}$, the associated uncertainty is $\Delta\hat{A} = \sqrt{\langle\hat{A}^2\rangle - \langle\hat{A}\rangle^2}$.

(c) Use the result from (b) to estimate the so-called "zero-point" energy of the particle (look up this concept if you are unfamiliar with it). Consider only the momentum uncertainty.

6.17 Uncertainty Principle

Calculate $\Delta p_r \,\Delta r$ for the hydrogen 1s orbital, where

$$\psi_{1s}(r) = \frac{1}{\sqrt{\pi}}\left(\frac{1}{a_0}\right)^{3/2} e^{-r/a_0}$$

and show that this satisfies the Heisenberg Uncertainty Principle ($\Delta p_r \,\Delta r \geq \frac{1}{2}\hbar$). Hint:

$$\hat{p}_r\psi = -i\hbar\left(\frac{1}{r}\right)\frac{d}{dr}(r\psi) \quad \text{and} \quad \int_0^\infty x^n e^{-x}\,dx = n!.$$

6.18 Postulates

A particle with mass m is in the state

$$\psi(x,t) = A\exp\left[-a\left(\frac{mx^2}{\hbar} + it\right)\right],$$

where A and a are constants. For what value of the potential $V(x)$ does this wavefunction satisfy the Schrödinger equation?

6.19 Postulates and probability

A particle has a wavefunction

$$\psi(x,t) = Ax^2 e^{-i\omega t},$$

where A is a constant. Find the potential $V(x)$ that it experiences.

6.20 Nondegenerate perturbation theory

Calculate the ground-state energy of a particle of mass m moving in an infinite one-dimensional potential well of length a with walls at $x = 0$ and $x = a$. The bottom of the potential is modified by the perturbation

$$V(x) = \lambda A \sin\frac{\pi x}{a},$$

with $\lambda \ll 1$. Use first-order nondegenerate perturbation theory. For more information, see Miller (1979).

6.21 Nondegenerate perturbation theory

A particle of mass m is in the asymmetric one-dimensional box depicted in **Figure 6.11** (i.e., the particle is confined to the region between $0 \leq x \leq a$ but with a perturbation tacked on to it). Use first-order nondegenerate perturbation theory to calculate the particle's energies.

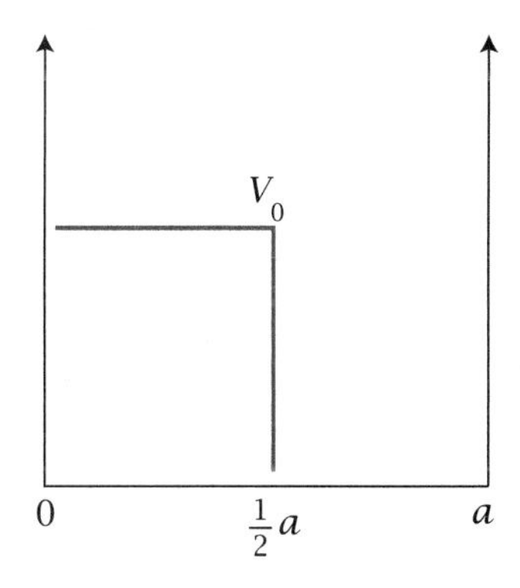

Figure 6.11 Potential for a particle in an asymmetric one-dimensional box.

6.22 Nondegenerate perturbation theory

An electron in a hydrogen atom is in its ground state. Its associated wavefunction and energy are

$$\psi_{1s}(r) = \frac{1}{\sqrt{\pi}}\left(\frac{1}{a_0}\right)^{3/2} e^{-r/a_0},$$
$$E_{1s} = -\frac{me^4}{8\epsilon_0^2 h^2}.$$

The atom is now exposed to a constant uniform electric field that points along the z direction of a Cartesian coordinate system. What is the first-order change to the electron's ground-state energy? The interaction Hamiltonian is given by

$$\hat{H}^{(1)} = eEr\cos\theta,$$

where e and E are constants.

6.18 REFERENCES

Miller GR (1979) The particle in the one-dimensional champagne bottle. *J. Chem. Ed.* **56**, 709.

Moore WJ (1989) *Schrödinger: Life and Thought*. Cambridge University Press, Cambridge, UK.

6.19 FURTHER READING

Introductory Quantum Mechanics Texts

Green NJB (1997) *Quantum Mechanics 1: Foundations*. Oxford University Press, Oxford, UK.

Green NJB (1998) *Quantum Mechanics 2: The Toolkit*. Oxford University Press, Oxford, UK.

Griffiths DJ (2004) *Introduction to Quantum Mechanics*, 2nd edn. Prentice Hall, Upper Saddle River, NJ.

Liboff RL (2003) *Introductory Quantum Mechanics*. Addison-Wesley, San Francisco, CA.

McMahon D (2005) *Quantum Mechanics Demystified*. McGraw-Hill, New York.

McQuarrie DA (2008) *Quantum Chemistry*, 2nd edn. University Science Books, Mill Valley, CA.

More Advanced Quantum Mechanics Texts

Cohen-Tannoudji C, Diu B, Laloe F (1977) *Quantum Mechanics* (2 volumes). Wiley, New York.

Zettili N (2009) *Quantum Mechanics, Concepts and Applications*, 2nd edn. Wiley, Chichester, UK.

Model Quantum Mechanics Problems

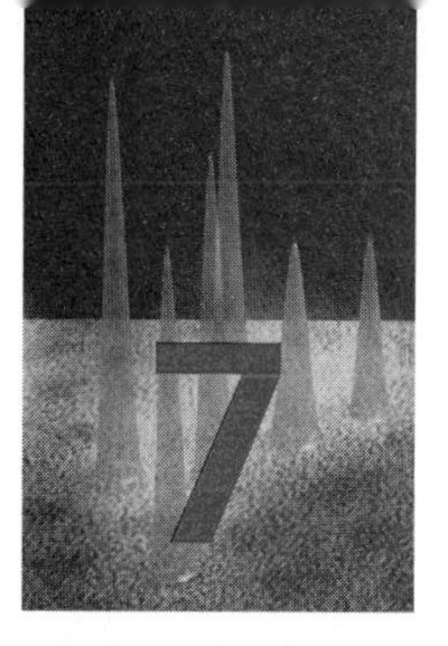

7.1 INTRODUCTION

Much of introductory quantum mechanics is spent solving so-called model systems that illustrate the confinement of a particle (e.g., an electron) in one, two, or three dimensions. In what follows, we work through these problems with suggestive names such as

- The particle in a box
- The particle in a cylinder
- The particle in a sphere

In all cases, we solve the Schrödinger equation to obtain the wavefunctions and energies of a confined particle. What distinguishes one problem from another are their boundary conditions. These constraints dictate where the carrier can or cannot be and reflect the underlying physical geometry as well as restrictions of the system. This is why one often refers to a box, a cylinder, and a sphere when modeling quantum wells, quantum wires (nanowires), and quantum dots, since these are their natural geometries, as first alluded to in Chapter 2 and seen in Chapter 4.

7.2 STANDARD MODEL PROBLEMS

7.2.1 Free Particle

Previously, we talked about a particle that had no constraints on it. This was the free particle or free electron problem. In one dimension, the free particle is described by a plane wave, $\psi = e^{ikx}$. Note that multiplying ψ by $e^{i\omega t}$ yields a traveling wave, as first discussed in Chapter 5.

Next, recall that the time-independent Schrödinger equation is

$$-\frac{\hbar^2}{2m}\nabla^2\psi = E\psi,$$

where ∇^2 is the Laplacian. In one dimension, if we consider only the x coordinate,

$$\nabla^2 = \frac{d^2}{dx^2}. \tag{7.1}$$

Therefore

$$-\frac{\hbar^2}{2m}\frac{d^2\psi}{dx^2} = E\psi.$$

To solve this, we rearrange the equation as follows:

$$\frac{d^2\psi}{dx^2} = -\frac{2mE}{\hbar^2}\psi,$$

and let $k^2 = 2mE/\hbar^2$, where k is called the wavevector. As a further aside, k is related to the particle's momentum via $p = \hbar k$. We can see this by referring to the de Broglie wave–particle duality, $\lambda = h/p$, first seen in Chapter 3. By rearranging the expression to $p = h/\lambda$ and multiplying both the numerator and denominator by 2π, one finds $p = \hbar/(\lambda/2\pi)$. This is just $p = \hbar/(1/k)$ which leads to our desired result, $p = \hbar k$.

On substituting k^2 into the above equation, we obtain the following second-order differential equation:

$$\frac{d^2\psi}{dx^2} + k^2\psi = 0. \tag{7.2}$$

It has general solutions, consisting of linear combinations of plane waves:

$$\psi = Ae^{ikx} + Be^{-ikx},$$

where A and B are constant coefficients to be determined.

Note that as the plane wave propagates, it repeats. Thus, the only constraint on it is periodicity (i.e., it has periodic boundary conditions, also called Born–von Kármán boundary conditions). The following therefore applies:

$$\psi(x) = \psi(x + \lambda),$$

with λ the wavelength or period of the wave.

Using our previous solution for ψ, we then have the following relationship for the first term:

$$\begin{aligned} Ae^{ikx} &= Ae^{ik(x+\lambda)} \\ &= Ae^{ikx}e^{ik\lambda}. \end{aligned}$$

A similar equality applies to the second term. Thus, for either to be true, $e^{ik\lambda} = 1$, and since the Euler relation says that $e^{ik\lambda} = \cos k\lambda + i \sin k\lambda$, we have

$$k\lambda = 2n\pi,$$

where $n = 0, \pm 1, \pm 2, \pm 3, \ldots$. We thus find

$$k = \frac{2n\pi}{\lambda} \tag{7.3}$$

as constraints on k that arise due to the periodic boundary conditions of the system.

Wavefunctions

Since there are no constraints on A or B, we are free to choose a solution. For convenience, let $B = 0$. This gives

$$\psi(x) = Ae^{ikx} \tag{7.4}$$

as the free particle wavefunction.

Energies

The accompanying energies are found using the definition of the wavevector. Namely, since $k = \sqrt{2mE/\hbar^2}$, we find that

$$E = \frac{\hbar^2 k^2}{2m}. \quad (7.5)$$

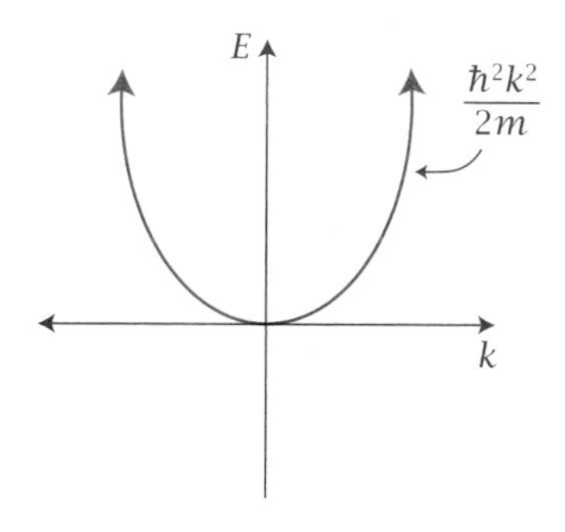

Figure 7.1 Parabolic dependence of the free particle energy with k.

This reveals a parabolic energy dependence on k, as illustrated in **Figure 7.1**. Note that Equation 7.5 could have also been derived by noting the classical relationship between momentum and energy, $E = p^2/2m$, given that $p = \hbar k$:

$$E = \frac{p^2}{2m} = \frac{\hbar^2 k^2}{2m}.$$

7.2.2 Particle in a One-Dimensional Box

Let us now discuss the first of several standard problems in introductory quantum mechanics. This is the problem of a particle in a one-dimensional box and is a way to introduce the effects of spatial confinement on carrier wavefunctions and energies. In the problem, we again solve the Schrödinger equation

$$\left[-\frac{\hbar^2}{2m}\frac{d^2}{dx^2} + V(x)\right]\psi = E\psi,$$

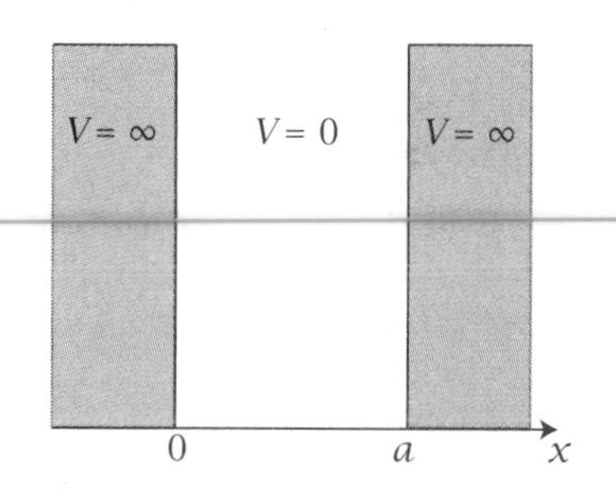

Figure 7.2 Potential for a particle in a one-dimensional box.

this time with the boundary conditions shown in **Figure 7.2**. Namely, the potential inside the box is zero and is infinite outside of it. As a consequence, the particle is always physically confined to the interval $0 < x < a$. We therefore have for the potential

$$V(x) = \begin{cases} 0 & \text{when } 0 < x < a, \\ \infty & \text{elsewhere,} \end{cases}$$

with boundary conditions

$$\psi(0) = 0,$$
$$\psi(a) = 0.$$

We now solve the Schrödinger equation in the region of interest. Since $V = 0$, we have

$$\frac{d^2\psi}{dx^2} = -\frac{2mE}{\hbar^2}\psi,$$

whereupon letting $k^2 = 2mE/\hbar^2$ yields the following second-order differential equation

$$\frac{d^2\psi}{dx^2} + k^2\psi = 0.$$

General solutions to this are linear combinations of plane waves

$$\psi(x) = Ae^{ikx} + Be^{-ikx},$$

where A and B are constant coefficients to be determined.

Let us now apply the problem's boundary conditions to specify A and B. Namely, the first says that the wavefunction must go to zero at the origin (i.e., $\psi(0) = 0$). We therefore have

$$\psi(0) = A + B = 0,$$

or $B = -A$. This gives the complex wavefunction

$$\psi(x) = A(e^{ikx} - e^{-ikx}),$$

which we can rewrite using the Euler relation:

$$\begin{aligned} \psi(x) &= 2iA\left(\frac{e^{ikx} - e^{-ikx}}{2i}\right) \\ &= 2iA \sin kx. \end{aligned}$$

Next, we apply the second boundary condition (i.e., $\psi(a) = 0$) to get

$$\psi(a) = 2iA \sin ka = 0.$$

At this point, to obtain a solution, we have either $A = 0$ (the trivial solution) or $ka = n\pi$, with n an integer. We pursue the second solution, since $A = 0$ just tells us that the wavefunction is zero. As a consequence, k values are restricted to

$$k = \frac{n\pi}{a}. \tag{7.6}$$

Notice the factor-of-two difference between these k values and those in the free particle case (Equation 7.3).

Putting everything together, we now have

$$\psi(x) = 2iA \sin \frac{n\pi x}{a},$$

or, more generally,

$$\psi(x) = N \sin \frac{n\pi x}{a},$$

where N is a normalization constant to be determined. This is the general form of the wavefunction for a particle in a one-dimensional box.

Normalization

As discussed in Chapter 6, the probabilistic interpretation of wavefunctions means that they must be normalized (i.e., the particle must be located somewhere within the box with unity probability):

$$\begin{aligned} 1 &= \int_0^a \psi(x)^* \psi(x)\, dx \\ &= \int_0^a N^2 \sin^2 \frac{n\pi x}{a}\, dx. \end{aligned}$$

We therefore solve the above integral to find N by invoking the trigonometric identity $\sin^2 m = \frac{1}{2}(1 - \cos 2m)$. This gives

$$\begin{aligned} 1 &= \frac{N^2}{2}\left(x|_0^a - \int_0^a \cos \frac{2n\pi x}{a}\, dx\right) \\ &= \frac{N^2}{2} a. \end{aligned}$$

Solving for N then yields

$$N = \sqrt{\frac{2}{a}}.$$

Wavefunctions

Normalized wavefunctions for a particle in a one-dimensional box are therefore

$$\psi(x) = \sqrt{\frac{2}{a}} \sin \frac{n\pi x}{a}, \tag{7.7}$$

which we referred to earlier in Chapter 6. **Figure 7.3** shows the first few of these wavefunctions. Also illustrated are the accompanying probability densities.

Energies

Associated energies arise from the equivalence just found:

$$k = \frac{n\pi}{a},$$

$$\sqrt{\frac{2mE}{\hbar^2}} = \frac{n\pi}{a},$$

yielding

$$E = \frac{n^2 h^2}{8ma^2}, \tag{7.8}$$

where $n = 1, 2, 3, \ldots$. Note that the quantum number n is never zero, as this would violate the Heisenberg Uncertainty Principle (Chapter 6). Finally, we can extend these results to two and three dimensions by simply varying the form of the Laplacian and assuming what are called separable solutions. This will be seen next.

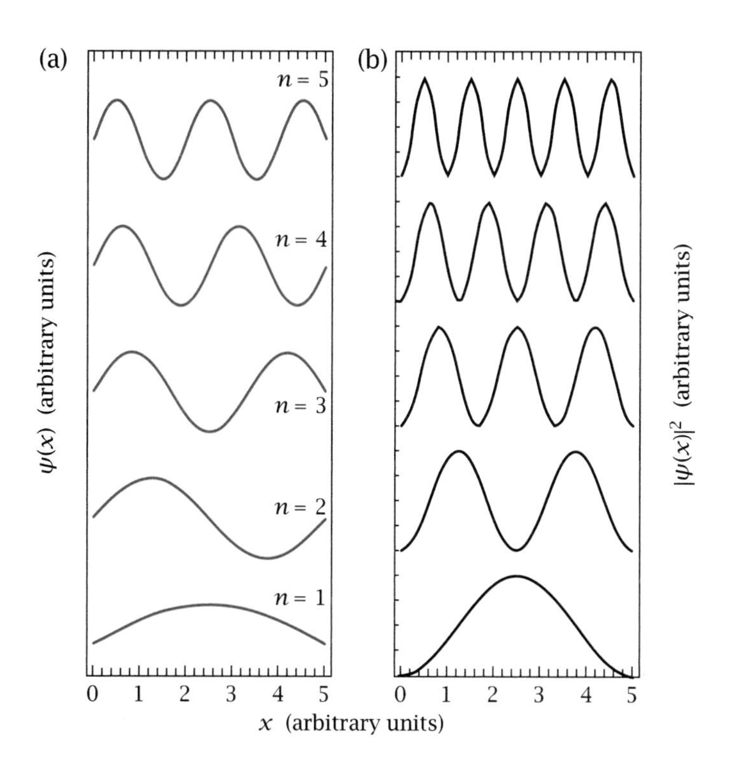

Figure 7.3 The first five wavefunctions for a particle in a one-dimensional box (a) and their associated probability densities (b). Traces are offset for clarity.

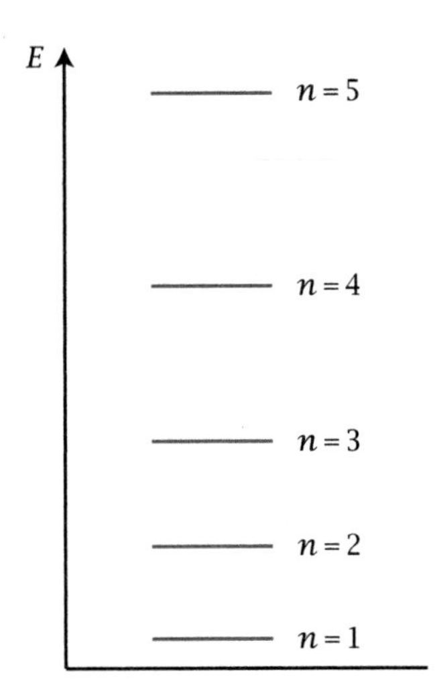

Figure 7.4 Energy levels for a particle in a one-dimensional box.

Figure 7.4 shows the first five energy levels of the particle in a box along with their corresponding quantum numbers. Note the absence of degeneracies (i.e., multiple levels having the same energy) in the one-dimensional box energy levels. It is also apparent that the states are closely spaced together at lower energies but become more distant as E increases. The reader may verify this by evaluating $E(n+1) - E(n)$.

7.2.3 Particle in a Two-Dimensional Box: Degeneracies

At this point, let us briefly illustrate how to handle problems in higher dimensions, since the systems we will ultimately deal with exhibit *three* physical dimensions. For simplicity, we shall solve the problem of a particle confined to a two-dimensional box in the (x, y) plane of a Cartesian coordinate system.

What results are wavefunctions and energies that have x and y dependences. Furthermore, a natural consequence of working in higher dimensions is the increased likelihood of finding degeneracies. Namely, there will be an increased propensity for finding states that have the same energy. This stems from built-in symmetries of the system.

The Schrödinger equation that we must solve is

$$\left[-\frac{\hbar^2}{2m}\nabla^2 + V(x, y)\right]\psi(x, y) = E\psi(x, y),$$

where, in two dimensions, the Laplacian takes the form

$$\nabla^2 = \frac{\partial^2}{\partial x^2} + \frac{\partial^2}{\partial y^2}. \tag{7.9}$$

As a consequence, we have

$$\left[-\frac{\hbar^2}{2m}\left(\frac{\partial^2}{\partial x^2} + \frac{\partial^2}{\partial y^2}\right) + V(x, y)\right]\psi(x, y) = E\psi(x, y).$$

We assume the potentials and boundary conditions illustrated in **Figure 7.5**. Namely, we have a symmetric box with a dimension a along each axis and with a potential that is zero inside the box but infinite outside. The particle is therefore confined to the interior of the box. Furthermore, its wavefunctions go to zero at the edges. We therefore

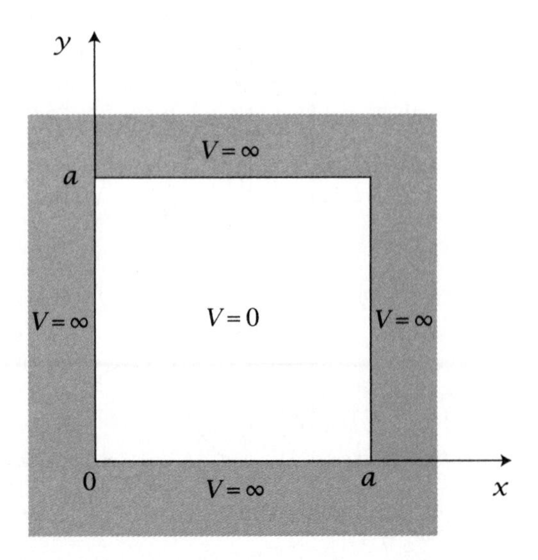

Figure 7.5 Potential for a particle in a two-dimensional box.

have for the potential

$$V(x,y)\begin{cases}0 & \text{when } 0<x<a \text{ and } 0<y<a,\\ \infty & \text{elsewhere,}\end{cases}$$

with accompanying boundary conditions

$$\begin{aligned}\psi(0,y)&=0 \quad \text{for } 0\le y\le a,\\ \psi(a,y)&=0 \quad \text{for } 0\le y\le a,\\ \psi(x,0)&=0 \quad \text{for } 0\le x\le a,\\ \psi(x,a)&=0 \quad \text{for } 0\le x\le a.\end{aligned}$$

We now solve the Schrödinger equation in the region of interest, where $V(x,y)=0$:

$$-\frac{\hbar^2}{2m}\left(\frac{\partial^2}{\partial x^2}+\frac{\partial^2}{\partial y^2}\right)\psi(x,y)=E\psi(x,y).$$

To simplify things, let us assume a separable solution of the form

$$\psi(x,y)=\psi_x(x)\psi_y(y)\equiv\psi_x\psi_y,$$

which is basically a product of two independent wavefunctions, each with its own x or y dependence. There is good reason for choosing such a solution, because what ultimately results are two separate one-dimensional particle-in-a-box problems. Since we have already solved this model system, the two-dimensional problem therefore simplifies greatly.

Introducing $\psi=\psi_x\psi_y$ into the Schrödinger equation then gives

$$\psi_y\left(-\frac{\hbar^2}{2m}\frac{d^2\psi_x}{dx^2}\right)+\psi_x\left(-\frac{\hbar^2}{2m}\frac{d^2\psi_y}{dy^2}\right)=E\psi_x\psi_y,$$

whereupon dividing both sides by $\psi_x\psi_y$ yields

$$\frac{1}{\psi_x}\left(-\frac{\hbar^2}{2m}\right)\frac{d^2\psi_x}{dx^2}+\frac{1}{\psi_y}\left(-\frac{\hbar^2}{2m}\right)\frac{d^2\psi_y}{dy^2}=E.$$

Notice that this expression remains valid for all values of x or y. As a consequence, each term on the left-hand side must equal a constant; these constants turn out to be the energy contributions from x and y. We thus separate the total energy term into two contributions, one from x and one from y:

$$E=E_x+E_y.$$

This in turn means that we can separate the full equation into two smaller ones:

$$\frac{1}{\psi_x}\left(-\frac{\hbar^2}{2m}\right)\frac{d^2\psi_x}{dx^2}=E_x,$$

$$\frac{1}{\psi_y}\left(-\frac{\hbar^2}{2m}\right)\frac{d^2\psi_y}{dy^2}=E_y,$$

which on rearranging become

$$-\frac{\hbar^2}{2m}\frac{d^2\psi_x}{dx^2}=E_x\psi_x$$

$$-\frac{\hbar^2}{2m}\frac{d^2\psi_y}{dy^2}=E_y\psi_y.$$

What results are two one-dimensional particle-in-a-box problems. This illustrates how assuming a separable solution greatly simplifies the original two-dimensional problem. We now solve both one-dimensional problems, and given that we have already demonstrated this earlier, we just list the results here.

The obtained (independently) normalized wavefunctions are

$$\psi_x = \sqrt{\frac{2}{a}} \sin \frac{n_x \pi x}{a},$$

$$\psi_y = \sqrt{\frac{2}{a}} \sin \frac{n_y \pi y}{a},$$

with corresponding energies

$$E_x = \frac{n_x^2 h^2}{8ma^2},$$

$$E_y = \frac{n_y^2 h^2}{8ma^2}.$$

In both cases, n_x and n_y are integers that reflect the quantization of the particle.

Wavefunctions

When everything is put together, the total wavefunction for the particle in a two-dimensional square box is

$$\psi(x, y) = \frac{2}{a} \sin \frac{n_x \pi x}{a} \sin \frac{n_y \pi y}{a}, \tag{7.10}$$

where again n_x and n_y are independent integers. **Figure 7.6** shows the first four wavefunctions of a particle in a two-dimensional box. The associated probability densities are shown in **Figure 7.7**. As an exercise, the reader may plot the $n_x = 5, n_y = 5$ wavefunction.

Energies

The corresponding energies are

$$E = \frac{h^2}{8ma^2}(n_x^2 + n_y^2), \tag{7.11}$$

with $n_x = 1, 2, 3, \ldots$ and $n_y = 1, 2, 3, \ldots$.

Finally, **Figure 7.8** illustrates the energies of the particle in a two-dimensional box. One sees that because the box is symmetric, there exist a number of degeneracies, where different states end up having the same energy. For example, it is evident that the state characterized by the quantum numbers $n_x = 1$, $n_y = 2$ possesses the same energy as the state with quantum numbers $n_x = 2$, $n_y = 1$. Note that such degeneracies can be "lifted" by making the box asymmetric. For example, if we were to elongate one side of the box, say the y dimension, to a length b, the particle's wavefunctions and energies would become

$$\psi(x, y) = \frac{2}{\sqrt{ab}} \sin \frac{n_x \pi x}{a} \sin \frac{n_y \pi y}{b}$$

and

$$E = \frac{h^2}{8m}\left(\frac{n_x^2}{a^2} + \frac{n_y^2}{b^2}\right).$$

The reader may verify this as an exercise.

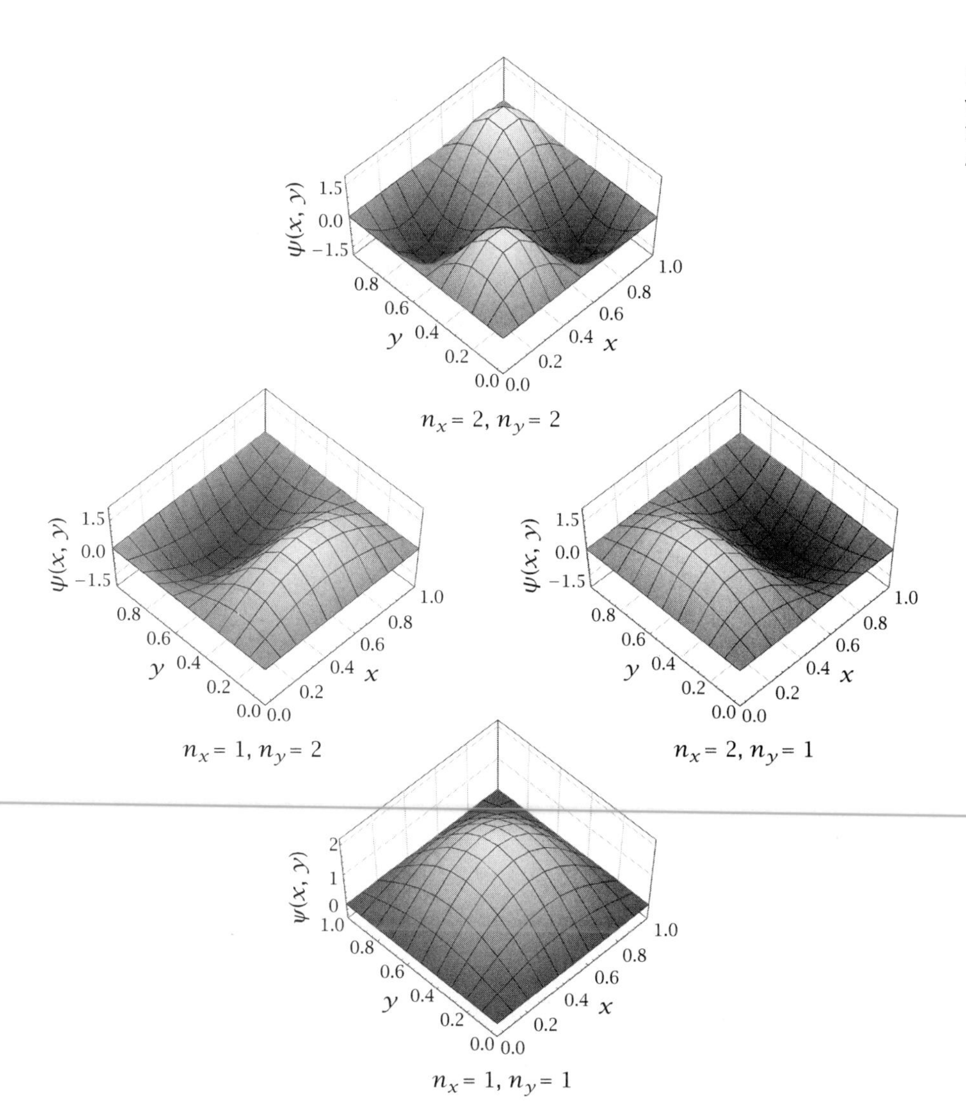

Figure 7.6 First four wavefunctions of a particle in a two-dimensional box. (Units of x, y, and $\psi(x, y)$ are arbitrary.)

7.2.4 Particle in a Three-Dimensional Box: Degeneracies

The problem can now be extended to three dimensions. Namely, we can consider a particle trapped inside a three-dimensional box having equal lengths on all sides (i.e., a cube). The relevant potentials and boundary conditions are shown in **Figure 7.9**, where we have a box with a length a on all three sides and a potential of zero (infinity) inside (outside). The reader may then use the same boundary conditions and overall procedure outlined above for the one- and two-dimensional cases to show that the resulting particle wavefunctions and energies are

$$\psi(x, y, z) = \left(\frac{2}{a}\right)^{3/2} \sin\frac{n_x \pi x}{a} \sin\frac{n_y \pi y}{a} \sin\frac{n_z \pi z}{a} \tag{7.12}$$

and

$$E = \frac{h^2}{8ma^2}(n_x^2 + n_y^2 + n_z^2). \tag{7.13}$$

In both expressions, a is the width of the box, while $n_x = 1, 2, 3, \ldots$, $n_y = 1, 2, 3, \ldots$, and $n_z = 1, 2, 3, \ldots$ are quantum numbers that reflect the confinement of the particle. **Figure 7.10** illustrates the first few energies of

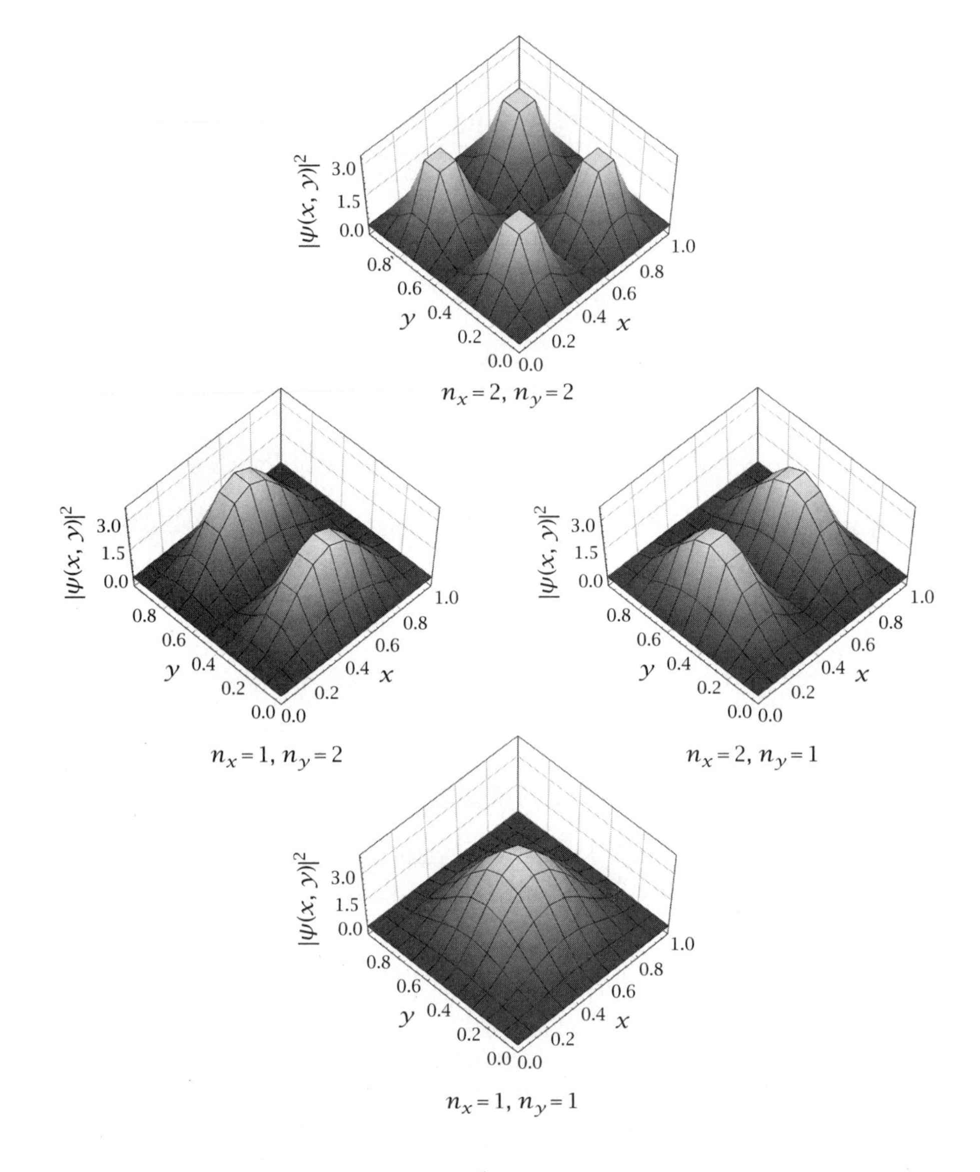

Figure 7.7 Probability densities associated with the first four wavefunctions of a particle in a two-dimensional box. (Units of x, y, and $|\psi(x,y)|^2$ are arbitrary.)

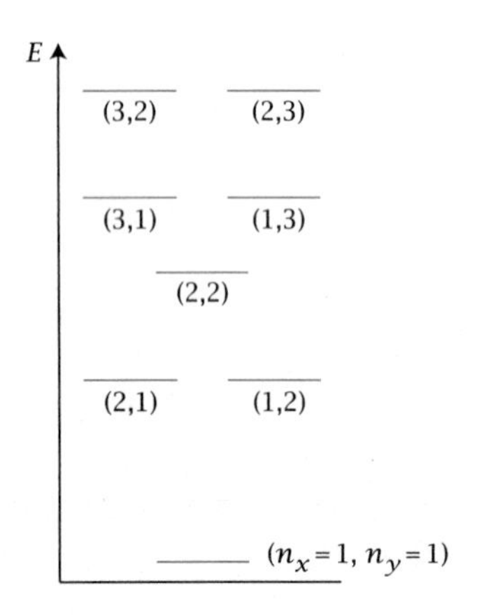

Figure 7.8 Energy levels for a particle in a two-dimensional box.

the particle in a three-dimensional box. Again, because of the system's high degree of symmetry, there are significant degeneracies, with multiple states having the same energy.

7.3 MODEL PROBLEMS FOR WELLS, WIRES, AND DOTS

Let us now discuss three quantum mechanical models that describe a carrier confined to a quantum well, a quantum wire, or a quantum dot. As mentioned earlier, all entail solving the Schrödinger equation in order to obtain valid particle wavefunctions and associated energies. The only differences between problems are their boundary conditions and the form of the Laplacian, since we will eventually work in cylindrical and spherical coordinates.

7.3.1 Particle in a Three-Dimensional Slab: An Eventual Envelope Function for the Quantum Well

We will first solve the quantum mechanical problem of a particle confined to a three-dimensional slab with a finite volume V_{tot}. The obtained

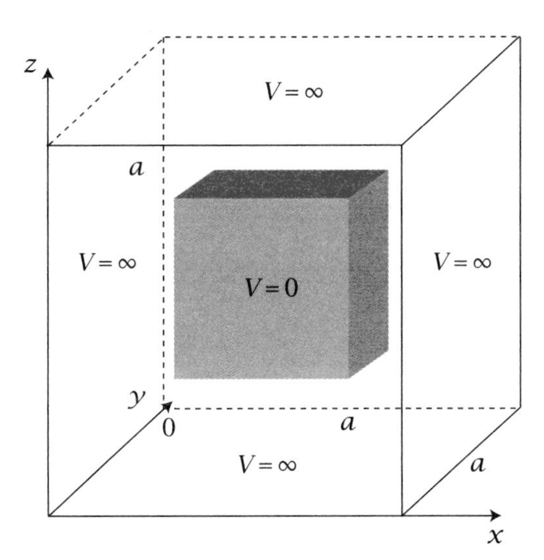

Figure 7.9 Potential for a particle in a three-dimensional box.

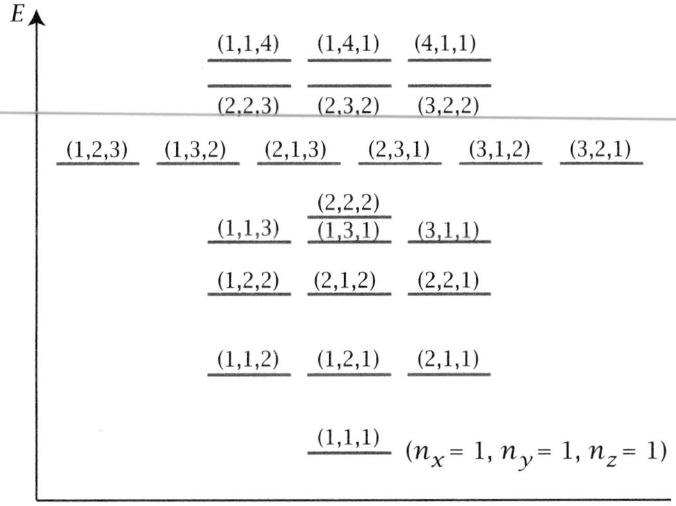

Figure 7.10 Energy levels for a particle in a three-dimensional box.

wavefunction will be part of what is eventually the full wavefunction of an electron or a hole in a semiconductor quantum well. Specifically, it provides the "envelope" part of the quantum well wavefunction for either the electron or the hole. For now, however, we shall just take this to be the whole story.

Figure 7.11 shows the potential energy and boundary conditions of the problem. In the quantum well, the particle experiences periodic boundary conditions within the (x, y) plane. This assumes that the well is fairly wide, with lengths L_x and L_y, respectively, such that no confinement exists along these directions. By contrast, along the z direction, boundary conditions, identical to those of the particle in a one-dimensional box, exist. Namely, between $0 < z < a$, the potential is zero. Elsewhere the potential is infinite. As a consequence, the particle is physically restricted to the region between 0 and a. We thus have for the potential

$$V(x, y, z) = \begin{cases} 0 & \text{for } 0 < z < a, \;\; 0 < x < L_x, \;\text{ and }\; 0 < y < L_y, \\ \infty & \text{elsewhere,} \end{cases}$$

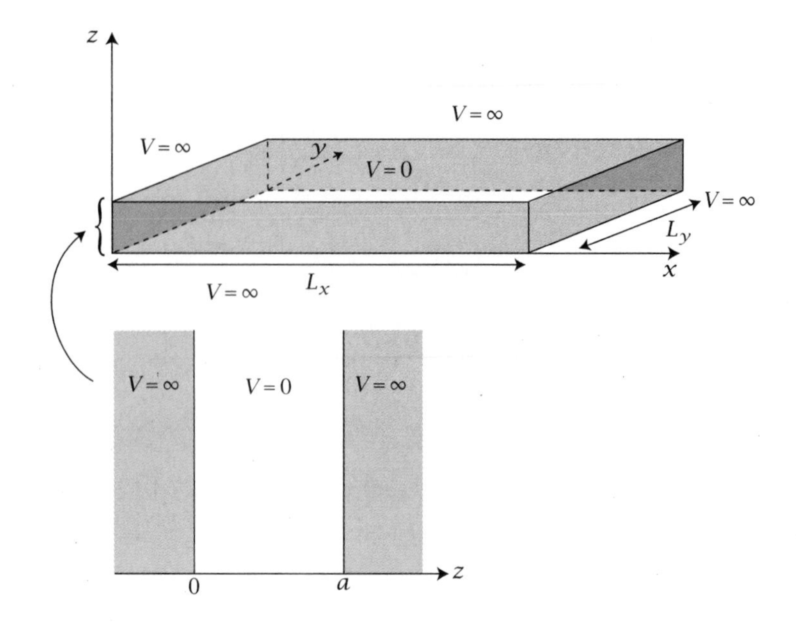

Figure 7.11 Illustration of a quantum well and its potential along the z direction of a Cartesian coordinate system.

with accompanying boundary conditions

$$\begin{aligned}
\psi(0, y, z) &= 0 \quad (0 \le y \le L_y) \quad (0 \le z \le a), \\
\psi(L_x, y, z) &= 0 \quad (0 \le y \le L_y) \quad (0 \le z \le a), \\
\psi(x, 0, z) &= 0 \quad (0 \le x \le L_x) \quad (0 \le z \le a), \\
\psi(x, L_y, z) &= 0 \quad (0 \le x \le L_x) \quad (0 \le z \le a), \\
\psi(x, y, 0) &= 0 \quad (0 \le x \le L_x) \quad (0 \le y \le L_y), \\
\psi(x, y, a) &= 0 \quad (0 \le x \le L_x) \quad (0 \le y \le L_y).
\end{aligned}$$

We now solve the Schrödinger equation in the region of interest (i.e., inside the box), where the potential is zero:

$$-\frac{\hbar^2}{2m_{\text{eff}}}\nabla^2\psi = E\psi.$$

Note that instead of m we use m_{eff} to describe an effective mass for carriers when in a semiconductor. In later chapters, when distinguishing electrons and holes, we will use m_e and m_h to represent their respective effective masses.

In three dimensions, the Laplacian in Cartesian coordinates takes the form

$$\nabla^2 = \frac{\partial^2}{\partial x^2} + \frac{\partial^2}{\partial y^2} + \frac{\partial^2}{\partial z^2}. \tag{7.14}$$

The Schrödinger equation therefore becomes

$$-\frac{\hbar^2}{2m_{\text{eff}}}\left(\frac{\partial^2}{\partial x^2} + \frac{\partial^2}{\partial y^2} + \frac{\partial^2}{\partial z^2}\right)\psi = E\psi.$$

To solve it, we assume a separable solution of the form

$$\psi(x, y, z) = \psi_x(x)\psi_y(y)\psi_z(z) \equiv \psi_x\psi_y\psi_z,$$

which is the product of three independent wavefunctions, each with its own x, y, or z dependence. This wavefunction is then inserted into the

Schrödinger equation to obtain

$$\psi_y\psi_z\left(-\frac{\hbar^2}{2m_{\text{eff}}}\right)\frac{d^2\psi_x}{dx^2}+\psi_x\psi_z\left(-\frac{\hbar^2}{2m_{\text{eff}}}\right)\frac{d^2\psi_y}{dy^2}+\psi_x\psi_y\left(-\frac{\hbar^2}{2m_{\text{eff}}}\right)\frac{d^2\psi_z}{dz^2} = E\psi_x\psi_y\psi_z.$$

Dividing both sides by $\psi_x\psi_y\psi_z$ then gives

$$\frac{1}{\psi_x}\left(-\frac{\hbar^2}{2m_{\text{eff}}}\right)\frac{d^2\psi_x}{dx^2}+\frac{1}{\psi_y}\left(-\frac{\hbar^2}{2m_{\text{eff}}}\right)\frac{d^2\psi_y}{dy^2}+\frac{1}{\psi_z}\left(-\frac{\hbar^2}{2m_{\text{eff}}}\right)\frac{d^2\psi_z}{dz^2}=E,$$

and since this equation is valid for all x, y, and z, each term on the left-hand side must equal a constant. They turn out to be the energy contributions from each dimension, E_i. The total energy therefore has three terms:

$$E = E_x + E_y + E_z.$$

As a consequence, we obtain three separate equations:

$$\frac{1}{\psi_x}\left(-\frac{\hbar^2}{2m_{\text{eff}}}\right)\frac{d^2\psi_x}{dx^2}=E_x,$$
$$\frac{1}{\psi_y}\left(-\frac{\hbar^2}{2m_{\text{eff}}}\right)\frac{d^2\psi_y}{dy^2}=E_y,$$
$$\frac{1}{\psi_z}\left(-\frac{\hbar^2}{2m_{\text{eff}}}\right)\frac{d^2\psi_z}{dz^2}=E_z,$$

which can alternatively be written as

$$-\frac{\hbar^2}{2m_{\text{eff}}}\frac{d^2\psi_x}{dx^2}=E_x\psi_x,$$
$$-\frac{\hbar^2}{2m_{\text{eff}}}\frac{d^2\psi_y}{dy^2}=E_y\psi_y,$$
$$-\frac{\hbar^2}{2m_{\text{eff}}}\frac{d^2\psi_z}{dz^2}=E_z\psi_z.$$

Mimicking what we did before for the particle in a two-dimensional box, each equation is now solved separately to find ψ_i and E_i.

x contribution

The equation we wish to solve here is

$$-\frac{\hbar^2}{2m_{\text{eff}}}\frac{d^2\psi_x}{dx^2}=E_x\psi_x,$$
$$\frac{d^2\psi_x}{dx^2}=-\frac{2m_{\text{eff}}E_x}{\hbar^2}\psi_x.$$

We let $k_x^2 = 2m_{\text{eff}}E_x/\hbar^2$ to obtain the following equation:

$$\frac{d^2\psi_x}{dx^2}+k_x^2\psi_x=0.$$

General solutions to this second-order differential equation are linear combinations of plane waves:

$$\psi_x = A_x e^{ik_x x} + B_x e^{-ik_x x},$$

with A_x and B_x constant coefficients to be determined.

Now, since there are no constraints on the wavefunction along x, other than periodicity, we choose the particular solution

$$\psi_x(x) = A_x e^{ik_x x},$$

where $k_x = 2n_x\pi/L_x$, $n_x = 0, \pm 1, \pm 2, \pm 3, \ldots$, and L_x is the length of the quantum well along the x direction. Note that we could independently normalize $\psi_x(x)$ by solving for

$$1 = \int_0^{L_x} |\psi_x|^2\, dx$$

to find $A_x = 1/\sqrt{L_x}$. Note further that although the integral uses the limits 0 and L_x, we could just as well have used the symmetric limits $-\frac{1}{2}L_x$ and $\frac{1}{2}L_x$. Let us normalize the above wavefunction later, together with the y contribution below.

y contribution

Let us now carry out the same analysis to solve

$$-\frac{\hbar^2}{2m_{\text{eff}}}\frac{d^2\psi_y}{dy^2} = E_y\psi_y.$$

In the same manner as above for the x contribution, we obtain

$$\psi_y(y) = A_y e^{ik_y y},$$

with $k_y = 2n_y\pi/L_y$, $n_y = 0, \pm 1, \pm 2, \pm 3, \ldots$, and L_y the length of the quantum well along the y direction. Again, the y contribution could be normalized separately by solving for

$$1 = \int_0^{L_y} |\psi_y|^2\, dy,$$

to obtain $A_y = 1/\sqrt{L_y}$, but we shall do this later along with the normalization for the x contribution above.

z contribution

To obtain the z contribution, we solve

$$-\frac{\hbar^2}{2m_{\text{eff}}}\frac{d^2\psi_z}{dz^2} = E_z\psi_z,$$

which is our standard problem of a particle in a one-dimensional box that describes carrier confinement along the z direction. As always, the equation is first rearranged to

$$\frac{d^2\psi_z}{dz^2} + k_z^2\psi_z = 0$$

with $k_z^2 = 2m_{\text{eff}}E_z/\hbar^2$, whereupon its general solutions are linear combinations of plane waves:

$$\psi_z(z) = A_z e^{ik_z z} + B_z e^{-ik_z z}.$$

We now apply the boundary conditions of the problem, namely that $\psi_z(0) = 0$ and $\psi_z(a) = 0$, to find explicit forms for A_z and B_z.

Applying the first boundary condition gives

$$\psi_z(0) = A_z + B_z = 0,$$

from which we conclude that $B_z = -A_z$. When this is introduced into the wavefunction, we obtain

$$\begin{aligned}\psi_z(z) &= A_z(e^{ik_z z} - e^{-ik_z z}) \\ &= 2iA_z \sin k_z z,\end{aligned}$$

through the Euler relation. Next, applying the second boundary condition gives

$$\psi_z(a) = 2iA_z \sin k_z a = 0.$$

Solutions to this are either $A_z = 0$ (the trivial solution) or $k_z a = n_z \pi$ with $n_z = 1, 2, 3, \ldots$. We choose the latter, since we never want the trivial solution (life would be too easy). As a consequence,

$$k_z = \frac{n_z \pi}{a}.$$

Notice again the factor-of-two difference that fixed or static boundary conditions impose on k_z relative to k_x and k_y, where periodic boundary conditions apply.

The z contribution to the overall wavefunction is therefore

$$\psi_z(z) = N \sin\left(\frac{n_z \pi}{a} z\right),$$

where N is a normalization constant that we can find by solving

$$1 = \int_0^a |\psi_z|^2 \, dz.$$

This yields

$$\psi_z = \sqrt{\frac{2}{a}} \sin\left(\frac{n_z \pi}{a} z\right)$$

as our z contribution to the total wavefunction.

Wavefunctions

At this point, let us lump all the wavefunction contributions together to obtain the particle's total wavefunction. Since

$$\psi = \psi_x \psi_y \psi_z,$$

we find that

$$\psi = N\sqrt{\frac{2}{a}} \sin\left(\frac{n_z \pi}{a} z\right) e^{ik_x x} e^{ik_y y},$$

where N is a normalization constant for the two (x, y)-plane wave contributions seen earlier.

Normalization

We find N as follows. Starting with

$$1 = \int_{V_{\text{tot}}} |\psi(x, y, z)|^2 \, d^3 r$$

we have

$$\begin{aligned}
1 &= \int_{V_{\text{tot}}} \left[N\sqrt{\frac{2}{a}} \sin\left(\frac{n_z\pi}{a}z\right) e^{ik_x x} e^{ik_y y} \right]^* \\
&\quad \times \left[N\sqrt{\frac{2}{a}} \sin\left(\frac{n_z\pi}{a}z\right) e^{ik_x x} e^{ik_y y} \right] d^3r \\
&= N^2 \int_{V_{\text{tot}}} \frac{2}{a} \sin^2\left(\frac{n_z\pi}{a}z\right) d^3r \\
&= N^2 \int_A dx\, dy \int_0^a \frac{2}{a} \sin^2\left(\frac{n_z\pi}{a}z\right) dz \\
&= N^2 A \frac{2}{a} \left(\frac{1}{2}\right) \int_0^a 1 - \cos\left(\frac{2n_z\pi}{a}z\right) dz \\
&= \frac{N^2 A}{a} \left[a - \frac{a}{2n_z\pi} \sin\left(\frac{2n_z\pi}{a}z\right)\Big|_0^a \right] \\
&= N^2 A,
\end{aligned}$$

where $A = L_x L_y$ is the quantum well's area in the (x, y) plane. Our desired normalization constant is therefore $N = 1/\sqrt{A} = 1/\sqrt{L_x L_y}$. As a consequence, the total wavefunction for a carrier in a given state within the quantum well is

$$\psi(x, y, z) = \frac{1}{\sqrt{A}} \sqrt{\frac{2}{a}} \sin\left(\frac{n_z\pi}{a}z\right) e^{ik_x x} e^{ik_y y}. \tag{7.15}$$

Energies

The particle's energy can now be found from $E = E_x + E_y + E_z$. Recall that the energy contribution from the x dimension is found using the equivalence

$$k_x^2 = \frac{2m_{\text{eff}} E_x}{\hbar^2},$$

which leads to

$$E_x = \frac{\hbar^2 k_x^2}{2m_{\text{eff}}}.$$

Similarly, the energy contribution from the y dimension is found using

$$k_y^2 = \frac{2m_{\text{eff}} E_y}{\hbar^2},$$

and results in

$$E_y = \frac{\hbar^2 k_y^2}{2m_{\text{eff}}}.$$

Finally, the z contribution is found from

$$k_z = \frac{n_z\pi}{a},$$

$$\sqrt{\frac{2m_{\text{eff}} E_z}{\hbar^2}} = \frac{n_z\pi}{a},$$

giving

$$E_z = \frac{n_z^2 h^2}{8m_{\text{eff}} a^2}.$$

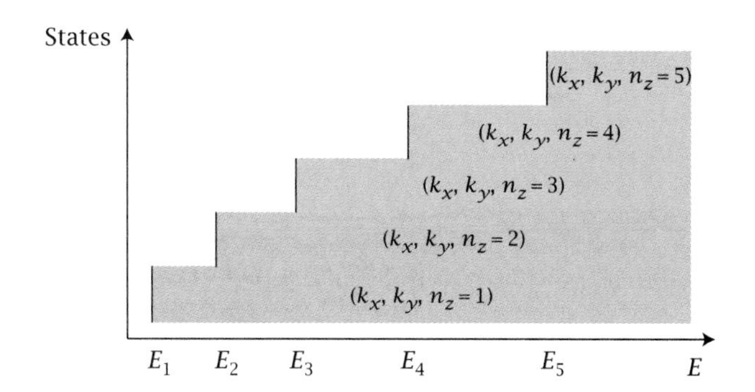

Figure 7.12 Illustration of the energy levels of a particle confined to a three-dimensional slab.

When all of these are put together, the particle's total energy is

$$E = \frac{\hbar^2}{2m_{\text{eff}}}(k_x^2 + k_y^2) + \frac{n_z^2 h^2}{8m_{\text{eff}} a^2}. \tag{7.16}$$

Note that we have implicitly assumed that the carrier's effective mass m_{eff} is the same in all directions (x, y, and z). Be aware that this assumption is generally not true in real life.

Figure 7.12 illustrates the energies of the quantum well. Solid vertical lines denote energies where $n_z = 1, 2, 3, \ldots$ and $k_x = k_y = 0$. The shaded regions denote the near continuum of energies where n_z has a given value and where $k_x \neq 0$ and $k_y \neq 0$ contributions add to this base energy. Values for different n_z are offset for clarity. Note that calculating this near continuum of energies is tedious and Chapter 9 will discuss a density-of-states calculation that will provide an overall function that encompasses all of these energies without having to solve for them one at a time.

7.3.2 Quantum Well Electron/Hole Wavefunctions and Overlap Integrals for Envelope Function Selection Rules

Having found the energies of the confined particle as well as the associated wavefunctions, we are now interested in evaluating various transition probabilities between energy levels upon the absorption of light. Although we will consider this in more detail in Chapter 12 when we talk about interband transitions, for now, let us evaluate what is referred to as an overlap integral.

In this regard, transitions in semiconductors occur by promoting an electron from a filled valence band into an empty conduction band. This can be recast into the presence of an electron in the conduction band and its absence (i.e., the presence of a hole) in the valence band. Note that the concept of bands will be more fully discussed in Chapter 10. Therefore, in this electron/hole picture, the likelihood of a given transition occurring between two states is described by an overlap integral between the final-state electron wavefunction and the initial-state hole wavefunction. Note that we have implicitly assumed working in the strong confinement regime (Chapter 3).

Based on the previous quantum well model, the total wavefunctions for the electron (e) and hole (h) are

$$\psi_e = \frac{1}{\sqrt{A}}\sqrt{\frac{2}{a}}\sin\left(\frac{n_{z,e}\pi}{a}z\right)e^{ik_{x,e}x}e^{ik_{y,e}y} \tag{7.17}$$

and

$$\psi_h = \frac{1}{\sqrt{A}}\sqrt{\frac{2}{a}}\sin\left(\frac{n_{z,h}\pi}{a}z\right)e^{ik_{x,h}x}e^{ik_{y,h}y}. \tag{7.18}$$

For notational convenience, we can re-express these functions as

$$\psi_e = \frac{1}{\sqrt{A}}\sqrt{\frac{2}{a}}\sin\left(\frac{n_{z,e}\pi}{a}z\right)e^{i\mathbf{k}_e\cdot\mathbf{r}_{xy}}$$

and

$$\psi_h = \frac{1}{\sqrt{A}}\sqrt{\frac{2}{a}}\sin\left(\frac{n_{z,h}\pi}{a}z\right)e^{i\mathbf{k}_h\cdot\mathbf{r}_{xy}},$$

where $\mathbf{k}_e = (k_{x,e}, k_{y,e}, k_{z,e})$, $\mathbf{k}_h = (k_{x,h}, k_{y,h}, k_{z,h})$, and $\mathbf{r}_{xy} = (x, y, 0)$ represents a coordinate within the (x, y) plane.

The electron and hole overlap integral is then written in Dirac bra–ket notation as $\langle\psi_e|\psi_h\rangle$ (see Chapter 6), where

$$\langle\psi_e|\psi_h\rangle = \int \psi_e^*\psi_h\, d^3r.$$

The integral then evaluates as follows:

$$\begin{aligned}
\langle\psi_e|\psi_h\rangle &= \int_{V_{\text{tot}}}\left[\frac{1}{\sqrt{A}}\sqrt{\frac{2}{a}}\sin\left(\frac{n_{z,e}\pi}{a}z\right)e^{i\mathbf{k}_e\cdot\mathbf{r}_{xy}}\right]^* \\
&\quad\times\left[\frac{1}{\sqrt{A}}\sqrt{\frac{2}{a}}\sin\left(\frac{n_{z,h}\pi}{a}z\right)e^{i\mathbf{k}_h\cdot\mathbf{r}_{xy}}\right]d^3r \\
&= \frac{1}{A}\int_{V_{\text{tot}}} e^{i(\mathbf{k}_h-\mathbf{k}_e)\cdot\mathbf{r}_{xy}}\frac{2}{a}\sin\left(\frac{n_{z,e}\pi}{a}z\right)\sin\left(\frac{n_{z,h}\pi}{a}z\right)d^3r \\
&= \frac{1}{A}\int_A e^{i(\mathbf{k}_h-\mathbf{k}_e)\cdot\mathbf{r}_{xy}}\,dx\,dy\int_0^a\frac{2}{a}\sin\left(\frac{n_{z,e}\pi}{a}z\right)\sin\left(\frac{n_{z,h}\pi}{a}z\right)dz.
\end{aligned}$$

First Integral

The first integral above is evaluated by assuming that $\mathbf{k}_h \neq \mathbf{k}_e$. This gives

$$\begin{aligned}
\frac{1}{A}\int_A e^{i(\mathbf{k}_h-\mathbf{k}_e)\cdot\mathbf{r}_{xy}}\,dx\,dy &= \frac{1}{A}\int_{-L_x/2}^{L_x/2} e^{i(k_{x,h}-k_{x,e})x}\,dx\int_{-L_y/2}^{L_y/2} e^{i(k_{y,h}-k_{y,e})y}\,dy \\
&= \frac{1}{A}\,\frac{1}{i(k_{x,h}-k_{x,e})}e^{i(k_{x,h}-k_{x,e})x}\Big|_{-L_x/2}^{L_x/2} \\
&\quad\times\frac{1}{i(k_{y,h}-k_{y,e})}e^{i(k_{y,h}-k_{y,e})y}\Big|_{-L_y/2}^{L_y/2} \\
&= \frac{1}{A}\left(\frac{1}{i\Delta k_x}\right)\left(\frac{1}{i\Delta k_y}\right)\left(e^{i\Delta k_x L_x/2} - e^{-i\Delta k_x L_x/2}\right) \\
&\quad\times\left(e^{i\Delta k_y L_y/2} - e^{-i\Delta k_y L_y/2}\right) \\
&= \frac{1}{A}\left(\frac{2i}{i\Delta k_x}\right)\left(\frac{2i}{i\Delta k_y}\right)\sin\frac{\Delta k_x L_x}{2}\sin\frac{\Delta k_y L_y}{2}
\end{aligned}$$

where

$$\Delta k_x = k_{x,h} - k_{x,e} = \frac{2\pi n_{x,h}}{L_x} - \frac{2\pi n_{x,e}}{L_x} = \frac{2\pi\Delta n_x}{L_z}$$

and likewise

$$\Delta k_y = \frac{2\pi\Delta n_y}{L_y},$$

with Δn_x and Δn_y both integers. As a consequence,

$$\frac{1}{A}\int_A e^{i(\mathbf{k}_h-\mathbf{k}_e)\cdot\mathbf{r}_{xy}}\,dx\,dy = \frac{4}{A\Delta k_x \Delta k_y}\sin\pi\Delta n_x \sin\pi\Delta n_y,$$

which evaluates to zero.

The entire overlap integral is therefore zero if $\mathbf{k}_h \neq \mathbf{k}_e$. The only case where the first term is nonzero occurs when

$$\mathbf{k}_h = \mathbf{k}_e.$$

In this case, the first integral becomes

$$\frac{1}{A}\int_A e^{i(\mathbf{k}_h-\mathbf{k}_e)\cdot\mathbf{r}_{xy}}\,dx\,dy = \frac{1}{A}\int_A dx\,dy$$
$$= \frac{1}{A}A = 1.$$

Later, in Chapter 12, we will see that we could have already predicted this, since the $\mathbf{k}_h = \mathbf{k}_e$ rule arises from momentum conservation during optical transitions. The $\Delta k = 0$ rule also suggests that optical transitions are "vertical" in k-space for direct gap semiconductors. More about this later.

Second Integral

Next, for the remaining integral, we want to solve

$$\frac{2}{a}\int_0^a \sin\left(\frac{n_{z,e}\pi}{a}z\right)\sin\left(\frac{n_{z,h}\pi}{a}z\right)dz.$$

To evaluate this, we first assume that $n_{z,e} \neq n_{z,h}$. Then, using the trigonometric identity $\sin a \sin b = \frac{1}{2}[\cos(a-b) - \cos(a+b)]$, we see that the integral becomes

$$\frac{2}{a}\left(\frac{1}{2}\right)\int_0^a \left\{\cos\left[\left(\frac{n_{z,e}-n_{z,h}}{a}\right)\pi z\right] - \cos\left[\left(\frac{n_{z,e}+n_{z,h}}{a}\right)\pi z\right]\right\}dz$$
$$= \frac{1}{a}\left\{\frac{a}{(n_{z,e}-n_{z,h})\pi}\sin\left[\left(\frac{n_{z,e}-n_{z,h}}{a}\right)\pi z\right]\Bigg|_0^a\right.$$
$$\left. - \frac{a}{(n_{z,e}+n_{z,h})\pi}\sin\left[\left(\frac{n_{z,e}+n_{z,h}}{a}\right)\pi z\right]\Bigg|_0^a\right\}$$
$$= \frac{1}{a}\left\{\frac{a}{(n_{z,e}-n_{z,h})\pi}\sin[(n_{z,e}-n_{z,h})\pi] - \frac{a}{(n_{z,e}+n_{z,h})\pi}\sin[(n_{z,e}+n_{z,h})\pi]\right\}.$$

Since $n_{z,e}$ and $n_{z,h}$ are both integers and their difference as well as their sum are integers, the term evaluates to zero.

As a consequence, the only case where the second integral does not yield zero is when

$$n_{z,e} = n_{z,h}.$$

We can illustrate this explicitly. For notational convenience, let $n_{z,e} = n_{z,h} = n$, to obtain

$$\begin{aligned}\frac{2}{a}\int_0^a \left(\sin\frac{n_{z,e}\pi}{a}z\right)\sin\left(\frac{n_{z,h}\pi}{a}z\right)dz &= \frac{2}{a}\int_0^a \sin^2\left(\frac{n\pi}{a}z\right)dz \\ &= \frac{2}{a}\left(\frac{1}{2}\right)\int_0^a 1-\cos\left(\frac{2n\pi}{a}z\right)dz \\ &= \frac{1}{a}\left[a - \frac{a}{2n\pi}\sin\left(\frac{2n\pi}{a}z\right)\Big|_0^a\right] \\ &= 1.\end{aligned}$$

Summary

To summarize, $\langle\psi_e|\psi_h\rangle \neq 0$ only if

$$\Delta\mathbf{k} = 0, \tag{7.19}$$

where $\Delta\mathbf{k} = \mathbf{k}_h - \mathbf{k}_e$ and

$$\Delta n_z = 0, \tag{7.20}$$

with $\Delta n_z = n_{z,e} - n_{z,h}$. These restrictions are so called selection rules for quantum well transitions.

The concept is more fully illustrated in **Figure 7.13**, which depicts the first few electron and hole energy levels, as well as the allowed transitions between them. Note that hole energies increase in the opposite direction to electron energies. There is a reason for this, which we will see later when we discuss bands in Chapter 10. For now, we simply take this for granted.

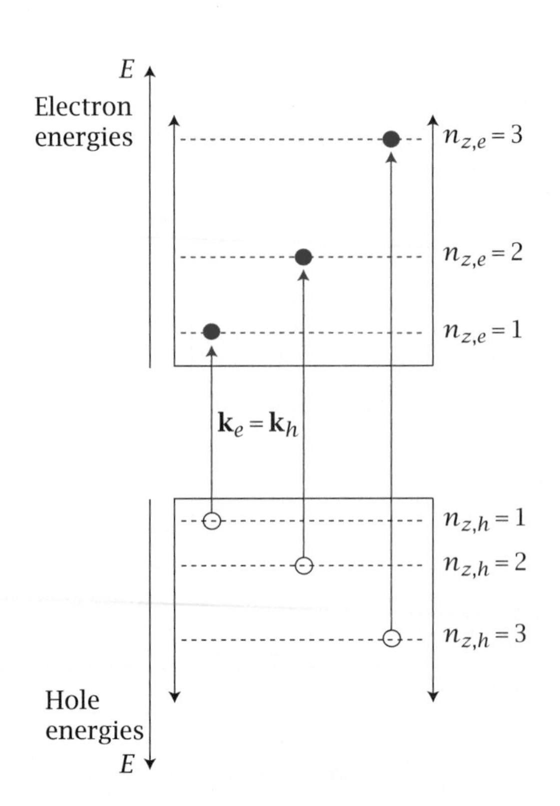

Figure 7.13 Optical transitions for a quantum well allowed by its envelope function selection rules.

7.3.3 Particle in a Cylinder: An Eventual Envelope Function for the Quantum Wire

Next, let us solve the model problem of a particle in a cylinder. This yields the envelope part of the electron (or hole) wavefunction in a semiconductor nanowire when its full wavefunction is considered. Again, more about this will be seen in Chapter 12. For now, let us assume that solving this problem yields the whole story, much like the earlier quantum well case.

In the cylinder, we have the geometry and associated boundary conditions shown in **Figure 7.14**. Namely, let us assume that the potential inside the cylinder is zero and that it is infinite outside. As a consequence, the carrier is trapped inside the finite-volume cylinder with a volume V_{tot}.

We solve the Schrödinger equation in the region of interest to obtain the particle's wavefunctions and energies. Since the Laplacian is written in cylindrical coordinates,

$$\nabla^2 = \frac{1}{r}\frac{\partial}{\partial r}\left(r\frac{\partial}{\partial r}\right) + \frac{1}{r^2}\frac{\partial^2}{\partial \theta^2} + \frac{\partial^2}{\partial z^2}, \tag{7.21}$$

the Schrödinger equation becomes

$$-\frac{\hbar^2}{2m_{\text{eff}}}\left[\frac{1}{r}\frac{\partial}{\partial r}\left(r\frac{\partial}{\partial r}\right) + \frac{1}{r^2}\frac{\partial^2}{\partial \theta^2} + \frac{\partial^2}{\partial z^2}\right]\psi = E\psi,$$

or

$$\left[\frac{1}{r}\frac{\partial}{\partial r}\left(r\frac{\partial}{\partial r}\right) + \frac{1}{r^2}\frac{\partial^2}{\partial \theta^2} + \frac{\partial^2}{\partial z^2}\right]\psi = -\frac{2m_{\text{eff}}E}{\hbar^2}\psi,$$

whereupon letting $k^2 = 2m_{\text{eff}}E/\hbar^2$ gives

$$\left[\frac{1}{r}\frac{\partial}{\partial r}\left(r\frac{\partial}{\partial r}\right) + \frac{1}{r^2}\frac{\partial^2}{\partial \theta^2} + \frac{\partial^2}{\partial z^2}\right]\psi = -k^2\psi.$$

Multiplying by r^2 and consolidating terms then yields

$$\left[r\frac{\partial}{\partial r}\left(r\frac{\partial}{\partial r}\right) + \frac{\partial^2}{\partial \theta^2} + r^2\frac{\partial^2}{\partial z^2}\right]\psi + (kr)^2\psi = 0.$$

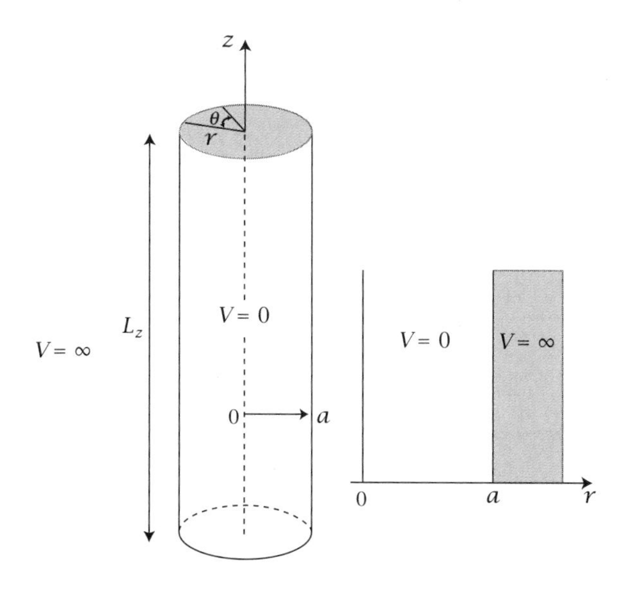

Figure 7.14 Illustration of a quantum wire, its coordinate system, and the potential along the radial direction.

At this point, assume a general, separable, wavefunction of the form

$$\psi = A(r,\theta)B(z) \equiv AB,$$

where the last equivalence just makes our notation easier in what follows.

We now introduce this wavefunction into the above equation to obtain

$$r\frac{\partial}{\partial r}\left(r\frac{\partial(AB)}{\partial r}\right) + \frac{\partial^2(AB)}{\partial\theta^2} + r^2\frac{\partial^2(AB)}{\partial z^2} + (kr)^2 AB = 0,$$

or

$$Br\frac{\partial}{\partial r}\left(r\frac{\partial A}{\partial r}\right) + B\frac{\partial^2 A}{\partial\theta^2} + Ar^2\frac{d^2 B}{dz^2} + (kr)^2 AB = 0.$$

Dividing both sides by AB then gives

$$\frac{1}{A}r\frac{\partial}{\partial r}\left(r\frac{\partial A}{\partial r}\right) + \frac{1}{A}\frac{\partial^2 A}{\partial\theta^2} + \frac{1}{B}r^2\frac{d^2 B}{dz^2} + (kr)^2 = 0,$$

whereupon further division by r^2 yields

$$\frac{1}{Ar}\frac{\partial}{\partial r}\left(r\frac{\partial A}{\partial r}\right) + \frac{1}{Ar^2}\frac{\partial^2 A}{\partial\theta^2} + \frac{1}{B}\frac{d^2 B}{dz^2} + k^2 = 0.$$

Finally, consolidating terms based on their (r,θ) versus z dependences gives

$$\frac{1}{Ar}\frac{\partial}{\partial r}\left(r\frac{\partial A}{\partial r}\right) + \frac{1}{Ar^2}\frac{\partial^2 A}{\partial\theta^2} + k^2 = -\frac{1}{B}\frac{d^2 B}{dz^2}.$$

It is apparent that both sides of the equation have different dependences yet are always equal. As a consequence, they must equal the same constant. Call it k_z^2 for convenience. We therefore have the following two equations, which we will solve separately:

$$\frac{1}{Ar}\frac{\partial}{\partial r}\left(r\frac{\partial A}{\partial r}\right) + \frac{1}{Ar^2}\frac{\partial^2 A}{\partial\theta^2} + k^2 = k_z^2,$$

$$-\frac{1}{B}\frac{d^2 B}{dz^2} = k_z^2.$$

We solve the easy one first:

$$\frac{d^2 B}{dz^2} = -k_z^2 B.$$

Rearrange this to

$$\frac{d^2 B}{dz^2} + k_z^2 B = 0$$

and recognize that its general solutions are linear combinations of plane waves:

$$B(z) = C_1 e^{ik_z z} + C_2 e^{-ik_z z},$$

where C_1 and C_2 are constant coefficients to be determined. Relevant boundary conditions are (**Figure 7.14**)

$$B(0) = 0,$$

$$B(L_z) = 0.$$

On applying the first, we find that $C_1 + C_2 = 0$, or $C_2 = -C_1$. When this is reintroduced into the above wavefunction, we have

$$B(z) = C_1\left(e^{ik_z z} - e^{-ik_z z}\right).$$

As we have seen previously, the complex expression in parentheses can be converted into a real function using the Euler relation. We find that

$$\begin{aligned} B(z) &= 2iC_1\left(\frac{e^{ik_z z} - e^{-ik_z z}}{2i}\right) \\ &= 2iC_1 \sin k_z z, \end{aligned}$$

or

$$B(z) = N \sin k_z z,$$

where N is a normalization constant determined by enforcing wavefunction normalization.

At this point, applying the second boundary condition gives

$$B(L_z) = N \sin(k_z L_z) = 0.$$

For this to be true, we have either the trivial solution $N = 0$ or $k_z L_z = n_z \pi$ with $n_z = 1, 2, 3, \ldots$. Since we do not want the trivial solution (ever), the following values of k_z apply:

$$k_z = \frac{n_z \pi}{L_z}.$$

Note that we have assumed a finite (but large) length L_z and have therefore not assumed periodic boundary conditions. Admittedly, what constitutes a large value for L_z remains a little subjective. Instead, we have assumed static boundary conditions (i.e., $B(0) = 0$ and $B(L_z) = 0$). One could alternatively assume periodicity, but this will not change the answer significantly since, in either case, k_z will be small.

We now enforce wavefunction normalization. This is done by solving

$$1 = \int_0^{L_z} |B(z)|^2 \, dz$$

to find $N = \sqrt{2/L_z}$. The particle's wavefunction along the z direction can therefore be written as

$$B(z) = \sqrt{\frac{2}{L_z}} \sin\left(\frac{n_z \pi}{L_z} z\right). \tag{7.22}$$

Note that in the case where periodic boundary conditions are assumed, the solution changes to $B(z) = (\sqrt{1/L_z})e^{ik_z z}$. We will see this in the next section and, in fact, we will use this alternative version in Chapter 12.

Let us now solve the second, more complicated, equation. We have

$$\left[\frac{1}{Ar}\frac{\partial}{\partial r}\left(r\frac{\partial A}{\partial r}\right) + \frac{1}{Ar^2}\frac{\partial^2 A}{\partial \theta^2} + k^2\right] = k_z^2.$$

Multiplying both sides by r^2 gives

$$\frac{r}{A}\frac{\partial}{\partial r}\left(r\frac{\partial A}{\partial r}\right) + \frac{1}{A}\frac{\partial^2 A}{\partial \theta^2} + (kr)^2 = (k_z r)^2,$$

whereupon consolidating terms yields

$$r\frac{\partial}{\partial r}\left(r\frac{\partial A}{\partial r}\right)+\frac{\partial^2 A}{\partial\theta^2}+A\left[(kr)^2-(k_z r)^2\right]=0.$$

At this point, assume a separable solution of the form

$$A=C(r)D(\theta)\equiv CD,$$

with the last equivalence to simplify our notation in what follows. When this is inserted into the above equation, we find that

$$r\frac{\partial}{\partial r}\left(r\frac{\partial(CD)}{\partial r}\right)+\frac{\partial^2(CD)}{\partial\theta^2}+CD\left[(kr)^2-(k_z r)^2\right]=0$$

$$Dr\frac{d}{dr}\left(r\frac{dC}{dr}\right)+C\frac{d^2D}{d\theta^2}+CD\left[(kr)^2-(k_z r)^2\right]=0.$$

Dividing by CD then gives

$$\frac{1}{C}r\frac{d}{dr}\left(r\frac{dC}{dr}\right)+\frac{1}{D}\frac{d^2D}{d\theta^2}+\left[(kr)^2-(k_z r)^2\right]=0,$$

and consolidating terms based on their r or θ dependences yields

$$\frac{1}{C}r\frac{d}{dr}\left(r\frac{dC}{dr}\right)+\left[(kr)^2-(k_z r)^2\right]=-\frac{1}{D}\frac{d^2D}{d\theta^2}.$$

This can subsequently be simplified (using the product rule of differentiation) to

$$\frac{1}{C}\left(r^2\frac{d^2C}{dr^2}+r\frac{dC}{dr}\right)+\left[(kr)^2-(k_z r)^2\right]=-\frac{1}{D}\frac{d^2D}{d\theta^2},$$

where, as seen before, both sides have different dependences yet are always equal. They therefore equal a constant, which we call m^2 for convenience. Note that m should not be confused with the particle's mass. Unfortunately, the same letters are frequently used to denote different things. We are now left with two separate equations to solve. They are

$$\frac{1}{C}\left(r^2\frac{d^2C}{dr^2}+r\frac{dC}{dr}\right)+\left[(kr)^2-(k_z r)^2\right]=m^2,$$

$$-\frac{1}{D}\frac{d^2D}{d\theta^2}=m^2.$$

At this point, let us address the easy one first. The equation is

$$\frac{d^2D}{d\theta^2}=-m^2D,$$

which can be rearranged into

$$\frac{d^2D}{d\theta^2}+m^2D=0.$$

The reader will recognize that its general solutions are linear combinations of plane waves:

$$D(\theta)=D_1e^{im\theta}+D_2e^{-im\theta},$$

where D_1 and D_2 are again constant coefficients. Since there are no strict boundary conditions on D_1 or D_2, we are free to choose a particular solution, namely

$$D(\theta) = D_1 e^{im\theta}.$$

Let us now enforce periodicity, so that $D(\theta) = D(\theta + 2\pi)$. As a consequence,

$$\begin{aligned} D_1 e^{im\theta} &= D_1 e^{im(\theta+2\pi)} \\ &= D_1 e^{im\theta} e^{im2\pi}, \end{aligned}$$

and for the equality to hold, $e^{im2\pi} = 1$. From the Euler relation, we see that $\cos 2m\pi + i \sin 2m\pi = 1$, and, as a consequence, $m = 0, \pm 1, \pm 2, \pm 3, \ldots$. We thus have

$$D(\theta) = D_1 e^{im\theta}. \tag{7.23}$$

We now address the radial part of the wavefunction. The relevant equation is

$$\frac{1}{C}\left(r^2 \frac{d^2C}{dr^2} + r\frac{dC}{dr}\right) + \left[(kr)^2 - (k_z r)^2\right] = m^2,$$

where we multiply both sides by C and consolidate terms to obtain

$$r^2 \frac{d^2C}{dr^2} + r\frac{dC}{dr} + C\left[(kr)^2 - (k_z r)^2 - m^2\right] = 0.$$

At this point, let $K^2 = k^2 - k_z^2$, to get

$$r^2 \frac{d^2C}{dr^2} + r\frac{dC}{dr} + C\left[(Kr)^2 - m^2\right] = 0.$$

An appropriate change of variables, namely, $\xi = Kr$ (so $r = \xi/K$ and $dr = d\xi/K$) and $C(r) \to C(\xi/K)$, then yields our final desired expression:

$$\xi^2 \frac{d^2C}{d\xi^2} + \xi\frac{dC}{d\xi} + C(\xi^2 - m^2) = 0. \tag{7.24}$$

Equation 7.24 is called the *Bessel equation.*

General solutions to it are linear combination of *Bessel functions* of the first (J_m) and second (Y_m) kinds:

$$C(r) = C_1 J_m(\xi) + C_2 Y_m(\xi),$$

where C_1 and C_2 are constant coefficients. Note that Bessel functions of the second kind diverge as $r \to 0$. This leads to unphysical solutions. As a consequence, $C_2 = 0$, and we are left with

$$C(r) = C_1 J_m(\xi).$$

Alternatively, when normal units are reintroduced,

$$C(r) = C_1 J_m\left(\frac{\alpha_{m,n}}{a} r\right). \tag{7.25}$$

The reader may be puzzled by the transformation. However, it will become apparent after going through the particle's associated energies. In the equation, a is the radius of the cylinder and $\alpha_{m,n}$ is the dual-index

Table 7.1 Roots of the Bessel function $J_m(x)$

Root number (n)	$J_0(x)$	$J_1(x)$	$J_2(x)$	$J_3(x)$	$J_4(x)$	$J_5(x)$
1	2.4048	3.8317	5.1356	6.3802	7.5883	8.7715
2	5.5201	7.0156	8.4172	9.7610	11.0647	12.3386
3	8.6537	10.1735	11.6198	13.0152	14.3725	15.7002
4	11.7915	13.3237	14.7960	16.2235	17.6160	18.9801
5	14.9309	16.4706	17.9598	19.4094	20.8269	22.2178

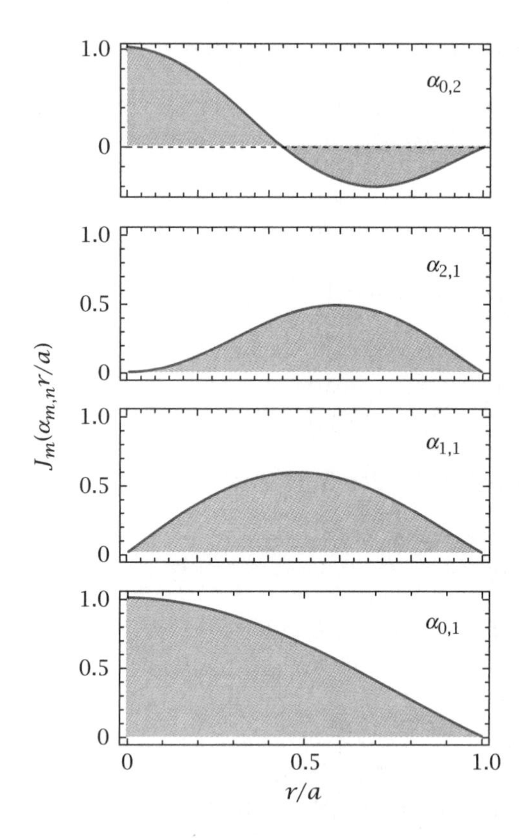

Figure 7.15 First four Bessel functions in order of increasing $\alpha_{m,n}$ value.

Bessel function root. The subscript m is the order of the Bessel function, with n its associated root number, also called a radial quantum number. Bessel function roots are listed in **Table 7.1**. Several Bessel functions are plotted in **Figure 7.15** to allow the reader to get a qualitative sense of their appearance.

Wavefunctions

When everything is put together, we obtain the following wavefunctions for the particle confined to a cylinder:

$$\psi(r,\theta,z) = NJ_m\left(\frac{\alpha_{m,n}}{a}r\right)e^{im\theta}\sqrt{\frac{2}{L_z}}\sin\left(\frac{n_z\pi}{L_z}z\right), \tag{7.26}$$

where $N = C_1D_1$ is a normalization constant that can be determined by solving

$$1 = \int_{V_{\text{tot}}} |\psi(r,\theta,\phi)|^2\, d^3r,$$

as seen earlier.

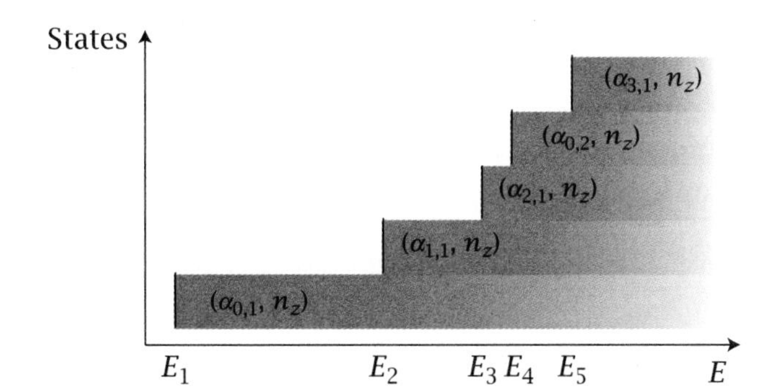

Figure 7.16 Energy levels of a particle in a three-dimensional cylinder.

Energies

In tandem, the particle's energies can be found by applying the problem's boundary conditions. Namely, when $r = a, J_m(\xi) = J_m(Ka) = 0$. As a consequence, Ka must equal a root of the Bessel function ($\alpha_{m,n}$). Thus, $K = \alpha_{m,n}/a$, and this should make Equation 7.25 more apparent. Next, returning to the energies, we have

$$Ka = \alpha_{m,n},$$

$$\sqrt{k^2 - k_z^2}\, a = \alpha_{m,n},$$

where we recall that $k^2 = 2m_{\text{eff}}E/\hbar^2$ and $k_z = n_z\pi/L_z$. Solving for E, which is contained in k, then yields

$$E = \frac{\hbar^2}{2m_{\text{eff}}}\left(\frac{\alpha_{m,n}^2}{a^2} + \frac{n_z^2\pi^2}{L_z^2}\right), \tag{7.27}$$

with the first term in the parentheses reflecting the radial contribution to the total energy and the second term representing the (small) longitudinal energy contribution. The reader may verify this as an exercise.

These energies are illustrated in **Figure 7.16**, where the solid vertical lines denote the energy for a given $\alpha_{m,n}$ value with $n_z = 0$. Shaded regions represent energies for a particular $\alpha_{m,n}$ with nonzero n_z contributions. The shading just depicts the fact that at lower energies the states are closer together whereas at higher energies they become sparser. We saw this earlier in **Figure 7.4**. Levels, associated with a given $\alpha_{m,n}$ value, are also offset for clarity. In Chapter 9 we will talk about the density of states of a nanowire that allows us to obtain an expression that encompasses all of these states in a more tractable manner.

7.3.4 Particle-in-a-Cylinder Electron/Hole Wavefunctions and Overlap Integrals for Envelope Function Selection Rules

As with the quantum well, we are now interested in evaluating the likelihood that an optical transition occurs between different nanowire energy levels upon the absorption of light. We will consider this in more detail in Chapter 12. However, for now let us evaluate our electron/hole overlap integral.

As discussed earlier, transitions in semiconductors occur by promoting an electron from a filled valence band into an empty conduction band. This can be recast as the absence of an electron (i.e., a hole) in the valence band and the presence of an electron in the conduction band. As a consequence, in this electron/hole picture, the likelihood of a given transition between two states is described by the overlap integral between the wavefunctions of the electron in the final state and that of the complementary hole in the initial state.

Electron and hole wavefunctions, using the above cylindrical nanowire model, are

$$\psi_e = N_e\left[J_{m_e}\left(\frac{\alpha_{m_e,n_e}}{a}r\right)e^{im_e\theta}\right]\sqrt{\frac{2}{L_z}}\sin\left(\frac{n_{z,e}\pi}{L_z}z\right) \tag{7.28}$$

and

$$\psi_h = N_h\left[J_{m_h}\left(\frac{\alpha_{m_h,n_h}}{a}r\right)e^{im_h\theta}\right]\sqrt{\frac{2}{L_z}}\sin\left(\frac{n_{z,h}\pi}{L_z}z\right), \tag{7.29}$$

where N_e and N_h are normalization constants. Note also that m_e and m_h here should not be confused with electron or hole masses.

In Dirac bra–ket notation, the electron and hole overlap integral is written as $\langle\psi_e|\psi_h\rangle$, where

$$\begin{aligned}\langle\psi_e|\psi_h\rangle &= \int_{V_{\text{tot}}}\psi_e^*\psi_h\,d^3r\\ &= \int_0^a\int_0^{2\pi}\int_0^{L_z}\psi_e^*\psi_h r\,dr\,d\theta\,dz.\end{aligned}$$

This expands to

$$\begin{aligned}&\int_0^a\int_0^{2\pi}\int_0^{L_z}\left[N_eJ_{m_e}\left(\frac{\alpha_{m_e,n_e}}{a}r\right)e^{im_e\theta}\sqrt{\frac{2}{L_z}}\sin\left(\frac{n_{z,e}\pi}{L_z}z\right)\right]^*\\ &\quad\times\left[N_hJ_{m_h}\left(\frac{\alpha_{m_h,n_h}}{a}r\right)e^{im_h\theta}\sqrt{\frac{2}{L_z}}\sin\left(\frac{n_{z,h}\pi}{L_z}z\right)\right]r\,dr\,d\theta dz\\ &= N_eN_h\int_0^a rJ_{m_e}\left(\frac{\alpha_{m_e,n_e}}{a}r\right)J_{m_h}\left(\frac{\alpha_{m_h,n_h}}{a}r\right)dr\int_0^{2\pi}e^{-im_e\theta}e^{im_h\theta}\,d\theta\\ &\quad\times\int_0^{L_z}\frac{2}{L_z}\sin\left(\frac{n_{z,e}\pi}{L_z}z\right)\sin\left(\frac{n_{z,h}\pi}{L_z}z\right)dz,\end{aligned}$$

whereupon we evaluate each integral separately.

To begin, the second integral is (note that we are proceeding out of order, but life is easier this way)

$$\int_0^{2\pi}e^{i(m_h-m_e)\theta}\,d\theta,$$

where m_e and m_h are both integers. Again, these indices are not to be confused with electron or hole effective masses. It is an unfortunate coincidence that the same letter is shared by our quantum numbers and carrier masses. Assuming that $m_e \neq m_h$, we integrate to find

$$\frac{1}{i(m_h-m_e)}e^{i(m_h-m_e)\theta}\bigg|_0^{2\pi} = \frac{1}{i\Delta m}(e^{i\Delta m2\pi}-e^{i\Delta m(0)}),$$

where $\Delta m = m_h - m_e$. We then have

$$\frac{1}{i\Delta m}\left(e^{i2\pi\Delta m}-1\right),$$

and since Δm is an integer, the entire term equals zero. Thus, the only case that leads to a nonzero value (i.e., 2π) is when

$$m_e = m_h,$$

or where

$$\Delta m = 0. \tag{7.30}$$

This is our first nanowire selection rule.

We now evaluate the third integral. It is

$$\frac{2}{L_z}\int_0^{L_z} \sin\left(\frac{n_{z,e}\pi}{L_z}z\right)\sin\left(\frac{n_{z,h}\pi}{L_z}z\right)dz,$$

where we first assume that $n_{z,e} \neq n_{z,h}$ and also let $\alpha = n_{z,e}\pi/L_z$ and $\beta = n_{z,h}\pi/L_z$. We then have

$$\begin{aligned}\frac{2}{L_z}\int_0^{L_z} \sin\alpha z \sin\beta z\, dz &= \frac{2}{L_z}\left(\frac{1}{2}\right)\int_0^{L_z}[\cos(\alpha z - \beta z) - \cos(\alpha z + \beta z)]\, dz\\ &= \frac{1}{L_z}\left\{\frac{L_z}{(n_{z,e}-n_{z,h})\pi}\sin\left[(n_{z,e}-n_{z,h})\frac{\pi z}{L_z}\right]\Bigg|_0^{L_z}\right.\\ &\quad\left. - \frac{L_z}{(n_{z,e}+n_{z,h})\pi}\sin\left[(n_{z,e}+n_{z,h})\frac{\pi z}{L_z}\right]\Bigg|_0^{L_z}\right\},\end{aligned}$$

which shows that the entire integral is zero, since the sum as well as the difference of $n_{z,e}$ and $n_{z,h}$ are integers. As a consequence, the only case where nonzero values occur requires that

$$n_{z,e} = n_{z,h},$$

or

$$\Delta n_z = 0. \tag{7.31}$$

We can see this explicitly since the third integral is then

$$\frac{2}{L_z}\int_0^{L_z} \sin^2\left(\frac{n\pi}{L_z}z\right)dz,$$

where $n = n_{z,e} = n_{z,h}$, and equals 1. Equation 7.31 is therefore our second nanowire selection rule.

Finally, let us evaluate the first integral, containing Bessel functions. It is

$$N_e N_h \int_0^a rJ_{m_e}\left(\frac{\alpha_{m_e,n_e}}{a}r\right)J_{m_h}\left(\frac{\alpha_{m_h,n_h}}{a}r\right)dr.$$

Note that from the second integral just evaluated, $m_e = m_h$, which we will denote by m for convenience. We thus have

$$N_e N_h \int_0^a rJ_m\left(\frac{\alpha_{m,n_e}}{a}r\right)J_m\left(\frac{\alpha_{m,n_h}}{a}r\right)dr.$$

This integrates to (the integral can be looked up in a table)

$$N_e N_h \frac{a^2}{2}\left[J_{m+1}(\alpha_{m,n})\right]\delta_{n_e,n_h}.$$

From this, we see that the only nonzero values arise when

$$n = n_e = n_h,$$

or when

$$\Delta n = 0. \tag{7.32}$$

This is the last of our nanowire selection rules.

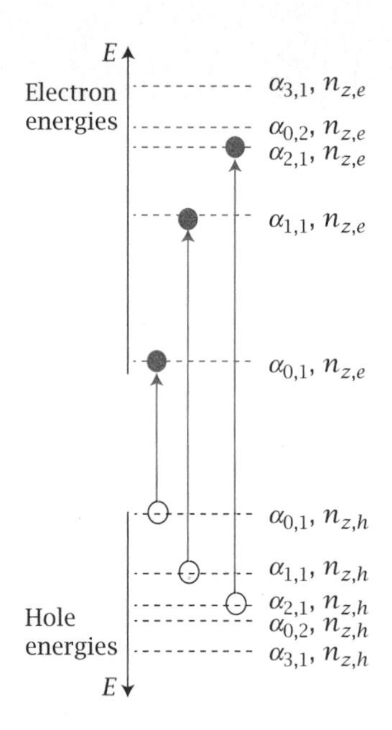

Figure 7.17 Optical transitions for a nanowire allowed by its envelope function selection rules.

Summary

To summarize, the overlap integral for nanowire electron/hole transitions, $\langle\psi_e|\psi_h\rangle$, is zero unless

$$\begin{aligned} \Delta n &= 0 \quad \text{(for the radial part)},\\ \Delta m &= 0 \quad \text{(for the angular part)},\\ \Delta n_z &= 0 \quad \text{(for the longitudinal part)}. \end{aligned}$$

These are our nanowire (envelope function) selection rules. They are illustrated in **Figure 7.17**, where allowed transitions between electron and hole levels are shown. Note that hole energies increase in the opposite direction to those of electrons. The reason for this has to do with the different curvature of the hole's valence band relative to the electron's conduction band and will be demonstrated in Chapter 10. Finally, note that additional selection rules arise later on when dealing with the full nanowire electron/hole wavefunctions. We will discuss these polarization-specific selection rules at that point.

7.3.5 An Alternative Model of a Particle in a Very Long Cylinder

Alternative ways exist for describing the confined particle in a nanowire. Namely, as alluded to earlier, we could assume that the wire is very long along the z direction, though not infinitely long. It still has a finite volume V_{tot}. However, having a very long wire allows us to use periodic boundary conditions for the z contribution to the total particle wavefunction instead of the static boundary conditions seen earlier. This will change the appearance of the overall wavefunction slightly, but not by much. Since we will assume this model later on in Chapter 12 when we talk about interband transitions, let us derive the particle's wavefunctions and energies under these conditions.

We start with the same approach as before by writing the Laplacian in cylindrical coordinates. The Schrödinger equation $-(\hbar^2/2m_{\text{eff}})\nabla^2\psi = E\psi$ becomes

$$-\frac{\hbar^2}{2m_{\text{eff}}}\left[\frac{1}{r}\frac{\partial}{\partial r}\left(r\frac{\partial}{\partial r}\right)+\frac{1}{r^2}\frac{\partial^2}{\partial\theta^2}+\frac{\partial^2}{\partial z^2}\right]\psi = E\psi.$$

As seen earlier, this expression rearranges to the following equation:

$$\left[r\frac{\partial}{\partial r}\left(r\frac{\partial}{\partial r}\right)+\frac{\partial^2}{\partial\theta^2}+r^2\frac{\partial^2}{\partial z^2}\right]\psi+(kr)^2\psi = 0,$$

where $k^2 = 2m_{\text{eff}}E/\hbar^2$.

Wavefunctions

Assuming a separable wavefunction $\psi = A(r,\theta)B(z) \equiv AB$ then leads to the following two equations we have solved previously:

$$\frac{1}{Ar}\frac{\partial}{\partial r}\left(r\frac{\partial A}{\partial r}\right)+\frac{1}{Ar^2}\frac{\partial^2 A}{\partial\theta^2}+k^2 = k_z^2,$$

$$-\frac{1}{B}\frac{d^2B}{dz^2} = k_z^2.$$

Recall that k_z^2 is a constant in either case.

In this example, we are interested in the second equation, since the solution to the first, namely

$$A(r,\theta) = NJ_m\left(\frac{\alpha_{m,n}}{a}r\right)e^{im\theta}, \tag{7.33}$$

remains the same.

Now, the solution to the second equation differs a little here only because we assume a very long wire and hence invoke periodic, as opposed to static, boundary conditions. As a consequence, we have

$$-\frac{1}{B}\frac{d^2B}{dz^2} = k_z^2,$$

or

$$\frac{d^2B}{dz^2} + k_z^2 B = 0.$$

General solutions to this equation are linear combinations of plane waves:

$$B(z) = B_1 e^{ik_z z} + B_2 e^{-ik_z z},$$

where B_1 and B_2 are constant coefficients to be determined. Since the only restriction on our wavefunctions is periodicity, we let $B_2 = 0$ to obtain

$$B(z) = B_1 e^{ik_z z}$$

as the longitudinal contribution to the overall wavefunction. We can normalize this separately,

$$1 = \int_0^{L_z} |B(z)|^2\, dz = \int_0^{L_z} B_1^2\, dz$$
$$= B_1^2 L_z,$$

to show that $B_1^2 = 1/L_z$, or $B_1 = 1/\sqrt{L_z}$. The total longitudinal part of the wavefunction is therefore

$$B(z) = \frac{1}{\sqrt{L_z}}e^{ik_z z}.$$

Putting everything together yields the following wavefunction:

$$\psi(r,\theta,z) = \frac{1}{\sqrt{L_z}}\left[NJ_m\left(\frac{\alpha_{m,n}}{a}r\right)e^{im\theta}\right]e^{ik_z z}, \tag{7.34}$$

which is the functional form we will use in Chapter 12 when discussing nanowire interband transitions.

Energies

The corresponding energies have two contributions: one from the radial/angular part of the wavefunction and the other from the longitudinal part. The former is the same as shown earlier and contributes

$$E_{r,\theta} = \frac{\hbar^2}{2m_{\text{eff}}}\frac{\alpha_{m,n}^2}{a^2}.$$

The latter is basically the energy contribution of a free particle and is $p^2/2m_{\rm eff}$, where $p = \hbar k$. As a consequence,

$$E_z = \frac{\hbar^2 k_z^2}{2m_{\rm eff}} = \frac{\hbar^2}{2m_{\rm eff}} \frac{(2n_z)^2\pi^2}{L_z^2},$$

since $k_z = 2n_z\pi/L_z$. The reader may recall that for periodic boundary conditions k_z will take values like this. However, if it is unclear, the result can be derived again by recalling that, due to periodicity, $B(z) = B(z + L_z)$. This leads to the equivalence $B_1 e^{ik_z z} = B_1 e^{ik_z z} e^{ik_z L_z}$, and from the Euler relation it should be apparent that for this to be true, $k_z = 2n_z\pi/L_z$.

The particle's total energy is the sum of these two contributions:

$$E = \frac{\hbar^2}{2m_{\rm eff}} \left[\frac{\alpha_{m,n}^2}{a^2} + \frac{(2n_z)^2\pi^2}{L_z^2} \right]. \tag{7.35}$$

Notice that the only real difference between these energies and those found earlier is the extra factor of two next to n_z in the numerator of the second term. Provided that L_z is large in either case, this does not represent a significant difference between the two expressions.

Overlap Integral

Let us repeat the evaluation of the nanowire envelope function electron/hole overlap integral for completeness. Starting with the initial and final wavefunctions

$$\psi_e = \frac{1}{\sqrt{L_z}} \left[N_e J_{m_e} \left(\frac{\alpha_{m_e,n_e}}{a} r \right) e^{im_e\theta} \right] e^{ik_{z,e}z}$$

and

$$\psi_h = \frac{1}{\sqrt{L_z}} \left[N_h J_{m_h} \left(\frac{\alpha_{m_h,n_h}}{a} r \right) e^{im_h\theta} \right] e^{ik_{z,h}z},$$

we evaluate $\langle \psi_e | \psi_h \rangle$:

$$\begin{aligned}
\langle \psi_e | \psi_h \rangle &= \int_{V_{\rm tot}} \left\{ \frac{1}{\sqrt{L_z}} \left[N_e J_{m_e} \left(\frac{\alpha_{m_e,n_e}}{a} r \right) e^{im_e\theta} \right] e^{ik_{z,e}z} \right\}^* \\
&\quad \times \left\{ \frac{1}{\sqrt{L_z}} \left[N_h J_{m_h} \left(\frac{\alpha_{m_h,n_h}}{a} r \right) e^{im_h\theta} \right] e^{ik_{z,h}z} \right\} d^3 r \\
&= \frac{N_e N_h}{L_z} \int_{L_z} e^{i(k_{z,h}-k_{z,e})z}\, dz \int_A J_{m_e}^* \left(\frac{\alpha_{m_e,n_e}}{a} r \right) \\
&\quad \times J_{m_h} \left(\frac{\alpha_{m_h,n_h}}{a} r \right) e^{i(m_h - m_e)\theta}\, r\, dr\, d\theta \\
&= \frac{N_e N_h}{L_z} \int_{L_z} e^{i(k_{z,h}-k_{z,e})z}\, dz \int_0^{2\pi} e^{i(m_h - m_e)\theta}\, d\theta \\
&\quad \times \int_0^a r J_{m_e} \left(\frac{\alpha_{m_e,n_e}}{a} r \right) J_{m_h} \left(\frac{\alpha_{m_h,n_h}}{a} r \right) dr.
\end{aligned}$$

We have evaluated the last two integrals previously. They result in $2\pi\delta_{m_e,m_h}$ and $\frac{1}{2}a^2 J_{m+1}(\alpha_{m,n})\delta_{n_e,n_h}$, respectively. We therefore have

$$= \frac{N_e N_h}{L_z} (2\pi\delta_{m_e,m_h}) \left[\frac{a^2}{2} J_{m+1}(\alpha_{m,n})\delta_{n_e,n_h} \right] \int_{L_z} e^{i(k_{z,h}-k_{z,e})z}\, dz.$$

The remaining integral is now easy to evaluate and yields

$$\begin{aligned}\int_{-L_z/2}^{L_z/2} e^{i(k_{z,h}-k_{z,e})z}\,dz &= \frac{1}{i(k_{z,h}-k_{z,e})} e^{i(k_{z,h}-k_{z,e})z}\Big|_{-L_z/2}^{L_z/2} \\ &= \frac{1}{i(k_{z,h}-k_{z,e})}(e^{i(k_{z,h}-k_{z,e})L_z/2} - e^{-i(k_{z,h}-k_{z,e})L_z/2}) \\ &= \frac{2}{k_{z,h}-k_{z,e}} \sin\left(\frac{k_{z,h}-k_{z,e}}{2} L_z\right).\end{aligned}$$

Note that in the evaluation one could also have used the limits 0 and L_z. The answer is the same either way. Since $k_{z,h} = 2n_{z,h}\pi/L_z$ and $k_{z,e} = 2n_{z,e}\pi/L_z$, we find $k_{z,h} - k_{z,e} = (2\pi/L_z)(n_{z,h} - n_{z,e})$, and thus

$$\int_{-L_z/2}^{L_z/2} e^{i(k_{z,h}-k_{z,e})z}\,dz = \frac{2}{k_{z,h}-k_{z,e}} \sin[\pi(n_{z,h}-n_{z,e})].$$

This equals zero since $n_{z,h} - n_{z,e}$ is an integer. The only nonzero value therefore occurs when $n_{z,h} = n_{z,e}$. In this case, it yields $L_z\delta_{n_{z,e},n_{z,h}}$, giving

$$\langle\psi_e|\psi_h\rangle = \frac{N_e N_h}{L_z}[2\pi\delta_{m_e,m_h}]\left[\frac{a^2}{2}J_{m+1}(\alpha_{m,n})\delta_{n_e,n_h}\right][L_z\delta_{n_{z,e},n_{z,h}}]$$

or

$$\langle\psi_e|\psi_h\rangle \propto \delta_{m_e,m_h}\delta_{n_e,n_h}\delta_{n_{z,e},n_{z,h}}.$$

From this, we see that the selection rules for nanowire optical transitions are

$$\Delta n = 0 \quad \text{(for the radial part),} \tag{7.36}$$

$$\Delta m = 0 \quad \text{(for the angular part),} \tag{7.37}$$

$$\Delta n_z = 0 \quad \text{(for the longitudinal part),} \tag{7.38}$$

which are identical to what we found previously.

7.3.6 Particle in a Sphere: An Eventual Envelope Function for the Quantum Dot

Finally, let us solve the problem of a particle in a sphere. This leads to the envelope part of the total wavefunction that describes a carrier confined to a quantum dot. We will see this later on in Chapter 12. **Figure 7.18** illustrates the potentials involved in the problem. In particular, the potential for the carrier inside the sphere is zero, while it is infinite outside. As a consequence, the particle is localized to the interior of the sphere. Furthermore, boundary conditions force the carrier wavefunction to go to zero at the sphere's edges. We therefore have the following constraints: for the potentials,

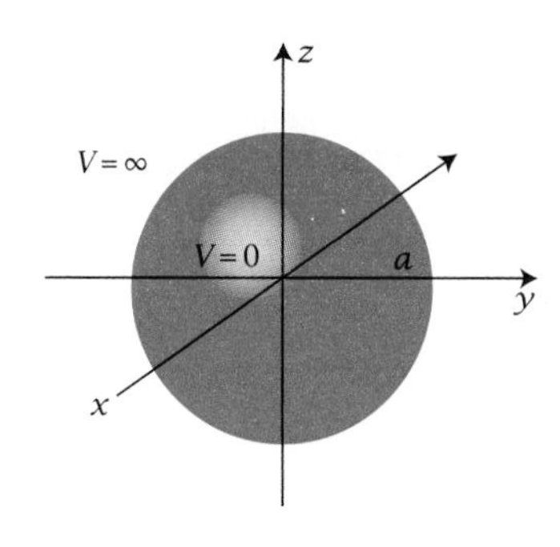

Figure 7.18 Illustration of a quantum dot and its potential.

$$V(r,\theta,\phi) = \begin{cases} 0 & \text{when } r < a, \\ \infty & \text{elsewhere,} \end{cases}$$

with corresponding boundary conditions

$$\psi(a,\theta,\phi) = 0.$$

This time, we solve the Schrödinger equation with the Laplacian written in spherical coordinates. Namely, we have

$$\nabla^2 = \frac{1}{r^2}\frac{\partial}{\partial r}\left(r^2\frac{\partial}{\partial r}\right) + \frac{1}{r^2\sin\theta}\frac{\partial}{\partial\theta}\left(\sin\theta\frac{\partial}{\partial\theta}\right) + \frac{1}{r^2\sin^2\theta}\frac{\partial^2}{\partial\phi^2}, \tag{7.39}$$

so the Schrödinger equation takes the form

$$-\frac{\hbar^2}{2m_{\text{eff}}}\left[\frac{1}{r^2}\frac{\partial}{\partial r}\left(r^2\frac{\partial}{\partial r}\right)+\frac{1}{r^2\sin\theta}\frac{\partial}{\partial\theta}\left(\sin\theta\frac{\partial}{\partial\theta}\right)+\frac{1}{r^2\sin^2\theta}\frac{\partial^2}{\partial\phi^2}\right]\psi=E\psi.$$

Multiply this by r^2 and rearrange terms to obtain

$$-\hbar^2\left[\frac{\partial}{\partial r}\left(r^2\frac{\partial}{\partial r}\right)+\frac{1}{\sin\theta}\frac{\partial}{\partial\theta}\left(\sin\theta\frac{\partial}{\partial\theta}\right)+\frac{1}{\sin^2\theta}\frac{\partial^2}{\partial\phi^2}\right]\psi=2m_{\text{eff}}r^2E\psi,$$

which can be further simplified by consolidating terms based on their r versus (θ,ϕ) dependences:

$$-\hbar^2\frac{\partial}{\partial r}\left(r^2\frac{\partial\psi}{\partial r}\right)-2m_{\text{eff}}r^2E\psi-\hbar^2\left[\frac{1}{\sin\theta}\frac{\partial}{\partial\theta}\left(\sin\theta\frac{\partial}{\partial\theta}\right)+\frac{1}{\sin^2\theta}\frac{\partial^2}{\partial\phi^2}\right]\psi=0.$$

Note that the third term is simply the quantum mechanical angular momentum operator ($\hat{L}^2$) (see Chapter 6, **Table 6.1**):

$$\hat{L}^2=-\hbar^2\left[\frac{1}{\sin\theta}\frac{\partial}{\partial\theta}\left(\sin\theta\frac{\partial}{\partial\theta}\right)-\frac{1}{\sin^2\theta}\frac{\partial^2}{\partial\phi^2}\right],$$

for which it turns out that $\hat{L}^2\psi=\hbar^2 l(l+1)\psi$, with l an integer.

As a consequence,

$$-\hbar^2\frac{\partial}{\partial r}\left(r^2\frac{\partial\psi}{\partial r}\right)-2m_{\text{eff}}r^2E\psi+\hat{L}^2\psi=0,$$

$$-\hbar^2\frac{\partial}{\partial r}\left(r^2\frac{\partial\psi}{\partial r}\right)-2m_{\text{eff}}Er^2\psi+\hbar^2 l(l+1)\psi=0,$$

$$\frac{\partial}{\partial r}\left(r^2\frac{\partial\psi}{\partial r}\right)+\frac{2m_{\text{eff}}E}{\hbar^2}r^2\psi-l(l+1)\psi=0.$$

Letting $k^2=2m_{\text{eff}}E/\hbar^2$ then gives

$$\frac{\partial}{\partial r}\left(r^2\frac{\partial\psi}{\partial r}\right)+\psi\left[(kr)^2-l(l+1)\right]=0.$$

We now solve this equation using a separable wavefunction of the form

$$\psi=x(r)y(\theta,\phi)\equiv xy.$$

The last equivalence just makes our subsequent notation simpler. When this is inserted into the above equation, we get

$$\frac{\partial}{\partial r}\left(r^2\frac{\partial(xy)}{\partial r}\right)+xy\left[(kr)^2-l(l+1)\right]=0,$$

or

$$y\frac{d}{dr}\left(r^2\frac{dx}{dr}\right)+xy\left[(kr)^2-l(l+1)\right]=0.$$

On applying the product rule for differentiation and dividing by y, we obtain

$$\left(r^2\frac{d^2x}{dr^2}+2r\frac{dx}{dr}\right)+x\left[(kr)^2-l(l+1)\right]=0.$$

We then change variables, letting $z = kr$ $(dr = dz/k)$, to find

$$(kr)^2 \frac{d^2x}{dz^2} + 2(kr)\frac{dx}{dz} + x\left[z^2 - l(l+1)\right] = 0.$$

This ultimately simplifies to

$$z^2 \frac{d^2x}{dz^2} + 2z\frac{dx}{dz} + x[z^2 - l(l+1)] = 0. \quad (7.40)$$

Equation 7.40 is called the *spherical Bessel equation*. Its solutions are linear combinations of *spherical Bessel functions* of the first (j_l) and second (y_l) kinds:

$$x(r) = Aj_l(z) + By_l(z),$$

where A and B are constant coefficients to be determined. Note that spherical Bessel functions are denoted by the lower-case letters j and y as opposed to their upper-case counterparts (J and Y), which represent normal Bessel functions. Furthermore, as with normal Bessel functions, spherical Bessel functions of the second kind diverge as $r \to 0$. As a consequence, to obtain physical wavefunctions for the particle, $B = 0$. We are therefore left with

$$x(r) = Aj_l(z).$$

When real units are put back in (i.e., $z = kr$ and $k = \alpha_{l,n}/a$), we find our radial wavefunction

$$x(r) = Aj_l\left(\frac{\alpha_{l,n}}{a} r\right), \quad (7.41)$$

where $\alpha_{l,n}$ is the root of the spherical Bessel function, l is its order, and n is the index of its particular root (also called a radial quantum number). As with the example of a particle in a cylinder, the conversion back to real units will become apparent when we discuss the particle's associated energies. **Table 7.2** lists the first few spherical Bessel functions and **Table 7.3** lists their roots. **Figure 7.19** also illustrates the first

Table 7.2 Spherical Bessel functions

Spherical Bessel function	Explicit expression
$j_0(z)$	$\frac{\sin z}{z}$
$j_1(z)$	$\frac{\sin z}{z^2} - \frac{\cos z}{z}$
$j_2(z)$	$\left(\frac{3}{z^2} - 1\right)\frac{\sin z}{z} - \frac{3\cos z}{z^2}$
$j_3(z)$	$\left(\frac{15}{z^3} - \frac{6}{z}\right)\frac{\sin z}{z} - \left(\frac{15}{z^2} - 1\right)\frac{\cos z}{z}$

Table 7.3 Roots of the spherical Bessel function $j_l(x)$

Root number n	$j_0(x)$	$j_1(x)$	$j_2(x)$	$j_3(x)$	$j_4(x)$
1	3.141593	4.493409	5.763459	6.87932	8.182561
2	6.283185	7.725252	9.095011	10.417119	11.704907
3	9.424778	10.904122	12.322941	13.698023	15.039665
4	12.566370	14.066194	15.514603	16.923621	18.301256
5	15.707963	17.220755	18.689036	20.121806	21.525418

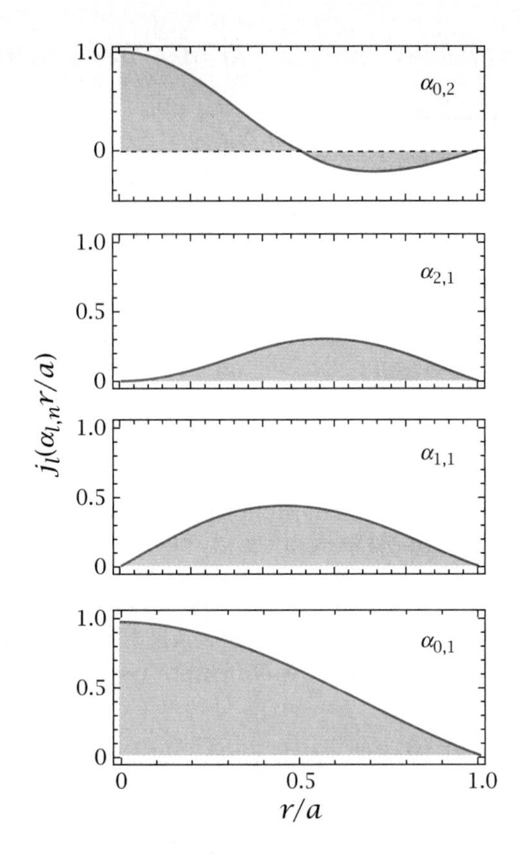

Figure 7.19 First four spherical Bessel functions in order of increasing $\alpha_{l,n}$ value.

few spherical Bessel functions to allow the reader to get a sense of what they look like.

Table 7.4 Spherical harmonic functions

Spherical harmonic function	Explicit expression
Y_0^0	$\frac{1}{\sqrt{4\pi}}$
Y_1^0	$\sqrt{\frac{3}{4\pi}} \cos\theta$
Y_1^1	$\sqrt{\frac{3}{8\pi}} \sin\theta\, e^{i\phi}$
Y_1^{-1}	$\sqrt{\frac{3}{8\pi}} \sin\theta\, e^{-i\phi}$
Y_2^0	$\sqrt{\frac{5}{16\pi}}(3\cos^2\theta - 1)$
Y_2^1	$\sqrt{\frac{15}{8\pi}} \sin\theta\cos\theta\, e^{i\phi}$
Y_2^{-1}	$\sqrt{\frac{15}{8\pi}} \sin\theta\cos\theta\, e^{-i\phi}$
Y_2^2	$\sqrt{\frac{15}{32\pi}} \sin^2\theta\, e^{2i\phi}$
Y_2^{-2}	$\sqrt{\frac{15}{32\pi}} \sin^2\theta\, e^{-2i\phi}$

Wavefunctions

The total wavefunction for the particle in a sphere is the product of the above radial term with an angular contribution. While the radial part is represented by spherical Bessel functions, angular contributions are spherical harmonics $Y_l^m(\theta, \phi)$. The reader may consult one of the quantum mechanics texts cited in Further Reading to see this if unfamiliar with the eigenfunctions of the square of the angular momentum operator, $\hat{L}^2$. **Table 7.4** lists the first few spherical harmonics.

The total wavefunction for a particle confined to a sphere is then

$$\psi(r, \theta, \phi) = N j_l\left(\frac{\alpha_{l,n}}{a} r\right) Y_l^m(\theta, \phi). \tag{7.42}$$

Note that the spherical harmonics are already normalized. Hence, N is really only a normalization constant for the spherical Bessel functions. It can be shown that the normalized, lowest-energy wavefunction for the particle in a sphere is

$$\psi(r, \theta, \phi) = \frac{1}{\sqrt{2\pi a}} \frac{\sin(\pi r/a)}{r}. \tag{7.43}$$

The reader may evaluate this as an exercise.

Energies

The particle's energies can be found from the relationship

$$k = \frac{\alpha_{l,n}}{a},$$

where $k = \sqrt{2m_{\text{eff}}E/\hbar^2}$ and $\alpha_{l,n}$ is the dual-index root of the spherical Bessel function. The equivalence occurs because the radial wavefunction must equal zero at the radius of the sphere. Thus, $j_l(ka) = j_l(\alpha_{l,n})$, and this makes it apparent how we converted from $j_l(z)$ to $j_l(\alpha_{l,n}r/a)$. Next, we find from the equivalence that

$$\sqrt{\frac{2m_{\text{eff}}E}{\hbar^2}} = \frac{\alpha_{l,n}}{a},$$

and solving for the energy yields

$$E = \frac{\hbar^2\alpha_{l,n}^2}{2m_{\text{eff}}a^2}. \tag{7.44}$$

In this expression, l is the order of the spherical Bessel function and n is the root's associated index (or radial quantum number).

Figure 7.20 is a schematic of the first few energies of a particle in a sphere along with their associated Bessel function roots. Note that, unlike the quantum well and nanowire, no near continuum of energies exists. This is because quantum dot energies are *discrete* atomic-like transitions, unlike the former two cases, where bands exist.

7.3.7 Particle-in-a-Sphere Electron/Hole Wavefunctions and Overlap Integrals for Selection Rules

Once again, we are interested in determining the likelihood of an optical transition between different quantum dot energy levels. This will be dealt with more extensively in Chapter 12, but for now let us simply evaluate the envelope function electron/hole overlap integral which reflects the promotion of an electron from one state to another, leaving behind a hole in the initial ground state. We have carried out such overlap integrals before in dealing with quantum wells and nanowires. The concept/implementation with quantum dots is no different.

The total wavefunctions for an electron and a hole, based on the above quantum dot model, are

$$\psi_e = N_e j_{l_e}\left(\frac{\alpha_{l_e,n_e}}{a}r\right) Y_{l_e}^{m_e}(\theta,\phi) \tag{7.45}$$

and

$$\psi_h = N_h j_{l_h}\left(\frac{\alpha_{l_h,n_h}}{a}r\right) Y_{l_h}^{m_h}(\theta,\phi). \tag{7.46}$$

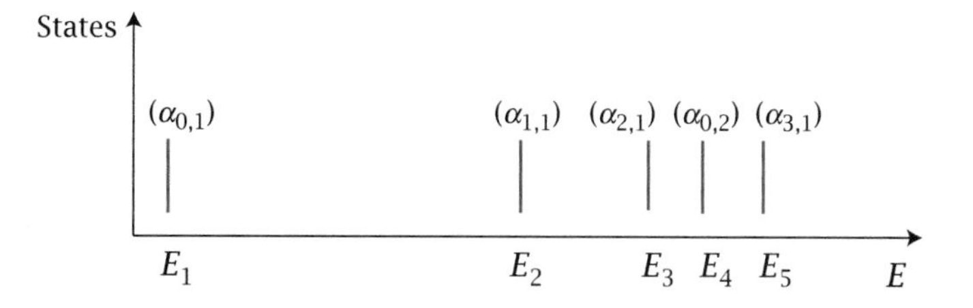

Figure 7.20 Energy levels of a particle confined to a three-dimensional sphere.

The overlap integral, written in Dirac bra–ket notation, is then

$$\langle \psi_e | \psi_h \rangle = \int_{V_{\text{tot}}} \psi_e^* \psi_h \, d^3 r.$$

This expands to

$$\begin{aligned}
&= N_e N_h \int_{V_{\text{tot}}} j_{l_e}^* Y_{l_e}^{m_e *} j_{l_h} Y_{l_h}^{m_h} \, d^3 r \\
&= N_e N_h \int_{V_{\text{tot}}} j_{l_e}^* Y_{l_e}^{m_e *} j_{l_h} Y_{l_h}^{m_h} r^2 \, dr \sin\theta \, d\theta \, d\phi \\
&= N_e N_h \int_0^a j_{l_e}^* j_{l_h} r^2 \, dr \int_0^{\pi} \int_0^{2\pi} Y_{l_e}^{m_e *} Y_{l_h}^{m_h} \sin\theta \, d\theta \, d\phi \\
&= \langle j_{l_e} | j_{l_h} \rangle \langle Y_{l_e}^{m_e} | Y_{l_h}^{m_h} \rangle
\end{aligned}$$

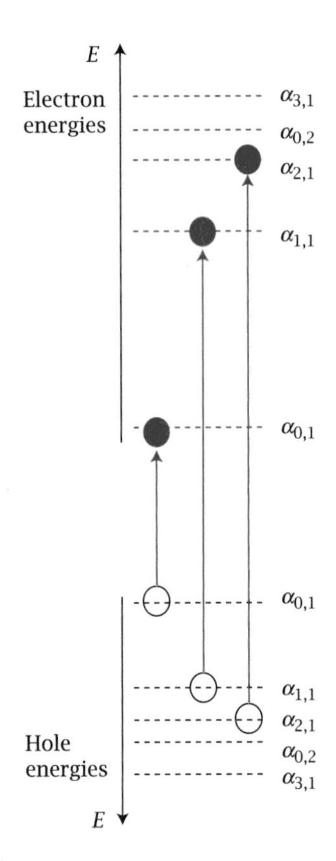

Figure 7.21 Optical transitions for a quantum dot allowed by its envelope function selection rules.

and readily evaluates to

$$\delta_{n_e, n_h} \delta_{l_e, l_h} \delta_{m_e, m_h},$$

given the orthogonality of the spherical Bessel functions (i.e., the spherical Bessel functions can be shown to be orthogonal over the interval between 0 and a) and their accompanying spherical harmonics. As a consequence, the quantum dot selection rules we seek can immediately be read off.

$\Delta n = 0$: the root indices (or radial quantum numbers) of the spherical Bessel functions must be equal;

$\Delta l = 0$: transitions are made between wavefunctions with spherical Bessel functions of the same order, i.e., s–s, p–p, d–d, recalling atomic orbital terminology associated with the orbital angular momentum quantum number l;

$\Delta m = 0$: the angular quantum number does not change.

Figure 7.21 shows a qualitative picture of these transitions between different quantum dot electron/hole energy levels. Note that, as with the quantum well and nanowire cases earlier, hole energies increase in the opposite direction of those of the electron. The reason for this will be discussed in Chapter 10.

The three lowest-energy transitions therefore occur between the following electron/hole states:

First Transition

$$\psi_e = N j_0 \left(\frac{\alpha_{01}}{a} r \right) Y_0^0(\theta, \phi),$$

$$\psi_h = N j_0 \left(\frac{\alpha_{01}}{a} r \right) Y_0^0(\theta, \phi),$$

with an energy

$$\begin{aligned}
E &= E_g + \frac{\hbar^2 \alpha_{01}^2}{2 m_{\text{eff},e} a^2} + \frac{\hbar^2 \alpha_{01}^2}{2 m_{\text{eff},h} a^2} \\
&= E_g + \frac{\hbar^2 \alpha_{01}^2}{2 \mu a^2},
\end{aligned}$$

where

$$\frac{1}{\mu} = \frac{1}{m_{\text{eff},e}} + \frac{1}{m_{\text{eff},h}}$$

is the electron/hole reduced mass, $m_{\text{eff},e}$ and $m_{\text{eff},h}$ are the electron and hole effective masses, and E_g is the bulk band gap (as will be seen later in Chapter 10). We refer to this as a 1s–1s transition.

Second Transition

$$\psi_e = Nj_1\left(\frac{\alpha_{11}}{a}r\right)Y_1^{1,0,-1}(\theta,\phi),$$
$$\psi_h = Nj_1\left(\frac{\alpha_{11}}{a}r\right)Y_1^{1,0,-1}(\theta,\phi),$$

where the superscript $1, 0, -1$ on the spherical harmonic just denotes that, for $l = 0$, m has a number of possible values: $m = 1, 0, -1$. Readers unfamiliar with this should consult one of the quantum mechanics texts in Further Reading, focusing on the sections dealing with the hydrogen atom problem. The associated transition energy is

$$\begin{aligned} E &= E_g + \frac{\hbar^2\alpha_{11}^2}{2m_{\text{eff},e}a^2} + \frac{\hbar^2\alpha_{11}^2}{2m_{\text{eff},h}a^2} \\ &= E_g + \frac{\hbar^2\alpha_{11}^2}{2\mu a^2}. \end{aligned}$$

The second transition can be referred to as a 1p–1p transition.

Third Transition

$$\psi_e = Nj_2\left(\frac{\alpha_{21}}{a}r\right)Y_2^{2,1,0,-1,-2}(\theta,\phi),$$
$$\psi_h = Nj_2\left(\frac{\alpha_{21}}{a}r\right)Y_2^{2,1,0,-1,-2}(\theta,\phi),$$

where again the superscript $2, 1, 0, -1, -2$ on the spherical harmonic just denotes that, for $l = 2$, m can take a number of possible values: $m = 2, 1, 0, -1, -2$. The associated transition energy is

$$\begin{aligned} E &= E_g + \frac{\hbar^2\alpha_{21}^2}{2m_{\text{eff},e}a^2} + \frac{\hbar^2\alpha_{21}^2}{2m_{\text{eff},h}a^2} \\ &= E_g + \frac{\hbar^2\alpha_{21}^2}{2\mu a^2}. \end{aligned}$$

This is referred to as a 1d–1d transition. As mentioned above, E_g represents the bulk band gap of the system. More about E_g will be seen in Chapter 10.

7.4 SUMMARY

In what follows, we will see a few more model quantum mechanics problems that can be solved to represent nanostructures and their confined carriers. However, what we are ultimately aiming for is Chapter 12 since interband transitions underlie the linear absorption and emission of semiconductor nanostructures. On the way there, we will need to cover three additional topics: the density of states (Chapter 9), the concept of bands (Chapter 10), and time-dependent perturbation theory (Chapter 11).

7.5 THOUGHT PROBLEMS

7.1 Particle in a symmetric one-dimensional box

Find the energies and normalized wavefunctions of a particle in a symmetric one-dimensional box with barriers at $x = -\frac{1}{2}a$ and $x = \frac{1}{2}a$. Assume the free electron mass m_0.

7.2 Particle in a one-dimensional box

Consider the standard problem of a particle in a symmetric one-dimensional box. Estimate the energy difference between the first and second excited states of an electron where $a = 5$ nm. Carry out the same evaluation for a much larger 1 kg object confined to a macroscopic $a = 1$ m box. Comment on the relevance of confinement effects in each case.

7.3 Degeneracies: two-dimensional box

Calculate the first four energy levels of a particle in a two-dimensional box where $a = b = 5$ nm. Compare these energies with those in the case where $a = 5$ nm and $b = 10$ nm.

(a) What happens to the degeneracies of the levels?

(b) Generally speaking, why are the energies of the second case lower than the first?

(c) When are degeneracies most likely to be present?

(d) Finally, compare these energies with the thermal energy at room temperature and explain which levels are most likely to be populated.

Express all answers in eV.

7.4 Degeneracies: three-dimensional box

Calculate the first seven energy levels of an electron in an $a =$ 5 nm three-dimensional box. Compare these results with those found for a particle in a sphere in Problem 7.5.

7.5 Degeneracies: particle in a sphere

Calculate the eigenenergies of a free electron (mass m_0) in a 5 nm diameter sphere for $l = 0, 1, 2, 3$ using

(a) the lowest root of the Bessel function

(b) the second-lowest root

(c) the third-lowest root.

Assume a particle-in-a-sphere model. Summarize all these energies in ascending order in one table.

7.6 Wavefunctions: particle in a sphere

Use a mathematical modeling program to draw the radial wavefunctions corresponding to the lowest three energies in the table that you drew up earlier as the answer to Problem 7.5. Make sure to normalize each wavefunction.

7.7 Normalization: three-dimensional box

Normalize the total wavefunction of a carrier in a three-dimensional box with infinite potentials on all sides and with an edge length a. The carrier's wavefunction is

$$\psi_{n_x,n_y,n_z}(x, y, z) = N \sin\frac{n_x \pi x}{a} \sin\frac{n_y \pi y}{a} \sin\frac{n_z \pi z}{a}$$

in the interval $0 \le x \le a$ and $0 \le y \le a$ and $0 \le z \le a$.

7.8 Normalization: particle in a cylinder

Find the normalization constant N for the particle-in-a-cylinder wavefunction (Equation 7.26)

$$\psi(r, \theta, \phi) = N J_m\left(\frac{\alpha_{m,n}}{a} r\right) e^{im\theta} \sqrt{\frac{2}{L_z}} \sin\frac{n_z \pi z}{L_z}.$$

Hint:

$$\int_0^a J_m^2\left(\frac{\alpha_{m,n}}{a} r\right) r\, dr = \frac{a^2}{2} J_{m+1}^2(\alpha_{m,n}).$$

7.9 Normalization: particle in a sphere

Find the normalization constant for the lowest particle-in-a-sphere wavefunction. Then show that the entire wavefunction is normalized.

7.10 Probabilities: two-dimensional box

For a particle confined to a two-dimensional infinite potential in the interval $0 \le x \le a$ and $0 \le y \le b$, calculate the probability that the particle is found within the region $0 \le y \le \frac{1}{2}b$ and $0 \le x \le \frac{1}{2}a$.

7.11 Probabilities: particle in a sphere

For an electron trapped in an infinite spherical well with radius a, calculate the probability of finding the electron inside an inner sphere of radius $\frac{1}{2}a$. Assume the lowest-energy wavefunction.

7.12 Transition probabilities: three-dimensional box

Determine the generic transition moment and transition probability P for a particle in a three-dimensional box and find the associated selection rules. Assume x-polarized light and recall that

$$P \propto |\langle n, j, k|\hat{\mu} \cdot \hat{\mathbf{e}}|1, 1, 1\rangle|^2,$$

where n, j, and k are quantum numbers associated with the x, y, and z directions.

7.13 Transition probabilities: three-dimensional box

For a three-dimensional box with $a = 100$ Å and the incident excitation light polarized along the x direction, calculate the energies of all possible transitions up to $n = 10$ and provide their intensities as defined by the transition probability ratio $P_{n,1,1}/P_{2,1,1}$. Express your energies in eV.

7.14 Confinement: general

Does the effective band gap of a low-dimensional semiconductor increase with degree of confinement (the band gap is akin to the HOMO–LUMO gap of a molecular system)? If so, what are the relative electron and hole contributions? Provide a mathematical expression. Which term dominates and why? We will see more about effective masses in Chapter 10.

7.15 Confinement and Coulomb

Often enough, there is a Coulomb attraction between the electron and hole in a confined system. As a consequence the energies one calculates via particle-in-a-box type expressions are often corrected for this Coulomb attraction. Think back to the basic expression for the Coulomb energy and qualitatively explain why this term is important or not important in a confined low-dimensional system. *Hint*: for simplicity, think of a spherically symmetric system and think of how these energies scale with r or a.

7.16 Confinement: nanorod

Imagine you are studying a single CdSe nanorod. The diameter of the nanorod is 5 nm and its aspect ratio is 10. The bulk exciton Bohr radius of CdSe is 5.6 nm. The bulk band gap at room temperature is 1.74 eV. The effective electron and heavy hole masses are $m_e = 0.13m_0$ and $m_h = 1.1m_0$.

Model the absorption spectrum of this nanorod by

- Providing the energies of the three lowest transitions in eV.
- Providing their corresponding wavelengths in nm.
- Providing their approximate colors.

Make any necessary approximations. State them clearly.

7.17 Confinement: nanowire

Consider InP nanowires with diameters of 10, 15, and 20 nm. See Gudiksen et al. (2002). Assume $m_e = 0.078m_0$ and $m_h = 0.4m_0$. Calculate the energy of the first three optical transitions. For simplicity, ignore the length of the wire. What colors do you expect these wires to emit: UV, visible, IR? How does this compare with the results in the paper by Gudiksen et al.?

7.18 Intrawire variations: nanowire

Consider the same 10 nm diameter InP nanowire as in Problem 7.17. Assume that the wire exhibits a 5% width variation along its length. In effect, one part of the wire is slightly thicker than another. Estimate the magnitude of the emission energy variation, assuming that the emission occurs from the lowest-energy transition in all parts of the same wire. As with Problem 7.17, assume that $m_e = 0.078m_0$ and $m_h = 0.4m_0$.

7.19 Polarization anisotropy: nanowire

Very crudely, model the absorption polarization anisotropy of a nanowire. Assume that the wire is oriented along the z direction with one end at the origin and the other at L_z. Furthermore, assume light traveling along the x direction and polarized along either the y or z directions. The polarization anisotropy in this case is defined as

$$\rho = \frac{I_\parallel - I_\perp}{I_\parallel + I_\perp},$$

where $I_\parallel \propto |\langle\psi_e|z|\psi_h\rangle|^2$ ($I_\perp \propto |\langle\psi_e|y|\psi_h\rangle|^2$) represents the amount of light absorbed whose polarization vector is parallel (perpendicular) to the nanowire growth axis. Assume that we are dealing with the lowest electron and hole excited states. For additional information on this topic, see Wang et al. (2001).

7.20 Colloidal quantum dots

New syntheses of colloidal quantum dots are being developed that exhibit narrow size distributions. Often, in addition to TEM characterization of the particles (Chapter 14), authors will characterize their size distribution in terms of their emission linewidth (see, e.g., Wu et al. 2005). For a 5 nm diameter CdSe quantum dot, estimate the impact a 5% diameter variation has on the emission energy spread. This ignores what is called homogeneous broadening but gives a sense of the inhomogeneous broadening that exists due to the residual size distribution of the ensemble. Model the dot using the particle-in-a-sphere model and assume that $m_e = 0.13m_0$ and $m_h = 1.1m_0$.

7.21 Phonon bottleneck

In quantum dots, it has been predicted that the relaxation of excitations can be slow due to the discreteness of excited states (see, e.g., Benisty et al. 1991). This is because relaxation between excited states often occurs by the emission of optical phonons. Given quantized phonon energies, this then requires that the energy spacing between quantum dot states, ΔE, satisfies

$$\Delta E = nE_{LO}$$

where $n = 1, 2, 3, \ldots$ and where E_{LO} is the system's LO (longitudinal optical) phonon energy. Often, this is not the case. Furthermore, even if the condition is met, the probability of a multiphonon process is small. As a consequence, a phonon "bottleneck" is said to exist in quantum dots.

Provide an expression for the number of phonons needed to account for the energy difference between the first and second excited states of a quantum dot modeled as a particle in a sphere. Then, for a 5 nm-diameter CdSe quantum dot where $m_e = 0.13m_0$, $m_h = 1.1m_0$, and $E_{LO} = 26$ meV, calculate n. Round values as needed.

7.22 Confinement

Qualitatively compare the size-dependent energies of a particle in a one-dimensional box (quantum well), an infinite circular potential (quantum wire), and a spherical box (quantum dot). Basically ask yourself how all these energies scale with r or a. Are there any similarities? Now take a look at Yu et al. (2003) and comment on the conclusions of the paper. See if you can reproduce the plots shown there.

7.23 Matching conditions

Core/shell semiconductor particles are popular these days since they were found to improve the quantum yield of many colloidal quantum dot systems. Read Dabbousi et al. (1997), Schooss et al. (1994), and Kim et al. (2003), with particular emphasis on the first article. Try to reproduce the plot of the radial wavefunction for both the electron and hole in the Dabbousi et al. paper. Do this for the first (CdSe/matrix) and second (CdSe/ZnS) cases.

7.6 REFERENCES

Benisty H, Sotomayor-Torrés CM, Weisbuch C (1991) Intrinsic mechanism for the poor luminescence properties of quantum-box systems. *Phys. Rev. B* **44**, 10 945.

Dabbousi BO, Rodriguez-Viejo J, Mikulec FV, et al. (1997) (CdSe)ZnS core-shell quantum dots: synthesis and characterization of a size series of highly luminescent nanocrystallites. *J. Phys. Chem. B* **101**, 9463.

Gudiksen MS, Wang J, Lieber CM (2002) Size-dependent photoluminescence from single indium phosphide nanowires. *J. Phys. Chem. B* **106**, 4036.

Kim S, Fisher B, Eisler HJ, Bawendi M (2003) Type-II quantum dots: CdTe/CdSe (core/shell) and CdSe/ZnTe (core/shell) heterostructures. *J. Am. Chem. Soc.* **125**, 11 466.

Schooss D, Mews A, Eychmuller A, Weller H (1994) Quantum-dot quantum well CdS/HgS/CdS: theory and experiment. *Phys. Rev. B* **49**, 17 072.

Wang J, Gudiksen MS, Duan X, et al. (2001) Highly polarized photoluminescence and photodetection from single indium phosphide nanowires. *Science* **293**, 1455.

Wu D, Kordesch ME, Van Patten PG (2005) A new class of capping ligands for CdSe nanocrystal synthesis. *Chem. Mater.* **17**, 6436.

Yu H, Li J, Loomis RA, et al. (2003) Two- versus three-dimensional quantum confinement in indium phosphide wires and dots. *Nature Mat.* **2**, 517.

7.7 FURTHER READING

Introductory Quantum Mechanics Texts

Green NJB (1997) *Quantum Mechanics 1: Foundations.* Oxford University Press, Oxford, UK.

Green NJB (1998) *Quantum Mechanics 2: The Toolkit.* Oxford University Press, Oxford, UK.

Griffiths DJ (2004) *Introduction to Quantum Mechanics*, 2nd edn. Prentice Hall, Upper Saddle River, NJ.

Liboff RL (2003) *Introductory Quantum Mechanics.* Addison-Wesley, San Francisco, CA.

McMahon D (2005) *Quantum Mechanics Demystified.* McGraw-Hill, New York.

McQuarrie DA (2008) *Quantum Chemistry*, 2nd edn. University Science Books, Mill Valley, CA.

More Advanced Quantum Mechanics Texts

Cohen-Tannoudji C, Diu B, Laloe F (1977) *Quantum Mechanics* (2 volumes). Wiley, New York.

Zettili N (2009) *Quantum Mechanics, Concepts and Applications*, 2nd edn. Wiley, Chichester, UK.

Additional Model Problems

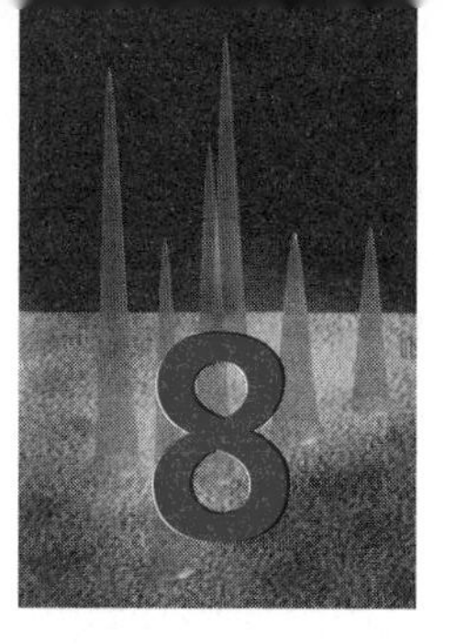

8.1 INTRODUCTION

In Chapter 7, we found the wavefunctions and energies of a particle confined to various geometries with infinite potentials surrounding their internal (potential-free) regions. However, in reality, surrounding potentials are never infinite. As a consequence, there exists some leakage of the wavefunction into these regions. This phenomenon is referred to as tunneling and will be important later when dealing with the problem of a particle in a periodic potential—a model that ultimately leads us to the concept of bands.

For now, though, let us determine the wavefunctions and energies of a particle confined to three additional model systems. Two of them—the particle in a finite one-dimensional box and the harmonic oscillator—illustrate the above tunneling phenomenon. The third is a subset of the previous particle-in-a-cylinder problem and can be used to describe so-called quantum corrals. Note that many other problems exist that we could deal with here. All seek to model the behavior of carriers localized within different physical and/or electronic geometries. However, we leave them as exercises for the reader.

8.2 PARTICLE IN A FINITE ONE-DIMENSIONAL BOX

In our first problem, we revisit the problem of a particle in a one-dimensional box. This time, instead of having an infinite potential for the carrier outside the region of interest, we let it be finite. **Figure 8.1** illustrates the situation. For the potential,

$$V(x) = \begin{cases} 0 & \text{where } 0 < x < a, \\ V & \text{elsewhere.} \end{cases}$$

For the boundary conditions, our wavefunction matching conditions (Chapter 6) require that

$$\begin{aligned} \psi_1(0) &= \psi_2(0), \\ \psi_1'(0) &= \psi_2'(0), \\ \psi_2(a) &= \psi_3(a), \\ \psi_2'(a) &= \psi_3'(a). \end{aligned}$$

Note that the wavefunction in the region $0 < x < a$ where $V = 0$ is wave-like, as we saw earlier in the problem of a particle in an infinite box. The only difference is that, since the confining potential is not infinite, there will be a nonzero probability of finding the particle in Regions 1 and 3 shown in **Figure 8.1**.

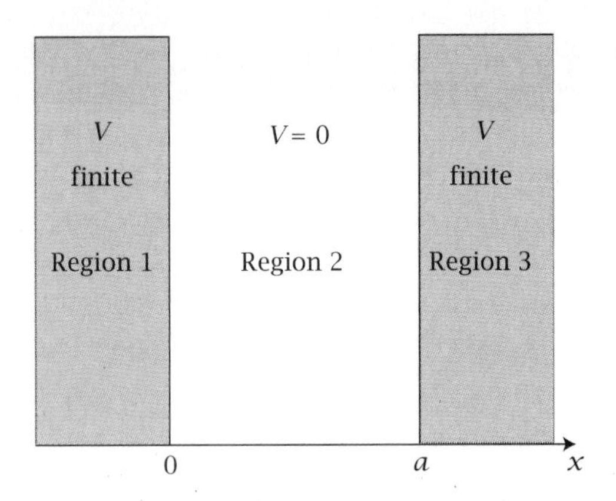

Figure 8.1 Potential for a particle in a finite one-dimensional box.

Once we have determined what the wavefunctions for all three regions look like, we will apply the above wavefunction matching conditions to ensure wavefunction continuity at the well/barrier interfaces. In all cases, we assume $V > E$ and that m_{eff}, the effective mass of the particle, is the same in all regions. Note that this latter assumption is not generally valid, since, in practice, the well and barrier regions are often made of different materials.

8.2.1 Wavefunction in Region 1

The Schrödinger equation in Region 1 has the form

$$-\frac{\hbar^2}{2m_{\text{eff}}}\frac{d^2\psi_1}{dx^2} + V\psi_1 = E\psi_1.$$

This can be rearranged to

$$\frac{d^2\psi_1}{dx^2} = \frac{2m_{\text{eff}}(V-E)}{\hbar^2}\psi_1$$

whereupon letting $\beta^2 = 2m_{\text{eff}}(V-E)/\hbar^2$ and consolidating terms gives

$$\frac{d^2\psi_1}{dx^2} - \beta^2\psi_1 = 0.$$

This equation is similar to what we have seen previously in Chapter 7. The exception is the negative sign in front of the second term. As a consequence, rather than general solutions being linear combinations of plane waves, they are linear combinations of growing and decaying exponentials:

$$\psi_1 = Ae^{\beta x} + Be^{-\beta x},$$

where A and B are constant coefficients to be determined, using the constraints/matching conditions of the problem. We will return to this in a moment.

8.2.2 Wavefunction in Region 2

The Schrödinger equation in Region 2 is

$$-\frac{\hbar^2}{2m_{\text{eff}}}\frac{d^2\psi_2}{dx^2} = E\psi_2,$$

since the potential between 0 and a is zero. We rearrange this to the form

$$\frac{d^2\psi_2}{dx^2} + \frac{2m_{\text{eff}}E}{\hbar^2}\psi_2 = 0$$

and let $k^2 = 2m_{\text{eff}}E/\hbar^2$ to obtain

$$\frac{d^2\psi_2}{dx^2} + k^2\psi_2 = 0.$$

As before, general solutions to this second-order differential equation are linear combinations of plane waves:

$$\psi_2 = Ce^{ikx} + De^{-ikx}.$$

8.2.3 Wavefunction in Region 3

The Schrödinger equation in Region 3 is identical to that in Region 1. So, rather than work this out explicitly, we can immediately see that the solution has the same form as in Region 1. We therefore have

$$\psi_3 = Fe^{\beta x} + Ge^{-\beta x},$$

where F and G are constant coefficients, determined using the constraints/matching conditions of the problem.

8.2.4 Summary

At this point, the wavefunctions for Regions 1, 2, and 3 are

$$\begin{aligned}\psi_1 &= Ae^{\beta x} + Be^{-\beta x},\\ \psi_2 &= Ce^{ikx} + De^{-ikx},\\ \psi_3 &= Fe^{\beta x} + Ge^{-\beta x}.\end{aligned}$$

We can now solve for the coefficients, A, B, C, D, F, and G by applying our quantum mechanical matching conditions. Namely,

$$\begin{aligned}\psi_1(0) &= \psi_2(0),\\ \psi_1'(0) &= \psi_2'(0)\end{aligned}$$

and

$$\begin{aligned}\psi_2(a) &= \psi_3(a),\\ \psi_2'(a) &= \psi_3'(a).\end{aligned}$$

However, before doing this, it is apparent that some of these coefficients can already be dropped since they lead to unreasonable wavefunctions. Specifically, the coefficients B and F can be excluded because Chapter 6 has shown that physically relevant wavefunctions must be finite and hence must not diverge as $x \to \infty$ or $x \to -\infty$ (i.e., both growing exponential terms are unphysical since this would mean that the wavefunction could not be normalized). As a consequence, we are left with the following wavefunctions in Regions 1, 2, and 3:

$$\begin{aligned}\psi_1 &= Ae^{\beta x},\\ \psi_2 &= Ce^{ikx} + De^{-ikx},\\ \psi_3 &= Ge^{-\beta x}.\end{aligned}$$

On applying the above matching conditions, we are left with a series of four simultaneous equations to solve:

$$\begin{aligned}
\psi_1(0) = \psi_2(0) &\Longrightarrow A = C + D, \\
\psi_1'(0) = \psi_2'(0) &\Longrightarrow \beta A = ikC - ikD, \\
\psi_2(a) = \psi_3(a) &\Longrightarrow Ce^{ika} + De^{-ika} = Ge^{-\beta a}, \\
\psi_2'(a) = \psi_3'(a) &\Longrightarrow ikCe^{ika} - ikDe^{-ika} = -\beta Ge^{-\beta a},
\end{aligned}$$

which can be rearranged to yield

$$\begin{aligned}
A - C - D &= 0, \\
\beta A - ikC + ikD &= 0, \\
Ce^{ika} + De^{-ika} - Ge^{-\beta a} &= 0, \\
ikCe^{ika} - ikDe^{-ika} + \beta Ge^{-\beta a} &= 0.
\end{aligned}$$

In matrix form, these linear equations can be written as

$$\begin{pmatrix} 1 & -1 & -1 & 0 \\ \beta & -ik & ik & 0 \\ 0 & e^{ika} & e^{-ika} & -e^{-\beta a} \\ 0 & ike^{ika} & -ike^{-ika} & \beta e^{-\beta a} \end{pmatrix} \begin{pmatrix} A \\ C \\ D \\ G \end{pmatrix} = \begin{pmatrix} 0 \\ 0 \\ 0 \\ 0 \end{pmatrix},$$

where the solution is either the trivial solution, $A = C = D = G = 0$, or the result from having the determinant of the first matrix be zero. Since we are not interested in the trivial solution, we choose the latter option. Note that in what follows we assume some knowledge of linear algebra. If necessary, the reader should consult one of the books listed in Further Reading for more information on solving a system of linear equations.

The determinant we wish to find is

$$\begin{vmatrix} 1 & -1 & -1 & 0 \\ \beta & -ik & ik & 0 \\ 0 & e^{ika} & e^{-ika} & -e^{-\beta a} \\ 0 & ike^{ika} & -ike^{-ika} & \beta e^{-\beta a} \end{vmatrix} = 0.$$

It can be evaluated a number of ways. In what follows, we first apply a technique called Gaussian elimination to make our determinant "lower diagonal." The Gaussian elimination approach simply entails taking linear combinations of rows (or of rows multiplied by some constant) in order to place zeros along the lower left corner of the determinant (i.e., the lower diagonal).

To start, we multiply the first row by $-\beta$ and add it to the second row. This gives us a new second row. We denote this with the following shorthand notation:

$$-\beta(\text{row } 1) + (\text{row } 2) \rightarrow (\text{row } 2).$$

This gives

$$\begin{vmatrix} 1 & -1 & -1 & 0 \\ 0 & \beta - ik & \beta + ik & 0 \\ 0 & e^{ika} & e^{-ika} & -e^{-\beta a} \\ 0 & ike^{ika} & -ike^{-ika} & \beta e^{-\beta a} \end{vmatrix} = 0.$$

Notice the new zero in the second row.

Next, we replace ike^{ika} (the second element of the last row) with a zero. This is achieved by multiplying the third row with $-ik$ and adding it to the fourth row. We denote this by

$$-ik(\text{row } 3) + (\text{row } 4) \to (\text{row } 4).$$

The result is

$$\begin{vmatrix} 1 & -1 & -1 & 0 \\ 0 & \beta - ik & \beta + ik & 0 \\ 0 & e^{ika} & e^{-ika} & -e^{-\beta a} \\ 0 & 0 & -2ike^{-ika} & (\beta + ik)e^{-\beta a} \end{vmatrix} = 0.$$

To further simplify things, let us multiply the second row by $-1/(\beta - ik)$ and call the result our new second row. We can also multiply the third row by e^{-ika} and call the result our new third row. These actions are denoted by the following shorthand notation:

$$-\frac{1}{\beta - ik}(\text{row } 2) \to (\text{row } 2),$$
$$e^{-ika}(\text{row } 3) \to (\text{row } 3),$$

and yield

$$\begin{vmatrix} 1 & -1 & -1 & 0 \\ 0 & -1 & -\dfrac{\beta + ik}{\beta - ik} & 0 \\ 0 & 1 & e^{-2ika} & -e^{-ika-\beta a} \\ 0 & 0 & -2ike^{-ika} & (\beta + ik)e^{-\beta a} \end{vmatrix} = 0.$$

At this point, we sum the second and third rows to obtain a new third row:

$$(\text{row } 2) + (\text{row } 3) \to (\text{row } 3).$$

In this way, we continue adding zeros along the determinant's lower diagonal. We thus find that

$$\begin{vmatrix} 1 & -1 & -1 & 0 \\ 0 & -1 & -\dfrac{\beta + ik}{\beta - ik} & 0 \\ 0 & 0 & e^{-2ika} - \dfrac{\beta + ik}{\beta - ik} & -e^{-ika-\beta a} \\ 0 & 0 & -2ike^{-ika} & (\beta + ik)e^{-\beta a} \end{vmatrix} = 0.$$

Finally, multiplying the second row by -1,

$$-(\text{row } 2) \to (\text{row } 2),$$

gives

$$\begin{vmatrix} 1 & -1 & -1 & 0 \\ 0 & 1 & \dfrac{\beta + ik}{\beta - ik} & 0 \\ 0 & 0 & e^{-2ika} - \dfrac{\beta + ik}{\beta - ik} & -e^{-ika-\beta a} \\ 0 & 0 & -2ike^{-ika} & (\beta + ik)e^{-\beta a} \end{vmatrix} = 0.$$

At this point, the method of subdeterminants can be invoked to show that solving the original 4×4 determinant is now equivalent to evaluating the following 2×2 determinant:

$$\begin{vmatrix} e^{-2ika} - \dfrac{\beta + ik}{\beta - ik} & -e^{-ika-\beta a} \\ -2ike^{-ika} & (\beta + ik)e^{-\beta a} \end{vmatrix} = 0.$$

The reader may again consult one of the books listed in Further Reading to learn more about the properties of determinants.

When this 2×2 determinant is evaluated, we obtain the following equation

$$\left(e^{-2ika} - \frac{\beta + ik}{\beta - ik}\right)(\beta + ik)e^{-\beta a} - 2ike^{-2ika-\beta a} = 0.$$

When expanded, this yields

$$(\beta + ik)e^{-2ika-\beta a} - \frac{(\beta + ik)^2}{\beta - ik}e^{-\beta a} - 2ike^{-2ika-\beta a} = 0.$$

Upon further simplification, we obtain

$$(\beta - ik)^2 e^{-2ika} - (\beta + ik)^2 = 0$$

whereupon multiplying both sides by e^{ika} gives

$$(\beta - ik)^2 e^{-ika} - (\beta + ik)^2 e^{ika} = 0.$$

Further simplification leads to

$$2i(\beta^2 - k^2)\frac{e^{ika} - e^{-ika}}{2i} + 2(2i\beta k)\frac{e^{ika} + e^{-ika}}{2} = 0,$$

or

$$2i(\beta^2 - k^2)\sin ka + 4i\beta k \cos ka = 0.$$

Finally, dividing the result by $2i \cos ka$ yields

$$(\beta^2 - k^2)\tan ka + 2\beta k = 0,$$

and results in our desired final expression:

$$\tan ka = \frac{2\beta k}{k^2 - \beta^2},$$

where

$$k = \sqrt{\frac{2m_{\text{eff}}E}{\hbar^2}},$$

$$\beta = \sqrt{\frac{2m_{\text{eff}}(V - E)}{\hbar^2}}.$$

We therefore have

$$\tan\left(\sqrt{\frac{2m_{\text{eff}}E}{\hbar^2}}a\right) = \frac{2\sqrt{E(V - E)}}{2E - V} \tag{8.1}$$

as the actual equation one solves numerically to find the allowed energies of the particle. In practice, this is done using a computer to find the roots of

$$\tan\left(\sqrt{\frac{2m_{\text{eff}}E}{\hbar^2}}a\right) - \frac{2\sqrt{E(V - E)}}{2E - V} = 0$$

(i.e., the values of E that make this relationship true). The one thing we immediately notice is that, for the same well width, the finite box yields slightly lower energies relative to results from the particle in an infinite box. This is because the finite barrier box well width is effectively larger and consequently leads to correspondingly smaller energies.

EXAMPLE 8.1

A Comparison Between an Infinite and a Finite Potential Box for a GaAs Quantum Well We can illustrate this result using a GaAs quantum well modeled as a simple one-dimensional box. Let us assume an electron effective mass $m_{\text{eff}} = 0.067m_0$ and a box well width $a = 1 \times 10^{-8}$ m. From before, the energies of the particle in an infinite box are $E = n^2h^2/8m_{\text{eff}}a^2$. The lowest three energies are therefore

$$E_1 = 0.056\,\text{eV},$$
$$E_2 = 0.224\,\text{eV},$$
$$E_3 = 0.504\,\text{eV}.$$

If we now assume a finite box with a barrier height of $V = 1$ eV, Equation 8.1 shows that the corresponding lowest three energies are

$$E_1 = 0.042\,\text{eV},$$
$$E_2 = 0.168\,\text{eV},$$
$$E_3 = 0.374\,\text{eV}.$$

By comparing these two sets of results, it is apparent that the finite-box energies are smaller. This illustrates how leakage of the wavefunction into the barrier makes the effective box larger.

Finally, note that one can go back and find what the particle's wavefunctions actually look like. This involves solving for the coefficients A, C, D, and G and enforcing wavefunction normalization. Although this can be done, it turns out to be tedious. So at this point, let us stop here and understand that such a task can be undertaken, leading to wavefunctions that qualitatively look like those shown in **Figure 8.2**.

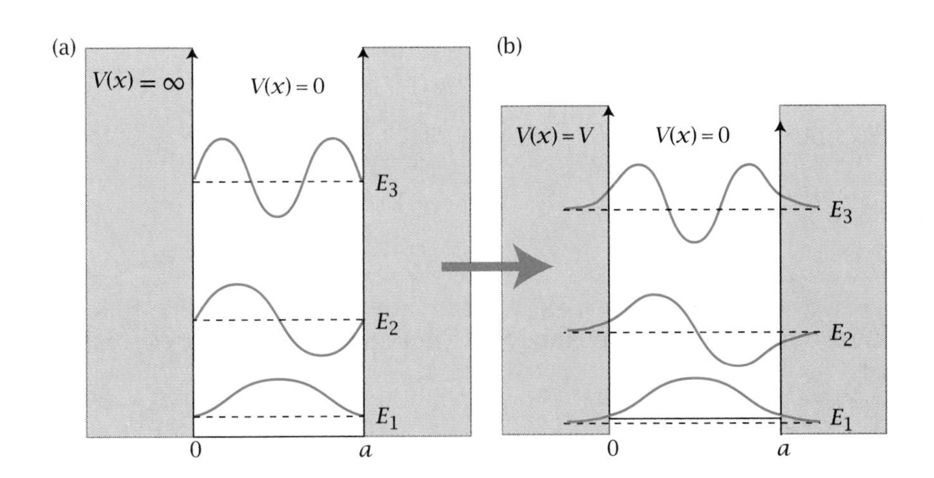

Figure 8.2 Wavefunctions for a particle in (a) an infinite and (b) a finite one-dimensional box.

8.3 PARTICLE IN AN INFINITE CIRCULAR BOX

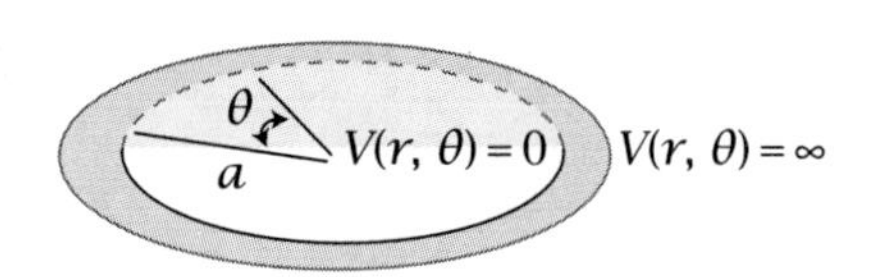

Figure 8.3 Potential for a particle confined to an infinite circular box.

We will now deal with a standard quantum mechanical problem useful for modeling the electronic properties of quantum corrals or even nanowires when longitudinal contributions to the carrier wavefunctions and energies are ignored. More about quantum corrals can be found in Crommie et al. (1993, 1995). This is essentially a two-dimensional problem where there exists an infinite potential for a particle trapped within a given radius a from the origin, as illustrated in **Figure 8.3**.

For this problem, the potential is

$$V(r,\theta) = \begin{cases} 0 & \text{where } 0 < r < a, \\ \infty & \text{elsewhere.} \end{cases}$$

The corresponding wavefunction boundary conditions are

$$\psi(r = a, \theta) = 0,$$

$$\psi(r,\theta) = \psi(r,\theta + 2\pi), \quad \text{where } 0 < r < a \text{ (i.e., periodicity).}$$

The Schrödinger equation is then solved with the above constraints to yield the confined particle's wavefunctions and energies.

Since the Laplacian in polar coordinates is

$$\nabla^2 = \frac{1}{r}\frac{\partial}{\partial r}\left(r\frac{\partial}{\partial r}\right) + \frac{1}{r^2}\frac{\partial^2}{\partial\theta^2}, \tag{8.2}$$

the Schrödinger equation is

$$-\frac{\hbar^2}{2m_{\text{eff}}}\left[\frac{1}{r}\frac{\partial}{\partial r}\left(r\frac{\partial}{\partial r}\right) + \frac{1}{r^2}\frac{\partial^2}{\partial\theta^2}\right]\psi = E\psi.$$

We simplify it as follows:

$$\left[\frac{1}{r}\frac{\partial}{\partial r}\left(r\frac{\partial}{\partial r}\right) + \frac{1}{r^2}\frac{\partial^2}{\partial\theta^2}\right]\psi = -\frac{2m_{\text{eff}}E}{\hbar^2}\psi,$$

letting $k^2 = 2m_{\text{eff}}E/\hbar^2$ and consolidating terms to get

$$\left[\frac{1}{r}\frac{\partial}{\partial r}\left(r\frac{\partial}{\partial r}\right) + \frac{1}{r^2}\frac{\partial^2}{\partial\theta^2}\right]\psi + k^2\psi = 0.$$

Next, multiplying by r^2 gives

$$r\frac{\partial}{\partial r}\left(r\frac{\partial\psi}{\partial r}\right) + (kr)^2\psi + \frac{\partial^2\psi}{\partial\theta^2} = 0,$$

where one notices that the first two terms depend only on r while the third term depends only on θ. We therefore assume a separable solution of the form

$$\psi = A(r)B(\theta) \equiv AB.$$

The last equivalence just makes our notation simpler in what follows.

Before going on, let us evaluate the first term in the above expression. When AB is inserted into it, we find that

$$r\frac{\partial}{\partial r}\left[r\frac{\partial(AB)}{\partial r}\right] = B\left(r^2A'' + rA'\right).$$

We therefore replace this into our full expression to get

$$Br^2\frac{d^2A}{dr^2} + Br\frac{dA}{dr} + (kr)^2AB + A\frac{d^2B}{d\theta^2} = 0.$$

At this point, dividing by AB and isolating terms based on their r or θ dependences gives

$$\frac{1}{A}r^2\frac{d^2A}{dr^2} + \frac{1}{A}r\frac{dA}{dr} + (kr)^2 = -\frac{1}{B}\frac{d^2B}{d\theta^2}.$$

Notice that the left-hand side depends only on r while the right-hand side depends only on θ. Since this equality holds for any value of θ or r, both sides must equal a constant. Call it m^2 (as usual, this is not to be confused with a mass).

We now have two equations to solve:

$$-\frac{1}{B}\frac{d^2B}{d\theta^2} = m^2$$

and

$$\frac{r^2}{A}\frac{d^2A}{dr^2} + \frac{r}{A}\frac{dA}{dr} + (kr)^2 = m^2.$$

Let us solve for B first. This turns out to be the easier of the two equations. We have

$$-\frac{1}{B}\frac{d^2B}{d\theta^2} = m^2,$$

which we rearrange into a familiar form,

$$\frac{d^2B}{d\theta^2} + m^2B = 0.$$

As before, general solutions to this second-order differential equation are linear combination of plane waves:

$$B(\theta) = C_1e^{im\theta} + C_2e^{-im\theta}.$$

Since there are no constraints defining what C_1 or C_2 are, we are free to choose a solution. For convenience, we let $C_2 = 0$, giving

$$B(\theta) = C_1e^{im\theta}.$$

Note that m is an integer. This is a consequence of the wavefunction having to satisfy periodicity. Namely, since $B(\theta) = B(\theta + 2\pi)$,

$$\begin{aligned} C_1e^{im\theta} &= C_1e^{im(\theta+2\pi)} \\ &= C_1e^{im\theta}e^{im2\pi} \end{aligned}$$

and $e^{im2\pi} = 1$. From the Euler relation, $e^{im2\pi} = \cos m2\pi + i\sin m2\pi = 1$. Thus, for this to be true, $m = 0, \pm1, \pm2, \pm3, \ldots$.

Next, let us solve the harder radial equation

$$\frac{r^2}{A}\frac{d^2A}{dr^2} + \frac{r}{A}\frac{dA}{dr} + (kr)^2 = m^2.$$

We rearrange this by dividing by r^2 and consolidating terms to get

$$\frac{d^2A}{dr^2} + \frac{1}{r}\frac{dA}{dr} + A\left(k^2 - \frac{m^2}{r^2}\right) = 0.$$

The resulting equation is called the Bessel equation, which we have already encountered in Chapter 7. To make this more apparent, let us change variables by letting $z = kr$ ($dr = dz/k$). This yields

$$k^2\frac{d^2A}{dz^2} + \frac{k}{r}\frac{dA}{dz} + A\left(k^2 - \frac{m^2}{r^2}\right) = 0,$$

which when multiplied by r^2 gives

$$(kr)^2\frac{d^2A}{dz^2} + (kr)\frac{dA}{dz} + A\left[(kr)^2 - m^2\right] = 0,$$

that is,

$$z^2\frac{d^2A}{dz^2} + z\frac{dA}{dz} + A(z^2 - m^2) = 0, \tag{8.3}$$

which is the common form of Bessel's equation (see Equation 7.24).

8.3.1 Wavefunctions

General solutions to the Bessel equation are linear combinations of Bessel functions of the first kind (J_m) and Bessel functions of the second kind (Y_m):

$$A(r) = CJ_m(z) + DY_m(z),$$

where C and D are constant coefficients to be determined.

Note that Bessel functions of the first kind are well behaved at the origin. However, Bessel functions of the second kind diverge as $r \to 0$. As a consequence, they must be excluded if we are to have physically relevant wavefunctions. Thus, $D = 0$, resulting in

$$A(r) = CJ_m(z). \tag{8.4}$$

The first four Bessel functions of the first kind are illustrated in **Figure 8.4**. One sees that they are oscillatory in nature and have numerous zero crossings.

The total wavefunction of the particle in an infinite circular box is therefore the product

$$\begin{aligned}\psi &= A(r)B(\theta)\\ &= NJ_m(z)e^{im\theta},\end{aligned}$$

where N is a normalization constant. Note that the zeros of the Bessel function occur when $ka = \alpha_{m,n}$ ($\alpha_{m,n}$ is the dual-index root of the Bessel function). Thus, $k = \alpha_{m,n}/a$ and, since $z = kr$, an alternative expression for the total wavefunction using real units is

$$\psi = NJ_m\left(\frac{\alpha_{m,n}}{a}r\right)e^{im\theta}. \tag{8.5}$$

8.3.2 Energies

Finally, since

$$k = \sqrt{\frac{2m_{\text{eff}}E}{\hbar^2}} = \frac{\alpha_{m,n}}{a},$$

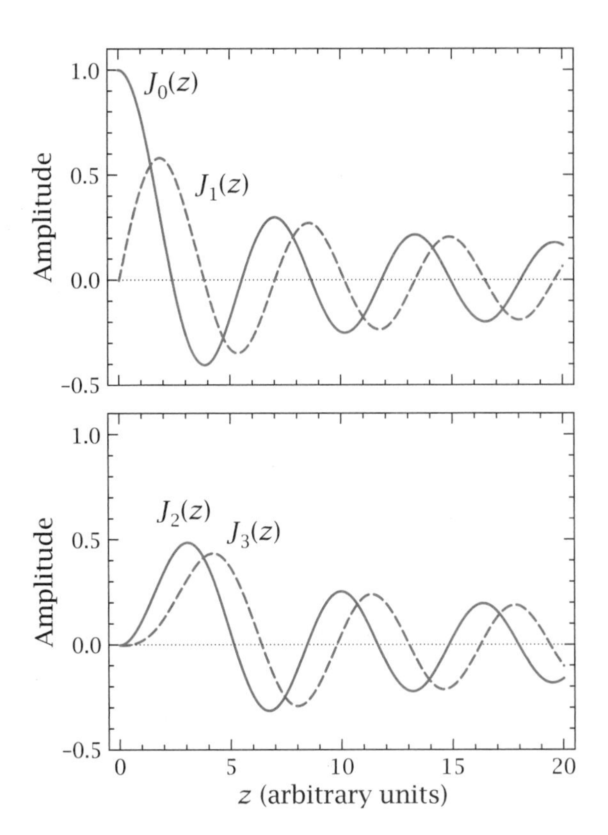

Figure 8.4 First four Bessel functions of the first kind.

we can use this relationship to find the energies of the confined particle. Solving for E gives

$$E = \frac{\hbar^2 \alpha_{m,n}^2}{2 m_{\text{eff}} a^2}, \tag{8.6}$$

where $\alpha_{m,n}$ is again the root of the Bessel function and a is the radius of the circular well. A table of Bessel function roots can be found in Chapter 7 (**Table 7.1**), and **Figure 8.5** shows the first five lowest energies of the particle. Corresponding roots $\alpha_{m,n}$ associated with each energy are also shown.

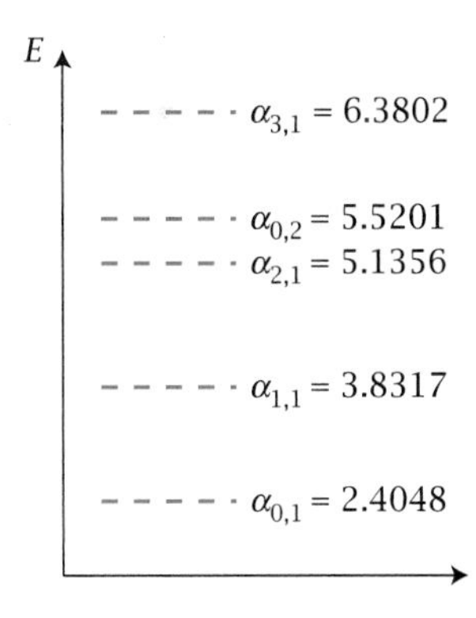

Figure 8.5 First five energies of a particle in a circular box.

8.4 HARMONIC OSCILLATOR

The last of our three model problems appears in many places, especially when discussing phonons (i.e., vibrations in solids). This, like all previous cases, involves solving the Schrödinger equation for a given potential. Here, $V(x)$ is quadratic and reflects the Hooke's law ($F = -k_{\text{Hooke}} x$) behavior of vibrating systems. In one dimension, we write

$$V(x) = \tfrac{1}{2} m_{\text{eff}} \omega^2 x^2, \tag{8.7}$$

where $\omega = \sqrt{k_{\text{Hooke}}/m_{\text{eff}}}$, with k_{Hooke} a force constant and m_{eff} the particle's mass. Note that in our problem the particle of interest is an atom as opposed to an electron or a hole. This once again illustrates the wave-like nature of matter. The harmonic oscillator potential is shown in **Figure 8.6**, along with a schematic depicting the ball-and-spring physical representation of the oscillator.

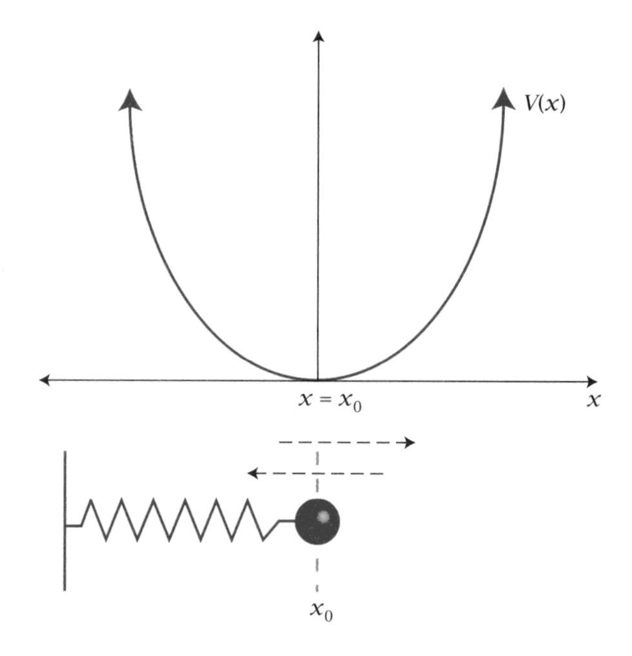

Figure 8.6 One-dimensional harmonic oscillator potential, along with a ball-and-spring physical description.

As usual, the goal is to solve the Schrödinger equation to find the particle's wavefunctions and energies. Given the potential in Equation 8.7, the Schrödinger equation is

$$\left(-\frac{\hbar^2}{2m_{\text{eff}}}\frac{d^2}{dx^2} + \frac{1}{2}m_{\text{eff}}\omega^2 x^2\right)\psi = E\psi,$$

and will be more complicated to solve since a nonzero potential exists in the region of interest.

Note that a number of ways exist for solving the above equation. One common approach involves using a power series solution. Another entails using so-called quantum mechanical raising and lowering operators. In this text, we will pursue the former, although the latter actually turns out to be more straightforward to carry out. It just requires a more thorough introduction to quantum mechanical operators beyond what we have outlined in Chapter 6. More information about raising and lowering operators (also called creation and annihilation operators) can be found in the quantum mechanics texts listed in Further Reading. The reader is therefore encouraged to see more about them.

To begin our solution of the harmonic oscillator problem, we start with

$$-\frac{\hbar^2}{2m_{\text{eff}}}\frac{d^2}{dx^2}\psi + V\psi = E\psi$$

and rearrange it to

$$\frac{d^2\psi}{dx^2} + \frac{2m_{\text{eff}}(E - V)}{\hbar^2}\psi = 0.$$

We simplify the result using dimensionless variables by (a) letting $y = x/\alpha$, where α has units of distance, (b) letting $\epsilon = E/\beta$, where β has units of energy, and (c) letting $V(x) = \frac{1}{2}m_{\text{eff}}\omega^2 x^2$. On inserting these expressions into the Schrödinger equation, we find that

$$\frac{1}{\alpha^2}\frac{d^2\psi}{dy^2} + \frac{2m_{\text{eff}}}{\hbar^2}\left(\beta\epsilon - \frac{1}{2}m_{\text{eff}}\omega^2 x^2\right)\psi = 0.$$

Multiplying by α^2 and expanding terms then gives

$$\frac{d^2\psi}{dy^2} + \frac{2m_{\text{eff}}\alpha^2\beta\epsilon}{\hbar^2}\psi - \frac{m_{\text{eff}}^2\omega^2\alpha^4 y^2}{\hbar^2}\psi = 0,$$

where the variables α and β are defined such that $2m_{\text{eff}}\alpha^2\beta/\hbar^2 = 1$ and $m_{\text{eff}}^2\omega^2\alpha^4/\hbar^2 = 1$. Thus,

$$\alpha = \sqrt{\frac{\hbar}{m_{\text{eff}}\omega}} \tag{8.8}$$

(obtained from the second equivalence) represents a natural length scale for the problem and

$$\beta = \tfrac{1}{2}\hbar\omega \tag{8.9}$$

(obtained from the first equivalence) is the corresponding natural energy scale. The reader may verify this as an exercise.

At this point, we are left with the equation

$$\left[\frac{d^2}{dy^2} + (\epsilon - y^2)\right]\psi = 0, \tag{8.10}$$

which we solve by assuming a general solution of the form

$$\psi(y) = f(y)e^{-y^2/2}. \tag{8.11}$$

The choice of this functional form can be motivated as follows. In the limit where $y \to \infty$, Equation 8.10 becomes

$$\frac{d^2\psi}{dy^2} - y^2\psi \simeq 0.$$

It therefore has general solutions of the form

$$\psi \sim Ae^{y^2/2} + Be^{-y^2/2}, \tag{8.12}$$

where A and B are coefficients to be determined. Since the wavefunction must satisfy a normalization condition, we exclude the first term given that it diverges as $y \to \infty$. This leaves

$$\psi \sim Be^{-y^2/2}. \tag{8.13}$$

At this point, we can check that the result is more or less a solution of the equation by taking its second derivative. Namely,

$$\psi'' \sim -B(-y^2e^{-y^2/2} + e^{-y^2/2}).$$

Note that in the limit where y is large, the first term dominates the second and we essentially have

$$\psi'' \sim By^2e^{-y^2/2}.$$

By inserting this into $\psi'' - y^2\psi \simeq 0$ it is clear that our effective (large-y) equation is satisfied.

The function $f(y)$ turns out to be what is called a Hermite polynomial. For now, though, let us simply insert Equation 8.11 into Equation 8.10

to see what happens. Looking ahead, we will need to evaluate some second derivatives. Let us do this ahead of time.

The first derivative of ψ is

$$\frac{d}{dy}[f(y)e^{-y^2/2}] = -yf(y)e^{-y^2/2} + f'(y)e^{-y^2/2}.$$

The second derivative is somewhat tedious to obtain and involves repeated application of the product rule of differentiation. However, doing so gives

$$\frac{d}{dy}\left[\frac{d}{dy}(f(y)e^{-y^2/2})\right] = [f''(y) - 2yf'(y) + f(y)(y^2 - 1)]e^{-y^2/2},$$

and, when inserted into Equation 8.10, yields

$$[f''(y) - 2yf'(y) + f(y)(y^2 - 1)]e^{-y^2/2} + (\epsilon - y^2)f(y)e^{-y^2/2} = 0.$$

This reduces to

$$f''(y) - 2yf'(y) + (\epsilon - 1)f(y) = 0,$$

whereupon defining $\epsilon - 1 = 2n$ gives

$$f''(y) - 2yf'(y) + 2nf(y) = 0. \tag{8.14}$$

We will discuss later what values the variable n can take. The equation is now solved by assuming a power series expansion of $f(y)$. Note that we could also use other approaches such as Laplace transforms if we wanted to.

For our power series approach, let us assume two classes of solutions: one with even symmetry and the other with odd symmetry. For notational simplicity, call them $f_1(y)$ and $f_2(y)$, respectively. Their power series representations are

$$\begin{aligned} f_1(y) &= \sum_{m=0}^{\infty} a_{2m}y^{2m} \qquad \text{(even)} \\ &= a_o + a_2y^2 + a_4y^4 + a_6y^6 + \cdots, \\ f_2(y) &= \sum_{m=0}^{\infty} a_{2m+1}y^{2m+1} \qquad \text{(odd)} \\ &= a_1y + a_3y^3 + a_5y^5 + a_7y^7 + \cdots. \end{aligned}$$

It is apparent that the first function just incorporates the even powers of y while the second incorporates the odd powers. The coefficients a_{2m} and a_{2m+1} just indicate how much of a given term is included in either expansion. Note that the letter m is just a dimensionless index and should not be confused with a mass. As alluded to above, both $f_1(y)$ and $f_2(y)$ ultimately turn out to be Hermite polynomials. For now, though, we start evaluating Equation 8.14 by assuming the even-symmetry solutions. This leads to a recurrence relationship that links successive even coefficients a_{2m}. We will then do the same with the odd solutions.

The even polynomials and their derivatives are

$$\begin{aligned} f_1(y) &= \sum_{m=0}^{\infty} a_{2m}y^{2m}, \\ f_1'(y) &= \sum_{m=0}^{\infty} a_{2m}(2m)y^{2m-1}, \\ f_1''(y) &= \sum_{m=0}^{\infty} a_{2m}(2m)(2m-1)y^{2m-2}. \end{aligned}$$

We recast the last expression in terms of powers of y^{2m}, leading to

$$f_1''(y) = \sum_{m=0}^{\infty} a_{2m+2}(2m+2)(2m+1)y^{2m}.$$

The reader can verify that these two versions of $f_1''(y)$ are equivalent. Next, inserting $f_1(y)$, $f_1'(y)$, and $f_1''(y)$ into Equation 8.14 gives

$$\sum_{m=0}^{\infty} a_{2m+2}(2m+2)(2m+1)y^{2m} - 2y\sum_{m=0}^{\infty} a_{2m}(2m)y^{2m-1} + 2n\sum_{m=0}^{\infty} a_{2m}y^{2m} = 0,$$

which simplifies to

$$\sum_{m=0}^{\infty} a_{2m+2}(2m+2)(2m+1)y^{2m} - 2\sum_{m=0}^{\infty} a_{2m}(2m)y^{2m} + 2n\sum_{m=0}^{\infty} a_{2m}y^{2m} = 0.$$

Consolidating coefficients of y^{2m} then yields

$$\sum_{m=0}^{\infty} [a_{2m+2}(2m+2)(2m+1) - 2a_{2m}(2m) + 2na_{2m}]\, y^{2m} = 0.$$

At this point, we either have the trivial solution (i.e., $y = 0$) or find that the term in square brackets equals zero. Since we never want the trivial solution,

$$a_{2m+2}(2m+2)(2m+1) - 2a_{2m}(2m) + 2na_{2m} = 0.$$

This simplifies to

$$a_{2m+2}(2m+2)(2m+1) - (4m-2n)a_{2m} = 0,$$

and, when all terms are consolidated, this yields

$$a_{2m+2} = \frac{2(2m-n)}{(2m+1)(2m+2)}a_{2m} \tag{8.15}$$

as a recurrence relationship for the even coefficients defining $f_1(y)$ (i.e., by knowing a_0, we have a_2. By knowing a_2, we have a_4, so on).

Next, we do the same analysis for the odd polynomials and their derivatives:

$$f_2(y) = \sum_{m=0}^{\infty} a_{2m+1}y^{2m+1},$$

$$f_2'(y) = \sum_{m=0}^{\infty} a_{2m+1}(2m+1)y^{2m},$$

$$f_2''(y) = \sum_{m=0}^{\infty} a_{2m+1}(2m+1)(2m)y^{2m-1}.$$

As before, we recast the last expression in terms of powers of $2m+1$:

$$f_2''(y) = \sum_{m=0}^{\infty} a_{2m+3}(2m+3)(2m+2)y^{2m+1}.$$

The equivalence between these two versions of $f_2''(y)$ can again be checked explicitly. We now replace these expressions into Equation 8.14 to obtain

$$\sum_{m=0}^{\infty} a_{2m+3}(2m+3)(2m+2)y^{2m+1} - 2y\sum_{m=0}^{\infty} a_{2m+1}(2m+1)y^{2m}$$
$$+ 2n\sum_{m=0}^{\infty} a_{2m+1}y^{2m+1} = 0.$$

This simplifies to

$$\sum_{m=0}^{\infty} a_{2m+3}(2m+3)(2m+2)y^{2m+1} - 2\sum_{m=0}^{\infty} a_{2m+1}(2m+1)y^{2m+1}$$
$$+ 2n\sum_{m=0}^{\infty} a_{2m+1}y^{2m+1} = 0,$$

and consolidating terms gives

$$\sum_{m=0}^{\infty} [a_{2m+3}(2m+3)(2m+2) - 2a_{2m+1}(2m+1) + 2na_{2m+1}]y^{2m+1} = 0.$$

Finally, we either have the trivial solution (i.e., $y = 0$) or find that the term in square brackets equals zero. Since we do not want the trivial solution,

$$a_{2m+3}(2m+3)(2m+2) - 2a_{2m+1}(2m+1) + 2na_{2m+1} = 0.$$

This simplifies to

$$a_{2m+3}(2m+3)(2m+2) - [2(2m+1) - 2n]a_{2m+1} = 0,$$

and, on consolidating terms, we obtain the desired recurrence relationship for the odd coefficients that define $f_2(y)$:

$$a_{2m+3} = \frac{2[(2m+1) - n)]}{(2m+3)(2m+2)} a_{2m+1}. \tag{8.16}$$

Equation 8.16 means that if we have a_1, we also have a_3. Likewise, if we have a_3 we have a_5, and so forth.

8.4.1 Wavefunctions

At this point, we have

$$f_1(y) = \sum_{m=0}^{\infty} a_{2m}y^{2m},$$
$$f_2(y) = \sum_{m=0}^{\infty} a_{2m+1}y^{2m+1},$$

with recurrence relationships linking successive coefficients in either series (Equations 8.15 and 8.16). In principle, we now know what ψ looks like. However, we are not done yet. To obtain a *normalizable* wavefunction, it turns out that the infinite series in $f_1(y)$ and $f_2(y)$ must eventually terminate (i.e., there must be some value of a_{2m+2} or a_{2m+3} that equals zero). We can demonstrate why as follows. First, note that the ratios between successive coefficients in the even and odd series are

$$\frac{a_{2m+2}}{a_{2m}} = \frac{2(2m-n)}{(2m+1)(2m+2)} \quad \text{(even)}$$

and

$$\frac{a_{2m+3}}{a_{2m+1}} = \frac{2[(2m+1)-n]}{(2m+3)(2m+2)} \quad \text{(odd)}.$$

In either case, if the index m is large, we find that

$$\frac{a_{2m+2}}{a_{2m}} \sim \frac{1}{m},$$
$$\frac{a_{2m+3}}{a_{2m+1}} \sim \frac{1}{m}.$$

This means that for either the even or odd series we can generically write its expansion as

$$f(y) \sim \sum_{m}^{\infty} \frac{1}{m!} y^{2m}$$

when m is large. What is interesting about this, then, is that, in this regime, $f(y)$ is identical to the series expansion for e^{y^2}:

$$e^{y^2} = \sum_{m=0}^{\infty} \frac{1}{m!} y^{2m}.$$

As a consequence, this becomes a problem, because if we recall that

$$\psi(y) = f(y)e^{-y^2/2},$$

then, when $f(y) \sim e^{y^2}$, we have $\psi(y) \sim e^{y^2/2}$, which diverges at large y. We must therefore terminate the series expansions of $f_1(y)$ and $f_2(y)$. To do this, note that in Equations 8.15 and 8.16, making n an integer enables the numerator to become zero at some point and thus stops either series. We therefore conclude that n must be an integer: $n = 0, 1, 2, 3, \ldots$.

Important consequences arise from this. First, we will find discrete polynomials for $f_1(y)$ and $f_2(y)$, leading to normalizable ψ. Next, with n an integer, our energies will be quantized, since $\epsilon - 1 = 2n$. Let us therefore look at the appearance of $f_1(y)$ and $f_2(y)$. This will be followed by the corresponding quantized energies in the next subsection.

For a given n, we expect an associated wavefunction ψ. Let us therefore see what the accompanying power series $f(y)$ looks like. We begin with the even series $f_1(y)$ and then alternate with the odd series $f_2(y)$.

For $n = 0$

For $n = 0$, we begin with $f_1(y)$, which we expect will stop after a few terms. Starting with

$$f_1(y) = \sum_{m=0}^{\infty} a_{2m} y^{2m} = a_0 + a_2 y^2 + a_4 y^4 + \cdots,$$

we will use Equation 8.15 to find a_2, a_4, a_6, and so forth from a_0. For a_2, we therefore find

$$a_2 = \frac{-2n}{(1)(2)} a_0 = -n a_0.$$

Since $n = 0$, we immediately see that $a_2 = 0$. Next, given the above recurrence relationship, we immediately see that $a_4 = 0$, $a_6 = 0$, and so on. We can stop now, since it is apparent that

$$f_1(y, n = 0) = a_0.$$

For $n = 1$

For our second wavefunction, let us move on to $f_2(y)$, which should also be a truncated polynomial. Starting with

$$f_2(y) = \sum_{m=0}^{\infty} a_{2m+1} y^{2m+1} = a_1 y + a_3 y^3 + a_5 y^5 + \cdots,$$

we use the recurrence relationship (Equation 8.16) to find a_3, a_5, $a_7, \ldots$ from a_1. For a_3, we therefore find

$$a_3 = \frac{2(-n+1)}{(3)(2)} a_1 = \frac{1}{3}(-n+1) a_1,$$

and, since $n = 1$, $a_3 = 0$. As a consequence, if $a_3 = 0$, so are a_5, a_7, and so on. We can stop now, since the power series has been truncated. We find

$$f_2(y, n = 1) = a_1 y.$$

For $n = 2$

We now revert back to $f_1(y)$ for the next wavefunction. We again expect to find a truncated polynomial. Using Equation 8.15, we find

$$a_2 = \frac{-2n}{(1)(2)} a_0 = -n a_0,$$

and, since $n = 2$, $a_2 = -2a_0$. A second application of the recurrence relationship gives a_4:

$$a_4 = \frac{2(2-n)}{(3)(4)} a_2 = \frac{2-n}{6} a_2.$$

Since $n = 2$, we have $a_4 = 0$. The series therefore stops, since every additional coefficient is zero. We are left with

$$\begin{aligned} f_1(y, n = 2) &= a_0 - 2a_0 y^2 \\ &= a_0(1 - 2y^2). \end{aligned}$$

For $n = 3$

In the same alternating fashion, let us revert back to $f_2(y)$ and find its coefficients. Equation 8.16 shows that

$$a_3 = \frac{2(-n+1)}{(3)(2)} a_1 = \frac{1}{3}(-n+1) a_0,$$

and since $n = 3$, $a_3 = -\frac{2}{3}a_1$. Next, a_5 is

$$a_5 = \frac{2(3-n)}{(5)(4)}a_3,$$

and since $n = 3$, $a_5 = 0$. The series thus stops. We therefore find

$$\begin{aligned} f_2(y, n=3) &= a_1 y - \tfrac{2}{3}a_1 y^3 \\ &= a_1(y - \tfrac{2}{3}y^3). \end{aligned}$$

In this manner, we can go on and on to find as many expressions for $f_1(y)$ and $f_2(y)$ as needed to compile the wavefunctions associated with a given n. A brief summary of the polynomials we have found so far shows that

$$\begin{aligned} f_1(y, n=0) &= a_0, \\ f_2(y, n=1) &= a_1 y, \\ f_1(y, n=2) &= a_0(1 - 2y^2), \\ f_2(y, n=3) &= a_1(y - \tfrac{2}{3}y^3). \end{aligned}$$

However, a cursory comparison of the above functions (noting that a_0 and a_1 need not be the same for different n values) with the so-called *Hermite polynomials* $H_n(y)$ reveals that they are identical apart from the scaling factors a_0 or a_1, along with potential sign differences. To illustrate, the first few even- and odd-symmetry Hermite polynomials are listed in **Table 8.1**. In fact, if we choose a_0 or a_1 so that the term with the largest power of y in each polynomial has a prefactor of 2^n, we can see that the above functions are identical with the Hermite polynomials. Thus, instead of writing ψ using $f_1(y)$ and $f_2(y)$, let us just refer to the Hermite polynomials directly. They are formally defined as

$$H_n(y) = (-1)^n e^{y^2} \frac{d^n}{dy^n} e^{-y^2}. \tag{8.17}$$

Table 8.1 Hermite polynomials

Hermite polynomial	Explicit expression
$H_0(y)$	1
$H_1(y)$	$2y$
$H_2(y)$	$4y^2 - 2$
$H_3(y)$	$8y^3 - 12y$
$H_4(y)$	$16y^4 - 48y^2 + 12$
$H_5(y)$	$32y^5 - 160y^3 + 120y$

We now bring our discussion of harmonic oscillator wavefunctions to a close. The total (generic) wavefunction we have been seeking is written as

$$\psi_n = N_n H_n(y) e^{-y^2/2},$$

where

$$N_n = \frac{1}{\sqrt{2^n n!}}\left(\frac{m_{\text{eff}}\omega}{\pi\hbar}\right)^{1/4}$$

is our usual normalization coefficient and $H_n(y)$ are the above Hermite polynomials. In terms of real variables ($y = \sqrt{m_{\text{eff}}\omega/\hbar}\,x$), the final wavefunctions, including normalization, look like

$$\psi_n(x) = \left[\frac{1}{\sqrt{2^n n!}}\left(\frac{m_{\text{eff}}\omega}{\pi\hbar}\right)^{1/4}\right] H_n\left(\sqrt{\frac{m_{\text{eff}}\omega}{\hbar}}x\right) \exp(-m_{\text{eff}}\omega x^2/2\hbar). \tag{8.18}$$

Table 8.2 Harmonic oscillator wavefunctions

Harmonic oscillator wavefunctions	Explicit expression
$\psi_0(x)$	$\left(\frac{m_{\text{eff}}\omega}{\pi\hbar}\right)^{1/4} \exp(-m_{\text{eff}}\omega x^2/2\hbar)$
$\psi_1(x)$	$\frac{\sqrt{2}}{\pi^{1/4}}\left(\frac{m_{\text{eff}}\omega}{\hbar}\right)^{3/4} x\exp(-m_{\text{eff}}\omega x^2/2\hbar)$
$\psi_2(x)$	$\frac{1}{\sqrt{2}}\left(\frac{m_{\text{eff}}\omega}{\pi\hbar}\right)^{1/4}\left[2\left(\frac{m_{\text{eff}}\omega}{\hbar}\right)x^2 - 1\right]\exp(-m_{\text{eff}}\omega x^2/2\hbar)$

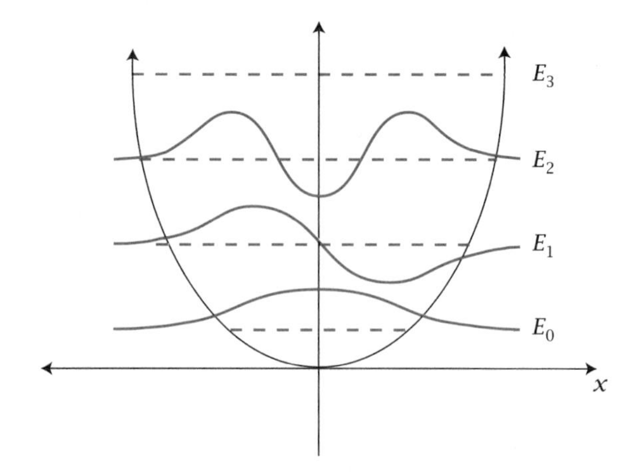

Figure 8.7 First few energies and wavefunctions for a one-dimensional harmonic oscillator.

Table 8.2 provides a summary of the first few of these harmonic oscillator wavefunctions.

8.4.2 Energies

Finally, for the energies, recall that we previously defined $\epsilon - 1 = 2n$. Solving for ϵ then gives $\epsilon = 2\left(n + \frac{1}{2}\right)$. Next, since $E = \beta\epsilon$ and $\beta = \frac{1}{2}\hbar\omega$, we find that

$$E = 2\beta\left(n + \tfrac{1}{2}\right).$$

This results in our desired expression for the harmonic oscillator energies:

$$E = \hbar\omega\left(n + \tfrac{1}{2}\right), \tag{8.19}$$

where $n = 0, 1, 2, \ldots$. These energies, their uniform spacing, and their accompanying wavefunctions are shown in **Figure 8.7**

8.5 SUMMARY

Having run through some more model quantum mechanics problems that describe particles confined to different representative nanostructures, we are now ready to move on. As alluded to earlier in Chapter 7, we now introduce the concept of a density of states. This expression compactly describes the many states present in a system for a given energy range of interest. This will be done without having to solve for each level independently and, as a consequence, represents a major convenience.

8.6 THOUGHT PROBLEMS

8.1 Particle in a finite-barrier box

Consider an InP quantum well whose thickness is $l = 10\,\text{nm}$. Now rather than assuming an infinite box model, consider a finite barrier height of 0.3 eV. Assume that the electron's effective mass $m_e = 0.077m_0$. Calculate the energies of the electron states trapped within the well. Next, increase the barrier height to 1 eV and repeat the same calculations. This can be done using your favorite mathematical modeling program (Mathcad, Matlab, Mathematica, etc.).

8.2 Particle in an infinite circular box

Confirm that the wavefunctions of the particle in an infinite circular box are orthogonal. *Hint*: Use the Euler relation.

8.3 Particle in an infinite circular box

Consider a particle in an infinite circular box. Its wavefunction happens to be

$$\psi(r,\theta) = A(r)\cos\theta + B(r)\sin 3\theta,$$

with $A(r)$ and $B(r)$ encompassing the radial part of the wavefunction as well as any appropriate normalization constants. What energies are found when measured?

8.4 Quantum corral

Consider a quantum corral, described in the classic paper by Crommie et al. (1993). Assume that the radius of the corral is 71.3 Å. In addition, assume that the electron mass is $m_{eff} = 0.38m_0$. Calculate the lowest three energy levels of the corral for $m = 0$ and $m = 1$ (note that these are indices, not masses). How do they compare with the tabulated values in the paper? To simplify things, consider the energy differences between states.

8.5 Quantum corral

For the same quantum corral as in Problem 8.4, plot the first three wavefunctions of the system.

8.6 Rigid rotor

The linear absorption spectrum of C_{60} shows transitions at 404, 328, 256, and 211 nm. Assume that there are 60 electrons in the molecule and that electronic transitions require a change in orbital angular momentum of ± 1 (i.e., $\Delta l = \pm 1$). The absorption of C_{60} can be modeled using another quantum mechanical model called the "rigid rotor" model. Alternatively, it is called the "particle-*on*-a-sphere" model. Read the article by Ball (1994). Assume the Pauli principle and describe which states likely participate in the 404 nm transition.

8.7 Particle on a ring

Consider the case of a particle constrained to move on the perimeter of a ring of radius r. This is the particle-on-a-ring problem. Show by an appropriate transformation to polar coordinates (i.e., $x = r\cos\theta$, $y = r\sin\theta$) that the Schrödinger equation

$$\left[-\frac{\hbar^2}{2m}\nabla^2 + V(r,\theta)\right]\psi = E\psi$$

becomes

$$-\frac{\hbar^2}{2I}\frac{\partial^2\psi}{\partial\theta^2} = E\psi.$$

In this expression, $I = mr^2$ and E is the particle's energy. The potential along the ring is assumed to be zero, but is infinite elsewhere.

Hints:

$$d\frac{\tan^{-1} u}{dx} = \frac{1}{1+u^2}\frac{du}{dx},$$

and, for a function $f(r,\theta)$,

$$\left(\frac{\partial f}{\partial x}\right)_y = \left(\frac{\partial f}{\partial r}\right)_\theta \left(\frac{\partial r}{\partial x}\right)_y + \left(\frac{\partial f}{\partial \theta}\right)_r \left(\frac{\partial \theta}{\partial x}\right)_y,$$

$$\left(\frac{\partial f}{\partial y}\right)_x = \left(\frac{\partial f}{\partial r}\right)_\theta \left(\frac{\partial r}{\partial y}\right)_x + \left(\frac{\partial f}{\partial \theta}\right)_r \left(\frac{\partial \theta}{\partial y}\right)_x.$$

Note that r is constant in this problem.

Finally, when the Schrödinger equation in polar coordinates has been derived, show that $\psi = Ae^{ik\theta}$ is a general solution to the problem. Find the normalization constant A and also the particle's associated energies. Think about the appropriate boundary conditions of the problem.

8.8 Triangular potential well

One occasionally encounters a situation where the carrier in a semiconductor experiences a potential like that depicted in **Figure 8.8**. This is the problem of a particle in a triangular well. The potential is

$$V(x) = eE\hat{x}$$

for $x > 0$, with E the magnitude of the electric field responsible for the potential. The system's boundary condition is $\psi(0) = 0$, since the potential is infinite when $x \leq 0$. Show that the Schrödinger equation reduces to the Airy equation, whose solutions are linear combinations of Airy functions. To simplify things, use scaled variables where

$$y = \frac{x}{x_0}$$

and

$$\epsilon = \frac{E}{E_0},$$

with x the coordinate, E the particle's energy,

$$x_0 = \left(\frac{\hbar^2}{2me|E|}\right)^{1/3} \quad \text{(a natural unit of length in the problem),}$$

and

$$E_0 = \left[\frac{(e|E|\hbar)^2}{2m}\right]^{1/3} \quad \text{(a natural unit of energy in the problem).}$$

The Airy equation expressed using these variables is then

$$\frac{d^2\psi}{dy^2} - (y - \epsilon)\psi = 0.$$

Provide an expression for the corresponding energies.

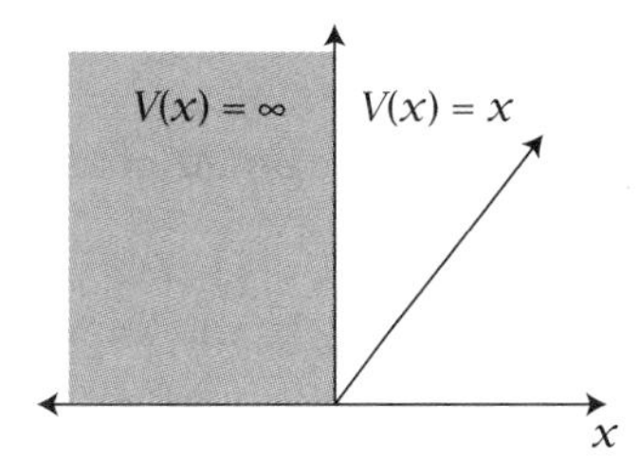

Figure 8.8 Triangular potential well.

8.9 Wavefunctions, triangular potential well

Assume for the situation described in Problem 8.8 that the particle's first and second excited-state wavefunctions can be approximated by

$$\psi_1(x) = N_1 x e^{-bx/2},$$
$$\psi_2(x) = N_2 x(A - bx)e^{-bx/2},$$

where $x > 0$, b is a constant, and N_1 (N_2) is the normalization constant of the first (second) excited state. Find the value of A that makes the two wavefunctions orthogonal. Then find the normalization constants N_1 and N_2.

8.10 Eigenvalues and eigenfunctions

The first four wavefunctions of the one-dimensional harmonic oscillator are

$$\psi_0(x) = \left(\frac{\gamma}{\pi}\right)^{1/4} e^{-\gamma x^2/2},$$

$$\psi_1(x) = \left(\frac{4\gamma^3}{\pi}\right)^{1/4} x e^{-\gamma x^2/2},$$

$$\psi_2(x) = \left(\frac{\gamma}{4\pi}\right)^{1/4} (2\gamma x^2 - 1) e^{-\gamma x^2/2},$$

$$\psi_3(x) = \left(\frac{\gamma^3}{9\pi}\right)^{1/4} (2\gamma x^3 - 3x) e^{-\gamma x^2/2},$$

where $\gamma = m_{\text{eff}}\omega/\hbar$

(a) Find the associated energies of each.

(b) Verify that the second wavefunction, $\psi_1(x)$, is normalized.

8.11 Two-dimensional harmonic oscillator

Find the energies and wavefunctions of an isotropic two-dimensional harmonic oscillator.

8.12 Degeneracies, three-dimensional harmonic oscillator

What are the degeneracies of the four lowest energy levels of an isotropic three-dimensional harmonic oscillator?

8.13 Probability, two-dimensional harmonic oscillator

Assume a two-dimensional harmonic oscillator where the force constant is the same along the x and y directions (i.e., $k_{\text{Hooke}} = k_x = k_y$).

(a) For the ground-state wavefunction, what is the probability of finding the particle in the first quadrant? Does the answer make sense?

(b) Next, assume that the wavefunction has a time dependence. What is the probability of finding the particle in the first quadrant? Does the answer make sense? Explain.

8.14 Transition probability, one-dimensional harmonic oscillator

Show that the transition probability of the one-dimensional harmonic oscillator can basically be written as

$$P \propto |\langle n|x|0\rangle|^2 \cos^2\theta,$$

where we have assumed light traveling along the z direction with the x axis being parallel to the vibrational coordinate and θ being the angle away from it. Assume transitions originating from the ground state to higher excited states.

8.15 Selection rules, one-dimensional harmonic oscillator

Show that for the one-dimensional harmonic oscillator

$$\langle 2|x|0\rangle = 0,$$

$$\langle 2|x|1\rangle \neq 0,$$

where

$$|0\rangle = \left(\frac{\gamma}{\pi}\right)^{1/4} e^{-\gamma x^2/2},$$

$$|1\rangle = \left(\frac{4\gamma^3}{\pi}\right)^{1/4} x e^{-\gamma x^2/2},$$

$$|2\rangle = \left(\frac{\gamma}{4\pi}\right)^{1/4} (2\gamma x^2 - 1) e^{-\gamma x^2/2},$$

with $\gamma = m_{\text{eff}}\omega/\hbar$. It turns out that the harmonic oscillator selection rules require $\Delta n = \pm 1$.

8.16 Transition probability, two-dimensional harmonic oscillator

Calculate the following transition dipole moments for the two-dimensional harmonic oscillator, assuming x-polarized light:

$$|0,0\rangle \to |0,1\rangle, \quad \text{i.e., } \langle 0,1|x|0,0\rangle,$$

$$|0,0\rangle \to |1,0\rangle, \quad \text{i.e., } \langle 1,0|x|0,0\rangle,$$

$$|0,0\rangle \to |1,1\rangle, \quad \text{i.e., } \langle 1,1|x|0,0\rangle.$$

In all cases, consider the following one-dimensional wavefunctions applicable to either x or y (the specific case for x is shown, but one gets the same functions for y, by replacing x with y in these expressions):

$$|0\rangle = \left(\frac{\gamma}{\pi}\right)^{1/4} e^{-\gamma x^2/2},$$

$$|1\rangle = \left(\frac{4\gamma^3}{\pi}\right)^{1/4} x e^{-\gamma x^2/2},$$

$$|2\rangle = \left(\frac{\gamma}{4\pi}\right)^{1/4} (2\gamma x^2 - 1) e^{-\gamma x^2/2}.$$

Note that $\gamma = m_{\text{eff}}\omega/\hbar$.

8.17 Boundary conditions, half harmonic oscillator

Consider the potential shown in Figure 8.9. Provide expressions for the resulting particle wavefunctions and energies. Assume knowledge of the one-dimensional harmonic oscillator wavefunctions and energies. This problem does not require much work.

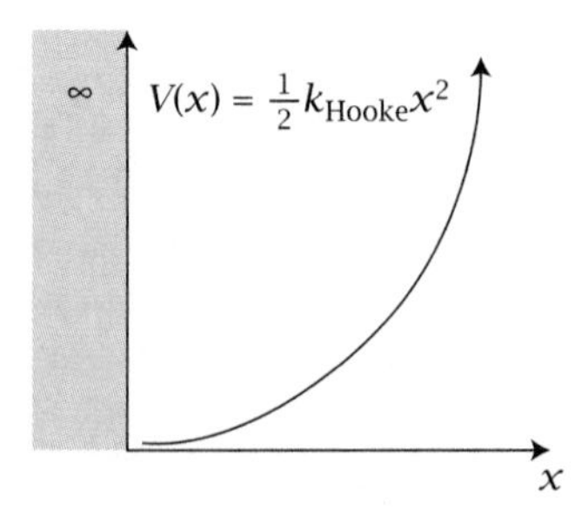

Figure 8.9 Half harmonic oscillator potential.

8.18 Uncertainty Principle

Show that the lowest-energy solution to the one-dimensional harmonic oscillator problem is consistent with the Heisenberg Uncertainty Principle. Recall that the ground-state wavefunction is

$$|0\rangle = \left(\frac{\gamma}{\pi}\right)^{1/4} e^{-\gamma x^2/2}.$$

8.19 Postulates

A particle in a one-dimensional harmonic oscillator potential has the following wavefunction at $t = 0$:

$$\psi(x, t=0) = A|0\rangle + \frac{1}{\sqrt{3}}|1\rangle + \frac{1}{\sqrt{5}}|2\rangle,$$

where

$$|0\rangle = \left(\frac{\gamma}{\pi}\right)^{1/4} e^{-\gamma x^2/2},$$

$$|1\rangle = \left(\frac{4\gamma^3}{\pi}\right)^{1/4} x e^{-\gamma x^2/2},$$

$$|2\rangle = \left(\frac{\gamma}{4\pi}\right)^{1/4} (2\gamma x^2 - 1) e^{-\gamma x^2/2}.$$

In these wavefunctions, $\gamma = \sqrt{k\mu/\hbar^2}$, where k_{Hooke} is the force constant and μ is the reduced mass of the system.

(a) Find A so that $\psi(x, 0)$ is normalized.

(b) If measurements of the particle's energy are made, what values will be found and with what probabilities? What is the average energy of the particle?

(c) Determine the probability of finding the particle at some later time, t, in the state

$$|\phi\rangle = \left(\frac{4\gamma^3}{\pi}\right)^{1/4} x e^{-\gamma x^2/2} e^{-i(3/2)\omega t}.$$

(d) Determine the probability of finding the particle at some later time t in the state

$$|\phi\rangle = \frac{1}{4\sqrt{3}}\left(\frac{\gamma}{\pi}\right)^{1/4} (8\gamma^{3/2}x^3 - 12\gamma^{1/2}x) e^{-\gamma x^2/2} e^{-i(7/2)\omega t}.$$

8.20 Nondegenerate perturbation theory

Consider a one-dimensional harmonic oscillator. Now assume that an external perturbation to the system is turned on. It consists of an electric field with a magnitude E. The system's Hamiltonian therefore changes from

$$H = \frac{p^2}{2m} + \frac{m\omega^2}{2}x^2$$

to

$$H = \frac{p^2}{2m} + \frac{m\omega^2}{2}x^2 - qEx,$$

where q is the system's charge. Find expressions for the new ground-state wavefunction and energy to first order. The particle's original wavefunction and energy are

$$|0\rangle = \left(\frac{\gamma}{\pi}\right)^{1/4} e^{-\gamma x^2/2},$$

$$E_0 = \tfrac{1}{2}\hbar\omega.$$

Note that there are many ways of working out this problem. One approach is straightforward and involves using nondegenerate perturbation theory. Another approach simply involves completing the square and changing variables.

8.21 Other potentials

Using your knowledge of the energy-level spacing for a one-dimensional particle in an infinite potential box as well as the energy-level spacing of a one-dimensional harmonic oscillator, predict whether the energy levels of the particle confined to the potentials shown in **Figure 8.10** converge or diverge. That is, as the energy gets larger, does the spacing between states increase or decrease?

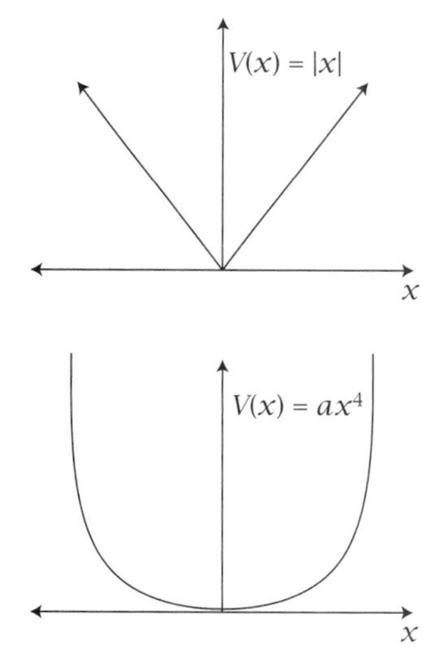

Figure 8.10 Potentials for Problem 8.21.

8.7 REFERENCES

Ball DW (1994) Electronic absorptions of C_{60}: a quantum mechanical model. *J. Chem. Ed.* **71**, 463.

Crommie MF, Lutz CP, Eigler DM (1993) Confinement of electrons to quantum corrals on a metal surface. *Science* **262**, 218.

Crommie MF, Lutz CP, Eigler DM, Heller EJ (1995) Quantum corrals. *Physica D* **83**, 98.

8.8 FURTHER READING

Linear algebra

Arfken GB, Weber HJ (2003) *Mathematical Methods for Physicists*, 6th edn. Academic Press, San Diego.

Kreyszig E (2011) *Advanced Engineering Mathematics* 10th edn. Wiley, New York.

McQuarrie DA (2003) *Mathematical Methods for Scientists and Engineers*. University Science Books, Sausalito, CA.

Introductory Quantum Mechanics Texts

Green NJB (1997) *Quantum Mechanics 1: Foundations.* Oxford University Press, Oxford, UK.

Green NJB (1998) *Quantum Mechanics 2: The Toolkit.* Oxford University Press, Oxford, UK.

Griffiths DJ (2004) *Introduction to Quantum Mechanics*, 2nd edn. Prentice Hall, Upper Saddle River, NJ.

Liboff RL (2003) *Introductory Quantum Mechanics.* Addison-Wesley, San Francisco, CA.

McMahon D (2005) *Quantum Mechanics Demystified.* McGraw-Hill, New York.

McQuarrie DA (2008) *Quantum Chemistry*, 2nd edn. University Science Books, Mill Valley, CA.

More Advanced Quantum Mechanics Texts

Cohen-Tannoudji C, Diu B, Laloe F (1977) *Quantum Mechanics* (2 volumes). Wiley, New York.

Zettili N (2009) *Quantum Mechanics, Concepts and Applications*, 2nd edn. Wiley, Chichester, UK.

Density of States

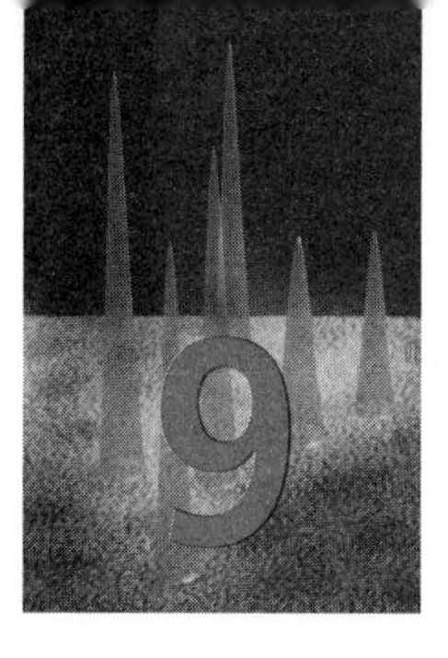

9.1 INTRODUCTION

We have just seen in Chapters 6–8 that one can solve the Schrödinger equation to find the energies and wavefunctions of a particle of interest. We have also alluded to the fact that the model systems we have introduced can be used to describe quantum wells, wires, and dots. Now to truly have complete information about a system of interest, we should know all of its possible energies and wavefunctions.

However, our past experience solving the Schrödinger equation shows that many valid energies and associated wavefunctions exist. Complicating this, in certain cases, degeneracies are possible, resulting in states that possess the same energy. As a consequence, the task of calculating each and every possible carrier wavefunction as well as its corresponding energy is impractical to say the least.

Fortunately, rather than solve the Schrödinger equation multiple times, we can instead find what is referred to as a density of states. This is basically a function that when multiplied by an interval of energy, provides the total concentration of available states in that energy range. Namely,

$$N_{\text{interval}} = \rho_{\text{energy}}(E)\, dE,$$

where N_{interval} is the carrier density present in the energy range dE and $\rho_{\text{energy}}(E)$ is our desired density of states function. Alternatively, one could write

$$N_{\text{tot}} = \int_{E_1}^{E_2} \rho_{\text{energy}}(E)\, dE$$

to represent the total concentration of available states in the system between the energies E_1 and E_2.

Obtaining $\rho_{\text{energy}}(E)$ is accomplished through what is referred to as a density-of-states calculation. We will illustrate this below and in later parts of the chapter. Apart from giving us a better handle on the state distribution of our system, the resulting DOS is important for many subsequent calculations, ranging from estimating the occupancy of states to calculating optical transition probabilities and/or transition rates upon absorbing and emitting light.

In what follows, we briefly recap model system solutions to the Schrödinger equation seen earlier in Chapter 7. This involves solutions for a particle in a one-, two-, or three-dimensional box, from which we will estimate the associated density of states. We will then calculate the density of states associated with model bulk systems, quantum wells, wires, and dots. This will be complemented by subsequent calculations of what is referred to as the joint density of states, which ultimately relates to the absorption spectrum of these systems.

Particle in a One-Dimensional Box

Let us start with the particle in a one-dimensional box, which we saw earlier in Chapter 7. Recall that we had a potential that was zero inside the box and infinite outside. We found that the wavefunctions and energies of the confined particle were (Equation 7.7)

$$\psi_n = \sqrt{\frac{2}{a}} \sin \frac{n\pi x}{a}$$

and (Equation 7.8)

$$E = \frac{n^2 h^2}{8 m_{\text{eff}} a^2} = \frac{\hbar^2 k^2}{2 m_{\text{eff}}},$$

where $n = 1, 2, 3, \ldots$ and $k = n\pi/a$, with a the length of the box. **Table 9.1** lists the first 10 energies of the particle. The first eight energy levels are also illustrated schematically in **Figure 9.1**. Note that we can go on listing these energies and wavefunctions indefinitely, since no limit exists to the value that n can take. But clearly this is impractical.

Table 9.1 Energies of a particle in a one-dimensional box

n	Energy
1	$E_1 = h^2/8m_{\text{eff}}a^2$
2	$4E_1$
3	$9E_1$
4	$16E_1$
5	$25E_1$
6	$36E_1$
7	$49E_1$
8	$64E_1$
9	$81E_1$
10	$100E_1$

We will now calculate the density of states for this system to illustrate how we can begin wrapping our arms around the problem of describing all of these states in a concise fashion. Before beginning, note that there will be a slight difference here with what we will see later on. In this section, we are calculating a density of energy levels. In the next section, the density that we will derive refers to a state density. As a consequence, there will be an extra factor of two in those latter expressions to account for spin degeneracy, stemming from the Pauli principle.

At this point, associated with each value of n is an energy. (In later sections, we will utilize k in our calculations since there exists a 1:1 correspondence between k and energy.) Let us define a density of levels as

$$g(E) = \frac{\Delta n}{\Delta E} \sim \frac{dn}{dE}$$

with units of number per unit energy. Note that the approximation dn/dE improves as our box becomes larger (large a) such that we begin to see a near-continuous distribution of energies with the discreteness due to quantization becoming less pronounced. Since $E = n^2 h^2/8m_{\text{eff}}a^2$, we find that $n = \sqrt{8m_{\text{eff}}a^2/h^2}\sqrt{E}$, whereupon evaluating dn/dE gives

$$g(E) \sim \frac{1}{2}\sqrt{\frac{8m_{\text{eff}}a^2}{h^2}}\frac{1}{\sqrt{E}}.$$

We see that the density of levels becomes sparser as E increases. This is apparent from the function's inverse square root dependence. Your intuition should also tell you this from simply looking at **Table 9.1** or at **Figure 9.1**, where the spacing between states grows noticeably larger as E increases.

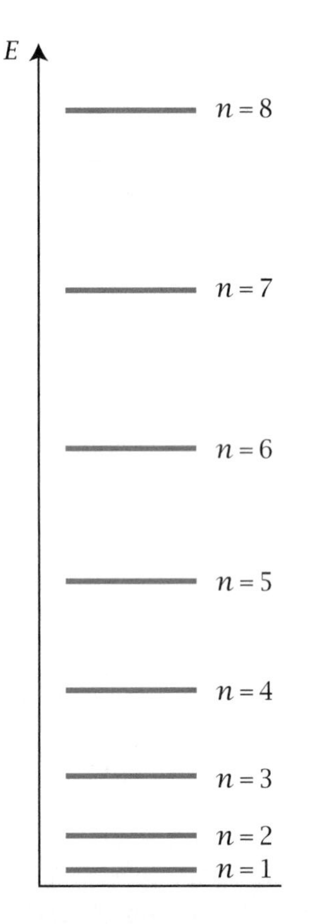

Figure 9.1 The first eight energy levels of a particle in a one-dimensional box. The spacing between states is to scale.

We can now obtain an approximate density of states with units of number per unit energy per unit length by dividing our previous expression by the length of the box, a. Thus, $\rho_{\text{energy}} = g(E)/a$ results in

$$\rho_{\text{energy, 1D box}} \sim \sqrt{\frac{2m_{\text{eff}}}{h^2}}\frac{1}{\sqrt{E}}. \tag{9.1}$$

Although not exact, we can illustrate Equation 9.1 for ourselves using the energies listed in **Table 9.1** and shown in **Figure 9.1**. Specifically,

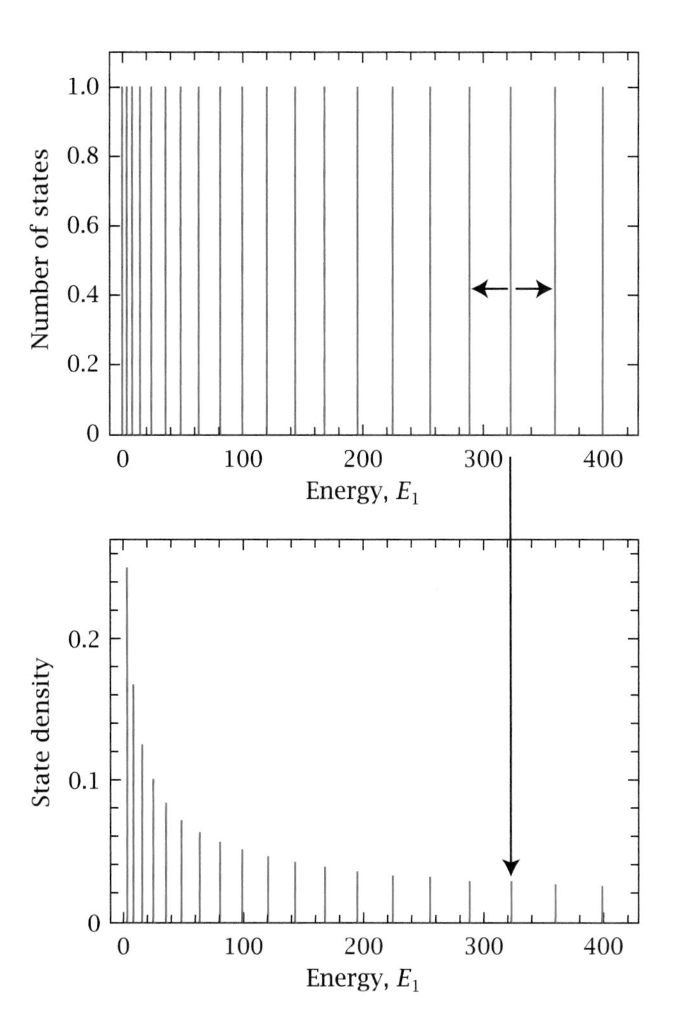

Figure 9.2 Histogram of the states of a particle in a one-dimensional box and the associated probability density obtained by weighting each value of the histogram by the average "distance" in energy to the next higher or lower energy state.

the top of **Figure 9.2** depicts a histogram of different energies possible for the particle in a one-dimensional box. Next, an effective probability density that mimics $\rho_{\text{energy, 1D box}}$ can be created by dividing each value of the histogram by the average "distance" in energy to the next state. On doing this, we see that the resulting probability density decays in agreement with one's intuition and with Equation 9.1.

Table 9.2 Energies of a particle in a two-dimensional box

n_x	n_y	Energy
1	1	$E_{11} = 2h^2/8m_{\text{eff}}a^2$
1	2	$2.5E_{11}$
2	1	
2	2	$4E_{11}$
1	3	$5E_{11}$
3	1	
2	3	$6.5E_{11}$
3	2	
1	4	$8.5E_{11}$
4	1	
3	3	$9E_{11}$
2	4	$10E_{11}$
4	2	
3	4	$12.5E_{11}$
4	3	
1	5	$13E_{11}$
5	1	

Particle in a Two-Dimensional Box

Let us now consider a particle in a two-dimensional box. This problem was also solved in Chapter 7 and assumes that a particle is confined to a square box with dimensions a within the (x, y) plane. The potential is zero inside the box and infinite outside. Solving the Schrödinger equation leads to the following wavefunctions (Equation 7.10):

$$\psi_{n_x,n_y} = \frac{2}{a}\sin\frac{n_x\pi x}{a}\sin\frac{n_y\pi y}{a}$$

with corresponding energies (Equation 7.11)

$$E = \frac{n^2h^2}{8m_{\text{eff}}a^2} = \frac{\hbar^2k^2}{2m_{\text{eff}}},$$

where $n^2 = n_x^2 + n_y^2$ ($n_x = 1, 2, 3, \ldots$; $n_y = 1, 2, 3, \ldots$) and $k^2 = k_x^2 + k_y^2$ ($k_x = n_x\pi/a$, $k_y = n_y\pi/a$). **Table 9.2** lists the first 10 energies of the particle and **Figure 9.3** illustrates the first eight energy levels with their degeneracies.

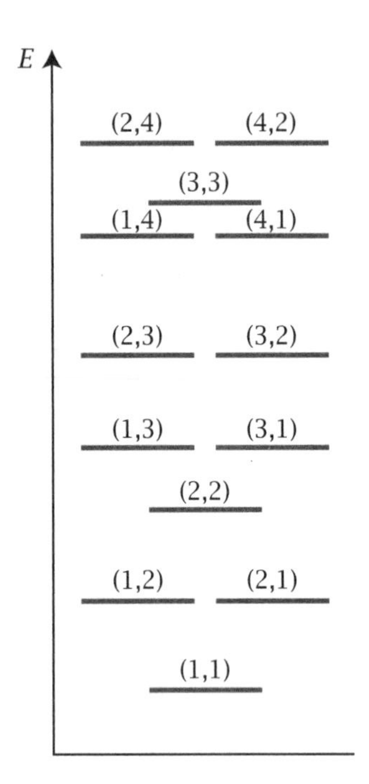

Figure 9.3 Depiction of the first eight energies of a particle in a two-dimensional box, including degeneracies. The numbers represent the indices n_x and n_y. The spacing between states is to scale.

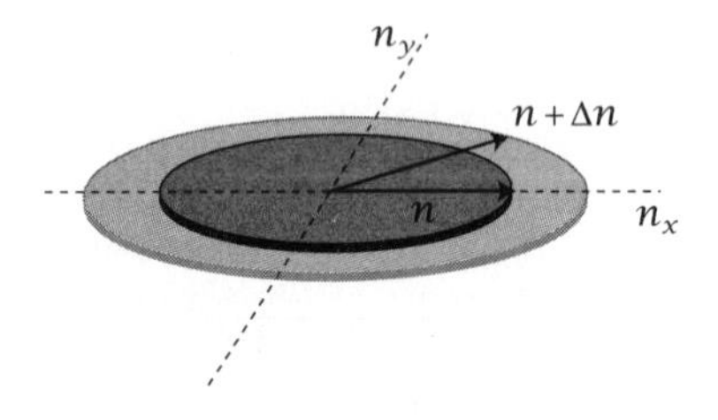

Figure 9.4 Constant-energy perimeters described by the indices n_x, n_y and the radius $n=\sqrt{n_x^2+n_y^2}$ for a particle in a two-dimensional box.

At this point, as with the previous one-dimensional box example, the density of levels associated with a unit change in energy is characterized by Δn. Since $E \propto n_x^2 + n_y^2$, where n_x and n_y are independent indices, we essentially have circles of constant energy in the (n_x, n_y) plane with a radius $n=\sqrt{n_x^2+n_y^2}$. This is illustrated in **Figure 9.4**. Thus a small change in radius, Δn, leads to a slightly larger circle (radius $n+\Delta n$) with a correspondingly larger energy.

We are interested in the states encompassed by these two circles, as represented by the area of the resulting annulus. As before, if the dimension of the box, a, becomes large, we may approximate $\Delta n \to dn$. We will assume that this holds from here on.

Now, the area of the annulus is $2\pi n\, dn$ (we will see a derivation of this shortly). It represents the total number of states present for a given change dn in energy. Since $n_x > 0$ and $n_y > 0$, we are only interested in the positive quadrant of the circle shown in **Figure 9.4**. As a consequence, the area of specific interest to us is $\frac{1}{4}(2\pi n\, dn)$. Then if $g(E)$ is our density of levels with units of number per unit energy, we can write

$$g_{\text{2D box}}(E)\, dE \sim \tfrac{1}{4}(2\pi n\, dn)$$

such that

$$g_{\text{2D box}}(E) \sim \frac{\pi}{2} n \frac{dn}{dE}.$$

Since $E = n^2h^2/8m_{\text{eff}}a^2$, $n=\sqrt{8m_{\text{eff}}a^2/h^2}\sqrt{E}$. As a consequence,

$$\frac{dn}{dE} = \frac{1}{2}\sqrt{\frac{8m_{\text{eff}}a^2}{h^2}}\frac{1}{\sqrt{E}}$$

and we find that

$$g_{\text{2D box}}(E) \sim \frac{2\pi\, m_{\text{eff}}a^2}{h^2}.$$

Finally, we can define a density of states $\rho_{\text{energy, 2D box}} = g_{\text{2D box}}(E)/a^2$, with units of number per unit energy per unit area by dividing $g_{\text{2D box}}(E)$ by the physical area of the box. This results in

$$\rho_{\text{energy, 2D box}} \sim \frac{2\pi\, m_{\text{eff}}}{h^2}, \tag{9.2}$$

which is a constant. In principle, one could have deduced this from looking at **Table 9.2** or **Figure 9.3**, although it is not immediately obvious.

Particle in a Three-Dimensional Box

Finally, let us consider a particle in a three-dimensional box. We did not work this problem out in Chapter 7 but instead listed the resulting wavefunctions and energies. However, as stated before, the reader can readily derive these results by applying the general strategies for finding the wavefunctions and energies of a particle in a box. The resulting wavefunctions and energies are (Equation 7.12)

$$\psi_{n_x,n_y,n_z} = \left(\frac{2}{a}\right)^{3/2} \sin\frac{n_x\pi x}{a} \sin\frac{n_y\pi y}{a} \sin\frac{n_z\pi z}{a}$$

and (Equation 7.13)

$$E_{n_x,n_y,n_z} = \frac{n^2 h^2}{8ma^2} = \frac{\hbar^2 k^2}{2m},$$

where $n^2 = n_x^2 + n_y^2 + n_z^2$ ($n_x = 1, 2, 3, \ldots$; $n_y = 1, 2, 3, \ldots$; $n_z = 1, 2, 3, \ldots$), $k^2 = k_x^2 + k_y^2 + k_z^2$ ($k_x = n_x\pi/a$, $k_y = n_y\pi/a$, $k_z = n_z\pi/a$) and a is the length of the box along all three sides. **Table 9.3** lists the first 10 energies of a particle in a three-dimensional box. The first eight energies are illustrated in **Figure 9.5** along with their degeneracies.

Now, just as with the one- and two-dimensional examples earlier, we will find the density of levels associated with a unit change in energy characterized by Δn. Since $E \propto n_x^2 + n_y^2 + n_z^2$, where n_x, n_y, n_z are all independent indices, we consider a sphere with a radius $n = \sqrt{n_x^2 + n_y^2 + n_z^2}$, possessing a constant-energy surface. This is illustrated in **Figure 9.6**. Thus, for a given change in radius, Δn, we encompass states represented by the volume of a thin shell between two spheres: one with radius n and the other with radius $n + \Delta n$. Furthermore, if a becomes large, we may approximate $\Delta n \to dn$, which we will assume from here on.

The volume of the shell is $4\pi n^2\, dn$ (we will see a derivation of this shortly). It represents the total number of states present for a given change dn in energy. Next, since $n_x > 0$, $n_y > 0$, and $n_z > 0$, we only consider the positive quadrant of the sphere's top hemisphere. As a consequence, the volume of interest to us is $\frac{1}{8}(4\pi n^2\, dn)$. Finally, if we define $g_{3\text{D box}}(E)$ as our energy density with units of number per unit energy, we can write

$$g_{3\text{D box}}(E)\, dE \sim \frac{1}{8}(4\pi n^2\, dn)$$

such that

$$g_{3\text{D box}}(E) \sim \frac{\pi n^2}{2}\left(\frac{dn}{dE}\right).$$

Next, since $E = n^2h^2/8m_{\text{eff}}a^2$ we have $n^2 = (8m_{\text{eff}}a^2/h^2)E$ and, as a consequence,

$$\frac{dn}{dE} = \frac{1}{2}\sqrt{\frac{8m_{\text{eff}}a^2}{h^2}}\frac{1}{\sqrt{E}}.$$

Table 9.3 Energies of a particle in a three-dimensional box

n_x	n_y	n_z	Energy
1	1	1	$E_{111} = 3h^2/8m_{\text{eff}}a^2$
1	1	2	$2E_{111}$
1	2	1	
2	1	1	
1	2	2	$3E_{111}$
2	1	2	
2	2	1	
1	1	3	$3.667E_{111}$
1	3	1	
3	1	1	
2	2	2	$4E_{111}$
1	2	3	$4.667E_{111}$
1	3	2	
2	1	3	
2	3	1	
3	1	2	
3	2	1	
2	2	3	$5.667E_{111}$
2	3	2	
3	2	2	
1	1	4	$6E_{111}$
1	4	1	
4	1	1	
1	3	3	$6.333E_{111}$
3	1	3	
3	3	1	
1	2	4	$7E_{111}$
1	4	2	
2	1	4	
2	4	1	
4	1	2	
4	2	1	

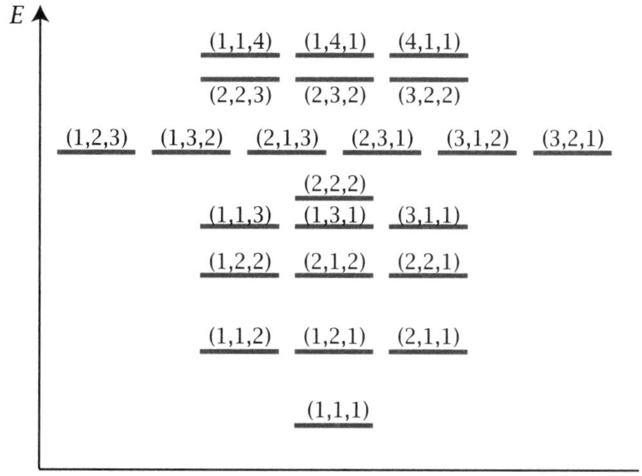

Figure 9.5 Depiction of the first eight energy levels of a particle in a three-dimensional box, including degeneracies. The numbers represent the indices n_x, n_y, n_z. The spacing between states is to scale.

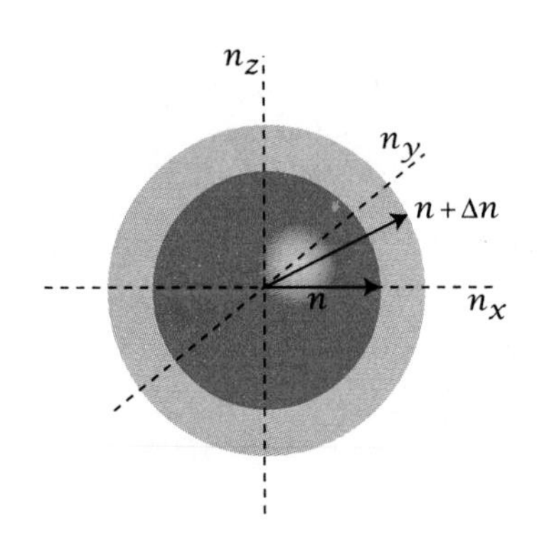

Figure 9.6 Constant-energy spherical surfaces described by the indices n_x, n_y, n_z and the radius $n = \sqrt{n_x^2 + n_y^2 + n_z^2}$ for a particle in a three-dimensional box.

We thus find that

$$g_{\text{3D box}}(E) \sim \frac{\pi}{4}\left(\frac{8m_{\text{eff}}a^2}{h^2}\right)^{3/2}\sqrt{E}.$$

We now define a density of states by dividing $g_{\text{3D box}}(E)$ by the physical volume of the box. This gives $\rho_{\text{energy, 3D box}} = g_{\text{3D box}}/a^3$, with units of number per unit energy per unit volume. The result

$$\rho_{\text{energy, 3D box}} \sim \frac{\pi}{4}\left(\frac{8m_{\text{eff}}}{h^2}\right)^{3/2}\sqrt{E} \tag{9.3}$$

possesses a characteristic $\sqrt{E}$ dependence. As a consequence, at higher energies, there are more available states in the system. In principle, one could have seen this trend by looking at **Table 9.3** or **Figure 9.5**. However, it is not obvious and highlights the usefulness of the density-of-states calculation.

9.2 DENSITY OF STATES FOR BULK MATERIALS, WELLS, WIRES, AND DOTS

Let us now switch to systems more relevant to us. Namely, we want to calculate the density of states for a model bulk system, quantum well, quantum wire, and quantum dot. We have already illustrated the basic approach in our earlier particle-in-a-box examples. However, there will be some slight differences that the reader will notice and should keep in mind.

For more information about these density-of-states calculations, the reader may consult the references cited in the Further Reading.

9.2.1 Bulk Density of States

First, let us derive the bulk density of states. There exist a number of approaches for doing this. In fact, we have just used one in the previous example of a particle in a three-dimensional box. The reader may verify this as an exercise, noting that the only difference is a factor of two that accounts for spin degeneracy.

In this section, we will demonstrate two common strategies, since one usually sees one or the other but not both. Either method leads to the same result. However, one may be conceptually easier to remember. In our first approximation, consider a sphere in k-space. Note that the actual geometry does not matter. Furthermore, while we will assume periodic boundary conditions, the previous example of a particle in a three-dimensional box did not (in fact, it assumed static boundary conditions).

Associated with this sphere is a volume

$$V_k = \tfrac{4}{3}\pi k^3,$$

where k is our "radius" and $k^2 = k_x^2 + k_y^2 + k_z^2$. In general,

$$k_x = \frac{2\pi n_x}{L_x}, \quad \text{where } n_x = 0, \pm 1, \pm 2, \pm 3, \ldots,$$

$$k_y = \frac{2\pi n_y}{L_y}, \quad \text{where } n_y = 0, \pm 1, \pm 2, \pm 3, \ldots,$$

$$k_z = \frac{2\pi n_z}{L_z}, \quad \text{where } n_z = 0, \pm 1, \pm 2, \pm 3, \ldots.$$

These k values arise from the periodic boundary conditions imposed on the carrier's wavefunction. We saw this earlier in Chapter 7 when we talked about the free particle problem. Namely, the free carrier wavefunction is represented by a traveling wave. As a consequence, it satisfies periodic boundary conditions in the extended solid and leads to the particular forms of k_x, k_y, and k_z shown. By contrast, the factor of 2 in the numerator is absent when static boundary conditions are assumed. Notice also that in a cubic solid, $L_x = L_y = L_z = Na$, where N is the number of unit cells along a given direction and a represents an interatomic spacing.

The above volume contains many energies, since

$$E = \frac{\hbar^2 k^2}{2m_{\text{eff}}} = \frac{p^2}{2m_{\text{eff}}}.$$

Furthermore, every point on the sphere's surface possesses the same energy. Spheres with smaller radii therefore have points on their surfaces with equivalent, but correspondingly smaller, energies. This is illustrated in **Figure 9.7**.

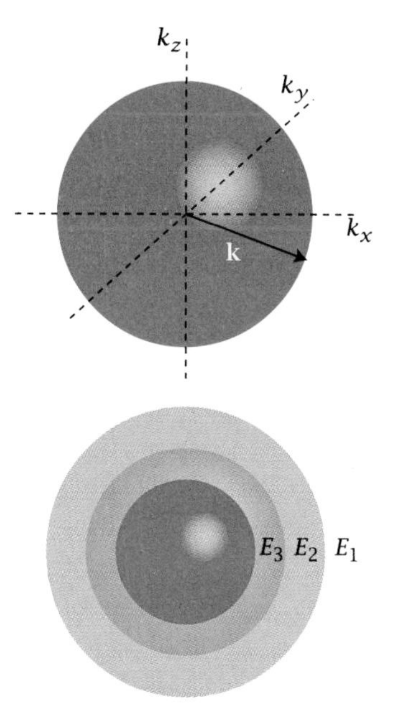

Figure 9.7 Constant-energy spherical surfaces in k-space. Different radii ($k = \sqrt{k_x^2 + k_y^2 + k_z^2}$) lead to different constant-energy surfaces denoted by E_1, E_2, and E_3 here, for example.

We now define a state by the smallest nonzero volume it possesses in k-space. This occurs when $k_x = 2\pi/L_x$, $k_y = 2\pi/L_y$, and $k_z = 2\pi/L_z$, so that

$$V_{\text{state}} = k_x k_y k_z = \frac{8\pi^3}{L_x L_y L_z}.$$

The concept is illustrated in **Figure 9.8**. Thus, within our imagined spherical volume of k-space, the total number of states present is

$$N_1 = \frac{V_k}{V_{\text{state}}} = \frac{\frac{4}{3}\pi k^3}{k_x k_y k_z} = \frac{k^3}{6\pi^2} L_x L_y L_z.$$

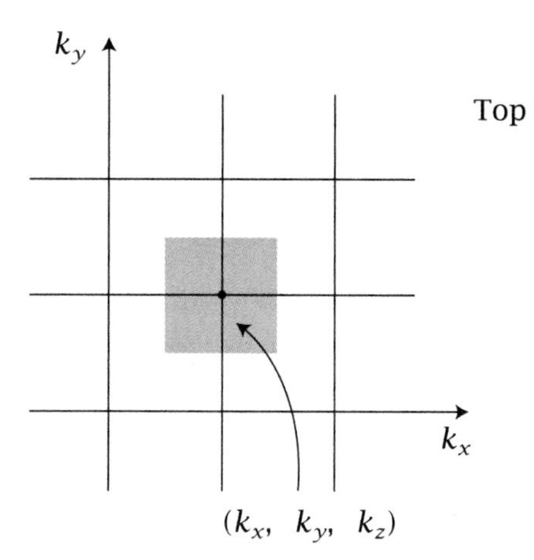

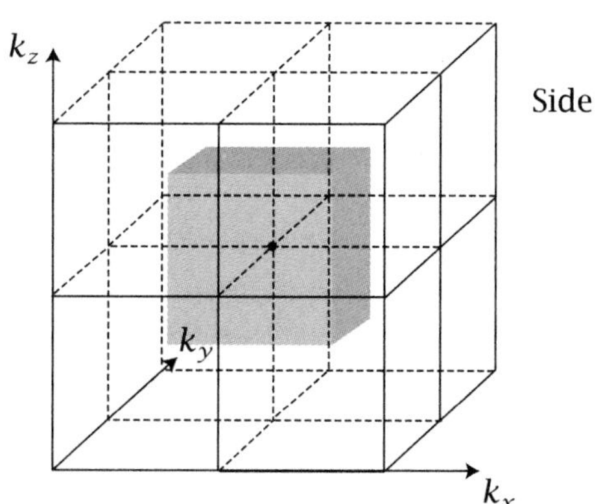

Figure 9.8 Volume (shaded) associated with a given state in k-space. Two views are shown: a top view onto the (k_x, k_y) plane and a three-dimensional side view.

Next, when dealing with electrons and holes, we must consider spin degeneracy, since two carriers, possessing opposite spin, can occupy the same state. As a consequence, we multiply the above expression by 2 to obtain

$$N_2 = 2N_1 = \frac{k^3}{3\pi^2} L_x L_y L_z.$$

This represents the total number of *available* states for carriers, accounting for spin.

We now define a density of states per unit volume, $\rho = N_2 / L_x L_y L_z$, with units of number per unit volume. This results in

$$\rho = \frac{k^3}{3\pi^2}.$$

Finally, considering an energy density, $\rho_{\text{energy}} = d\rho/dE$, with units number per unit energy per unit volume, we obtain

$$\rho_{\text{energy}}(E) = \frac{1}{3\pi^2} \frac{d}{dE} \left(\frac{2m_{\text{eff}} E}{\hbar^2} \right)^{3/2},$$

which simplifies to

$$\rho_{\text{energy}}(E) = \frac{1}{2\pi^2} \left(\frac{2m_{\text{eff}}}{\hbar^2} \right)^{3/2} \sqrt{E}. \tag{9.4}$$

This is our desired density-of-states expression for a bulk three-dimensional solid. Note that the function possesses a characteristic square root energy dependence.

Optional: An Alternative Derivation of the Bulk Density of States

This section is optional and the reader can skip it if desired. We discuss this alternative derivation because it is common to see a system's density of states derived, but not along with the various other approaches that exist.

We can alternatively derive the bulk density of states by starting with a uniform radial probability density in k-space,

$$w(k) = 4\pi k^2.$$

The corresponding (sphere) shell volume element is then

$$\begin{aligned} V_{\text{shell}} &= w(k)\, dk \\ &= 4\pi k^2\, dk. \end{aligned}$$

The latter expression can be derived by taking the volume difference between two spheres in k-space, one of radius $k + dk$ and the other of radius k. In this case, we have

$$\begin{aligned} V_{\text{shell}} &= V(k + dk) - V(k) \\ &= \tfrac{4}{3}\pi(k + dk)^3 - \tfrac{4}{3}\pi k^3, \end{aligned}$$

which when expanded gives

$$\begin{aligned} V_{\text{shell}} &= \tfrac{4}{3}\pi(k^3 + k^2\, dk + 2k^2\, dk + 2k\, dk^2 + k\, dk^2 + dk^3 - k^3) \\ &= \tfrac{4}{3}\pi(3k^2\, dk + 3k\, dk^2 + dk^3). \end{aligned}$$

Keeping terms only up to order dk, since dk^2 and dk^3 are much smaller values, yields

$$V_{\text{shell}} = 4\pi k^2 dk$$

as our desired shell volume. Now, the point of finding the shell volume element between two spheres is that the density-of-states calculation aims to find the number of available states per unit energy difference. This unit change is then dk, since there exists a direct connection between E and k.

Next, we have previously seen that the "volume" of a given state is

$$V_{\text{state}} = k_x k_y k_z = \frac{8\pi^3}{L_x L_y L_z}.$$

The number of states present in the shell is therefore the ratio $N_1 = V_{\text{shell}}/V_{\text{state}}$:

$$N_1 = \frac{k^2}{2\pi^2} L_x L_y L_z \, dk.$$

If spin degeneracy is considered, the result is multiplied by 2 to obtain

$$N_2 = 2N_1 = \frac{k^2}{\pi^2} L_x L_y L_z \, dk.$$

At this point, we can identify

$$g(k) = \frac{k^2}{\pi^2} L_x L_y L_z$$

as a probability density with units of number per unit k, since

$$N_2 = g(k)\, dk.$$

Furthermore, because of the 1:1 correspondence between k and E, we can make an additional link to the desired energy probability density $g(E)$ through

$$g(k)\, dk = g(E)\, dE.$$

As a consequence,

$$g(E) = g(k)\frac{dk}{dE}.$$

Recalling that

$$k = \sqrt{\frac{2m_{\text{eff}}E}{\hbar^2}} \quad \text{and} \quad \frac{dk}{dE} = \frac{1}{2}\sqrt{\frac{2m_{\text{eff}}}{\hbar^2}}\frac{1}{\sqrt{E}}$$

then gives

$$g(E) = \frac{L_x L_y L_z}{2\pi^2}\left(\frac{2m_{\text{eff}}}{\hbar^2}\right)^{3/2}\sqrt{E}$$

as the expression for the number of available states in the system per unit energy. Dividing this result by the real space volume $L_x L_y L_z$ yields our desired density of states, $\rho_{\text{energy}}(E) = g(E)/L_x L_y L_z$, with units of number per unit energy per unit volume:

$$\rho_{\text{energy}}(E) = \frac{1}{2\pi^2}\left(\frac{2m_{\text{eff}}}{\hbar^2}\right)^{3/2}\sqrt{E}.$$

The reader will notice that it is identical to what we found earlier (Equation 9.4).

9.2.2 Quantum Well Density of States

The density of states of a quantum well can also be evaluated. As with the bulk density-of-states evaluation, there are various approaches for doing this. We begin with one and leave the second derivation as an optional section that the reader can skip if desired.

First consider a circular area with radius $k = \sqrt{k_x^2 + k_y^2}$. Note that symmetric circular areas are considered, since two degrees of freedom exist within the (x, y) plane. The one direction of confinement that occurs along the z direction is excluded. The perimeters of the circles shown in **Figure 9.9** therefore represent constant-energy perimeters much like the constant-energy surfaces seen earlier in the bulk example.

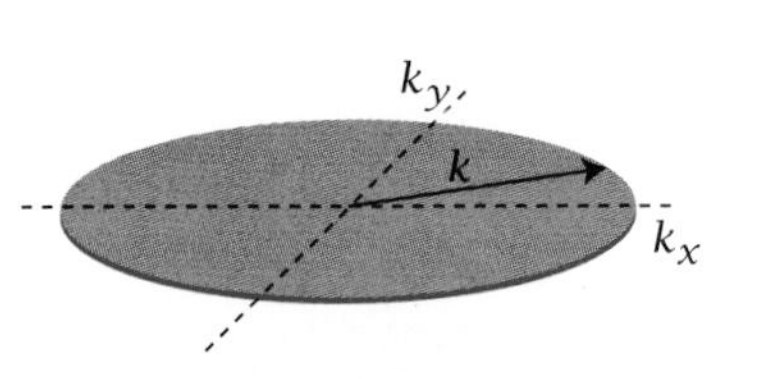

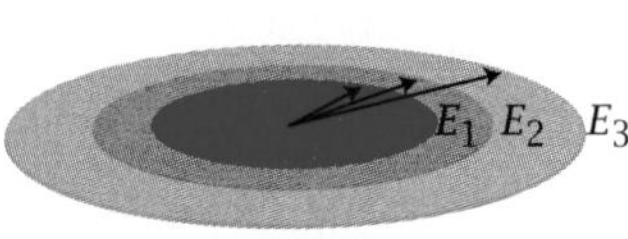

Figure 9.9 Constant-energy perimeters in k-space. The associated radius is $k = \sqrt{k_x^2 + k_y^2}$. Perimeters characterized by smaller radii have correspondingly smaller energies and are denoted by E_1, E_2, and E_3 here as an example.

The associated circular area in k-space is then

$$A_k = \pi k^2,$$

and encompasses many states having different energies. A given state within this circle occupies an area of

$$A_{\text{state}} = k_x k_y,$$

with $k_x = 2\pi/L_x$ and $k_y = 2\pi/L_y$, as illustrated in **Figure 9.10**. Recall that $L_x = L_y = Na$, where N is the number of unit cells along a given direction and a represents an interatomic spacing. Thus,

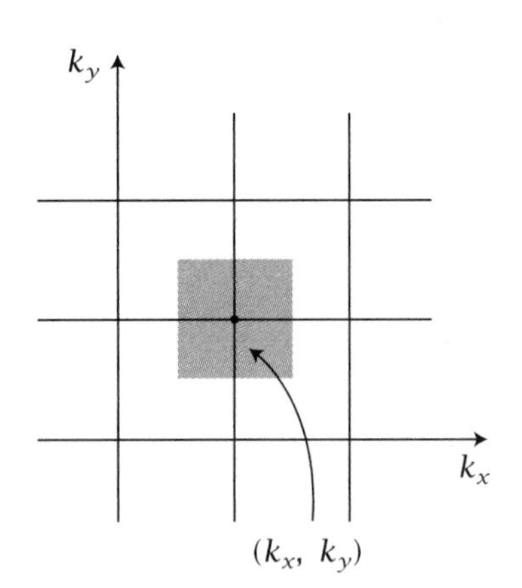

Figure 9.10 Area (shaded) associated with a given state in k-space for a two-dimensional system.

$$A_{\text{state}} = \frac{(2\pi)^2}{L_x L_y}.$$

The total number of states encompassed by this circular area is therefore $N_1 = A_k/A_{\text{state}}$, resulting in

$$N_1 = \frac{k^2}{4\pi} L_x L_y.$$

If we account for spin degeneracy, this value is further multiplied by 2,

$$N_2 = 2N_1 = \frac{k^2}{2\pi} L_x L_y,$$

giving the total number of available states for carriers, including spin. At this point, we can define an area density

$$\rho = \frac{N_2}{L_x L_y} = \frac{k^2}{2\pi} = \frac{m_{\text{eff}} E}{\pi \hbar^2}$$

with units of number per unit area, since $k = \sqrt{2m_{\text{eff}} E/\hbar^2}$. Our desired energy density is then $\rho_{\text{energy}} = d\rho/dE$ and yields

$$\rho_{\text{energy}} = \frac{m_{\text{eff}}}{\pi \hbar^2}, \tag{9.5}$$

with units of number per unit energy per unit area. Notice that it is a constant.

Notice also that this density of states in (x, y) accompanies states associated with each value of k_z (or n_z). As a consequence, each k_z (or n_z)

value is accompanied by a "subband" and one generally expresses this through

$$\rho_{\text{energy}}(E) = \frac{m_{\text{eff}}}{\pi\hbar^2}\sum_{n_z}\Theta(E - E_{n_z}), \qquad (9.6)$$

where n_z is the index associated with the confinement energy along the z direction and $\Theta(E - E_{n_z})$ is the Heaviside unit step function, defined by

$$\Theta(E - E_{n_z}) = \begin{cases} 0 & \text{if } E < E_{n_z} \\ 1 & \text{if } E > E_{n_z}. \end{cases} \qquad (9.7)$$

Optional: An Alternative Derivation of the Quantum Well Density of States

As before, this section is optional and the reader may skip it if desired.

We can alternatively derive the above density-of-states expression. First, consider a uniform radial probability density within a plane of k-space. It has the form

$$w(k) = 2\pi k$$

such that a differential area A_{annulus} can be defined as

$$\begin{aligned} A_{\text{annulus}} &= w(k)\,dk \\ &= 2\pi k\,dk. \end{aligned}$$

The latter expression can be derived by simply taking the difference in area between two circles having radii differing by dk. What results is

$$\begin{aligned} A_{\text{annulus}} &= A(k + dk) - A(k) \\ &= \pi(k + dk)^2 - \pi k^2 \\ &= \pi(k^2 + 2k\,dk + dk^2 - k^2) \\ &= \pi(2k\,dk + dk^2). \end{aligned}$$

For small dk, we can ignore dk^2. This yields

$$A_{\text{annulus}} = 2\pi k\,dk.$$

We are now interested in calculating the number of states within this differential area, since it ultimately represents the number of available states per unit energy. Since the smallest nonzero "area" occupied by a state is

$$A_{\text{state}} = k_x k_y = \frac{4\pi^2}{L_x L_y},$$

the number of states present in the annulus is

$$N_1 = \frac{A_{\text{annulus}}}{A_{\text{state}}} = \frac{k\,dk}{2\pi} L_x L_y.$$

To account for spin degeneracy, we multiply this value by 2 to obtain

$$N_2 = 2N_1 = \frac{k\,dk}{\pi} L_x L_y.$$

This represents the number of states available to carriers in a ring of thickness dk, accounting for spin degeneracy.

At this point, we identify

$$g(k) = \frac{k}{\pi} L_x L_y$$

as a probability density with units of number per unit k, since

$$N_2 = g(k)\, dk.$$

An equivalence with the analogous energy probability density $g(E)$ then occurs because of the 1:1 correspondence between k and E:

$$g(k)\, dk = g(E)\, dE.$$

Solving for $g(E)$ then yields

$$g(E) = g(k) \frac{dk}{dE},$$

where

$$k = \sqrt{\frac{2m_{\text{eff}} E}{\hbar^2}} \quad \text{and} \quad \frac{dk}{dE} = \frac{1}{2}\sqrt{\frac{2m_{\text{eff}}}{\hbar^2}} \frac{1}{\sqrt{E}}.$$

We thus have

$$g(E) = \left(\frac{m_{\text{eff}}}{\pi \hbar^2}\right) L_x L_y.$$

This energy density describes the number of available states per unit energy. Finally, dividing by $L_x L_y$ results in our desired density of states,

$$\rho_{\text{energy}} = \frac{g(E)}{L_x L_y} = \frac{m_{\text{eff}}}{\pi \hbar^2},$$

with units of number per unit energy per unit area. This is identical to Equation 9.5. Since an analogous expression exists for every k_z (or n_z) value, we subsequently generalize this result to

$$\rho_{\text{energy}}(E) = \frac{m_{\text{eff}}}{\pi \hbar^2} \sum_{n_z} \Theta(E - E_{n_z}),$$

where $\Theta(E - E_{n_z})$ is the Heaviside unit step function.

9.2.3 Nanowire Density of States

The derivation of the nanowire density of states proceeds in an identical manner. The only change is the different dimensionality. Whereas we discussed volumes and areas for bulk systems and quantum wells, we refer to lengths here. As before, there are also alternative derivations. One of these is presented in an optional section and can be skipped if desired.

For a nanowire, consider a symmetric line about the origin in k-space having a length $2k$:

$$L_k = 2k.$$

The associated "width" occupied by a given state is

$$L_{\text{state}} = k,$$

where k represents any one of three directions in k-space: k_x, k_y, or k_z. For convenience, choose the z direction. This will represent the single degree of freedom for carriers in the wire. We then have possible k_z values of

$$k_z = \frac{2\pi n_z}{L_z}, \quad \text{with } n_z = 0, \pm 1, \pm 2, \pm 3, \ldots,$$

where $L_z = Na$, N represents the number of unit cells along the z direction, and a is an interatomic spacing. The smallest positive nonzero length occurs when $n_z = 1$. As a consequence, the number of states found within L_k is

$$N_1 = \frac{L_k}{L_{\text{state}}} = \frac{k}{\pi} L_z.$$

If spin degeneracy is considered,

$$N_2 = 2N_1 = \frac{2k}{\pi} L_z$$

and describes the total number of available states for carriers.

We now define a density

$$\rho = \frac{N_2}{L_z} = \frac{2k}{\pi} = \frac{2}{\pi}\sqrt{\frac{2m_{\text{eff}}E}{\hbar^2}} \tag{9.8}$$

that describes the number of states per unit length, including spin. This leads to an expression for the DOS defined as $\rho_{\text{energy}} = d\rho/dE$:

$$\rho_{\text{energy}}(E) = \frac{2}{\pi}\left(\frac{dk}{dE}\right), \tag{9.9}$$

giving

$$\rho_{\text{energy}}(E) = \frac{1}{\pi}\sqrt{\frac{2m_{\text{eff}}}{\hbar^2}} \frac{1}{\sqrt{E}}. \tag{9.10}$$

Equation 9.10 is our desired expression, with units of number per unit energy per unit length. More generally, since this distribution is associated with confined energies along the other two directions, x and y, we write

$$\rho_{\text{energy}}(E) = \frac{1}{\pi}\sqrt{\frac{2m_{\text{eff}}}{\hbar^2}} \sum_{n_x, n_y} \frac{1}{\sqrt{E - E_{n_x, n_y}}} \Theta(E - E_{n_x, n_y}), \tag{9.11}$$

where E_{n_x, n_y} are the confinement energies associated with the x and y directions and $\Theta(E - E_{n_x, n_y})$ is the Heaviside unit step function. Notice the characteristic inverse square root dependence of the nanowire one-dimensional density of states.

Optional: An Alternative Derivation of the Nanowire Density of States

Alternatively, we can rederive the nanowire density-of-states expression by simply considering a differential length element dk between k and $k + dk$, with the length occupied by a given state being $k_z = 2\pi/L_z$. The number of states present is then

$$N_1 = \frac{dk}{k_z} = \frac{dk}{2\pi} L_z.$$

If spin degeneracy is considered, this expression is multiplied by two, giving

$$N_2 = 2N_1 = \frac{dk}{\pi}L_z.$$

We now define a probability density having the form

$$g(k) = \frac{L_z}{\pi},$$

since $N_2 = g(k)\,dk$. Furthermore, from the equivalence

$$g(k)\,dk = g(E)\,dE,$$

we find that

$$g(E) = g(k)\frac{dk}{dE}$$

with

$$k = \sqrt{\frac{2m_{\text{eff}}E}{\hbar^2}} \quad \text{and} \quad \frac{dk}{dE} = \frac{1}{2}\sqrt{\frac{2m_{\text{eff}}}{\hbar^2}}\frac{1}{\sqrt{E}}.$$

This yields

$$g(E) = \frac{L_z}{2\pi}\sqrt{\frac{2m_{\text{eff}}}{\hbar^2}}\frac{1}{\sqrt{E}}$$

as our density, with units of number per unit energy. Finally, we can define a density of states

$$\rho_{\text{energy}}(E) = \frac{2g(E)}{L_z},$$

where the extra factor of 2 accounts for the energy degeneracy between positive and negative k values. Note that we took this into account in our first derivation since we considered a symmetric length $2k$ about the origin. We then have

$$\rho_{\text{energy}}(E) = \frac{1}{\pi}\sqrt{\frac{2m_{\text{eff}}}{\hbar^2}}\frac{1}{\sqrt{E}},$$

which is identical to Equation 9.10. As before, the density of states can be generalized to

$$\rho_{\text{energy}}(E) = \frac{1}{\pi}\sqrt{\frac{2m_{\text{eff}}}{\hbar^2}}\sum_{n_x,n_y}\frac{1}{\sqrt{E - E_{n_x,n_y}}}\Theta(E - E_{n_x,n_y}),$$

since the distribution is associated with each confined energy E_{n_x,n_y}.

9.2.4 Quantum Dot Density of States

Finally, in a quantum dot, the density of states is just a series of delta functions, given that all three dimensions exhibit carrier confinement:

$$\rho_{\text{energy}}(E) = 2\delta(E - E_{n_x,n_y,n_z}). \tag{9.12}$$

In Equation 9.12, E_{n_x,n_y,n_z} are the confined energies of the carrier, characterized by the indices n_x, n_y, n_z. The factor of 2 accounts for spin degeneracy. To generalize the expression, we write

$$\rho_{\text{energy}}(E) = 2\sum_{n_x,n_y,n_z}\delta(E - E_{n_x,n_y,n_z}), \tag{9.13}$$

which accounts for all of the confined states in the system.

9.3 POPULATION OF THE CONDUCTION AND VALENCE BANDS

At this point, we have just calculated the density of states for three-, two-, one-, and zero-dimensional systems. While we have explicitly considered conduction band electrons in these derivations, such calculations also apply to holes in the valence band. The topic of bands will be discussed shortly in Chapter 10. Thus, in principle, with some slight modifications, one also has the valence band density of states for all systems of interest.

In either case, whether for the electron or the hole, the above density-of-states expressions just tell us the density of available states. They say nothing about whether or not such states are occupied. For this, we need the probability $P(E)$ that an electron or hole resides in a given state with an energy E. This will therefore be the focus of the current section and is also our first application of the density-of-states function to find where most carriers reside in a given band. Through this, we will determine both the carrier concentration and the position of the so-called Fermi level in each system.

9.3.1 Bulk

Let us first consider a bulk solid. We begin by evaluating the occupation of states in the conduction band, followed by the same calculation for holes in the valence band.

Conduction Band

For the conduction band, we have the following expression for the number of occupied states for a given unit energy difference per unit volume (alternatively, the concentration of electrons for a given unit energy difference):

$$n_e(E) = P_e(E)\rho_{\text{energy}}(E)\, dE.$$

In this expression, $P_e(E)$ is the probability that an electron possesses a given energy E and is called the Fermi–Dirac distribution. It is one of three distribution functions that we will encounter throughout this text:

$$P_e(E) = \frac{1}{1 + e^{(E-E_F)/kT}}. \tag{9.14}$$

Recall that we saw the other two (the Boltzmann distribution and the Bose–Einstein distribution) earlier in Chapter 5. There are several things to note. First, in Equation 9.14, E_F refers to the Fermi level, which should be distinguished from the Fermi energy seen earlier in Chapter 3. The two are equal only at 0 K. Next, note that the Fermi level E_F is sometimes referred to as the "chemical potential μ." However, we will stick to E_F here for convenience. The Fermi level or chemical potential is also the energy where $P_e(E = E_F) = 0.5$ (i.e., the probability of a carrier having this energy is 50%) and, as such, represents an important reference energy. Finally, **Figure 9.11** shows that the product of $P_e(E)$ and $\rho_{\text{energy}}(E)$ implies that most electrons reside near the conduction band edge.

The total concentration of electrons in the conduction band, n_c, is then the integral of $n_e(E)$ over all available energies:

$$n_c = \int_{E_c}^{\infty} P_e(E)\rho_{\text{energy}}(E)\, dE.$$

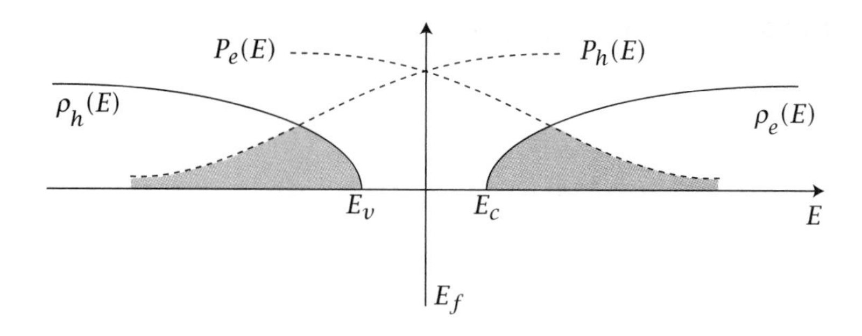

Figure 9.11 Conduction band and valence band occupation of states. The shaded regions denote where the carriers are primarily located in each band.

Notice that the lower limit of the integral is E_c, which represents the starting energy of the conduction band.

Next, we have previously found for bulk materials (Equation 9.4) that

$$\rho_{\text{energy}}(E) = \frac{1}{2\pi^2}\left(\frac{2m_{\text{eff}}}{\hbar^2}\right)^{3/2}\sqrt{E}.$$

Note that since we now explicitly refer to the electron in the conduction band, we replace m_{eff} with m_e to denote the electron's effective mass there. We also modify the expression to have a nonzero origin to account for the conduction band starting energy. We thus find

$$\rho_{\text{energy}}(E) = \frac{1}{2\pi^2}\left(\frac{2m_e}{\hbar^2}\right)^{3/2}\sqrt{E - E_c}.$$

Inserting this into our integral then yields

$$n_c = \int_{E_c}^{\infty}\left(\frac{1}{1+e^{(E-E_F)/kT}}\right)\left(\frac{1}{2\pi^2}\right)\left(\frac{2m_e}{\hbar^2}\right)^{3/2}\sqrt{E-E_c}\,dE$$

or

$$n_c = A\int_{E_c}^{\infty}\left(\frac{1}{1+e^{(E-E_F)/kT}}\right)\sqrt{E-E_c}\,dE, \tag{9.15}$$

where

$$A = \frac{1}{2\pi^2}\left(\frac{2m_e}{\hbar^2}\right)^{3/2}.$$

We can subsequently simplify this as follows:

$$n_c = A\int_{E_c}^{\infty}\left(\frac{1}{1+e^{(E-E_c)/kT}e^{(E_c-E_F)/kT}}\right)\sqrt{E-E_c}\,dE,$$

and let $\eta = (E - E_c)/kT$ as well as $\mu = (-E_c + E_F)/kT$ to obtain

$$n_c = A(kT)^{3/2}\int_0^{\infty}\frac{\sqrt{\eta}}{1+e^{\eta-\mu}}\,d\eta.$$

The resulting integral has been solved and can be looked up. It is called the Fermi–Dirac integral and is defined as follows:

$$F_{1/2}(\mu) = \int_0^{\infty}\frac{\sqrt{\eta}}{1+e^{\eta-\mu}}\,d\eta. \tag{9.16}$$

However, to stay instructive, let us just consider the situation where $E - E_F \gg kT$. In this case, the exponential term in the denominator of Equation 9.15 dominates, so that

$$\frac{1}{1+e^{(E-E_F)/kT}} \simeq e^{-(E-E_F)/kT}.$$

We therefore have

$$n_c \simeq A \int_{E_c}^{\infty} e^{-(E-E_F)/kT} \sqrt{E - E_c}\, dE,$$

which we can again modify to simplify things

$$n_c \simeq A \int_{E_c}^{\infty} e^{-[(E-E_c)+(E_c-E_F)]/kT} \sqrt{E - E_c}\, dE$$

$$\simeq A e^{-(E_c-E_F)/kT} \int_{E_c}^{\infty} e^{-(E-E_c)/kT} \sqrt{E - E_c}\, dE.$$

At this point, changing variables by letting $x = (E - E_c)/kT$, $E = E_c + xkT$, and $dE = kT\, dx$ as well as changing the limits of integration, gives

$$n_c \simeq A(kT)^{3/2} e^{-(E_c-E_F)/kT} \int_0^{\infty} e^{-x} \sqrt{x}\, dx.$$

To evaluate the integral here, we note that the gamma function $\Gamma(n)$ is defined as

$$\Gamma(n) = \int_0^{\infty} e^{-x} x^{n-1}\, dx. \tag{9.17}$$

Values of Γ are tabulated in the literature (see, e.g., Abramowitz and Stegun 1972; Beyer 1991). In our case, we have $\Gamma\left(\frac{3}{2}\right)$. As a consequence,

$$n_c \simeq A(kT)^{3/2} e^{-(E_c-E_F)/kT} \Gamma\left(\tfrac{3}{2}\right),$$

and we recall that $A = (1/2\pi^2)(2m_e/\hbar^2)^{3/2}$. Consolidating terms by letting $N_c = A(kT)^{3/2}\Gamma\left(\frac{3}{2}\right)$ then results in

$$n_c \simeq N_c e^{-(E_c-E_F)/kT} \tag{9.18}$$

and expresses the concentration of carriers in the bulk semiconductor conduction band.

Valence Band

We repeat the same calculation for the valence band. The approach is similar, but requires a few changes since we are now dealing with holes. As before, we first need a function $P_h(E)$ describing the probability that a hole possesses a given energy. Likewise, we need the previous bulk density of states (Equation 9.4).

Conceptually, the number of holes per unit volume for a given unit energy difference (i.e., the concentration of holes for a given unit energy difference) is

$$n_h(E) = P_h(E) \rho_{\text{energy}}(E)\, dE,$$

where the the probability that a hole occupies a given state can be expressed in terms of the absence of an electron. Since we have previously invoked the Fermi–Dirac distribution to express the probability that an electron occupies a given state, we can find an analogous probability for an electron *not* occupying it by subtracting $P_e(E)$ from 1. In this regard, the absence of an electron in a given valence band state is the same as having a hole occupy it. As a consequence, we write $P_h(E) = 1 - P_e(E)$ to obtain

$$P_h(E) = 1 - \frac{1}{1 + e^{(E-E_F)/kT}}. \tag{9.19}$$

Next, the hole density of states can be written as

$$\rho_{\text{energy}}(E) = \frac{1}{2\pi^2}\left(\frac{2m_h}{\hbar^2}\right)^{3/2}\sqrt{E_v - E},$$

which is slightly different from the corresponding electron density of states. Note that we have used m_h to denote the hole's effective mass. The different square root term also accounts for the fact that hole energies increase in the opposite direction to those of electrons since their effective mass has a different sign. We alluded to this earlier in Chapter 7 and will show it later in Chapter 10. Note that E_v is the valence band starting energy and, in many instances, is defined as the zero energy of the x axis in **Figure 9.11**. As a consequence, hole energies become larger towards more negative values in this picture. **Figure 9.11** likewise illustrates that most holes live near the valence band edge.

The total hole concentration is therefore the following integral over all energies:

$$\begin{aligned} n_v &= \int_{-\infty}^{E_v} P_h(E)\rho_{\text{energy}}(E)\,dE \\ &= B\int_{-\infty}^{E_v}\left(1 - \frac{1}{1 + e^{(E-E_F)/kT}}\right)\sqrt{E_v - E}\,dE, \end{aligned}$$

where

$$B = \frac{1}{2\pi^2}\left(\frac{2m_h}{\hbar^2}\right)^{3/2}.$$

Generally speaking, since $E < E_F$ in the valence band (recall that in this scheme, where we are using a single energy axis for both electrons and holes, hole energies increase towards larger negative values), we write

$$n_v = B\int_{-\infty}^{E_v}\left(1 - \frac{1}{1 + e^{-(E_F-E)/kT}}\right)\sqrt{E_v - E}\,dE.$$

We can now approximate the term in parentheses through the binomial expansion, keeping only the first two terms:

$$\frac{1}{1+x} \simeq 1 - x + x^2 - x^3 + x^4 + \cdots.$$

We therefore find that

$$\frac{1}{1 + e^{-(E_F-E)/kT}} \simeq 1 - e^{-(E_F-E)/kT} + \cdots.$$

On substituting this into n_v, we obtain

$$n_v \simeq B\int_{-\infty}^{E_v} e^{-(E_F-E)/kT}\sqrt{E_v - E}\,dE.$$

Changing variables, by letting $x = (E_v - E)/kT$, $E = E_v - xkT$, and $dE = -kT\,dx$, and remembering to change the limits of integration, then gives

$$\begin{aligned} n_v &\simeq B\int_{+\infty}^{0} e^{-[(E_F-E_v)+(E_v-E)]/kT}\sqrt{kTx}(-kT)\,dx \\ &\simeq B(kT)^{3/2}e^{-(E_F-E_v)/kT}\int_0^{\infty} e^{-x}\sqrt{x}\,dx. \end{aligned}$$

The last integral is again the gamma function $\Gamma\left(\frac{3}{2}\right)$. As a consequence, we find

$$n_v \simeq B(kT)^{3/2}e^{-(E_F-E_v)/kT}\Gamma\left(\frac{3}{2}\right).$$

Letting $N_v = B(kT)^{3/2}\Gamma\left(\frac{3}{2}\right)$ then yields

$$n_v \simeq N_v e^{-(E_F - E_v)/kT}, \tag{9.20}$$

which describes the hole concentration in the valence band.

Fermi Level

We can now determine the location of the Fermi level E_F from n_v and n_c, assuming that the material is intrinsic (i.e., not deliberately doped with impurities to have extra electrons or holes in it). We will also assume that the material is not being excited optically, so that the system remains in thermal equilibrium. Under these conditions, the following equivalence holds, since associated with every electron is a hole:

$$\begin{aligned} n_c &= n_v, \\ N_c e^{-(E_c - E_F)/kT} &= N_v e^{-(E_F - E_v)/kT}, \\ A(kT)^{3/2}\Gamma\left(\tfrac{3}{2}\right) e^{-(E_c - E_F)/kT} &= B(kT)^{3/2}\Gamma\left(\tfrac{3}{2}\right) e^{-(E_F - E_v)/kT}, \\ \frac{1}{2\pi^2}\left(\frac{2m_e}{\hbar^2}\right)^{3/2} e^{-(E_c - E_F)/kT} &= \frac{1}{2\pi^2}\left(\frac{2m_h}{\hbar^2}\right)^{3/2} e^{-(E_F - E_v)/kT}. \end{aligned}$$

Solving for E_F then yields

$$E_F = \frac{E_c + E_v}{2} + \frac{3}{4}kT\ln\left(\frac{m_h}{m_e}\right), \tag{9.21}$$

which says that at $T = 0\,\text{K}$, the Fermi level of an intrinsic semiconductor at equilibrium occurs halfway between the conduction band and the valence band. However, a weak temperature dependence exists, moving the Fermi level closer to the conduction band when $m_h > m_e$. Alternatively, if $m_e > m_h$, the Fermi level moves closer to the valence band. To a good approximation, the Fermi level lies midway between the two bands, as illustrated in **Figure 9.12**.

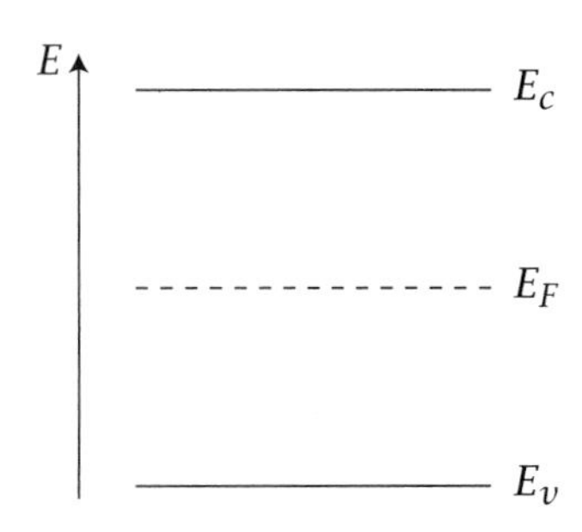

Figure 9.12 Position of the Fermi level relative to the conduction and valence band edges of an intrinsic semiconductor. At 0 K, it lies exactly halfway between the two bands.

9.3.2 Quantum Well

Let us repeat these calculations for a quantum well. The procedure is identical, and the reader can jump ahead to the results if desired.

Conduction Band

We start with our derived quantum well density of states (Equation 9.6),

$$\rho_{\text{energy}} = \frac{m_{\text{eff}}}{\pi\hbar^2}\sum_{n_{z,e}}\Theta(E - E_{n_e}),$$

where $n_{z,e}$ is the index associated with quantization along the z direction. Let us consider only the lowest $n_{z,e} = 1$ subband, as illustrated in **Figure 9.13**. Note further that since we are dealing with an electron in the conduction band, we replace m_{eff} with m_e. As a consequence,

$$\rho_{\text{energy}} = \frac{m_e}{\pi\hbar^2}.$$

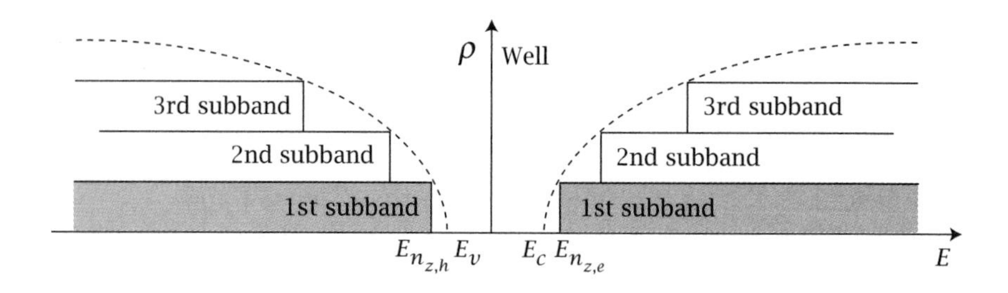

Figure 9.13 First subband (shaded) of the conduction and valence bands in a quantum well. For comparison purposes, the dashed lines represent the bulk density of states for both the electron and the hole.

Next, recall from the previous bulk example that the concentration of states for a given unit energy difference can be expressed as

$$n_e(E) = P_e(E)\rho_{\text{energy}}(E)\, dE.$$

The total concentration of electrons in the first subband is therefore the integral over all available energies:

$$n_C = \int_{E_{nz,e}}^{\infty} P_e(E)\rho_{\text{energy}}(E)\, dE,$$

where the lower limit to the integral appears because the energy of the first subband begins at $E_{n_{z,e}}$ not E_C. See **Figure 9.13**. Note that $P_e(E) = (1 + e^{(E-E_F)/kT})^{-1}$ is again our Fermi–Dirac distribution. We therefore find that

$$n_C = \frac{m_e}{\pi \hbar^2} \int_{E_{nz,e}}^{\infty} \frac{1}{1 + e^{(E-E_F)/kT}}\, dE.$$

To be instructive, we continue with an analytical evaluation of the integral by assuming that $E - E_F \gg kT$. This leads to

$$n_C \simeq \frac{m_e}{\pi \hbar^2} \int_{E_{nz,e}}^{\infty} e^{-(E-E_F)/kT}\, dE,$$

or

$$n_C \simeq \frac{m_e kT}{\pi \hbar^2} \exp\left(-\frac{E_{nz,e} - E_F}{kT}\right), \tag{9.22}$$

which is our desired carrier concentration in the first subband.

Valence Band

Let us repeat the above calculation to find the associated hole density of states. As with the conduction band, we need the probability that the hole occupies a given state within the valence band. The required probability distribution is then $P_h(E)$, which can be expressed in terms of the absence of an electron in a given valence band state (i.e., through $P_h(E) = 1 - P_e(E)$). We obtain

$$P_h(E) = 1 - \frac{1}{1 + e^{(E-E_F)/kT}}.$$

The concentration of states occupied for a given unit energy difference is then

$$n_h(E) = P_h(E)\rho_{\text{energy}}(E)\, dE,$$

where we consider only the first hole subband ($n_{z,h} = 1$) in Equation 9.6:

$$\rho_{\text{energy}} = \frac{m_h}{\pi \hbar^2}.$$

Notice that m_h is used instead of m_{eff} since we are explicitly considering a hole in the valence band.

Since the total concentration of holes in the first subband is the integral over all energies, we find that

$$n_v = \int_{-\infty}^{E_{n_{z,h}}} P_h(E)\rho_{\text{energy}}(E)\,dE.$$

The integral's upper limit is $E_{n_{z,h}}$ since the subband begins there and not at the bulk valence band edge E_v (see **Figure 9.13**). One therefore has

$$n_v = \frac{m_h}{\pi\hbar^2}\int_{-\infty}^{E_{n_{z,h}}}\left(1 - \frac{1}{1+e^{(E-E_F)/kT}}\right)dE$$

and since $E < E_F$ in the valence band, we alternatively write

$$n_v = \frac{m_h}{\pi\hbar^2}\int_{-\infty}^{E_{n_{z,h}}}\left(1 - \frac{1}{1+e^{-(E_F-E)/kT}}\right)dE.$$

Applying the binomial expansion to the term in parentheses

$$\frac{1}{1+e^{-(E_F-E)/kT}} \simeq 1 - e^{-(E_F-E)/kT} + \cdots,$$

then yields

$$n_v \simeq \frac{m_h}{\pi\hbar^2}\int_{-\infty}^{E_{n_{z,h}}} e^{-(E_F-E)/kT}\,dE.$$

This ultimately results in our desired expression for the valence band hole concentration:

$$n_v \simeq \frac{m_h kT}{\pi\hbar^2}\exp\left(-\frac{E_F - E_{n_{z,h}}}{kT}\right). \tag{9.23}$$

Fermi Level

Finally, as in the bulk example, let us evaluate the Fermi level of the quantum well. Again we assume that the material is intrinsic and that it remains at equilibrium. In this case,

$$n_c = n_v,$$

$$\frac{m_e kT}{\pi\hbar^2}\exp\left(-\frac{E_{n_{z,e}} - E_F}{kT}\right) = \frac{m_h kT}{\pi\hbar^2}\exp\left(-\frac{E_F - E_{n_{z,h}}}{kT}\right).$$

Eliminating common terms and solving for E_F then gives

$$E_F = \left(\frac{E_{n_{z,e}} + E_{n_{z,h}}}{2}\right) + \frac{kT}{2}\ln\left(\frac{m_h}{m_e}\right). \tag{9.24}$$

This shows that the Fermi level of an intrinsic two-dimensional material at equilibrium occurs halfway between its conduction band and valence band. The slight temperature dependence means that the Fermi level moves closer to the conduction (valence) band with increasing temperature if $m_h > m_e$ ($m_h < m_e$).

9.3.3 Nanowire

Finally, let us repeat the calculation one last time for a one-dimensional system. The procedure is identical to that in the bulk and quantum well examples above. The reader can therefore skip ahead to the results if desired.

Conduction Band

Let us start with the conduction band and use the Fermi–Dirac distribution in conjunction with the following one-dimensional density of states (Equation 9.11):

$$\rho_{\text{energy}}(E) = \frac{1}{\pi}\sqrt{\frac{2m_e}{\hbar^2}} \sum_{n_{x,e},n_{y,e}} \frac{1}{\sqrt{E - E_{n_{x,e},n_{y,e}}}} \Theta(E - E_{n_{x,e},n_{y,e}}).$$

$n_{x,e}$ and $n_{y,e}$ are integer indices associated with carrier confinement along the two confined directions x and y of the wire and m_{eff} is replaced with m_e to indicate an electron in the conduction band. We also consider only the lowest subband with energies starting at $E_{n_{x,e},n_{y,e}}$ with $n_{x,e} = n_{y,e} = 1$. This is illustrated in **Figure 9.14**.

The concentration of states for a given unit energy difference is therefore

$$n_e(E) = P_e(E)\rho_{\text{energy}}(E)\, dE.$$

The associated carrier concentration within the first subband is then the integral over all possible energies:

$$\begin{aligned} n_c &= \int_{E_{n_{x,e},n_{y,e}}}^{\infty} P_e(E)\rho_{\text{energy}}(E)\, dE \\ &= \frac{1}{\pi}\sqrt{\frac{2m_e}{\hbar^2}} \int_{E_{n_{x,e},n_{y,e}}}^{\infty} \frac{1}{1 + e^{(E-E_F)/kT}} \frac{1}{\sqrt{E - E_{n_{x,e},n_{y,e}}}}\, dE. \end{aligned}$$

If $E - E_F \gg kT$, then

$$\begin{aligned} n_c &\simeq \frac{1}{\pi}\sqrt{\frac{2m_e}{\hbar^2}} \int_{E_{n_{x,e},n_{y,e}}}^{\infty} e^{-(E-E_F)/kT} \frac{1}{\sqrt{E - E_{n_{x,e},n_{y,e}}}}\, dE \\ &\simeq \frac{1}{\pi}\sqrt{\frac{2m_e}{\hbar^2}} \int_{E_{n_{x,e},n_{y,e}}}^{\infty} \exp\left[-\frac{(E - E_{n_{x,e},n_{y,e}}) + (E_{n_{x,e},n_{y,e}} - E_F)}{kT}\right] \\ &\quad\times \frac{1}{\sqrt{E - E_{n_{x,e},n_{y,e}}}}\, dE \\ &\simeq \frac{1}{\pi}\sqrt{\frac{2m_e}{\hbar^2}} \exp\left(-\frac{E_{n_{x,e},n_{y,e}} - E_F}{kT}\right) \\ &\quad\times \int_{E_{n_{x,e},n_{y,e}}}^{\infty} \exp\left(-\frac{E - E_{n_{x,e},n_{y,e}}}{kT}\right) \frac{1}{\sqrt{E - E_{n_{x,e},n_{y,e}}}}\, dE, \end{aligned}$$

which we can simplify by letting $x = (E - E_{n_{x,e},n_{y,e}})/kT$, $E = E_{n_{x,e},n_{y,e}} + xkT$, and $dE = kT\, dx$, remembering to change the limits of integration.

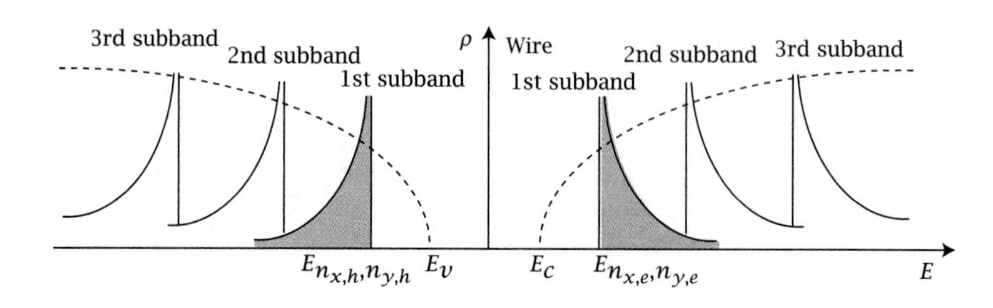

Figure 9.14 First subband (shaded) of both the conduction band and valence band of a one-dimensional system. For comparison purposes, the dashed lines represent the bulk electron and hole density of states.

This yields

$$n_c \simeq \frac{\sqrt{kT}}{\pi}\sqrt{\frac{2m_e}{\hbar^2}}\exp\left(-\frac{E_{n_{x,e},n_{y,e}} - E_F}{kT}\right)\int_0^\infty e^{-x}x^{-1/2}\,dx.$$

The last integral is again the gamma function, in this case $\Gamma\left(\frac{1}{2}\right)$. As a consequence, the desired expression for the electron concentration in the first subband of a one-dimensional material is

$$n_c \simeq \frac{1}{\pi}\sqrt{\frac{2m_e kT}{\hbar^2}}\exp\left(-\frac{E_{n_{x,e},n_{y,e}} - E_F}{kT}\right)\Gamma\left(\frac{1}{2}\right). \tag{9.25}$$

Valence Band

A similar calculation can be done for the holes in the valence band. From before, the hole probability distribution function is

$$P_h(E) = 1 - \frac{1}{1 + e^{(E-E_F)/kT}},$$

with a corresponding one-dimensional hole density of states (Equation 9.11)

$$\rho_{\text{energy}}(E) = \frac{1}{\pi}\sqrt{\frac{2m_h}{\hbar^2}}\sum_{n_{x,h},n_{y,h}}\frac{1}{\sqrt{E_{n_{x,h},n_{y,h}} - E}}\Theta(E_{n_{x,h},n_{y,h}} - E).$$

Note the use of m_h instead of m_{eff} for the hole's effective mass. The indices $n_{x,h}$ and $n_{y,h}$ represent integers associated with carrier confinement along the x and y directions of the system. Notice also the slight modification to the square root term and to the Heaviside unit step function. These changes account for hole energies increasing in the opposite direction to those of electrons due to their different effective mass signs. We alluded to this earlier in Chapter 7 and will see why in Chapter 10.

Considering only the lowest-energy subband, where $n_{x,h} = n_{y,h} = 1$, we have

$$\rho_{\text{energy}}(E) = \frac{1}{\pi}\sqrt{\frac{2m_h}{\hbar^2}}\frac{1}{\sqrt{E_{n_{x,h},n_{y,h}} - E}},$$

whereupon the concentration of states for a given unit energy difference is

$$n_h(E) = P_h(E)\rho_{\text{energy}}(E)\,dE.$$

The total carrier concentration is then the integral over all energies in the subband. We thus have

$$\begin{aligned} n_v &= \int_{-\infty}^{E_{n_{x,h},n_{y,h}}} P_h(E)\rho_{\text{energy}}(E)\,dE \\ &= \frac{1}{\pi}\sqrt{\frac{2m_h}{\hbar^2}}\int_{-\infty}^{E_{n_{x,h},n_{y,h}}}\left(1 - \frac{1}{1 + e^{(E-E_F)/kT}}\right)\frac{1}{\sqrt{E_{n_{x,h},n_{y,h}} - E}}\,dE, \end{aligned}$$

where the upper limit reflects the fact that the subband begins at $E_{n_{x,h},n_{y,h}}$ and not at E_v like in the bulk case. Since $E \ll E_F$, we may also write

$$n_v = \frac{1}{\pi}\sqrt{\frac{2m_h}{\hbar^2}}\int_{-\infty}^{E_{n_{x,h},n_{y,h}}}\left(1 - \frac{1}{1 + e^{-(E_F-E)/kT}}\right)\frac{1}{\sqrt{E_{n_{x,h},n_{y,h}} - E}}\,dE.$$

At this point, using the binomial expansion, keeping only the first two terms,

$$\frac{1}{1+e^{-(E_F-E)/kT}} \simeq 1 - e^{-(E_F-E)/kT},$$

gives

$$\begin{aligned}
n_v &\simeq \frac{1}{\pi}\sqrt{\frac{2m_h}{\hbar^2}} \int_{-\infty}^{E_{n_{x,h},n_{y,h}}} e^{-(E_F-E)/kT} \frac{1}{\sqrt{E_{n_{x,h},n_{y,h}} - E}}\, dE \\
&\simeq \frac{1}{\pi}\sqrt{\frac{2m_h}{\hbar^2}} \int_{-\infty}^{E_{n_{x,h},n_{y,h}}} \exp\left[-\frac{(E_F - E_{n_{x,h},n_{y,h}}) + (E_{n_{x,h},n_{y,h}} - E)}{kT}\right] \\
&\quad \times \frac{1}{\sqrt{E_{n_{x,h},n_{y,h}} - E}}\, dE \\
&\simeq \frac{1}{\pi}\sqrt{\frac{2m_h}{\hbar^2}} \exp\left(-\frac{E_F - E_{n_{x,h},n_{y,h}}}{kT}\right) \\
&\quad \times \int_{-\infty}^{E_{n_{x,h},n_{y,h}}} \exp\left(-\frac{E_{n_{x,h},n_{y,h}} - E}{kT}\right) \frac{1}{\sqrt{E_{n_{x,h},n_{y,h}} - E}}\, dE.
\end{aligned}$$

To simplify things, let $x = (E_{n_{x,h},n_{y,h}} - E)/kT$, $E = E_{n_{x,h},n_{y,h}} - xkT$, and $dE = -kT\,dx$, remembering to change the limits of integration. This yields

$$n_v \simeq \frac{1}{\pi}\sqrt{\frac{2m_h}{\hbar^2}}(kT)^{1/2} \exp\left(-\frac{E_F - E_{n_{x,h},n_{y,h}}}{kT}\right) \int_0^\infty e^{-x} x^{-1/2}\, dx,$$

where the last integral is again the gamma function $\Gamma\left(\frac{1}{2}\right)$. The final expression for the total hole concentration in the valence band is therefore

$$n_v \simeq \frac{1}{\pi}\sqrt{\frac{2m_h kT}{\hbar^2}} \exp\left(-\frac{E_F - E_{n_{x,h},n_{y,h}}}{kT}\right) \Gamma\left(\tfrac{1}{2}\right). \tag{9.26}$$

Fermi Level

Finally, we can find the Fermi level position in the same fashion as done previously for two- and three-dimensional materials. Namely, assuming that the material is intrinsic and remains at equilibrium, we have

$$\begin{aligned}
n_c &= n_v, \\
\frac{1}{\pi}\sqrt{\frac{2m_e kT}{\hbar^2}}\Gamma\left(\tfrac{1}{2}\right) \exp\left(-\frac{E_{n_{x,e},n_{y,e}} - E_F}{kT}\right) &= \frac{1}{\pi}\sqrt{\frac{2m_h kT}{\hbar^2}}\Gamma\left(\tfrac{1}{2}\right) \\
&\quad \times \exp\left(-\frac{E_F - E_{n_{x,h},n_{y,h}}}{kT}\right).
\end{aligned}$$

Solving for E_F then gives

$$E_F = \frac{E_{n_{x,e},n_{y,e}} + E_{n_{x,h},n_{y,h}}}{2} + \frac{kT}{4}\ln\left(\frac{m_h}{m_e}\right), \tag{9.27}$$

where we again find that the Fermi level lies midway between the conduction band and the valence band of the system. There is a slight temperature dependence, with E_F moving closer to the conduction band edge with increasing temperature provided that $m_h > m_e$.

9.4 QUASI-FERMI LEVELS

As a brief aside, we have just found the Fermi levels of various intrinsic systems under equilibrium. However, there are many instances where this is not the case. For example, the material could be illuminated so as to promote electrons from the valence band into the conduction band. This results in a nonequilibrium population of electrons and holes. If, however, we assume that both types of carriers achieve equilibrium among themselves before any additional relaxation processes occur, then we can refer to a *quasi*-Fermi level for both species. Thus, instead of referring to a single (common) Fermi level, we have $E_{F,e}$ and $E_{F,h}$ for electrons and holes, respectively, and only at equilibrium are they equal,

$$E_{F,e} = E_{F,h} = E_F. \tag{9.28}$$

Furthermore, if we were to look at the occupation probability for states in the conduction band under these conditions, we would use a slightly altered version of $P_e(E)$, namely

$$P_e(E) = \frac{1}{1 + e^{(E-E_{F,e})/kT}}, \tag{9.29}$$

where the quasi-Fermi level $E_{F,e}$ is used instead of E_F. Likewise, we would use

$$P_h(E) = 1 - \frac{1}{1 + e^{(E-E_{F,h})/kT}} = \frac{1}{1 + e^{(E_{F,h}-E)/kT}} \tag{9.30}$$

for holes.

9.5 JOINT DENSITY OF STATES

To conclude our derivation and discussion of the use of the density-of-states function, let us calculate the *joint* density of states for the conduction band and valence band of low-dimensional systems. Our primary motivation for doing this is that interband transitions occur between the two owing to the absorption and emission of light. This will be the subject of Chapter 12. As a consequence, we wish to consolidate our prior conduction band and valence band density-of-states calculations to yield the abovementioned joint density of states, which ultimately represents the absorption spectrum of the material. Namely, we will find that it is proportional to the system's absorption coefficient α, which is a topic first introduced in Chapter 5.

9.5.1 Bulk Case

First, let us redo our density-of-states calculation. This time, we will consider the conduction and valence bands in tandem. As before, we can perform the derivation in a number of ways. Let us just take the first approach used earlier where we began our density-of-states calculation with a sphere in k-space.

The volume of the sphere is

$$V_k = \tfrac{4}{3}\pi k^3,$$

where $k = \sqrt{k_x^2 + k_y^2 + k_z^2}$. Along the same lines, the corresponding volume of a given state is $V_{\text{state}} = k_x k_y k_z = 8\pi^3/L_x L_y L_z$. The number of states enclosed by the sphere is therefore

$$N_1 = \frac{V_k}{V_{\text{state}}} = \frac{k^3}{6\pi^2} L_x L_y L_z.$$

We now multiply N_1 by 2 to account for spin degeneracy,

$$N_2 = 2N_1 = \frac{k^3}{3\pi^2} L_x L_y L_z,$$

and consider a volume density

$$\rho = \frac{N_2}{L_x L_y L_z} = \frac{k^3}{3\pi^2}$$

with units of number per unit volume. Since we are dealing with interband transitions between states, our expression for k will differ from before. Pictorially, what occur in k-space are "vertical" transitions between the conduction band and valence band. This is depicted in **Figure 9.15**. In addition, recall from Chapter 7 that relevant envelope function selection rules generally led to a $\Delta k = 0$ requirement for optical transitions (for instance, see the quantum well case).

Let us therefore call the initial valence band k-value k_n and the final conduction band k-value k_k. The associated energy of the conduction band edge is then

$$E_{cb} = E_c + \frac{\hbar^2 k_k^2}{2m_e},$$

where E_c is the bulk conduction band edge energy and m_e is the electron's effective mass. For the valence band hole, we have

$$E_{vb} = E_v - \frac{\hbar^2 k_n^2}{2m_h},$$

where E_v is the bulk valence band edge energy and m_h is the hole's effective mass.

Next, the transition must observe conservation of momentum ($p = \hbar k$). Thus, the total final and initial momenta must be equal. We thus have

$$p_k = p_n + p_{\text{photon}}$$
$$\hbar k_k = \hbar k_n + \hbar k_{\text{photon}},$$

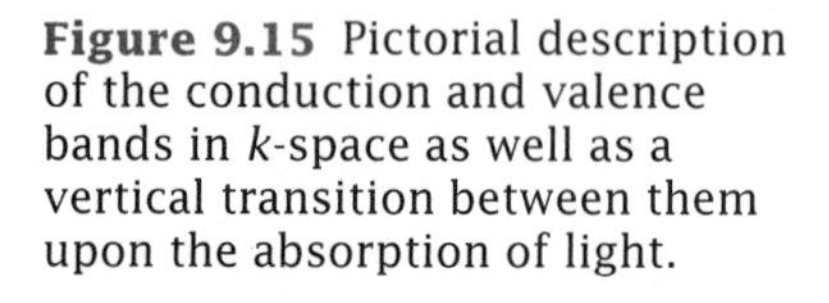

Figure 9.15 Pictorial description of the conduction and valence bands in k-space as well as a vertical transition between them upon the absorption of light.

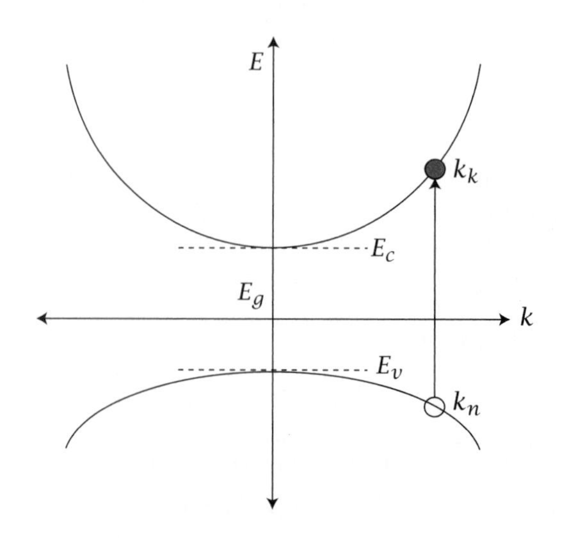

where p_{photon} represents the momentum carried by the photon. As a consequence,

$$k_k = k_n + k_{photon}. \tag{9.31}$$

We will see in Chapter 12 that k_k and k_n for interband optical transitions are both proportional to π/a, where a represents an interatomic spacing. By contrast, $k_{photon} \propto 1/\lambda$, where λ is the wavelength of light. As a consequence, $\lambda \gg a$ and we find that

$$k_k \simeq k_n. \tag{9.32}$$

This is why optical transitions between the valence band and conduction band in k-space are often said to be vertical.

Next, let us use this information to obtain an expression for the optical transition energy, from which we will obtain a relationship between k and E. Namely,

$$\begin{aligned} E &= E_{cb} - E_{vb} \\ &= (E_c - E_v) + \frac{\hbar^2}{2}\left(\frac{k_k^2}{m_e} + \frac{k_n^2}{m_h}\right). \end{aligned}$$

Note that the band gap E_g is the energy difference between the conduction band edge and the valence band energy. It is analogous to the HOMO–LUMO gap in molecular systems. As a consequence, $E_g = E_c - E_v$. In addition, we have just seen from the conservation of momentum that $k_k \simeq k_n \equiv k$. This results in

$$E = E_g + \frac{\hbar^2 k^2}{2}\left(\frac{1}{m_e} + \frac{1}{m_h}\right),$$

whereupon defining a reduced mass

$$\frac{1}{\mu} = \frac{1}{m_e} + \frac{1}{m_h}$$

yields

$$E = E_g + \frac{\hbar^2 k^2}{2\mu}. \tag{9.33}$$

Solving for k then gives

$$k = \sqrt{\frac{2\mu(E - E_g)}{\hbar^2}} \tag{9.34}$$

and if we return to our density $\rho = k^3/3\pi^2$, we find that

$$\rho = \frac{1}{3\pi^2}\left(\frac{2\mu(E - E_g)}{\hbar^2}\right)^{3/2}.$$

The desired bulk joint density of states is then $\rho_{joint} = d\rho/dE$, yielding

$$\rho_{joint}(E) = \frac{1}{2\pi^2}\left(\frac{2\mu}{\hbar^2}\right)^{3/2}\sqrt{E - E_g}, \tag{9.35}$$

which has a characteristic square root energy dependence. This is illustrated in **Figure 9.16**. It is also similar to the experimental bulk absorption spectrum of many semiconductors, although this lineshape is sometimes modified by additional Coulomb interactions between the electron and hole.

Figure 9.16 Summary of the joint density of states for a bulk system, a quantum well, a quantum wire, and a quantum dot.

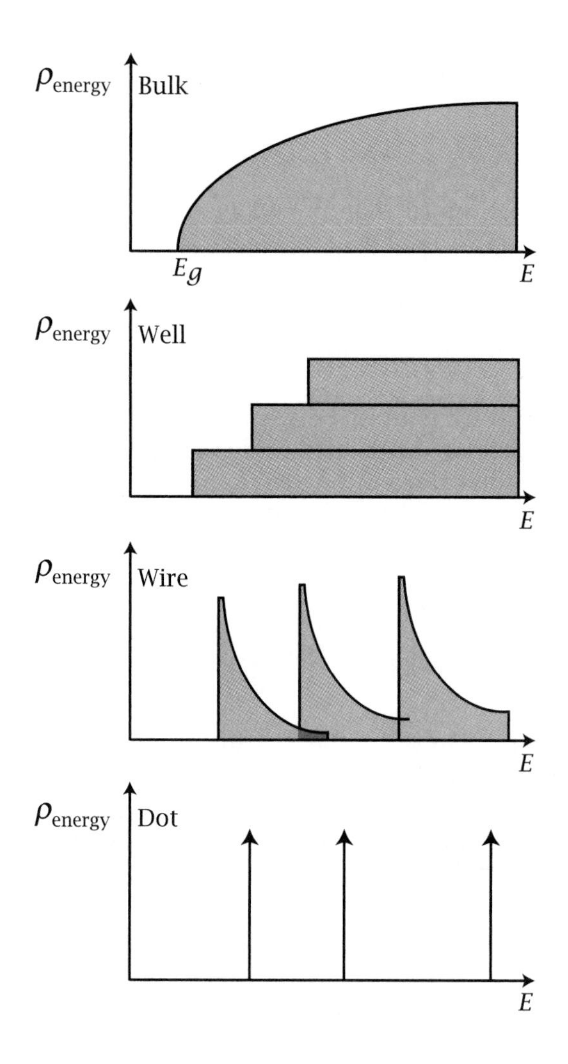

9.5.2 Quantum Well

We can now repeat the same calculation for a quantum well. As before, there are a number of strategies for doing this. Let us just take the first approach, where we begin with a circular area in k-space,

$$A_k = \pi k^2.$$

The corresponding area occupied by a given state is

$$A_{\text{state}} = k_x k_y,$$

where $k_x = 2\pi/L_x$ and $k_y = 2\pi/L_y$. The number of states encompassed by the circular area is therefore

$$N_1 = \frac{A_k}{A_{\text{state}}} = \frac{k^2}{4\pi} L_x L_y.$$

We multiply N_1 by 2 to account for spin degeneracy,

$$N_2 = 2N_1 = \frac{k^2}{2\pi} L_x L_y,$$

and subsequently define a density

$$\rho = \frac{N_2}{L_x L_y} = \frac{k^2}{2\pi},$$

with units of number per unit area.

At this point, let us find an expression for k that accounts for the interband transition energy. Recall that the energy of the conduction band electron is

$$E_{cb} = E_c^{'} + \frac{\hbar^2 k_k^2}{2m_e},$$

while that of the hole is

$$E_{vb} = E_v^{'} - \frac{\hbar^2 k_n^2}{2m_h}.$$

Note that instead of E_c and E_v, as seen in our earlier bulk example, we have $E_c^{'}$ and $E_v^{'}$. This is because these energies now contain contributions from quantization along the well's z direction, leading to discrete k_z values for the electron and hole:

$$k_{z_e} = \frac{\pi n_{z_e}}{a},$$

$$k_{z_h} = \frac{\pi n_{z_h}}{a}.$$

In both expressions, a is the thickness of the quantum well, with $n_{z_e} = 1, 2, 3, \ldots$ and $n_{z_h} = 1, 2, 3, \ldots$. So, for both the electron and hole, E_c and E_v have the following offsets:

$$E_c^{'} = E_c + \frac{\hbar^2 k_{z_e}^2}{2m_e},$$

$$E_v^{'} = E_v - \frac{\hbar^2 k_{z_h}^2}{2m_h}.$$

The transition energy is then

$$\begin{aligned} E &= E_{cb} - E_{vb} \\ &= (E_c^{'} - E_v^{'}) + \frac{\hbar^2}{2}\left(\frac{k_k^2}{m_e} + \frac{k_n^2}{m_h}\right), \end{aligned}$$

where $E_c^{'} - E_v^{'}$ is an effective energy $E_{n_{z,e},n_{z,h}}$ that includes both the bulk band gap energy and confinement contributions. Note also that $k_k \simeq k_n \equiv k$ due to momentum conservation, as demonstrated earlier. This results in

$$E = E_{n_{z,e},n_{z,h}} + \frac{\hbar^2 k^2}{2}\left(\frac{1}{m_e} + \frac{1}{m_h}\right).$$

We can again define a reduced mass $1/\mu = 1/m_e + 1/m_h$ to obtain

$$E = E_{n_{z,e},n_{z,h}} + \frac{\hbar^2 k^2}{2\mu}.$$

Solving for k then yields

$$k = \sqrt{\frac{2\mu(E - E_{n_{z,e},n_{z,h}})}{\hbar^2}}$$

and returning to our density gives

$$\rho = \frac{k^2}{2\pi} = \frac{\mu}{\pi}\frac{(E - E_{n_{z,e},n_{z,h}})}{\hbar^2}.$$

At this point, we define the joint density of states as $\rho_{\text{joint}} = d\rho/dE$ to obtain

$$\rho_{\text{joint}} = \frac{\mu}{\pi\hbar^2}. \tag{9.36}$$

Since this distribution is associated with each value of n_z, we can generalize the expression to

$$\rho_{\text{joint}} = \frac{\mu}{\pi\hbar^2} \sum_{n_{z,e},n_{z,h}} \Theta(E - E_{n_{z,e},n_{z,h}}). \tag{9.37}$$

The resulting joint density of states is illustrated schematically in **Figure 9.16** and represents the idealized linear absorption spectrum of the quantum well. As with the bulk example earlier, this lineshape is often modified by additional Coulomb interactions between the electron and hole.

9.5.3 Nanowire

Finally, let us repeat the above calculation for a nanowire. Consider a symmetric length in k-space about the origin,

$$L_k = 2k.$$

The width occupied by a given state is then $k_z = 2\pi/L_z$, so that the number of states contained within L_k is

$$N_1 = \frac{L_k}{k_z} = \frac{k}{\pi} L_z.$$

The result is multiplied by 2 to account for spin degeneracy

$$N_2 = 2N_1 = \frac{2k}{\pi} L_z.$$

We now define a density

$$\rho = \frac{N_2}{L_x} = \frac{2k}{\pi},$$

where we require an expression relating k to the transition energy in order to evaluate our final joint density of states.

In the case of a one-dimensional wire, the energy of the conduction band electron is

$$E_{cb} = E_c' + \frac{\hbar^2 k_k^2}{2m_e},$$

while for the hole it is

$$E_{vb} = E_v' - \frac{\hbar^2 k_n^2}{2m_h}.$$

E_c' and E_v' are again the effective electron and hole state energies offset from their bulk values by radial confinement along the x and y directions:

$$E_c' = E_c + \frac{\hbar^2 k_{xy,e}^2}{2m_e},$$

$$E_v' = E_v - \frac{\hbar^2 k_{xy,h}^2}{2m_h}.$$

In the above expressions, $k_{xy,e} = \sqrt{k_{x,e}^2 + k_{y,e}^2}$ and $k_{xy,h} = \sqrt{k_{x,h}^2 + k_{y,h}^2}$, where $k_{x,e} = n_{x,e}\pi/L_x$, $k_{y,e} = n_{y,e}\pi/L_y$, $k_{x,h} = n_{x,h}\pi/L_x$, $k_{y,h} = n_{y,h}\pi/L_y$

with $n_{x,e} = 1, 2, 3, \ldots,$ $n_{y,e} = 1, 2, 3, \ldots,$ $n_{x,h} = 1, 2, 3, \ldots,$ and $n_{y,h} = 1, 2, 3, \ldots$.

The transition energy is then

$$\begin{aligned} E &= E_{cb} - E_{vb} \\ &= E_{n_{x,e},n_{y,e},n_{x,h},n_{y,h}} + \frac{\hbar^2}{2}\left(\frac{k_k^2}{m_e} + \frac{k_n^2}{m_h}\right) \end{aligned}$$

with an effective transition energy $E_{n_{x,e},n_{y,e},n_{x,h},n_{y,h}} = E_c' - E_v'$ and $k_k \simeq k_n \equiv k$ owing to momentum conservation. We therefore have

$$E = E_{n_{x,e},n_{y,e},n_{x,h},n_{y,h}} + \frac{\hbar^2 k^2}{2}\left(\frac{1}{m_e} + \frac{1}{m_h}\right).$$

Defining a reduced mass $1/\mu = 1/m_e + 1/m_h$, yields

$$E = E_{n_{x,e},n_{y,e},n_{x,h},n_{y,h}} + \frac{\hbar^2 k^2}{2\mu}$$

and solving for k results in

$$k = \sqrt{\frac{2\mu(E - E_{n_{x,e},n_{y,e},n_{x,h},n_{y,h}})}{\hbar^2}}.$$

On returning to our density, we find that

$$\rho = \frac{2k}{\pi} = \frac{2}{\pi}\left(\frac{2\mu}{\hbar^2}\right)^{1/2} \sqrt{E - E_{n_{x,e},n_{y,e},n_{x,h},n_{y,h}}}.$$

We can thus define a joint density of states $\rho_{\text{joint}} = d\rho/dE$, to obtain

$$\rho_{\text{joint}}(E) = \frac{1}{\pi}\left(\frac{2\mu}{\hbar^2}\right)^{1/2} \frac{1}{\sqrt{E - E_{n_{x,e},n_{y,e},n_{x,h},n_{y,h}}}}. \tag{9.38}$$

This is our desired one-dimensional joint density of states and, as before, is associated with the values $n_{x,e}$, $n_{y,e}$, $n_{x,h}$, and $n_{y,h}$ due to carrier confinement along the x and y directions of the wire. The expression can thus be generalized to

$$\rho_{\text{joint}}(E) = \frac{1}{\pi}\sqrt{\frac{2\mu}{\hbar^2}} \sum_{n_{x,e},n_{y,e},n_{x,h},n_{y,h}} \frac{1}{\sqrt{E - E_{n_{x,e},n_{y,e},n_{x,h},n_{y,h}}}} \Theta(E - E_{n_{x,e},n_{y,e},n_{x,h},n_{y,h}}). \tag{9.39}$$

Equation 9.39 is illustrated schematically in **Figure 9.16**. In principle, it resembles the actual linear absorption spectrum of the system. However, in practice, additional electron–hole Coulomb interactions often (dramatically) modify the nanowire absorption spectrum.

9.5.4 Quantum Dot

Finally, for a quantum dot, there is no need to do this exercise since the joint density of states is a series of discrete atomic-like transitions. The zero-dimensional joint-density-of-states expression is therefore

$$\rho_{\text{joint}}(E) = 2\delta(E - E_{n_{x,e},n_{y,e},n_{z,e},n_{x,h},n_{y,h},n_{z,h}}), \tag{9.40}$$

where the energy $E_{n_{x,e},n_{y,e},n_{z,e},n_{x,h},n_{y,h},n_{z,h}}$ contains confinement contributions along the x, y, and z directions for both the electron and hole. Note that, as in the previous cases of the quantum well and nanowire, we implicitly assume that we are operating in the strong confinement regime with both the electron and hole confined independently. We then have

$$E_{n_{x,e},n_{y,e},n_{z,e},n_{x,h},n_{y,h},n_{z,h}} = E_g + \left(\frac{\hbar^2 k_{x,e}^2}{2m_e} + \frac{\hbar^2 k_{y,e}^2}{2m_e} + \frac{\hbar^2 k_{z,e}^2}{2m_e}\right) + \left(\frac{\hbar^2 k_{x,h}^2}{2m_h} + \frac{\hbar^2 k_{y,h}^2}{2m_h} + \frac{\hbar^2 k_{z,h}^2}{2m_h}\right),$$

where $k_{x,e} = n_{x,e}\pi/L_x$, $k_{y,e} = n_{y,e}\pi/L_y$, $k_{z,e} = n_{z,e}\pi/L_z$, $k_{x,h} = n_{x,h}\pi/L_x$, $k_{y,h} = n_{y,h}\pi/L_y$, and $k_{z,h} = n_{z,h}\pi/L_z$ ($n_{x,e} = 1, 2, 3, \ldots$, $n_{y,e} = 1, 2, 3, \ldots$, $n_{z,e} = 1, 2, 3, \ldots$, $n_{x,h} = 1, 2, 3, \ldots$, $n_{y,h} = 1, 2, 3, \ldots$, and $n_{z,h} = 1, 2, 3, \ldots$). This can be generalized to the following linear combination that accounts for all states present:

$$\rho_{\text{joint}}(E) = 2 \sum_{n_{x,e},n_{y,e},n_{z,e},n_{x,h},n_{y,h},n_{z,h}} \delta(E - E_{n_{x,e},n_{y,e},n_{z,e},n_{x,h},n_{y,h},n_{z,h}}). \tag{9.41}$$

The resulting joint density of states is illustrated schematically in **Figure 9.16**. In principle, it resembles the actual linear absorption spectrum of the dots. In practice, though, the inhomogeneous size distribution of nanocrystal ensembles often causes significant broadening of these transitions. Furthermore, note that Coulomb contributions are often not as significant here due to symmetry and strong confinement considerations. A more in-depth discussion of Coulomb effects in nanostructures is, however, beyond the scope of this text.

9.6 SUMMARY

This concludes our introduction to the density of states and to two early uses: (a) evaluating the occupation of states in bands and (b) evaluating the accompanying joint density of states. The latter is related to the absorption spectrum of the material and is therefore important in our quest to better understand interband transitions in low-dimensional systems. In the next chapter, we discuss bands and illustrate how they develop owing to the periodic potential experienced by carriers in actual solids. We follow this with a brief discussion of time-dependent perturbation theory (Chapter 11). This ultimately allows us to quantitatively model interband transitions in Chapter 12.

9.7 THOUGHT PROBLEMS

9.1 Density of states: bulk

Provide a numerical evaluation of ρ_{energy} for electrons in a bulk solid with $m_e = m_0$ and with an energy 100 meV above the system's conduction band edge. Express your answer in units of $\text{eV}^{-1}\ \text{cm}^{-3}$.

9.2 Density of states: bulk

Estimate the total number of states present per unit energy in a microscopic $1\ \mu\text{m}$ CdTe cube. The effective electron mass is $m_e = 0.1m_0$ and the electron energy of interest is 100 meV above the bulk conduction band edge. Express your result in units of eV^{-1}.

9.3 Effective conduction and valence band density of states: bulk

In Equations 9.18 and 9.20,

$$n_c \simeq N_c e^{-(E_c - E_F)/kT}$$
$$n_v \simeq N_v e^{-(E_F - E_v)/kT},$$

the prefactors N_c and N_v are often referred to as the effective conduction and valence band density of states. Provide numerical values for both in GaAs, where $m_e = 0.067m_0$ and $m_h = 0.64m_0$ at 300 K. Express you answers in units of cm^{-3}.

9.4 Density of states: quantum well

Numerically evaluate ρ_{energy} for electrons in a quantum well with $m_e = m_0$ and with an energy 100 meV above E_{n_e}, the first subband edge. Consider only the lowest subband. Express your answer in units of $eV^{-1}\,cm^{-2}$.

9.5 Density of states: quantum well

Numerically evaluate the prefactors in Equations 9.22 and 9.23:

$$n_c \simeq \frac{m_e kT}{\pi\hbar^2}\exp\left(-\frac{E_{n_{z,e}} - E_F}{kT}\right)$$
$$n_v \simeq \frac{m_h kT}{\pi\hbar^2}\exp\left(-\frac{E_F - E_{n_{z,h}}}{kT}\right).$$

They are referred to as critical densities. Consider GaAs, where $m_e = 0.067m_0$ and $m_h = 0.64m_0$ at 300 K. Express all your answers in units of cm^{-2}.

9.6 Density of states: nanowire

Provide a numerical evaluation of ρ_{energy} for electrons in a nanowire with $m_e = m_0$ and with an energy 100 meV above $E_{n_{x,e},n_{y,e}}$, the first subband edge. Consider only the lowest subband. Express your answer in units of $eV^{-1}\,cm^{-1}$.

9.7 Density of states: nanowire

Evaluate numerical values for the prefactors in Equations 9.25 and 9.26:

$$n_c = \frac{1}{\pi}\sqrt{\frac{2m_e kT}{\hbar^2}}\exp\left(-\frac{E_{n_{x,e},n_{y,e}} - E_F}{kT}\right)\Gamma\left(\tfrac{1}{2}\right)$$
$$= \sqrt{\frac{2m_e kT}{\pi\hbar^2}}\exp\left(-\frac{E_{n_{x,e},n_{y,e}} - E_F}{kT}\right),$$
$$n_v = \frac{1}{\pi}\sqrt{\frac{2m_h kT}{\hbar^2}}\exp\left(-\frac{E_F - E_{n_{x,h},n_{y,h}}}{kT}\right)\Gamma\left(\tfrac{1}{2}\right)$$
$$= \sqrt{\frac{2m_h kT}{\pi\hbar^2}}\exp\left(-\frac{E_F - E_{n_{x,h},n_{y,h}}}{kT}\right).$$

Consider GaAs where $m_e = 0.067m_0$ and $m_h = 0.64m_0$ at 300 K. Express all your answers in units of cm^{-1}.

9.8 Fermi–Dirac distribution

Consider the Fermi–Dirac distribution

$$f(E) = \frac{1}{1 + e^{(E-E_F)/kT}}.$$

Plot $f(E)$ for three different temperatures, say 10 K, 300 K, and 1000 K. Assume $E_F = 0.5$ eV. Show for yourself that the width of the distribution from where $f(E) \sim 1$ to where $f(E) \sim 0$ is approximately $4kT$. Thus, at room temperature, the width of the Fermi–Dirac distribution is of order 100 meV.

9.9 Fermi–Dirac distribution

The Fermi–Dirac distribution for electrons is

$$f_e(E) = \frac{1}{1 + e^{(E-E_F)/kT}}$$

and describes the probability that an electron occupies a given state. For holes, we have seen that one writes

$$f_h(E) = 1 - \frac{1}{1 + e^{(E-E_F)/kT}},$$

where this can be rationalized since the second term just reflects the probability that an electron occupies a given state and 1 minus this is the probability that it remains unoccupied (and hence is occupied by a hole). In some cases, however, one sees written

$$f_h(E) = \frac{1}{1 + e^{(E_F-E)/kT}}.$$

Show that the two expressions are identical. Then show in the limit where $E - E_F \gg kT$ for electrons and $E_F - E \gg kT$ for holes that the following Boltzmann approximations exist for both $f_e(E)$ and $f_h(E)$:

$$f_e(E) \simeq e^{-(E-E_F)/kT}$$
$$f_h(E) \simeq e^{-(E_F-E)/kT}.$$

9.10 Fermi level: bulk

Calculate the Fermi level of bulk silicon at 0 K, 10 K, 77 K, 300 K, and 600 K. Note that $E_g = E_c + E_v$ and assume that $m_e = 1.08m_0$ and $m_h = 0.55m_0$. Leave the answer in terms of E_g or, if you desire, look up the actual value of E_g and express all your answers in units of eV.

9.11 Fermi level: bulk

Evaluate the position of the Fermi level in bulk GaAs under equilibrium conditions (i.e. no optical excitation) relative to its valence band edge. Assume that $m_e = 0.067m_0$, $m_h = 0.64m_0$, and $E_g = 1.42$ eV at 300 K.

9.12 Fermi–Dirac integral

Show that the concentration of either electrons or holes in a bulk semiconductor can be given by the following expression:

$$n_{c(v)} = \frac{1}{2\pi^2}\left(\frac{2m_{e(h)}kT}{\hbar^2}\right)^{3/2}\int_0^\infty \frac{\sqrt{\eta}\,e^{\mu_{e(h)}-\eta}}{1 + e^{\mu_{e(h)}-\eta}}\,d\eta,$$

where the last integral is the Fermi–Dirac integral $F_{1/2}(\mu_{e(h)})$ written in such a way as to make its numerical evaluation more convenient. $m_{e(h)}$ is the electron (hole) effective mass, η is a variable of integration, and $\mu_{e(h)} = (E_F - E_{c(v)})/kT$.

9.13 Quasi-Fermi level: bulk, nonequilibrium conditions

Provide an estimate for the quasi-Fermi level of electrons in the conduction band of a CdS nanowire, modeled as a bulk system. Recall from Problem 9.12 that one can express the conduction band electron concentration in terms of the Fermi–Dirac integral as

$$n_c = \frac{1}{2\pi^2}\left(\frac{2m_e kT}{\hbar^2}\right)^{3/2}\int_0^\infty \frac{\sqrt{\eta}e^{\mu_e-\eta}}{1+e^{\mu_e-\eta}}\,d\eta,$$

where η is a variable of integration and $\mu_e = (E_F^{'} - E_c)/kT$, with E_c the conduction band edge and $E_F^{'}$ the desired quasi-Fermi level under nonequilibrium conditions. Assume an intrinsic material where all excess carriers are photogenerated through pulsed laser excitation. Recall from Chapter 5 that the number of carriers generated in such an experiment is

$$n = \left(\frac{I\sigma}{h\nu}\right)\tau_p,$$

where I is the laser intensity, σ is the nanowire absorption cross section, $h\nu$ is the photon energy, and τ_p is the laser pulse width. To carry out the numerical evaluation of $E_F^{'}$, use the following parameters:

$m_e = 0.2m_0$,

$T = 300\,\text{K}$,

$I = 10^{12}\,\text{W/m}^2$,

$\sigma = 3.13\times10^{-10}\,\text{cm}^2$,

387 nm excitation,

$\tau_p = 150\,\text{fs}$,

a nanowire radius $r = 7\,\text{nm}$,

a nanowire length $l = 10\,\mu\text{m}$.

9.14 Alternative bulk concentration

Show that the concentration of electrons in a bulk semiconductor can be described by the expression

$$n_c = \frac{1}{4}\left(\frac{2m_e kT}{\pi\hbar^2}\right)^{3/2} e^{\mu_e},$$

where $\mu_e = (E_{F,e} - E_c)/kT$ and $E_{F,e}$ is the quasi-Fermi level of carriers in the conduction band. Assume $E - E_{F,e} \gg kT$ and make an appropriate choice of variables. Hint: let $z = \sqrt{\eta}$, where $\eta = (E - E_c)/kT$. This approach has been used by Agarwal et al. (2005).

9.15 Quasi-Fermi level: bulk, nonequilibrium conditions

Evaluate the electron quasi-Fermi level in a bulk system under different photogenerated carrier concentrations n_c. Show that under low pump fluences where carrier densities are relatively small, the Fermi–Dirac distribution can be approximated by the Boltzmann distribution. Conversely, show that at high pump fluences where n_c is large, it is best to not make this approximation. First show that when $E - E_{F,e} \gg kT$,

$$n_c = \frac{1}{4}\left(\frac{2m_e kT}{\pi\hbar^2}\right)^{3/2} e^{\mu_e},$$

where $\mu_e = (E_{F,e} - E_c)/kT$ and $E_{F,e}$ is the quasi-Fermi level of electrons under illumination. Next, solve for μ_e and compare it with the same value found numerically by evaluating

$$n_c = \frac{1}{2\pi^2}\left(\frac{2m_e kT}{\hbar^2}\right)^{3/2}\int_0^\infty \frac{\sqrt{\eta}e^{\mu_e-\eta}}{1+e^{\mu_e-\eta}}\,d\eta,$$

where the integral is that of the full Fermi–Dirac function and η is a variable of integration. Consider the specific case of bulk GaAs, where $m_e = 0.067m_0$ and $m_h = 0.64m_0$, both at 300 K, and with excitation intensities leading to either $n_c = 10^{20}\,\text{m}^{-3}$ or $n_c = 10^{25}\,\text{m}^{-3}$.

9.16 Peak concentration

Consider the bulk density-of-states expression for electrons derived in the main text,

$$\rho_{\text{energy}}(E) = \frac{1}{2\pi^2}\left(\frac{2m_e}{\hbar^2}\right)^{3/2}\sqrt{E - E_c},$$

with E_c the conduction band energy. Assume that $E - E_F \gg kT$ in the Fermi–Dirac distribution. Find the energy associated with the peak of the resulting electron distribution function $n_e = \rho_{\text{energy}}(E)f(E)$.

9.17 Fermi energy

Consider the electron concentration of the conduction band in a bulk semiconductor. Show that at $T = 0\,\text{K}$, the associated Fermi energy is

$$E_F = \frac{\hbar^2}{2m_e}(3\pi^2 n_c)^{2/3}.$$

Hint: Consider the behavior of the Fermi–Dirac distribution at 0 K for $E > E_F$ versus $E < E_F$. Note also that the resulting expression was previously seen in Chapter 3 when discussing metals.

9.18 Bulk quasi-Fermi level: temperature dependence

In Problem 9.15, we compared two expressions for the carrier density in the conduction band of a bulk semiconductor:

$$n_c = \frac{1}{4}\left(\frac{2m_e kT}{\pi\hbar^2}\right)^{3/2} e^{\mu_e}$$

and

$$n_c = \frac{1}{2\pi^2}\left(\frac{2m_e kT}{\hbar^2}\right)^{3/2}\int_0^\infty \frac{\sqrt{\eta}e^{\mu_e-\eta}}{1+e^{\mu_e-\eta}}\,d\eta,$$

where $\mu_e = (E_{F,e} - E_c)/kT$ and $E_{F,e}$ is the quasi-Fermi level of electrons under illumination. For the specific case of GaAs where $m_e = 0.067m_0$ and $n_c = 10^{24}\,\text{m}^{-3}$, compare the quasi-Fermi levels from both expressions at 300 K and at 3000 K. Assume that m_e is temperature-independent. Explain any observed trends.

9.19 Quantum well carrier concentration

Evaluate the total carrier concentration n_c in a given conduction band subband of a quantum well. This time, however, do not assume $E - E_F \gg kT$. Consider the integral

$$\int \frac{1}{1+e^x}\,dx = x - \ln(1+e^x)$$

and assume lower and upper limits of E_1 and E_2, respectively.

9.8 REFERENCES

Abramowitz M, Stegun IA (1972) *Handbook of Mathematical Functions.* Dover Publications, New York.

Agarwal R, Barrelet CJ, Lieber CM (2005) Lasing in single cadmium sulfide nanowire optical cavities. *Nano Lett.* **5**, 917 (see Supporting Information).

Beyer WH (1991) *Standard Mathematical Tables and Formulae*, 29th edn. CRC Press, Boca Raton, FL.

9.9 FURTHER READING

Chuang SL (1995) *Physics of Optoelectronic Devices.* Wiley, New York.

Davies JH (1998) *The Physics of Low-Dimensional Semiconductors: An Introduction.* Cambridge University Press, Cambridge, UK.

Fox M (2010) *Optical Properties of Solids*, 2nd edn. Oxford University Press, Oxford, UK.

Liboff RL (2003) *Introductory Quantum Mechanics*, 4th edn. Addison-Wesley, San Francisco.

Bands

10.1 INTRODUCTION

Earlier, when dealing with quantum mechanics, the goal was to evaluate the wavefunctions and energies of a particle (e.g., an electron) under various potentials. The simplest case considered was the "free" particle problem where $V(r) = 0$ was introduced into the Schrödinger equation. This led to the equation

$$-\frac{\hbar^2}{2m}\nabla^2\psi = E\psi,$$

whose one-dimensional solutions are plane waves of the form

$$\psi = Ae^{ikx}.$$

The resulting energies are

$$E = \frac{p^2}{2m} = \frac{\hbar^2 k^2}{2m},$$

with $k = 2n\pi/L$ the wavevector, n an integer, and L a characteristic repeat length for the wave.

However, in real solids, the potential experienced by carriers is not zero owing to the regular periodic arrangement of positively charged nuclei. In this chapter, we therefore find the energies of a particle in a *periodic* potential, leading to a simple, but illustrative, model for electrons and holes in actual solids. The most important conclusion resulting from this exercise is that there exist allowed energies for the carrier, forming so-called bands, while others are forbidden, leading to band "gaps." These latter discontinuities in energy ultimately allow us to distinguish metals from semiconductors from insulators.

Three illustrative models for bands are introduced: the Kronig–Penney model, the tight binding approximation, and the nearly free electron approximation. The latter two models represent limiting cases where assumptions are made about how tightly or loosely bound electrons are to their parent atoms. Note that other more complicated approximations exist, such as the $k \cdot p$ model, but for simplicity we will not introduce them here. The reader may therefore refer to other more advanced texts cited in the Further Reading for descriptions of these models.

10.2 THE KRONIG–PENNEY MODEL

The Kronig–Penney model (Kronig and Penney 1931) is an approximation to the periodic atomic potential of a crystalline solid, arising from

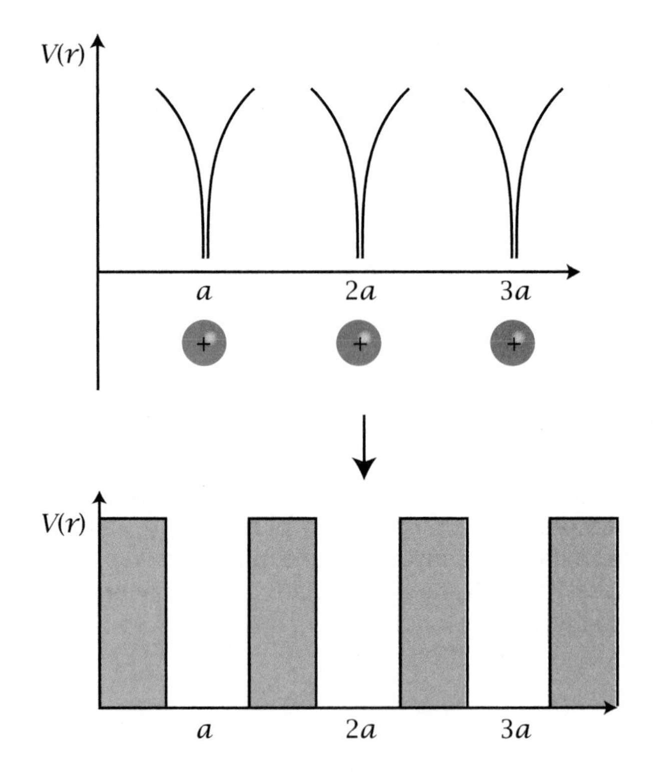

Figure 10.1 Illustration of how the periodic potential in a crystal is modeled by a series of rectangular barriers in the Kronig–Penney model.

the attractive $1/r$ Coulomb potential between electrons and positively charged nuclei. However, instead of employing the actual functional form of the potential, a series of rectangular barriers, having atomic spacings of a, $2a$, $3a$, etc., are used. These barriers represent the region of space between atomic cores, as illustrated schematically in **Figure 10.1**.

To find the energies that an electron possesses in this periodic potential, we solve the Schrödinger equation in a piecewise fashion much like we did earlier when solving the problem of a particle in a finite box in Chapter 8. We also employ standard quantum mechanical matching conditions for wavefunctions at a boundary, first discussed in Chapter 6. Namely, the wavefunctions are continuous at these junctions and so too are their first derivatives:

$$\psi_1(x) = \psi_2(x),$$
$$\psi_1'(x) = \psi_2'(x).$$

The three regions of interest are illustrated in **Figure 10.2**, where the potential in Region 1 is zero and the potential in Regions 2 and 3 is

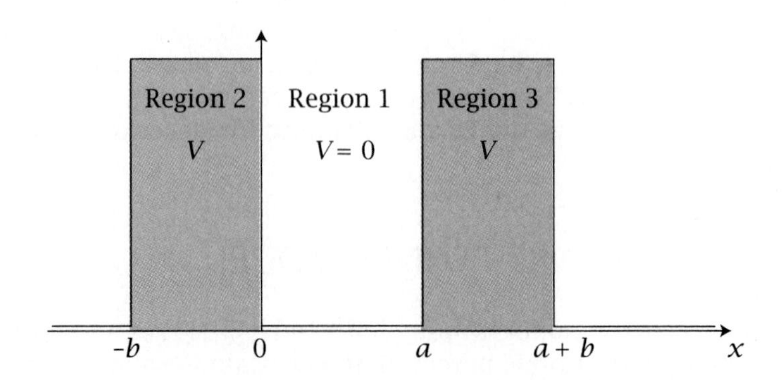

Figure 10.2 Schematic of the three regions of interest in the Kronig–Penney model.

V (i.e., a finite value). Furthermore, all the barriers have a finite width b with an accompanying interatomic spacing a. For convenience, we position one end of the well at the origin. However, this is not necessary, and the problem can be solved with an offset axis if desired.

Solving the Schrödinger Equation in Region 1

Let us first solve the Schrödinger equation in Region 1. Since the potential in this region is zero (see **Figure 10.2**), the Schrödinger equation can be written as

$$-\frac{\hbar^2}{2m_0}\frac{d^2\psi_1}{dx^2} = E\psi.$$

Note the use of the electron mass m_0 here and in what follows. Letting $\alpha^2 = 2m_0E/\hbar^2$ and consolidating terms yields the following equation:

$$\frac{d^2\psi_1}{dx^2} + \alpha^2\psi = 0.$$

As seen many times previously, general solutions are plane waves of the form

$$\psi_1(x) = Ae^{i\alpha x} + Be^{-i\alpha x},$$

with A and B constant coefficients.

Solving the Schrödinger Equation in Region 2

Next, in Region 2 (see **Figure 10.2**), the potential is nonzero. As a consequence, the Schrödinger equation is

$$-\frac{\hbar^2}{2m_0}\frac{d^2\psi_2}{dx^2} + V\psi_2 = E\psi_2.$$

This equation is then rearranged to

$$-\frac{\hbar^2}{2m_0}\frac{d^2\psi_2}{dx^2} = -(V - E)\psi_2,$$

where we have implicitly assumed that $V > E$. Note that solutions also exist if $V < E$, but this is not our primary concern here. The reader may consider this limit as an exercise.

On consolidating terms, we obtain

$$\frac{d^2\psi_2}{dx^2} = \frac{2m_0(V - E)}{\hbar^2}\psi_2,$$

whereupon letting $\beta^2 = 2m_0(V - E)/\hbar^2$ gives

$$\frac{d^2\psi_2}{dx^2} - \beta^2\psi_2 = 0.$$

General solutions to this second-order differential equation are then linear combinations of growing and decaying exponentials:

$$\psi_2(x) = Ce^{\beta x} + De^{-\beta x}.$$

We saw this earlier in Chapter 8. The prefactors C and D, like A and B, are constant coefficients.

Solving the Schrödinger Equation in Region 3

To determine the wavefunction in Region 3, we employ Bloch's theorem.

Bloch's Theorem Many years ago, Felix Bloch found that the wavefunction of an electron in a periodic potential could be written as

$$\psi(x) = u(x)e^{ikx}, \tag{10.1}$$

where $u(x)$ is a function possessing the same periodicity as the underlying crystal lattice and k is a wavevector oriented in the direction of carrier propagation. Bloch's solution is essentially the free electron plane wave solution modulated by a new function $u(x)$. Note that for simplicity we have assumed the one-dimensional case, using the x axis. More generally, we could write $\psi(\mathbf{r}) = u(\mathbf{r})e^{i\mathbf{k}\cdot\mathbf{r}}$ as will be seen later in Chapter 12.

An alternative version of Bloch's wavefunction that is useful to us is

$$\begin{aligned}\psi(x+a) &= u(x+a)e^{ik(x+a)} \\ &= u(x)e^{ikx}e^{ika}.\end{aligned}$$

In this expression, we have recognized that $u(x) = u(x+a)$ by periodicity. As a consequence, we obtain a general expression that we will use shortly. Specifically,

$$\psi(x+a) = \psi(x)e^{ika}, \tag{10.2}$$

which was obtained by recognizing that $\psi(x) = u(x)e^{ikx}$.

At this point, we can now relate the wavefunction in one part of the potential to that in another when offset by the potential's underlying periodicity. We thus find that

$$\psi_3(x) = \psi_2(x)e^{ik(a+b)},$$

where $a+b$ is the potential's period. Alternatively, we write

$$\psi_3(x) = (Ce^{\beta x} + De^{-\beta x})e^{ik(a+b)}.$$

Putting it All Together

We now have expressions for the carrier's wavefunction in Regions 1, 2 and 3. To join them together at their interfaces, we apply the standard quantum mechanical matching conditions first discussed in Chapter 6. The two interfaces of interest to us are circled in **Figure 10.3**.

Recall that the matching conditions simply require that

$$\begin{aligned}\psi_1(0) &= \psi_2(0), \\ \psi_1'(0) &= \psi_2'(0)\end{aligned}$$

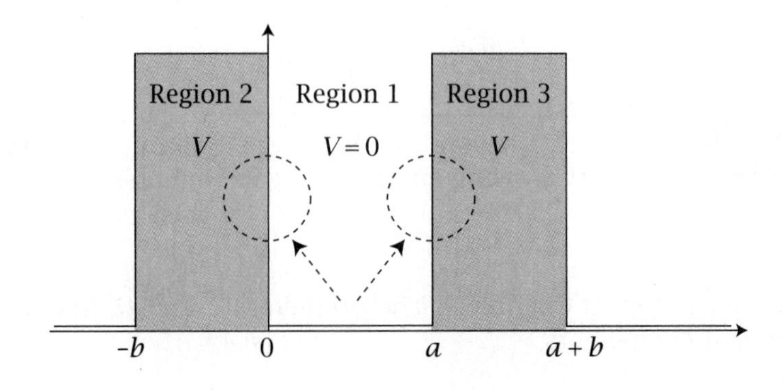

Figure 10.3 Regions of interest for applying wavefunction matching conditions. The interfaces between Regions 1 and 2 and between Regions 1 and 3 are circled.

and

$$\psi_1(a) = \psi_3(a),$$
$$\psi_1'(a) = \psi_3'(a).$$

In addition, note from Bloch's theorem that the following is true:

$$\psi_3(a) = \psi_2(-b)e^{ik(a+b)},$$
$$\psi_3'(a) = \psi_2'(-b)e^{ik(a+b)}.$$

We are therefore left with the following four equations:

$$\psi_1(0) = \psi_2(0),$$
$$\psi_1'(0) = \psi_2'(0),$$
$$\psi_1(a) = \psi_2(-b)e^{ik(a+b)},$$
$$\psi_1'(a) = \psi_2'(-b)e^{ik(a+b)},$$

which we can alternatively write as

$$\begin{aligned} A + B &= C + D, \\ i\alpha A - i\alpha B &= \beta C - \beta D, \\ Ae^{i\alpha a} + Be^{-i\alpha a} &= (Ce^{-\beta b} + De^{\beta b})e^{ik(a+b)}, \\ i\alpha Ae^{i\alpha a} - i\alpha Be^{-i\alpha a} &= (\beta Ce^{-\beta b} - \beta De^{\beta b})e^{ik(a+b)}. \end{aligned}$$

Since there are four equations and four unknowns, this system of linear equations can be recast into matrix form as

$$\begin{pmatrix} 1 & 1 & -1 & -1 \\ i\alpha & -i\alpha & -\beta & \beta \\ e^{i\alpha a} & e^{-i\alpha a} & -e^{-\beta b+ik(a+b)} & -e^{\beta b+ik(a+b)} \\ i\alpha e^{i\alpha a} & -i\alpha e^{-i\alpha a} & -\beta e^{-\beta b+ik(a+b)} & \beta e^{\beta b+ik(a+b)} \end{pmatrix} \begin{pmatrix} A \\ B \\ C \\ D \end{pmatrix} = \begin{pmatrix} 0 \\ 0 \\ 0 \\ 0 \end{pmatrix}$$

and may subsequently be solved in a number of ways. Here we will employ a technique referred to as Gaussian elimination. A detailed description of the method can be found in most linear algebra texts and in the references cited in Further Reading. However, briefly, the matrix equation is first condensed into what is referred to as augmented matrix form:

$$\left(\begin{array}{cccc|c} 1 & 1 & -1 & -1 & 0 \\ i\alpha & -i\alpha & -\beta & \beta & 0 \\ e^{i\alpha a} & e^{-i\alpha a} & -e^{-\beta b+ik(a+b)} & -e^{\beta b+ik(a+b)} & 0 \\ i\alpha e^{i\alpha a} & -i\alpha e^{-i\alpha a} & -\beta e^{-\beta b+ik(a+b)} & \beta e^{\beta b+ik(a+b)} & 0 \end{array}\right).$$

Next, one takes linear combinations of rows (but not columns) of this augmented matrix with the aim of making it "lower diagonal" (i.e., with all zeros on the lower diagonal).

As an illustration, let us multiply the first row by $-i\alpha$ and add it to the second row, calling the result row 2. Likewise, we can multiply the third row by $-i\alpha$ and add it to the fourth row, yielding a new fourth row. We denote these steps using the following shorthand notation first introduced in Chapter 8. Note that this is only one of many possible approaches.

$$-i\alpha(\text{row } 1) + \text{row } 2 \longrightarrow \text{row } 2,$$
$$-i\alpha(\text{row } 3) + \text{row } 4 \longrightarrow \text{row } 4.$$

On doing this, we obtain the following augmented matrix:

$$\left(\begin{array}{cccc|c} 1 & 1 & -1 & -1 & 0 \\ 0 & -2i\alpha & i\alpha-\beta & i\alpha+\beta & 0 \\ e^{i\alpha a} & e^{-i\alpha a} & -e^{-\beta b+ik(a+b)} & -e^{\beta b+ik(a+b)} & 0 \\ 0 & -2i\alpha e^{-i\alpha a} & (i\alpha-\beta)e^{-\beta b+ik(a+b)} & (i\alpha+\beta)e^{\beta b+ik(a+b)} & 0 \end{array}\right).$$

Next, we multiply the first row of the newly generated augmented matrix by $-e^{i\alpha a}$ and add it to the existing third row to obtain a new third row. This is denoted by

$$-e^{i\alpha a}(\text{row } 1) + \text{row } 3 \rightarrow \text{row } 3$$

and yields

$$\left(\begin{array}{cccc|c} 1 & 1 & -1 & -1 & 0 \\ 0 & -2i\alpha & i\alpha-\beta & i\alpha+\beta & 0 \\ 0 & -(e^{i\alpha a}-e^{-i\alpha a}) & e^{i\alpha a}-e^{-\beta b+ik(a+b)} & e^{i\alpha a}-e^{\beta b+ik(a+b)} & 0 \\ 0 & -2i\alpha e^{-i\alpha a} & (i\alpha-\beta)e^{-\beta b+ik(a+b)} & (i\alpha+\beta)e^{\beta b+ik(a+b)} & 0 \end{array}\right).$$

Finally, multiplying the second row by $-e^{-i\alpha a}$ and adding it to the fourth row gives us a new fourth row:

$$-e^{-i\alpha a}(\text{row } 2) + \text{row } 4 \rightarrow \text{row } 4.$$

The result is

$$\left(\begin{array}{cccc|c} 1 & 1 & -1 & -1 & 0 \\ 0 & -2i\alpha & i\alpha-\beta & i\alpha+\beta & 0 \\ 0 & -(e^{i\alpha a}-e^{-i\alpha a}) & e^{i\alpha a}-e^{-\beta b+ik(a+b)} & e^{i\alpha a}-e^{\beta b+ik(a+b)} & 0 \\ 0 & 0 & (i\alpha-\beta)(-e^{-i\alpha a}+e^{-\beta b+ik(a+b)}) & (i\alpha+\beta)(-e^{-i\alpha a}+e^{\beta b+ik(a+b)}) & 0 \end{array}\right).$$

We can go on in this fashion to achieve a true lower diagonal expression. However, let us stop here and re-expand the augmented matrix into

$$\begin{pmatrix} 1 & 1 & -1 & -1 \\ 0 & -2i\alpha & i\alpha-\beta & i\alpha+\beta \\ 0 & -(e^{i\alpha a}-e^{-i\alpha a}) & e^{i\alpha a}-e^{-\beta b+ik(a+b)} & e^{i\alpha a}-e^{\beta b+ik(a+b)} \\ 0 & 0 & (i\alpha-\beta)(-e^{-i\alpha a}+e^{-\beta b+ik(a+b)}) & (i\alpha+\beta)(-e^{-i\alpha a}+e^{\beta b+ik(a+b)}) \end{pmatrix}\begin{pmatrix} A \\ B \\ C \\ D \end{pmatrix} = \begin{pmatrix} 0 \\ 0 \\ 0 \\ 0 \end{pmatrix}.$$

Now, for this matrix equation to have a solution, we have either the trivial solution ($A = B = C = D = 0$) or the situation where the determinant of the first matrix equals zero. Since we are never really interested in the trivial solution, we find that

$$\begin{vmatrix} 1 & 1 & -1 & -1 \\ 0 & -2i\alpha & i\alpha-\beta & i\alpha+\beta \\ 0 & -(e^{i\alpha a}-e^{-i\alpha a}) & e^{i\alpha a}-e^{-\beta b+ik(a+b)} & e^{i\alpha a}-e^{\beta b+ik(a+b)} \\ 0 & 0 & (i\alpha-\beta)(-e^{-i\alpha a}+e^{-\beta b+ik(a+b)}) & (i\alpha+\beta)(-e^{-i\alpha a}+e^{\beta b+ik(a+b)}) \end{vmatrix} = 0.$$

Solving this 4×4 determinant can, in principle, be hard. However, we can readily evaluate it using the method of subdeterminants. As with Gaussian elimination, the method of subdeterminants can be found in most linear algebra texts and in the references in Further Reading, which the reader should consult if unfamiliar with the approach.

Briefly, given that we have three zeros down the first column, the method of subdeterminants says that this 4×4 determinant is equivalent to solving the following (smaller) 3×3 determinant:

$$\begin{vmatrix} -2i\alpha & i\alpha-\beta & i\alpha+\beta \\ -(e^{i\alpha a}-e^{-i\alpha a}) & e^{i\alpha a}-e^{-\beta b+ik(a+b)} & e^{i\alpha a}-e^{\beta b+ik(a+b)} \\ 0 & (i\alpha-\beta)(-e^{-i\alpha a}+e^{-\beta b+ik(a+b)}) & (i\alpha+\beta)(-e^{-i\alpha a}+e^{\beta b+ik(a+b)}) \end{vmatrix} = 0.$$

To further simplify the notation, let

$$J = (i\alpha - \beta)\left(-e^{-i\alpha a} + e^{-\beta b+ik(a+b)}\right),$$

$$K = (i\alpha + \beta)\left(-e^{-i\alpha a} + e^{\beta b+ik(a+b)}\right).$$

We then obtain the following determinant, which we again continue to solve via the method of subdeterminants:

$$\begin{vmatrix} -2i\alpha & i\alpha-\beta & i\alpha+\beta \\ -(e^{i\alpha a}-e^{-i\alpha a}) & e^{i\alpha a}-e^{-\beta b+ik(a+b)} & e^{i\alpha a}-e^{\beta b+ik(a+b)} \\ 0 & J & K \end{vmatrix} = 0.$$

As before, we proceed down the first column since a zero exists there that simplifies things. Note though that this is not necessary, since there are other ways to evaluate the determinant using different rows or columns. The reader may explore these other possibilities as an exercise. The above 3×3 determinant therefore equals the following linear combination of smaller 2×2 determinants:

$$-2i\alpha \begin{vmatrix} e^{i\alpha a}-e^{-\beta b+ik(a+b)} & e^{i\alpha a}-e^{\beta b+ik(a+b)} \\ J & K \end{vmatrix} + (e^{i\alpha a}-e^{-i\alpha a}) \begin{vmatrix} (i\alpha-\beta) & (i\alpha+\beta) \\ J & K \end{vmatrix} = 0.$$

Both are trivial to solve and result in the following expression:

$$-2i\alpha\big[\big(e^{i\alpha a}-e^{-\beta b+ik(a+b)}\big)K-\big(e^{i\alpha a}-e^{\beta b+ik(a+b)}\big)J\big] + (e^{i\alpha a}-e^{-i\alpha a})\,[(i\alpha-\beta)K-(i\alpha+\beta)J] = 0.$$

Simplifying the second term gives

$$-2i\alpha\big[\big(e^{i\alpha a}-e^{-\beta b+ik(a+b)}\big)K-\big(e^{i\alpha a}-e^{\beta b+ik(a+b)}\big)J\big] + 2i(-\alpha^2-\beta^2)\sin(\alpha a)\big(e^{\beta b+ik(a+b)}-e^{-\beta b+ik(a+b)}\big) = 0,$$

which subsequently reduces to

$$-2i\alpha\big[\big(e^{i\alpha a}-e^{-\beta b+ik(a+b)}\big)K-\big(e^{i\alpha a}-e^{\beta b+ik(a+b)}\big)J\big] + 2ie^{ik(a+b)}(-\alpha^2-\beta^2)\sin(\alpha a)\,\frac{e^{\beta b}-e^{-\beta b}}{2}(2) = 0.$$

This is equivalent to the following equation:

$$-2i\alpha\big[\big(e^{i\alpha a}-e^{-\beta b+ik(a+b)}\big)K-\big(e^{i\alpha a}-e^{\beta b+ik(a+b)}\big)J\big] + 4ie^{ik(a+b)}(-\alpha^2-\beta^2)\sin(\alpha a)\sinh(\beta b) = 0.$$

At this point, we focus on the first part of the expression by evaluating the two terms inside the square brackets.

The first term in the square brackets is

$$\big(e^{i\alpha a}-e^{-\beta b+ik(a+b)}\big)\big(i\alpha+\beta\big)\big(-e^{-i\alpha a}+e^{\beta b+ik(a+b)}\big),$$

which simplifies to

$$(i\alpha+\beta)[-1-e^{2ik(a+b)}+e^{ik(a+b)}(e^{i\alpha a+\beta b}+e^{-i\alpha a-\beta b})].$$

Next, the second term in the square brackets is

$$-\big(e^{i\alpha a}-e^{\beta b+ik(a+b)}\big)\big(i\alpha-\beta\big)\big(-e^{-i\alpha a}+e^{-\beta b+ik(a+b)}\big),$$

which likewise reduces to

$$-(i\alpha-\beta)[-1-e^{2ik(a+b)}+e^{ik(a+b)}(e^{i\alpha a-\beta b}+e^{-i\alpha a+\beta b})].$$

Consolidating the first and second terms then gives

$$(i\alpha+\beta)[-1-e^{2ik(a+b)}+e^{ik(a+b)}(e^{i\alpha a+\beta b}+e^{-i\alpha a-\beta b})]$$
$$-(i\alpha-\beta)[-1-e^{2ik(a+b)}+e^{ik(a+b)}(e^{i\alpha a-\beta b}+e^{-i\alpha a+\beta b})],$$

which simplifies to

$$-2\beta-2\beta e^{2ik(a+b)}+e^{ik(a+b)}\left(e^{i\alpha a}2i\alpha\frac{e^{\beta b}-e^{-\beta b}}{2}-e^{-i\alpha a}2i\alpha\frac{e^{\beta b}-e^{-\beta b}}{2}\right.$$
$$\left.+2\beta e^{i\alpha a}\frac{e^{\beta b}+e^{-\beta b}}{2}+2\beta e^{-i\alpha a}\frac{e^{\beta b}+e^{-\beta b}}{2}\right).$$

This can subsequently be written as

$$-2\beta-2\beta e^{2ik(a+b)}+e^{ik(a+b)}[2i\alpha e^{i\alpha a}\sinh(\beta b)-2i\alpha e^{-i\alpha a}\sinh(\beta b)$$
$$+2\beta e^{i\alpha a}\cosh(\beta b)+2\beta e^{-i\alpha a}\cosh(\beta b)]$$

or

$$-2\beta-2\beta e^{2ik(a+b)}$$
$$+2e^{ik(a+b)}\left[(i\alpha)(2i)\sinh(\beta b)\,\frac{e^{i\alpha a}-e^{-i\alpha a}}{2i}+2\beta\cosh(\beta b)\,\frac{e^{i\alpha a}+e^{-i\alpha a}}{2}\right].$$

In turn, one obtains

$$-2\beta-2\beta e^{2ik(a+b)}+2e^{ik(a+b)}[-2\alpha\sinh(\beta b)\sin(\alpha a)+2\beta\cosh(\beta b)\cos(\alpha a)].$$

At this point, putting the resulting expression back into our main result (just before evaluating its terms in the square brackets) and further consolidating terms gives

$$4i\alpha\beta+4i\alpha\beta e^{2ik(a+b)}-8i\alpha e^{ik(a+b)}[\beta\cosh(\beta b)\cos(\alpha a)-\alpha\sinh(\beta b)\sin(\alpha a)]$$
$$+4ie^{ik(a+b)}(-\alpha^2-\beta^2)\sin(\alpha a)\sinh(\beta b)=0.$$

Multiplying by $e^{-ik(a+b)}$ and dividing by $4i$ then results in

$$2\left(\alpha\beta\frac{e^{-ik(a+b)}+e^{ik(a+b)}}{2}\right)-2\alpha[\beta\cosh(\beta b)\cos(\alpha a)-\alpha\sinh(\beta b)\sin(\alpha a)]$$
$$+(-\alpha^2-\beta^2)\sin(\alpha a)\sinh(\beta b)=0,$$

which simplifies to

$$\cos k(a+b)+\frac{\alpha^2-\beta^2}{2\alpha\beta}\sinh(\beta b)\sin(\alpha a)-\cosh(\beta b)\cos(\alpha a)=0.$$

Our desired relationship is then

$$\frac{\beta^2-\alpha^2}{2\alpha\beta}\sinh(\beta b)\sin(\alpha a)+\cosh(\beta b)\cos(\alpha a)=\cos k(a+b),\quad(10.3)$$

which is the common textbook Kronig–Penney result.

Equation 10.3 is then solved *numerically* to obtain all valid energies E for the carrier in the crystal. To do this, recall that

$$\alpha=\sqrt{\frac{2m_0E}{\hbar^2}},$$
$$\beta=\sqrt{\frac{2m_0(V-E)}{\hbar^2}}.$$

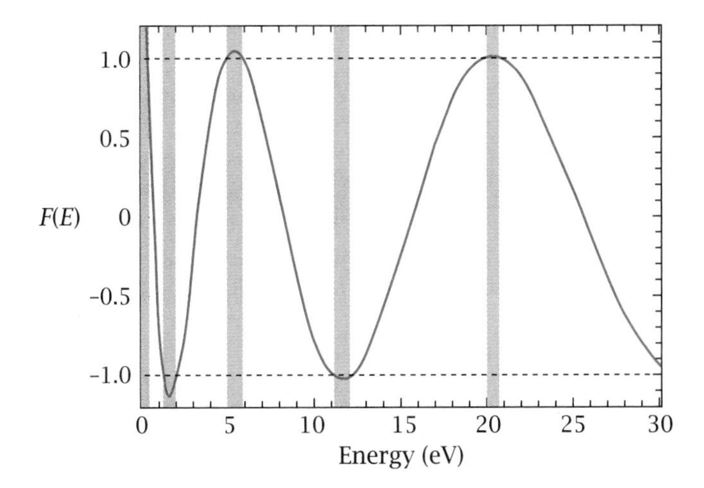

Figure 10.4 Left-hand side of the Kronig–Penney expression:

$$\frac{\beta^2 - \alpha^2}{2\alpha\beta}\sinh(\beta b)\sin(\alpha a) + \cosh(\beta b)\cos(\alpha a)$$

Regions where values of the left-hand side exceed 1 or −1 are shaded and represent band gaps for the carrier in the solid. This plot was constructed assuming $a = 5$ Å, $b = 0.1a$ and $V = 5$ eV.

Furthermore, note that $\cos k(a + b)$ on the right-hand side of the equation oscillates between −1 and 1. As a consequence, E has restrictions on it when in the left-hand side of the equation. These restrictions, in turn, result in regions of allowed energies referred to as bands and regions of unallowed energies called band gaps. **Figure 10.4** illustrates this by plotting the left-hand side of Equation 10.3 as a function of E.

In practice, one evaluates the allowed energies in Equation 10.3 by solving for the roots of the following expression, given a value of k

$$\frac{\beta^2 - \alpha^2}{2\alpha\beta}\sinh(\beta b)\sin(\alpha a) + \cosh(\beta b)\cos(\alpha a) - \cos k(a + b) = 0.$$

This enables one to plot an E versus k diagram, which resembles the familiar parabolic dispersion relationship for free electrons seen earlier in Chapter 7 (**Figure 7.1**). However, a major difference is that there exist

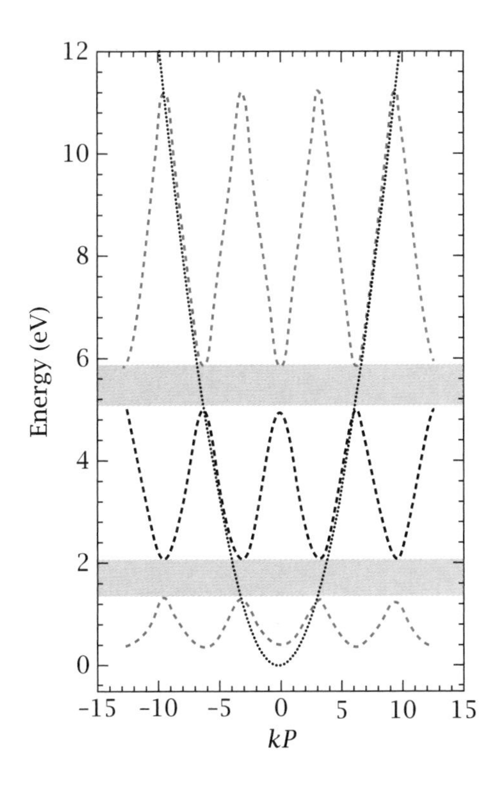

Figure 10.5 Kronig–Penney bands in the extended zone scheme. The first three bands are shown (dashed lines). The dotted line is the free particle parabolic energy relationship and the horizontal shaded regions represent band gaps. $P = a + b$ is the crystal's period.

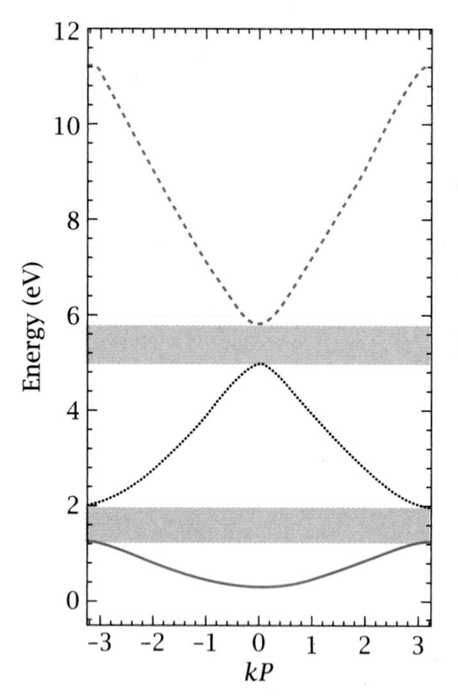

Figure 10.6 Kronig–Penney bands in the reduced zone scheme. The first three bands are shown (dashed, dotted, and solid lines). Horizontal shaded regions represent band gaps. $P = a + b$ is the crystal's period.

regions of unallowed energies that lead to band gaps, as illustrated by the horizontal shaded regions in **Figures 10.5** and **10.6**.

From looking at the right-hand side of Equation 10.3, one notices that these energy discontinuities occur at points where $k = n\pi/(a+b)$, with $n = 0, \pm1, \pm2, \pm3, \ldots$. These points define so-called zone edges. **Figure 10.5** therefore represents the allowed energies in what is called the "extended zone scheme" where multiple zones are shown. However, it is also common to see bands represented in the "reduced zone scheme" where the energies are folded back into the first Brillouin zone (i.e., the region between $-\pi/(a+b)$ and $\pi/(a+b)$). One then obtains plots like **Figure 10.6**.

Finally, we can check to see whether our derived relationship possesses the correct limits. Namely,

$$\frac{\beta^2 - \alpha^2}{2\alpha\beta}\sinh(\beta b)\sin(\alpha a) + \cosh(\beta b)\cos(\alpha a) = \cos k(a+b)$$

should give energies consistent with the free particle solution when $V \to 0$ (or $b \to 0$). It should also give the standard particle-in-a-box energies when $V \to \infty$.

Let us evaluate these limiting cases to see what we get.

Free Particle Limit ($V = 0$)

In the limit where $V = 0$,

$$\alpha = \sqrt{\frac{2m_0E}{\hbar^2}}$$

and

$$\beta = \sqrt{\frac{-2m_0E}{\hbar^2}}$$

such that $\beta = i\alpha$. This leads to the following Kronig–Penney expression:

$$\frac{-\alpha^2 - \alpha^2}{2i\alpha^2}\sinh(i\alpha b)\sin(\alpha a) + \cosh(i\alpha b)\cos(\alpha a) = \cos k(a+b).$$

Recalling that

$$\sinh(ix) = i\sin x,$$
$$\cosh(ix) = \cos x$$

then gives

$$-\sin(\alpha b)\sin(\alpha a) + \cos(\alpha b)\cos(\alpha a) = \cos k(a+b).$$

For this expression to be true, $k = \alpha$ (this can be seen from a trigonometric identity viewpoint). As a consequence,

$$E = \frac{\hbar^2k^2}{2m_0},$$

since $\alpha = \sqrt{2m_0E/\hbar^2}$. These are the free particle solutions derived earlier in Chapter 7.

Particle-in-a-Box Limit ($V = \infty$)

Next, let us consider the other limit where $V \to \infty$ and where the electron is strictly confined to the finite region between the barriers. To simplify things, rearrange everything in terms of $\xi = E/V$. This gives

$$\alpha^2 = \frac{2m_0 V}{\hbar^2}\xi,$$
$$\beta^2 = \frac{2m_0 V}{\hbar^2}(1-\xi),$$
$$\frac{\beta^2 - \alpha^2}{2\alpha\beta} = \frac{1-2\xi}{2\sqrt{\xi(1-\xi)}}.$$

Likewise, the βb terms in Equation 10.3 become

$$\beta b = \sqrt{\frac{2m_0 V}{\hbar^2}}\, a\left(\frac{b}{a}\right)\sqrt{1-\xi}$$

and letting $A = \sqrt{2m_0 V/\hbar^2}\, a$ and $r = b/a$ yields

$$\beta b = Ar\sqrt{1-\xi}.$$

The αa terms in Equation 10.3 become

$$\alpha a = A\sqrt{\xi}.$$

When all are replaced back into Equation 10.3, we find that

$$\frac{1-2\xi}{2\sqrt{\xi(1-\xi)}}\sinh(Ar\sqrt{1-\xi})\sin(A\sqrt{\xi}) + \cosh(Ar\sqrt{1-\xi})\cos(A\sqrt{\xi}) = \cos k(a+b).$$

As $\xi \to 0$, $V \to \infty$ and thus $A \to \infty$. The above expression is therefore approximately equivalent to

$$\frac{1}{2\sqrt{\xi}}\sinh(Ar)\sin(A\sqrt{\xi}) + \cosh(Ar) = \cos k(a+b).$$

Now, as $\xi \to 0$, the first term, $(1/2\sqrt{\xi})\sinh(Ar)\sin(A\sqrt{\xi})$, dominates the second, $\cosh(Ar)$. Furthermore, for the solution to remain bound by -1 and 1 (the limits on the right-hand side of the equation), $\sin(A\sqrt{\xi}) \to 0$. We can see this more clearly as follows. Given

$$\frac{1}{2\sqrt{\xi}}\sinh(Ar)\sin(A\sqrt{\xi}) \simeq \cos k(a+b),$$

the expression is equivalent to

$$\frac{1}{2\sqrt{\xi}}\left(\frac{e^{Ar} - e^{-Ar}}{2}\right)\sin(A\sqrt{\xi}) \simeq \cos k(a+b)$$

and as $A \to \infty$, we find that it is approximately equal to

$$\frac{1}{2\sqrt{\xi}}\left(\frac{e^{Ar}}{2}\right)\sin(A\sqrt{\xi}) \simeq \cos k(a+b).$$

The result clearly possesses a diverging e^{Ar} term. As a consequence, $\sin(A\sqrt{\xi}) \to 0$ to keep things bound. Thus, in this limit,

$$\sin(A\sqrt{\xi}) = 0$$

and $A\sqrt{\xi} = n\pi$, or

$$\sqrt{\frac{2m_0E}{\hbar^2}}a = n\pi,$$

with $n = 1, 2, 3, \ldots$. Solving for the energy E then yields

$$E = \frac{n^2h^2}{8m_0a^2},$$

which are the desired energies of a particle in an infinite one-dimensional box (Chapter 7). We thus find that both limits are satisfied by our Kronig–Penney solutions. More about the Kronig–Penney model can be found in Further Reading.

Summary

We see that the Kronig–Penney model lies in between two limiting cases for the behavior of an electron. At one end is the free electron limit. At the other is the limit of a confined particle in a one-dimensional box. In between lies a regime where bands of allowed energies exist separated by band gaps. The magnitude of these gaps then allows us to distinguish metals from semiconductors from insulators, as we will see later.

10.3 KRONIG–PENNEY MODEL WITH DELTA-FUNCTION BARRIERS

Another version of the Kronig–Penney model exists, employing Dirac δ-function barriers instead of the rectangular barriers shown in **Figure 10.1**. We introduce it here since it is also very illustrative and is commonly found in many texts. The potential of interest,

$$V(x) = V_0 \sum_{n=-\infty}^{\infty} \delta(x - na), \tag{10.4}$$

is shown in **Figure 10.7**, where V_0 represents the strength of the barrier with units of energy times length (for example, eV · Å).

As before, we solve the Schrödinger equation in a piecewise fashion to obtain the particle's energies. In this case, though, there are only two regions to consider and these are shown in **Figure 10.7**.

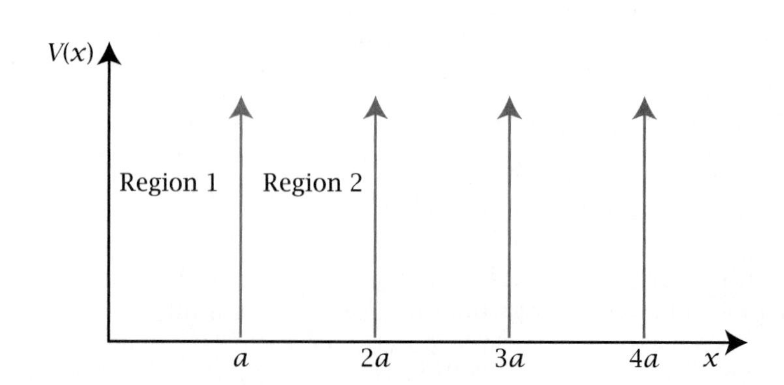

Figure 10.7 Illustration of the Dirac δ-function periodic potential.

10.3.1 Region 1 ($V = 0$)

In Region 1, the potential is zero. As a consequence, the Schrödinger equation can be written as

$$-\frac{\hbar^2}{2m_0}\frac{d^2\psi_1}{dx^2} = E\psi_1.$$

Letting $\alpha^2 = 2m_0E/\hbar^2$ then gives

$$\frac{d^2\psi_1}{dx^2} + \alpha^2\psi_1 = 0,$$

with plane wave solutions

$$\psi_1(x) = Ae^{i\alpha x} + Be^{-i\alpha x},$$

where A and B are constant coefficients.

10.3.2 Region 2 ($V = 0$)

In Region 2, $V = 0$ and we obtain similar solutions. However, rather than solve for them explicitly, let us employ Bloch's theorem, which states that (Equation 10.2)

$$\psi(x + a) = \psi(x)e^{ika}.$$

As a consequence, after considering a coordinate shift to utilize the same axis for both $\psi_1(x)$ and $\psi_2(x)$, we find that

$$\psi_2(x) = \psi_1(x - a)e^{ika},$$

or alternatively

$$\psi_2(x) = (Ae^{i\alpha(x-a)} + Be^{-i\alpha(x-a)})e^{ika}.$$

Next, from the wavefunction matching conditions used earlier, we know that

$$\psi_1(a) = \psi_2(a),$$
$$(Ae^{i\alpha a} + Be^{-i\alpha a}) = (A + B)e^{ika},$$

and therefore obtain

$$A(e^{i\alpha a} - e^{ika}) + B(e^{-i\alpha a} - e^{ika}) = 0,$$

after rearranging terms. Note that continuity of the first derivative is not satisfied when $V(x)$ consists of δ-functions, unlike our earlier Kronig–Penney model with rectangular barriers. A discontinuity exists, which we will determine. It, in turn, will provide a second expression relating A or B. This essentially follows the derivation by Wolfe (1978), where the idea is to integrate the Schrödinger equation about a from $a - \delta a$ to $a + \delta a$ and then consider the limit $\delta a \to 0$.

To begin, the Schrödinger equation with the potential present is

$$\left[-\frac{\hbar^2}{2m_0}\frac{d^2}{dx^2} + V(x)\right]\psi(x) = E\psi(x)$$

and can be rearranged to

$$\frac{d^2\psi(x)}{dx^2} = \frac{2m_0}{\hbar^2}(V - E)\psi(x).$$

At this point, integrating the left-hand side of the expression about $x = a$, with δa being an infinitesimally small offset, gives (left-hand side)

$$\int_{a-\delta a}^{a+\delta a} \frac{d^2\psi}{dx^2}\,dx = \int_{a-\delta a}^{a+\delta a} \frac{d}{dx}\left(\frac{d\psi(x)}{dx}\right)dx$$
$$= \left.\frac{d\psi(x)}{dx}\right|_{x=a+\delta a} - \left.\frac{d\psi(x)}{dx}\right|_{x=a-\delta a}.$$

This also equals the integral of the right-hand side,

$$\int_{a-\delta a}^{a+\delta a} \frac{d^2\psi(x)}{dx^2} = \int_{a-\delta a}^{a+\delta a} \frac{2m_0}{\hbar^2}(V-E)\psi(x)\,dx,$$

with $V = V_0\delta(x-a)$. We therefore have

$$\int_{a-\delta a}^{a+\delta a} \frac{d^2\psi(x)}{dx^2} = \int_{a-\delta a}^{a+\delta a} \frac{2m_0}{\hbar^2}\left[V_0\delta(x-a) - E\right]\psi(x)\,dx$$
$$= \frac{2m_0V_0}{\hbar^2}\int_{a-\delta a}^{a+\delta a} \delta(x-a)\psi(x)\,dx - \frac{2m_0E}{\hbar^2}\int_{a-\delta a}^{a+\delta a} \psi(x)\,dx.$$

As $\delta a \to 0$, the second term on the right-hand side of the equation goes to zero and we have

$$\int_{a-\delta a}^{a+\delta a} \frac{d^2\psi(x)}{dx^2} = \frac{2m_0V_0}{\hbar^2}\psi_2(a)$$

(in the expression, we have used $\psi_2(a)$; however, $\psi_2(a) = \psi_1(a)$ so $\psi_1(a)$ could be used as well), leading to

$$\left.\frac{d\psi(x)}{dx}\right|_{x=a+\delta a} - \left.\frac{d\psi(x)}{dx}\right|_{x=a-\delta a} = \frac{2m_0V_0}{\hbar^2}\psi_2(a),$$
$$\left.\frac{d\psi_2(x)}{dx}\right|_{x=a+\delta a} - \left.\frac{d\psi_1(x)}{dx}\right|_{x=a-\delta a} = \frac{2m_0V_0}{\hbar^2}\psi_2(a).$$

This difference illustrates the discontinuity in the wavefunction's first derivative when dealing with a delta-function potential. Rearranging and consolidating terms now gives

$$\left.\frac{d\psi_2(x)}{dx}\right|_{x=a+\delta a} - \frac{2m_0V_0}{\hbar^2}\psi_2(a) = \left.\frac{d\psi_1(x)}{dx}\right|_{x=a-\delta a}.$$

Let us look at the first term on the left-hand side of the above expression. Since

$$\psi_2(x) = (Ae^{i\alpha(x-a)} + Be^{-i\alpha(x-a)})e^{ika},$$

we find that

$$\frac{d\psi_2(x)}{dx} = (i\alpha Ae^{i\alpha(x-a)} - i\alpha Be^{-i\alpha(x-a)})e^{ika},$$

and this results in

$$\left.\frac{d\psi_2(x)}{dx}\right|_{x=a} = (i\alpha A - i\alpha B)e^{ika}$$

when we consider the limit $\delta a \to 0$.

Next, for the right-hand side, since

$$\psi_1(x) = Ae^{i\alpha x} + Be^{-i\alpha x},$$

we have

$$\frac{d\psi_1(x)}{dx} = i\alpha A e^{i\alpha x} - i\alpha B e^{-i\alpha x},$$

and therefore, in the limit $\delta a \to 0$,

$$\left.\frac{d\psi_1(x)}{dx}\right|_{x=a} = i\alpha A e^{i\alpha a} - i\alpha B e^{-i\alpha a}.$$

Putting everything together then yields

$$(i\alpha A - i\alpha B)e^{ika} - \frac{2m_0 V_0}{\hbar^2}[(A+B)e^{ika}] = i\alpha A e^{i\alpha a} - i\alpha B e^{-i\alpha a},$$

whereupon consolidating terms in A and B results in

$$\left[i\alpha(e^{ika} - e^{i\alpha a}) - \frac{2m_0 V_0}{\hbar^2}e^{ika}\right]A + \left[-i\alpha(e^{ika} - e^{-i\alpha a}) - \frac{2m_0 V_0}{\hbar^2}e^{ika}\right]B = 0.$$

Summary

To summarize, we now have two expressions describing the relationship between A and B, the first from continuity of the wavefunction at a and the second from the resulting discontinuity of the first derivative when dealing with delta functions:

$$(e^{i\alpha a} - e^{ika})A + (e^{-i\alpha a} - e^{ika})B = 0$$

and

$$\left[-i\alpha(e^{i\alpha a} - e^{ika}) - \frac{2m_0 V_0}{\hbar^2}e^{ika}\right]A + \left[i\alpha(e^{-i\alpha a} - e^{ika}) - \frac{2m_0 V_0}{\hbar^2}e^{ika}\right]B = 0.$$

These two equations can be expressed in matrix form as

$$\begin{pmatrix} e^{i\alpha a} - e^{ika} & e^{-i\alpha a} - e^{ika} \\ -i\alpha(e^{i\alpha a} - e^{ika}) - \frac{2m_0 V_0}{\hbar^2}e^{ika} & i\alpha(e^{-i\alpha a} - e^{ika}) - \frac{2m_0 V_0}{\hbar^2}e^{ika} \end{pmatrix}\begin{pmatrix} A \\ B \end{pmatrix} = \begin{pmatrix} 0 \\ 0 \end{pmatrix}.$$

At this point, we either have the trivial solution ($A = B = 0$) or find that the determinant of the first matrix equals zero. Since we never want the trivial solution,

$$\begin{vmatrix} e^{i\alpha a} - e^{ika} & e^{-i\alpha a} - e^{ika} \\ -i\alpha(e^{i\alpha a} - e^{ika}) - \frac{2m_0 V_0}{\hbar^2}e^{ika} & i\alpha(e^{-i\alpha a} - e^{ika}) - \frac{2m_0 V_0}{\hbar^2}e^{ika} \end{vmatrix} = 0.$$

This 2×2 determinant is easy to find and yields the following expression:

$$i\alpha(e^{i\alpha a} - e^{ika})(e^{-i\alpha a} - e^{ika}) - \frac{2m_0 V_0}{\hbar^2}e^{ika}(e^{i\alpha a} - e^{ika})$$
$$+ i\alpha(e^{i\alpha a} - e^{ika})(e^{-i\alpha a} - e^{ika}) + \frac{2m_0 V_0}{\hbar^2}e^{ika}(e^{-i\alpha a} - e^{ika}) = 0.$$

Simplifying it gives

$$i\alpha(1 - e^{i(k+\alpha)a} - e^{i(k-\alpha)a} + e^{2ika}) - \frac{2m_0 V_0}{\hbar^2}e^{ika}(e^{i\alpha a} - e^{ika})$$
$$+ i\alpha(1 - e^{i(k+\alpha)a} - e^{i(k-\alpha)a} + e^{2ika}) + \frac{2m_0 V_0}{\hbar^2}e^{ika}(e^{-i\alpha a} - e^{ika}) = 0,$$

or

$$2i\alpha(1 - e^{ika}e^{i\alpha a} - e^{ika}e^{-i\alpha a} + e^{2ika}) - \frac{2m_0 V_0}{\hbar^2} e^{ika}\left(e^{i\alpha a} - e^{-i\alpha a}\right) = 0.$$

Multiplying by e^{-ika} then yields

$$2i\alpha(e^{-ika} - e^{i\alpha a} - e^{-i\alpha a} + e^{ika}) - \frac{2m_0 V_0}{\hbar^2}(e^{i\alpha a} - e^{-i\alpha a}) = 0$$

$$2i\alpha\left[2\frac{(e^{ika} + e^{-ika})}{2} - 2\frac{(e^{i\alpha a} + e^{-i\alpha a})}{2}\right] - \frac{2m_0 V_0}{\hbar^2}(2i)\frac{e^{i\alpha a} - e^{-i\alpha a}}{2i} = 0$$

$$2i\alpha[2\cos(ka) - 2\cos(\alpha a)] - 2i\frac{2m_0 V_0}{\hbar^2}\sin(\alpha a) = 0$$

$$\cos(ka) - \cos(\alpha a) - \frac{2m_0 V_0}{\hbar^2}\frac{\sin(\alpha a)}{2\alpha} = 0,$$

which can be rearranged to

$$\left(\frac{m_0 V_0 a}{\hbar^2}\right)\frac{\sin(\alpha a)}{\alpha a} + \cos(\alpha a) = \cos(ka).$$

At this point, letting $P = m_0 V_0 a/\hbar^2$ gives

$$P\frac{\sin(\alpha a)}{\alpha a} + \cos(\alpha a) = \cos(ka), \tag{10.5}$$

which is our desired Kronig–Penney expression analogous to Equation 10.3 seen earlier. Recall that $\alpha = \sqrt{2m_0 E/\hbar^2}$ and that a is the lattice constant of the crystal.

One now solves the above expression numerically to find the values of E that make the expression true. However, we can already see the limiting cases of a free particle ($V = 0$) and of a confined particle ($V = \infty$), as illustrated below.

10.3.3 Free Particle Limit ($V = 0$)

If $V_0 = 0$ then $P = 0$ and we find that

$$\cos(\alpha a) = \cos(ka).$$

Therefore,

$$k = \alpha = \sqrt{\frac{2m_0 E}{\hbar^2}}.$$

As a consequence,

$$E = \frac{\hbar^2 k^2}{2m_0},$$

and these are the energies of a free particle shown previously.

10.3.4 Particle-in-a-Box Limit ($V = \infty$)

If $V_0 = \infty$, then $P \to \infty$ and $\sin(\alpha a) \to 0$ if the solutions are to remain bound by 1 and -1. Thus, in this limit,

$$\alpha a = n\pi,$$

$$\sqrt{\frac{2m_0 E}{\hbar^2}}a = n\pi.$$

Solving for E then gives

$$E = \frac{n^2 h^2}{8 m_0 a^2},$$

which are the standard solutions for a particle in a one-dimensional box as seen earlier.

10.3.5 General Case

In the general case, one finds the roots of the equation

$$P\frac{\sin(\alpha a)}{\alpha a} + \cos(\alpha a) - \cos(ka) = 0$$

for a given value of k. One can then construct the E versus k diagrams seen previously. Examples are shown in **Figures 10.8** and **10.9**. As in

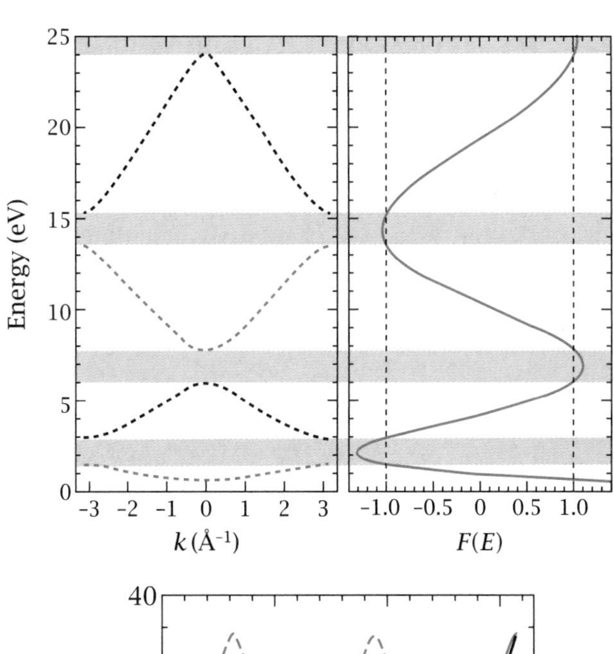

Figure 10.8 Kronig–Penney bands for the periodic δ-function potential in the reduced zone scheme (dashed lines). Horizontal shaded regions represent band gaps. The plot on the right is the left-hand side of Equation 10.5:

$$P\frac{\sin(\alpha a)}{\alpha a} + \cos(\alpha a).$$

Regions where it is less than -1 or greater than $+1$ are shaded and again represent band gaps for the electron.

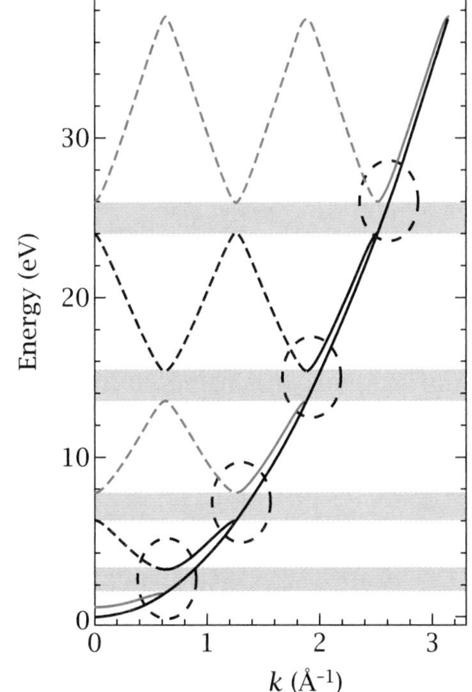

Figure 10.9 Band gaps (circled) in the Kronig–Penney δ-function potential model. They are also represented by the horizontal shaded regions. The solid parabola represents the free particle parabolic energy relationship.

the rectangular barrier Kronig–Penney model, we see the development of band gaps, especially at characteristic zone edges.

We can also find the magnitude of these energy gaps by numerically evaluating the Kronig–Penney expression at the zone edges of different bands. Results from this procedure are illustrated in **Table 10.1**, where band gaps between the first and second bands are tabulated for a given set of conditions. It is apparent that the simple Kronig–Penney model readily leads to realistic energy gaps found in many semiconductors. The reader may repeat these calculations with different conditions as an exercise.

Table 10.1 Calculated Kronig–Penney band gaps

	k (Å^{-1})	ΔE (eV)
	0.6283	1.471
$V_0 = 5$ eV Å,	1.2570	1.762
$a = 5$ Å	1.8850	1.872
	2.5130	1.922
	0.6283	2.280
$V_0 = 10$ eV Å,	1.2570	3.007
$a = 5$ Å	1.8850	3.388
	2.5130	3.596

10.3.6 Density of States

Note that the Kronig–Penney model can be quite illustrative. This is because it ultimately yields the allowed energies of a carrier in a crystal and hence the energies of the corresponding band. Because there exists a direct relationship between these energies and the effective mass of the carrier as well as its corresponding density of states, we can use the above Kronig–Penney results to estimate what these values or functions look like.

In what follows, we therefore calculate the Kronig–Penney density of states using the approach outlined in Wolfe (1978). The result of this procedure may then be compared with those derived earlier in Chapter 9. In particular, when we previously calculated a generic one-dimensional density of states, we found that it had a characteristic $1/\sqrt{E}$ dependence, leading to so-called van Hove singularities. Specifically, the equation found for the first subband was (Equation 9.10)

$$\rho_{\text{energy}}(E) = \frac{1}{\pi}\sqrt{\frac{2m_{\text{eff}}}{\hbar^2}}\frac{1}{\sqrt{E}}.$$

Let us therefore re-derive the expected one-dimensional density of states of a carrier, this time within the context of a one-dimensional Kronig–Penney model. Starting with Equation 10.5,

$$\cos(ka) = P\frac{\sin(\alpha a)}{\alpha a} + \cos(\alpha a) = f(\alpha),$$

where a is the lattice constant, we take the first derivative of both sides with respect to k and apply the chain rule. This yields

$$\frac{d[\cos(ka)]}{dk} = \frac{df(\alpha)}{d\alpha}\left(\frac{d\alpha}{dk}\right).$$

Subsequent rearrangement of terms gives

$$\frac{d\alpha}{dk} = -\frac{a\sin(ka)}{df(\alpha)/d\alpha}.$$

Note that $\alpha = \sqrt{2m_0E/\hbar^2}$. When substituted into the left-hand side of the above equation, followed by an application of the chain rule, we find

$$\frac{d\alpha}{dk} = \sqrt{\frac{2m_0}{\hbar^2}}\frac{dE^{1/2}}{dk} = \sqrt{\frac{2m_0}{\hbar^2}}\frac{dE^{1/2}}{dE}\left(\frac{dE}{dk}\right)$$
$$= \sqrt{\frac{2m_0}{\hbar^2}}\frac{1}{2}E^{-1/2}\left(\frac{dE}{dk}\right).$$

This, in turn, leads to the equivalence

$$\sqrt{\frac{2m_0}{\hbar^2}}\frac{1}{2\sqrt{E}}\left(\frac{dE}{dk}\right) = -\frac{a\sin(ka)}{df(\alpha)/d\alpha}.$$

Rearranging terms to find the energy gradient in k-space then yields

$$\frac{dE}{dk} = -\frac{2\sqrt{E}a\sin(ka)\sqrt{\hbar^2/2m_0}}{df(\alpha)/d\alpha},$$

or, better yet,

$$\frac{dk}{dE} = \frac{\sqrt{2m_0}[df(\alpha)/d\alpha]}{-2\sqrt{E}a\sin(ka)\sqrt{\hbar^2}}.$$

This is useful since we previously saw in Chapter 9 the following generic expression for the one-dimensional density of states (Equation 9.9):

$$\rho_{\text{energy}}(E) = \frac{2}{\pi}\left(\frac{dk}{dE}\right).$$

Introducing our expression for dk/dE then gives

$$\rho_{\text{energy}}(E) = -\sqrt{\frac{2m_0}{E}}\frac{1}{\pi\hbar}\left[\frac{df(\alpha)/d\alpha}{a\sin(ka)}\right],$$

and since

$$\begin{aligned}\frac{df(\alpha)}{d\alpha} &= \frac{d}{d\alpha}\left[P\frac{\sin(\alpha a)}{\alpha a} + \cos(\alpha a)\right] \\ &= P\left[-\frac{\sin(\alpha a)}{\alpha^2 a} + \frac{\cos(\alpha a)}{\alpha}\right] - a\sin(\alpha a),\end{aligned}$$

we obtain

$$\rho_{\text{energy}}(E) = \frac{1}{\pi}\sqrt{\frac{2m_0}{\hbar^2}}\left\{\frac{P\left[\frac{\sin(\alpha a)}{(\alpha a)^2} - \frac{\cos(\alpha a)}{\alpha a}\right] + \sin(\alpha a)}{\sin ka}\right\}\frac{1}{\sqrt{E}}. \quad (10.6)$$

We see that the above density-of-states expression is very similar to what we found earlier in Chapter 9 (Equation 9.10). The primary difference is the additional term in curly brackets.

Note that an alternative expression for Equation 10.6 exists. It is obtained by further rearranging the denominator of the above expression. Namely, instead of $\sin(ka)$, we utilize the fact that (Equation 10.5)

$$\cos(ka) = P\frac{\sin(\alpha a)}{\alpha a} + \cos(\alpha a)$$

to equivalently write

$$\begin{aligned}\sin(ka) &= \sqrt{1 - \cos^2(ka)} \\ &= \left\{1 - \left[\frac{P\sin(\alpha a)}{\alpha a} + \cos(\alpha a)\right]^2\right\}^{1/2}.\end{aligned}$$

Thus, our derived Kronig–Penney one-dimensional density of states can alternatively be written as

$$\rho_{\text{energy}}(E) = \frac{1}{\pi}\sqrt{\frac{2m_0}{\hbar^2}}\left\{\frac{P\left[\frac{\sin(\alpha a)}{(\alpha a)^2} - \frac{\cos(\alpha a)}{\alpha a}\right] + \sin(\alpha a)}{\left\{1 - \left[\frac{P\sin(\alpha a)}{\alpha a} + \cos(\alpha a)\right]^2\right\}^{1/2}}\right\}\frac{1}{\sqrt{E}}. \quad (10.7)$$

The result is illustrated in **Figure 10.10**, where vertical shaded regions denote the van Hove singularities present.

Figure 10.10 Kronig–Penney δ-function potential density of states. Shaded regions indicate where van Hove singularities exist.

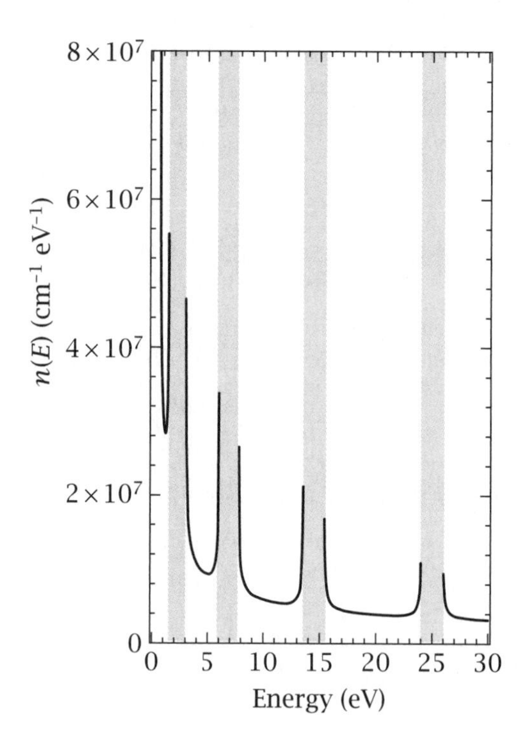

10.3.7 Effective Masses

Just as one can use the derived Kronig–Penney energies to find the carrier density of states, we can also use these results to find their effective masses. This applies to both electrons and holes.

The Electron Effective Mass m_e

Recall that the effective mass m_{eff} of a carrier is related to the curvature of its band in k-space through

$$\frac{1}{m_{\text{eff}}} = \frac{1}{\hbar^2}\left(\frac{d^2E}{dk^2}\right).$$

Thus, by knowing $E(k)$, we can find what m_{eff} is. A nice consequence of this calculation, when applied to the Kronig–Penney model, is that it naturally predicts that the effective mass of the electron, m_e, differs from that of the free electron, m_0. Namely,

$$m_e = \gamma_e m_0. \tag{10.8}$$

In this expression, γ_e is a unitless number. We will also see that when applied to holes, their effective masses m_h differ from m_e or m_0.

To illustrate, we proceed in much the same way as above when deriving the Kronig–Penney density of states (Wolfe 1978). Specifically, using

$$\cos(ka) = P\frac{\sin(\alpha a)}{\alpha a} + \cos(\alpha a) = f(\alpha),$$

we take the first derivative of both sides and apply the chain rule. This leads to

$$\frac{d\cos(ka)}{dk} = \frac{df(\alpha)}{d\alpha}\left(\frac{d\alpha}{dk}\right),$$

$$-a\sin(ka) = \frac{df(\alpha)}{d\alpha}\left(\frac{d\alpha}{dk}\right),$$

from which

$$\frac{d\alpha}{dk} = -\frac{a\sin(ka)}{df(\alpha)/d\alpha}.$$

Next, taking the second derivative of both sides gives

$$\frac{d^2\alpha}{dk^2} = -a\left[\sin(ka)\frac{d}{dk}\left(\frac{df(\alpha)}{d\alpha}\right)^{-1} + \frac{a\cos(ka)}{df(\alpha)/d\alpha}\right],$$

so that when we eventually consider the situation where $k \simeq 0$,

$$\frac{d^2\alpha}{dk^2} = -\frac{a^2\cos(ka)}{df(\alpha)/d\alpha}.$$

For what follows, recall that $\alpha = \sqrt{2m_0E/\hbar^2}$. The left-hand side of the above expression can now be expanded as follows:

$$\begin{aligned}
\frac{d}{dk}\left(\frac{d\alpha}{dk}\right) &= \sqrt{\frac{2m_0}{\hbar^2}}\frac{d}{dk}\left(\frac{dE^{1/2}}{dk}\right) \\
&= \sqrt{\frac{2m_0}{\hbar^2}}\frac{d}{dk}\left[\frac{dE^{1/2}}{dE}\left(\frac{dE}{dk}\right)\right] \\
&= \sqrt{\frac{m_0}{2\hbar^2}}\frac{d}{dk}\left[E^{-1/2}\left(\frac{dE}{dk}\right)\right] \\
&= \sqrt{\frac{m_0}{2\hbar^2}}\left(E^{-1/2}\frac{d^2E}{dk^2} - \frac{1}{2}E^{-3/2}\frac{dE}{dk}\right).
\end{aligned}$$

If we recognize that $dE/dk \simeq 0$ when $k \simeq 0$, we find

$$\frac{d^2\alpha}{dk^2} = \sqrt{\frac{m_0}{2\hbar^2 E}}\frac{d^2E}{dk^2} = -\frac{a^2\cos(ka)}{df(\alpha)/d\alpha}.$$

Rearranging terms then gives

$$\frac{d^2E}{dk^2} = \sqrt{\frac{2\hbar^2E}{m_0}}(-a^2)\frac{\cos(ka)}{df(\alpha)/d\alpha}.$$

Finally, since

$$\begin{aligned}
\frac{1}{m_e} &= \frac{1}{\hbar^2}\left(\frac{d^2E}{dk^2}\right) \\
&= -\frac{a^2}{\hbar}\sqrt{\frac{2E}{m_0}}\frac{\cos(ka)}{df(\alpha)/d\alpha},
\end{aligned}$$

inversion of the expression yields

$$m_e = -\frac{\hbar}{a^2}\sqrt{\frac{m_0}{2E}}\frac{df(\alpha)/d\alpha}{\cos(ka)},$$

from which we conclude that

$$m_e = \left[-\frac{1}{a^2}\sqrt{\frac{\hbar^2}{2m_0E}}\frac{df(\alpha)/d\alpha}{\cos(ka)}\right]m_0.$$

Note that this is already in the form of our desired effective mass expression.

All that is left for us to do is to find a better expression for $df(\alpha)/d\alpha$. The nice thing is that this was done previously when we found the Kronig–Penney density of states. Namely, we showed that

$$\frac{df(\alpha)}{d\alpha} = P\left[-\frac{\sin(\alpha a)}{\alpha^2 a} + \frac{\cos(\alpha a)}{\alpha}\right] - a\sin(\alpha a).$$

This allows us to complete our electron effective mass calculation, yielding

$$m_e = \frac{1}{a}\sqrt{\frac{\hbar^2}{2m_0E}}\frac{1}{\cos(ka)}\left\{P\left[\frac{\sin(\alpha a)}{(\alpha a)^2} - \frac{\cos(\alpha a)}{\alpha a}\right] + \sin(\alpha a)\right\} m_0, \quad (10.9)$$

where

$$\gamma_e = \frac{1}{a}\sqrt{\frac{\hbar^2}{2m_0E}}\frac{1}{\cos(ka)}\left\{P\left[\frac{\sin(\alpha a)}{(\alpha a)^2} - \frac{\cos(\alpha a)}{\alpha a}\right] + \sin(\alpha a)\right\}$$

in Equation 10.8. Notice that the effective mass depends on energy.

The Hole Effective Mass m_h

We can repeat the above calculation to find an identical expression for the effective mass of the hole,

$$m_h = \gamma_h m_0. \quad (10.10)$$

γ_h is again a unitless number. Although the resulting expression is essentially identical to that for m_e, it differs since it possesses a negative sign. This originates from using different energies in Equation 10.9, and one can see that this is the reason why hole energies increase in the opposite direction to those of the electron. In particular, we evaluate hole masses at the top of a band, while electron masses are found at the bottom of a band. As a consequence, while the hole's effective mass expression is identical to the electron's, it uses different input energies and hence yields not only different masses but an opposite sign as well. In practice, one generally ignores this sign difference when referring to hole masses. Our resulting hole effective mass expression is therefore

$$m_h = \frac{1}{a}\sqrt{\frac{\hbar^2}{2m_0E}}\frac{1}{\cos(ka)}\left\{P\left[\frac{\sin(\alpha a)}{(\alpha a)^2} - \frac{\cos(\alpha a)}{\alpha a}\right] + \sin(\alpha a)\right\} m_0. \quad (10.11)$$

From this, one sees that

$$\gamma_h = \frac{1}{a}\sqrt{\frac{\hbar^2}{2m_0E}}\frac{1}{\cos(ka)}\left\{P\left[\frac{\sin(\alpha a)}{(\alpha a)^2} - \frac{\cos(\alpha a)}{\alpha a}\right] + \sin(\alpha a)\right\}$$

in Equation 10.10.

Table 10.2 lists calculated m_e and m_h values using Equations 10.9 and 10.11 for the conditions listed. As an exercise, the reader may calculate analogous masses, using different input conditions.

10.4 OTHER BAND MODELS

Note that although the Kronig–Penney model is an illustrative approximation for the behavior of electrons in a crystal, other models exist, such as the tight binding approximation, the nearly free electron model,

Table 10.2 Kronig–Penney electron and hole effective masses

	k (Å^{-1})	m_e/m_0	m_h/m_0
	0.6283	0.205	0.333
$V_0 = 5$ eV Å,	1.2570	0.071	0.083
$a = 5$ Å	1.8850	0.034	0.037
	2.5130	0.02	0.021
	0.6283	0.315	0.666
$V_0 = 10$ eV Å,	1.2570	0.125	0.166
$a = 5$ Å	1.8850	0.064	0.074
	2.5130	0.038	0.042

and the $k \cdot p$ approximation. In what follows, we illustrate two of these models, namely the tight binding approximation and the nearly free electron model. Both represent limiting cases that describe the carrier's energy in a solid.

10.4.1 The Tight Binding Approximation

In the tight binding model, we assume that electrons, rather than being loosely held, are tightly bound to their parent atoms. As a consequence, their wavefunctions are localized mostly around individual atoms. However, a given carrier can still interact with another carrier when they are localized on nearest-neighbor atoms. Thus, what results is a description of electron wavefunctions in an extended solid through a linear combination of atomic orbitals. (The term "orbital" refers to the wavefunction of the carrier in its native state without other atoms present.) We have seen such descriptions before in our prior studies, and illustrate the general concept in **Figure 10.11**.

At this point, we want to solve the Schrödinger equation to determine the electron's energy in the crystal. But before doing this, note that each atomic wavefunction (denoted by the ket $|\phi_a\rangle$) satisfies the Schrödinger equation

$$\hat{H}_0|\phi_a\rangle = E_a|\phi_a\rangle,$$

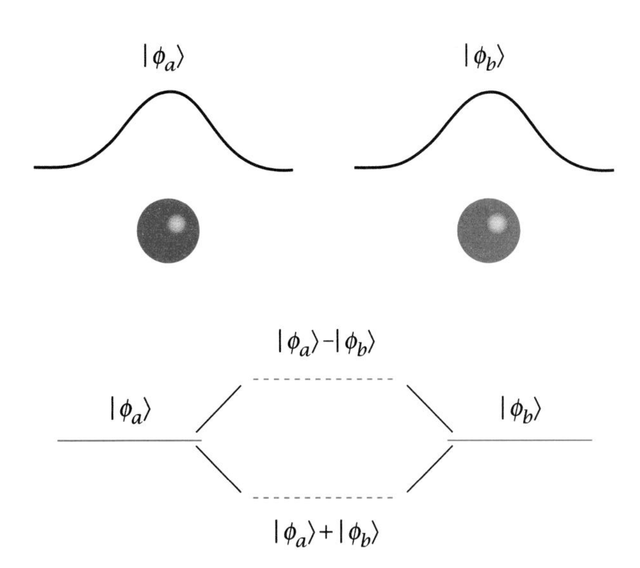

Figure 10.11 Illustration of the linear combination of atomic orbitals in a crystal.

where $\hat{H}_0 = -(\hbar^2/2m_0)\nabla^2 + V(\mathbf{r})$ is the carrier's Hamiltonian in the absence of all other atoms and $|\phi_a\rangle$ is the native atomic wavefunction or orbital.

Next, assume that the Schrödinger equation can be used to describe the wavefunction of the electron throughout the *entire* crystal

$$\hat{H}_{\text{tot}}|\psi\rangle = E|\psi\rangle.$$

Note that E is the energy of the electron in the crystal and that the total Hamiltonian for the electron in the solid is

$$\hat{H}_{\text{tot}} = -\frac{\hbar^2}{2m_0}\nabla^2 + \sum_n^N V(\mathbf{r} - \mathbf{r}_n).$$

Thus, unlike $\hat{H}_0$, it encompasses the full periodic potential of the crystal:

$$V(\mathbf{r}) = \sum_n^N V(\mathbf{r} - \mathbf{r}_n), \tag{10.12}$$

where N represents the total number of unit cells present and n is an integer.

Finally, the electron wavefunction is simply a Bloch-type linear combination of atomic orbitals:

$$|\psi\rangle = \sum_n^N e^{i\mathbf{k}\cdot\mathbf{r}_n}|\phi_a(\mathbf{r} - \mathbf{r}_n)\rangle. \tag{10.13}$$

For convenience, in what follows, we drop the subscript a in our expression since it is understood that these terms just represent atomic orbitals. Furthermore, we use Dirac bra–ket notation (Chapter 6) to make the overall notation simpler.

At this point, let us find our desired energies by multiplying both sides of the Schrödinger equation with $\langle\psi|$ to obtain

$$\langle\psi|\hat{H}_{\text{tot}}|\psi\rangle = E\langle\psi|\psi\rangle$$

or

$$E = \frac{\langle\psi|\hat{H}_{\text{tot}}|\psi\rangle}{\langle\psi|\psi\rangle}. \tag{10.14}$$

We now need an explicit expression for E.

First, let us evaluate the denominator. We have

$$\begin{aligned}\langle\psi|\psi\rangle &= \left[\sum_m^N e^{-i\mathbf{k}\cdot\mathbf{r}_m}\langle\phi(\mathbf{r} - \mathbf{r}_m)|\right]\left[\sum_n^N e^{i\mathbf{k}\cdot\mathbf{r}_n}|\phi(\mathbf{r} - \mathbf{r}_n)\rangle\right] \\ &= \sum_m^N\sum_n^N e^{i\mathbf{k}\cdot(\mathbf{r}_n - \mathbf{r}_m)}\langle\phi(\mathbf{r} - \mathbf{r}_m)|\phi(\mathbf{r} - \mathbf{r}_n)\rangle,\end{aligned}$$

where m and n represent indices for a given atom (note that m should not be confused with a mass here). At this point, we split the sum into two parts: one with the term with $m = n$ and the other containing all terms where $m \neq n$. We thus have

$$\begin{aligned}\langle\psi|\psi\rangle = &\sum_n^N\langle\phi(\mathbf{r} - \mathbf{r}_n)|\phi(\mathbf{r} - \mathbf{r}_n)\rangle \\ &+ \sum_{m\neq n}^N\sum_n^N e^{i\mathbf{k}\cdot(\mathbf{r}_n - \mathbf{r}_m)}\langle\phi(\mathbf{r} - \mathbf{r}_m)|\phi(\mathbf{r} - \mathbf{r}_n)\rangle.\end{aligned}$$

To simplify things, let us assume and invoke the orthonormality of wavefunctions on separate atoms. Namely,

$$\langle \phi(\mathbf{r}-\mathbf{r}_n)|\phi(\mathbf{r}-\mathbf{r}_n)\rangle = 1, \tag{10.15}$$

$$\langle \phi(\mathbf{r}-\mathbf{r}_m)|\phi(\mathbf{r}-\mathbf{r}_n)\rangle = 0. \tag{10.16}$$

This implies that wavefunctions centered primarily about one atom are orthogonal to those on other atoms and that each is normalized.

We therefore have

$$\langle \psi|\psi\rangle = \sum_n^N \langle \phi(\mathbf{r}-\mathbf{r}_n)|\phi(\mathbf{r}-\mathbf{r}_n)\rangle = N,$$

which is our desired denominator.

Next, we deal with the numerator. Here, we have

$$\begin{aligned}\langle \psi|\hat{H}_{\text{tot}}|\psi\rangle &= \left[\sum_m^N e^{-i\mathbf{k}\cdot\mathbf{r}_m}\langle \phi(\mathbf{r}-\mathbf{r}_m)|\right]\hat{H}_{\text{tot}}\left[\sum_n^N e^{i\mathbf{k}\cdot\mathbf{r}_n}|\phi(\mathbf{r}-\mathbf{r}_n)\rangle\right]\\ &= \sum_m^N\sum_n^N e^{i\mathbf{k}\cdot(\mathbf{r}_n-\mathbf{r}_m)}\langle \phi(\mathbf{r}-\mathbf{r}_m)|\hat{H}_{\text{tot}}|\phi(\mathbf{r}-\mathbf{r}_n)\rangle.\end{aligned}$$

We again split the term into two parts: one with $m = n$ and the other containing all terms where $m \neq n$:

$$\begin{aligned}\langle \psi|\hat{H}_{\text{tot}}|\psi\rangle &= \sum_n^N \langle \phi(\mathbf{r}-\mathbf{r}_n)|\hat{H}_{\text{tot}}|\phi(\mathbf{r}-\mathbf{r}_n)\rangle\\ &\quad+\sum_{m\neq n}^N\sum_n^N e^{i\mathbf{k}\cdot(\mathbf{r}_n-\mathbf{r}_m)}\langle \phi(\mathbf{r}-\mathbf{r}_m)|\hat{H}_{\text{tot}}|\phi(\mathbf{r}-\mathbf{r}_n)\rangle.\end{aligned}$$

Note that this system of linear equations can be considered from a matrix perspective where the first part represents elements along the diagonal while the second represents off-diagonal elements.

At this point, let us consider only nearest-neighbor interactions in the latter off-diagonal terms. What this says is that in the tight binding approximation, a wavefunction localized mostly around a given atom will only interact with the wavefunctions of its nearest neighbors. Contributions from carrier wavefunctions localized on atoms further away are not significant and can be ignored. Mathematically, this simplifies the above system of linear equations.

To illustrate, let $\mathbf{R} = \mathbf{r}_n - \mathbf{r}_m$ denote the vector between nearest-neighbor atoms. We can then write the numerator as

$$\begin{aligned}\langle \psi|\hat{H}_{\text{tot}}|\psi\rangle &= \sum_n^N \langle \phi(\mathbf{r}-\mathbf{r}_n)|\hat{H}_{\text{tot}}|\phi(\mathbf{r}-\mathbf{r}_n)\rangle\\ &\quad+\sum_{\mathbf{R}\neq 0}\sum_n^N e^{i\mathbf{k}\cdot\mathbf{R}}\langle \phi(\mathbf{r}-\mathbf{r}_n+\mathbf{R})|\hat{H}_{\text{tot}}|\phi(\mathbf{r}-\mathbf{r}_n)\rangle.\end{aligned}$$

Recall now that the potential can be written as

$$V(\mathbf{r}) = \sum_i^N V(\mathbf{r}-\mathbf{r}_i) = V(\mathbf{r}-\mathbf{r}_n) + \sum_{i\neq n}^N V(\mathbf{r}-\mathbf{r}_i)$$

so that the full Hamiltonian becomes

$$\hat{H}_{tot} = -\frac{\hbar^2}{2m_0}\nabla^2 + V(\mathbf{r} - \mathbf{r}_n) + \sum_{i\neq n}^{N} V(\mathbf{r} - \mathbf{r}_i)$$

or

$$\hat{H}_{tot} = \hat{H}_0 + \sum_{i\neq n}^{N} V(\mathbf{r} - \mathbf{r}_i), \tag{10.17}$$

where $\hat{H}_0$ is the unperturbed atomic orbital Hamiltonian.

Our numerator is then

$$\begin{aligned}\langle\psi|\hat{H}_{tot}|\psi\rangle &= \sum_{n}^{N}\langle\phi(\mathbf{r} - \mathbf{r}_n)|\left[\hat{H}_0 + \sum_{i\neq n}^{N} V(\mathbf{r} - \mathbf{r}_i)\right]|\phi(\mathbf{r} - \mathbf{r}_n)\rangle \\ &+ \sum_{\mathbf{R}\neq 0}\sum_{n}^{N} e^{i\mathbf{k}\cdot\mathbf{R}}\langle\phi(\mathbf{r} - \mathbf{r}_n + \mathbf{R})|\left[\hat{H}_0 + \sum_{i\neq n}^{N} V(\mathbf{r} - \mathbf{r}_i)\right]|\phi(\mathbf{r} - \mathbf{r}_n)\rangle,\end{aligned}$$

whereupon the four resulting terms simplify owing to the orthogonality of wavefunctions. Namely,

$$\begin{aligned}\langle\psi|\hat{H}_{tot}|\psi\rangle &= \sum_{n}^{N}\langle\phi(\mathbf{r} - \mathbf{r}_n)|\hat{H}_0|\phi(\mathbf{r} - \mathbf{r}_n)\rangle \\ &+ \sum_{n}^{N}\langle\phi(\mathbf{r} - \mathbf{r}_n)|\sum_{i\neq n}^{N} V(\mathbf{r} - \mathbf{r}_i)|\phi(\mathbf{r} - \mathbf{r}_n)\rangle \\ &+ \sum_{\mathbf{R}\neq 0}\sum_{n}^{N} e^{i\mathbf{k}\cdot\mathbf{R}}\langle\phi(\mathbf{r} - \mathbf{r}_n + \mathbf{R})|\hat{H}_0|\phi(\mathbf{r} - \mathbf{r}_n)\rangle \\ &+ \sum_{\mathbf{R}\neq 0}\sum_{n}^{N} e^{i\mathbf{k}\cdot\mathbf{R}}\langle\phi(\mathbf{r} - \mathbf{r}_n + \mathbf{R})|\sum_{i\neq n}^{N} V(\mathbf{r} - \mathbf{r}_i)|\phi(\mathbf{r} - \mathbf{r}_n)\rangle,\end{aligned}$$

where the third term disappears because of the previously invoked orthogonality of distant wavefunctions (i.e., $\langle\phi(\mathbf{r} - \mathbf{r}_n + \mathbf{R})|\phi(\mathbf{r} - \mathbf{r}_n)\rangle = 0$). This leaves

$$\begin{aligned}\langle\psi|\hat{H}_{tot}|\psi\rangle &= \sum_{n}^{N} E_a + \sum_{n}^{N}\langle\phi(\mathbf{r} - \mathbf{r}_n)|\sum_{i\neq n}^{N} V(\mathbf{r} - \mathbf{r}_i)|\phi(\mathbf{r} - \mathbf{r}_n)\rangle \\ &+ \sum_{\mathbf{R}\neq 0}\sum_{n}^{N} e^{i\mathbf{k}\cdot\mathbf{R}}\langle\phi(\mathbf{r} - \mathbf{r}_n + \mathbf{R})|\sum_{i\neq n}^{N} V(\mathbf{r} - \mathbf{r}_i)|\phi(\mathbf{r} - \mathbf{r}_n)\rangle.\end{aligned}$$

Recall that E_a is the electron's original atomic orbital energy and that we have assumed wavefunction normalization, namely $\langle\phi(\mathbf{r} - \mathbf{r}_n)|\phi(\mathbf{r} - \mathbf{r}_n)\rangle = 1$. The numerator thus becomes

$$\begin{aligned}\langle\psi|\hat{H}_{tot}|\psi\rangle &= \sum_{n}^{N} E_a + \sum_{n}^{N}\sum_{i\neq n}^{N}\langle\phi(\mathbf{r} - \mathbf{r}_n)|V(\mathbf{r} - \mathbf{r}_i)|\phi(\mathbf{r} - \mathbf{r}_n)\rangle \\ &+ \sum_{n}^{N}\sum_{\mathbf{R}\neq 0}\sum_{i\neq n}^{N} e^{i\mathbf{k}\cdot\mathbf{R}}\langle\phi(\mathbf{r} - \mathbf{r}_n + \mathbf{R})|V(\mathbf{r} - \mathbf{r}_i)|\phi(\mathbf{r} - \mathbf{r}_n)\rangle.\end{aligned}$$

Since the system exhibits translational symmetry, we let $\mathbf{r}_n \to 0$ (i.e., we center everything about the origin) to obtain

$$\langle\psi|\hat{H}_{\text{tot}}|\psi\rangle = NE_a + N\sum_{i\neq 0}^{N}\langle\phi(\mathbf{r})|V(\mathbf{r}-\mathbf{r}_i)|\phi(\mathbf{r})\rangle$$

$$+ N\sum_{\mathbf{R}\neq 0} e^{i\mathbf{k}\cdot\mathbf{R}}\sum_{i\neq 0}^{N}\langle\phi(\mathbf{r}+\mathbf{R})|V(\mathbf{r}-\mathbf{r}_i)|\phi(\mathbf{r})\rangle.$$

At this point, let

$$-A = \sum_{i\neq 0}^{N}\langle\phi(\mathbf{r})|V(\mathbf{r}-\mathbf{r}_i)|\phi(\mathbf{r})\rangle$$

$$-B = \sum_{i\neq 0}^{N}\langle\phi(\mathbf{r}+\mathbf{R})|V(\mathbf{r}-\mathbf{r}_i)|\phi(\mathbf{r})\rangle$$

to simplify our expression and to highlight that these interactions with other states lower the overall energy of the state in question. The resulting numerator is then

$$\langle\psi|\hat{H}_{\text{tot}}|\psi\rangle = NE_a - NA - N\sum_{\mathbf{R}\neq 0} Be^{i\mathbf{k}\cdot\mathbf{R}}.$$

Summary

When everything is put together, we find that

$$E = \frac{NE_a - NA - N\sum_{\mathbf{R}\neq 0} Be^{i\mathbf{k}\cdot\mathbf{R}}}{N},$$

giving our desired tight binding approximation for the electron's energy in a solid:

$$E = E_a - A - \sum_{\mathbf{R}\neq 0} Be^{i\mathbf{k}\cdot\mathbf{R}}. \tag{10.18}$$

This is a general result that can be applied to find the energies of an electron in a crystal, possessing a given unit cell, or, even more simply, to a linear chain of atoms much like in our earlier discussion of the Kronig–Penney model.

For comparison purposes, let us therefore consider this latter one-dimensional case. In a linear chain, each atom only has two nearest neighbors. As a consequence, the only values of R to consider, assuming that things are centered about the origin, are $R = a$ and $R = -a$, where a represents the lattice constant. The electron energy is therefore

$$E = E_a - A - (Be^{ika} + Be^{-ika}),$$

or, alternatively,

$$E = E_a - A - 2B\cos ka. \tag{10.19}$$

Finally, just like in the Kronig–Penney model, we can use these band energies to find the electron or hole effective mass as well as their associated density of states. For the former, recall that an expression for the effective mass was

$$\frac{1}{m_{\text{eff}}} = \frac{1}{\hbar^2}\left(\frac{d^2E}{dk^2}\right).$$

Thus, on applying our tight binding result (Equation 10.19) with $k = 0$, we find

$$\frac{1}{m_{\text{eff}}} = \frac{2Ba^2}{\hbar^2}. \tag{10.20}$$

The reader may verify this as an exercise.

We can also obtain the associated carrier density of states by recalling that (Equation 9.9)

$$\rho_{\text{energy}}(E) = \frac{2}{\pi}\left(\frac{dk}{dE}\right).$$

Since $E = E_a - A - 2B\cos ka$, we obtain

$$\frac{dk}{dE} = \frac{1}{2Ba\sin ka}.$$

This, in turn, yields our desired density of states,

$$\rho_{\text{energy}} = \frac{1}{\pi Ba}\frac{1}{\sin ka}. \tag{10.21}$$

As an exercise, the reader may plot the above expression, noting the associated van Hove singularities.

10.4.2 The Nearly Free Electron Model

Our next model represents the opposite limit of the tight binding approximation. Namely, it assumes that the electron is nearly free in the crystal although it experiences a slight perturbation in energy due to the presence of a periodic potential. As we will see, the model, like the tight binding and Kronig–Penney models before it, predicts the existence of bands and band gaps.

We start with the time-independent Schrödinger equation written in bra–ket notation:

$$\hat{H}_{\text{tot}}|\psi(\mathbf{r})\rangle = E|\psi(\mathbf{r})\rangle.$$

$\hat{H}_{\text{tot}}$ is the total Hamiltonian for the electron in the crystal, E is its associated energy, and $|\psi(\mathbf{r})\rangle$ is the carrier's full wavefunction.

This time, instead of describing the potential as a linear combination of distinct atomic potentials, let us simply describe it as a Fourier series owing to its periodicity.

We then have

$$V(\mathbf{r}) = \sum_{\mathbf{G}} V_{\mathbf{G}} e^{i\mathbf{G}\cdot\mathbf{r}}, \tag{10.22}$$

where $\mathbf{G}$ are reciprocal lattice vectors in k-space and $V_{\mathbf{G}}$ are Fourier coefficients reflecting the "weight" of each plane wave contribution. Now, ultimately, we assume that our potential is well defined and is highly periodic. As a consequence, mathematically all this means is that many of the $V_{\mathbf{G}}$ coefficients in our sum are zero. In the limiting case where we have a single perfect sinusoidal potential, all but one $V_{\mathbf{G}}$ value will be zero. But let us invoke this argument later.

Now, for the wavefunction, we previously assumed in the tight binding approximation that ψ could be described as a linear combination of atomic orbitals. This time, we will not do this and will simply assert that the wavefunction, whatever it is, is highly periodic. As a consequence, just as with the potential, we can describe the wavefunction using a Fourier series. We have

$$\psi(\mathbf{r}) = \sum_{\mathbf{k}} C_{\mathbf{k}} e^{i\mathbf{k}\cdot\mathbf{r}}, \qquad (10.23)$$

where $|\mathbf{k}| = k = 2n\pi/Na = 2n\pi/L$, n is an integer, N is the total number of unit cells in the crystal, a is the lattice constant, $Na = L$ reflects the length of the finite crystal, and $C_{\mathbf{k}}$ are plane wave weights.

From a conceptual standpoint, one can think of this linear combination of plane waves as a linear combination of different states each represented by an energy dictated by a given k value. This will be important for our eventual perturbation theory analysis, since we will eventually suggest that the perturbation is small such that only states close to each other in energy (i.e., states with similar but not identical k values) interact significantly.

Inserting everything into the Schrödinger equation gives

$$\left[-\frac{\hbar^2}{2m_0}\nabla^2 + \sum_{\mathbf{G}} V_{\mathbf{G}} e^{i\mathbf{G}\cdot\mathbf{r}}\right] \sum_{\mathbf{k}} C_{\mathbf{k}} e^{i\mathbf{k}\cdot\mathbf{r}} = E \sum_{\mathbf{k}} C_{\mathbf{k}} e^{i\mathbf{k}\cdot\mathbf{r}}$$

$$\sum_{\mathbf{k}} \left(\frac{\hbar^2 k^2}{2m_0}\right) C_{\mathbf{k}} e^{i\mathbf{k}\cdot\mathbf{r}} + \left[\sum_{\mathbf{G}} V_{\mathbf{G}} e^{i\mathbf{G}\cdot\mathbf{r}}\right]\left[\sum_{\mathbf{k}} C_{\mathbf{k}} e^{i\mathbf{k}\cdot\mathbf{r}}\right] = E \sum_{\mathbf{k}} C_{\mathbf{k}} e^{i\mathbf{k}\cdot\mathbf{r}}$$

$$\sum_{\mathbf{k}} \left[\left(\frac{\hbar^2 k^2}{2m_0}\right) - E\right] C_{\mathbf{k}} e^{i\mathbf{k}\cdot\mathbf{r}} + \sum_{\mathbf{G}} \sum_{\mathbf{k}} V_{\mathbf{G}} C_{\mathbf{k}} e^{i(\mathbf{G}+\mathbf{k})\cdot\mathbf{r}} = 0,$$

whereupon the last term can be rearranged as follows, since the sum is over all **G** and **k** values. This is essentially akin to changing variables by letting $\mathbf{k} \to \mathbf{k} - \mathbf{G}$. We thus obtain

$$\sum_{\mathbf{k}} \left(\frac{\hbar^2 k^2}{2m_0} - E\right) C_{\mathbf{k}} e^{i\mathbf{k}\cdot\mathbf{r}} + \sum_{\mathbf{G}} \sum_{\mathbf{k}} V_{\mathbf{G}} C_{\mathbf{k}-\mathbf{G}} e^{i\mathbf{k}\cdot\mathbf{r}} = 0.$$

Next, multiplying both sides by $e^{i\mathbf{k}'\cdot\mathbf{r}}$, where $\mathbf{k}'$ is another reciprocal lattice vector, and integrating over all space results in

$$\sum_{\mathbf{k}} \left[\left(\frac{\hbar^2 k^2}{2m_0} - E\right) C_{\mathbf{k}} + \sum_{\mathbf{G}} V_{\mathbf{G}} C_{\mathbf{k}-\mathbf{G}}\right] = 0, \qquad (10.24)$$

since the only nonzero terms occur when $\mathbf{k}' = \mathbf{k}$.

Equation 10.24 is essentially a long series of linear equations in the coefficients $C_{\mathbf{k}}$, which describe the wavefunction of interest. In this regard, one may alternatively think of it in matrix form where the first term within the square brackets represents the unperturbed energies of the various states and the second represents contributions to their energy by other states with differing energies (i.e., with different k values). The second term is therefore our perturbation, which we will assume is small.

Let us demonstrate this matrix view explicitly. For simplicity, consider only three states and furthermore assume at this point that all $V_{\mathbf{G}}$ terms are zero. We therefore have from Equation 10.24 that

$$\sum_{\mathbf{k}} \left(\frac{\hbar^2 k^2}{2m_0} - E\right) C_{\mathbf{k}} = 0.$$

To solve this series of linear equations, the determinant of the array described by $(\hbar^2 k^2/2m_0) - E$ must be zero. To provide a visual impression of what this looks like, let us write out the individual equations for k_1, k_2, and k_3:

Written out, the expression looks like

$$\left(\frac{\hbar^2 k_1^2}{2m_0} - E\right) C_{\mathbf{k}_1} = 0,$$

$$\left(\frac{\hbar^2 k_2^2}{2m_0} - E\right) C_{\mathbf{k}_2} = 0,$$

$$\left(\frac{\hbar^2 k_3^2}{2m_0} - E\right) C_{\mathbf{k}_3} = 0.$$

In matrix form, one writes

$$\begin{pmatrix} \frac{\hbar^2 k_1^2}{2m_0} - E & 0 & 0 \\ 0 & \frac{\hbar^2 k_2^2}{2m_0} - E & 0 \\ 0 & 0 & \frac{\hbar^2 k_3^2}{2m_0} - E \end{pmatrix} \begin{pmatrix} C_{\mathbf{k}_1} \\ C_{\mathbf{k}_2} \\ C_{\mathbf{k}_3} \end{pmatrix} = \begin{pmatrix} 0 \\ 0 \\ 0 \end{pmatrix},$$

which we solve to obtain the desired energies that make the expression true. The associated weights $C_{\mathbf{k}}$ can also be found. This is then nothing more than a standard eigenvalue/eigenvector problem. Since the above coefficients are not all zero (the trivial solution), the determinant of the first array equals zero:

$$\begin{vmatrix} \frac{\hbar^2 k_1^2}{2m_0} - E & 0 & 0 \\ 0 & \frac{\hbar^2 k_2^2}{2m_0} - E & 0 \\ 0 & 0 & \frac{\hbar^2 k_3^2}{2m_0} - E \end{vmatrix} = 0.$$

From this, we recover the free particle energies of the three states described by $\mathbf{k}_1$, $\mathbf{k}_2$, and $\mathbf{k}_3$:

$$E_1 = \frac{\hbar^2 k_1^2}{2m_0},$$

$$E_2 = \frac{\hbar^2 k_2^2}{2m_0},$$

$$E_3 = \frac{\hbar^2 k_3^2}{2m_0}.$$

At this point, let us return to the more interesting and realistic case where $V_{\mathbf{G}} \neq 0$. Using our three-state example, we then obtain from Equation 10.24 the following:

$$\left(\frac{\hbar^2 k_1^2}{2m_0} - E\right) C_{\mathbf{k}_1} + V_{11} C_{\mathbf{k}_1} + V_{12} C_{\mathbf{k}_2} + V_{13} C_{\mathbf{k}_3} = 0,$$

$$\left(\frac{\hbar^2 k_2^2}{2m_0} - E\right) C_{\mathbf{k}_2} + V_{21} C_{\mathbf{k}_1} + V_{22} C_{\mathbf{k}_2} + V_{23} C_{\mathbf{k}_3} = 0,$$

$$\left(\frac{\hbar^2 k_3^2}{2m_0} - E\right) C_{\mathbf{k}_3} + V_{31} C_{\mathbf{k}_1} + V_{32} C_{\mathbf{k}_2} + V_{33} C_{\mathbf{k}_3} = 0.$$

Here V_{11}, V_{22}, and V_{33} represent the potential contributions (V_G) due to states 1, 2, and 3. Note also that

- $V_{12} = V_{21}$ represents the potential due to an interaction between states 1 and 2.
- $V_{13} = V_{31}$ represents the potential due to an interaction between states 1 and state 3.
- $V_{23} = V_{32}$ represents the potential due to an interaction between states 2 and state 3.

The resulting matrix expression is then

$$\left[\begin{pmatrix} \frac{\hbar^2 k_1^2}{2m_0} - E & 0 & 0 \\ 0 & \frac{\hbar^2 k_2^2}{2m_0} - E & 0 \\ 0 & 0 & \frac{\hbar^2 k_3^2}{2m_0} - E \end{pmatrix} + \begin{pmatrix} V_{11} & V_{12} & V_{13} \\ V_{21} & V_{22} & V_{23} \\ V_{31} & V_{32} & V_{33} \end{pmatrix}\right]\begin{pmatrix} C_{\mathbf{k}_1} \\ C_{\mathbf{k}_2} \\ C_{\mathbf{k}_3} \end{pmatrix} = \begin{pmatrix} 0 \\ 0 \\ 0 \end{pmatrix},$$

or, alternatively,

$$\begin{pmatrix} \frac{\hbar^2 k_1^2}{2m_0} - E + V_{11} & V_{12} & V_{13} \\ V_{21} & \frac{\hbar^2 k_2^2}{2m_0} - E + V_{22} & V_{23} \\ V_{31} & V_{32} & \frac{\hbar^2 k_3^2}{2m_0} - E + V_{33} \end{pmatrix}\begin{pmatrix} C_{\mathbf{k}_1} \\ C_{\mathbf{k}_2} \\ C_{\mathbf{k}_3} \end{pmatrix} = \begin{pmatrix} 0 \\ 0 \\ 0 \end{pmatrix}.$$

For convenience, let $V_{11} = V_{22} = V_{33} = 0$, since these potentials only lead to a constant offset for the energies of states 1, 2, and 3. We therefore find

$$\begin{pmatrix} \frac{\hbar^2 k_1^2}{2m_0} - E & V_{12} & V_{13} \\ V_{21} & \frac{\hbar^2 k_2^2}{2m_0} - E & V_{23} \\ V_{31} & V_{32} & \frac{\hbar^2 k_3^2}{2m_0} - E \end{pmatrix}\begin{pmatrix} C_{\mathbf{k}_1} \\ C_{\mathbf{k}_2} \\ C_{\mathbf{k}_3} \end{pmatrix} = \begin{pmatrix} 0 \\ 0 \\ 0 \end{pmatrix}.$$

Note from before that the off-diagonal elements V_{12}, V_{13}, etc. represent contributions to the energy of a given state from interactions with other states. The magnitudes of these contributions, from a perturbation theory perspective (Chapter 6), then depend on the energy difference between the states involved. Thus for state 1, we see that V_{12} and V_{13} are weighted by $1/(E_{k_1} - E_{k_2})$ and $1/(E_{k_1} - E_{k_3})$, respectively.

At this point, let us assume that states 1 and 2 are much further away, energetically, from state 3. This allows us to ignore its contributions to their energies. As a consequence, we set $V_{13} = V_{31} = 0$ and $V_{23} = V_{32} = 0$

in our matrix expression. This leads to

$$\begin{pmatrix} \frac{\hbar^2 k_1^2}{2m_0} - E & V_{12} & 0 \\ V_{21} & \frac{\hbar^2 k_2^2}{2m_0} - E & 0 \\ 0 & 0 & \frac{\hbar^2 k_3^2}{2m_0} - E \end{pmatrix} \begin{pmatrix} C_{\mathbf{k}_1} \\ C_{\mathbf{k}_2} \\ C_{\mathbf{k}_3} \end{pmatrix} = \begin{pmatrix} 0 \\ 0 \\ 0 \end{pmatrix}.$$

We then solve the following determinant:

$$\begin{vmatrix} \frac{\hbar^2 k_1^2}{2m_0} - E & V_{12} & 0 \\ V_{21} & \frac{\hbar^2 k_2^2}{2m_0} - E & 0 \\ 0 & 0 & \frac{\hbar^2 k_3^2}{2m_0} - E \end{vmatrix} = 0$$

to obtain E. This 3×3 determinant is easy to evaluate since it turns out to be block-diagonal. Thus, in practice, all we need to do is solve the smaller 2×2 subdeterminant found in the upper left quadrant. That is, we must find the solutions to

$$\begin{vmatrix} \frac{\hbar^2 k_1^2}{2m_0} - E & V_{12} \\ V_{21} & \frac{\hbar^2 k_2^2}{2m_0} - E \end{vmatrix} = 0,$$

which will, in turn, give us the energies for states 1 and 2. For notational convenience, let $V_{12} = V_{21} = V$, $E_1 = \hbar^2 k_1^2/2m_0$ and $E_2 = \hbar^2 k_2^2/2m_0$. We then obtain the following equation:

$$(E_1 - E)(E_2 - E) - V^2 = 0,$$

which further reduces to

$$E^2 - (E_1 + E_2)E + (E_1 E_2 - V^2) = 0.$$

Finally, applying the familiar solution to a quadratic equation gives

$$E = \left(\frac{E_1 + E_2}{2}\right) \pm \frac{\sqrt{(E_1 + E_2)^2 - 4(E_1 E_2 - V^2)}}{2}$$

and ultimately yields the electron's energy in either state 1 or 2:

$$E = \left(\frac{E_1 + E_2}{2}\right) \pm \sqrt{\left(\frac{E_1 - E_2}{2}\right)^2 + V^2}. \tag{10.25}$$

Note that on examining the resulting expression, we find, for the case where we can ignore V, that the resulting energies are

$$E \simeq \left(\frac{E_1 + E_2}{2}\right) \pm \left(\frac{E_1 - E_2}{2}\right),$$

which reduce to E_1 or E_2. These are simply the free particle solutions shown earlier. If we do not completely ignore V, we can obtain electron energies, corrected to second order, using second-order nondegenerate perturbation theory (see Chapter 6 for more details). The resulting energies of states 1 and 2 are therefore

$$E_1 \simeq E_1^0 + \frac{|V|^2}{E_1^0 - E_2^0}, \tag{10.26}$$

$$E_2 \simeq E_2^0 + \frac{|V|^2}{E_2^0 - E_1^0}, \tag{10.27}$$

where we again see that the potential contribution is weighted by the inverse energy difference between interacting states (E_1^0 and E_2^0 are the unperturbed energies of the first and second states, respectively). In addition, because of the sign difference between expressions, the energies of each state change in opposite directions.

Near a degeneracy between E_1 and E_2, the potential term dominates within the square root term of Equation 10.25 and leads to approximate energies for either state of

$$E \simeq \left(\frac{E_1 + E_2}{2}\right) \pm V. \tag{10.28}$$

What this means is that an energy gap of magnitude $2V$ opens up between states. This gap has previously been seen in **Figures 10.5** and **10.6** and is ultimately the energy gap between bands that develops owing to the weak interaction between states in a crystal.

10.5 METALS, SEMICONDUCTORS, AND INSULATORS

Having established the concept of bands, we now distinguish metals from semiconductors and insulators as follows. Namely, when all of the electrons in a solid occupy the different electronic states available to them, metals find themselves with only partially filled bands. As a consequence, there are many empty states available for these carriers and no band gap exists. By contrast, both semiconductors and insulators have completely filled bands at 0 K and, as a consequence, exhibit band gap effects in their optical and electrical properties.

To illustrate, at higher temperatures, thermal energy can promote the transition of an electron from a full band (the valence band) to the next lowest-lying, empty (conduction) band. The probability that this occurs depends on the magnitude of the system's band gap and, in turn, helps to distinguish semiconductors from insulators. Namely, semiconductors generally have smaller gaps than insulators, making the event more likely. For example, the band gap of CdSe, a binary semiconductor, is 1.74 eV at room temperature while that of an insulator such as TiO_2 is 3.2 eV. As a consequence, insulators are essentially nonconducting whereas semiconductors are slightly better conductors. Absorption edges of the former also occur much further to the blue, towards the ultraviolet, whereas semiconductor absorption edges are often in the visible or near-infrared.

10.6 SUMMARY

With this, we conclude our introduction to bands in crystalline materials. We now move on to a quantum mechanical description of the interaction between carriers in a solid and light, resulting in absorption and emission processes in semiconductors. This is followed by a detailed description of interband transitions in Chapter 12.

10.7 THOUGHT PROBLEMS

10.1 Plots

Consider the rectangular barrier Kronig–Penney model. Use Mathcad, Matlab, Mathematica, or your favorite mathematical modeling program to visualize the actual bands. As a starting point, choose a barrier height of 5 eV, an atomic spacing $a = 5$ Å, and a barrier width $b = 0.1a$. Plot the free electron energies on top of the bands you have drawn. Show the plot in both the periodic zone scheme and the reduced zone scheme.

10.2 Plots

Consider the δ-function modification of the Kronig–Penney model. Again use a mathematical modeling program to draw the bands in the periodic and reduced zone schemes. As a starting point, choose $V_0 = 5$ eV Å and $a = 5$ Å.

10.3 Band gap

What temperature in kelvin is necessary to promote an electron from a semiconductor's valence band to its conduction band given a band gap $E_g = 1.17$ eV? How about for a semiconductor with a band gap $E_g = 2.5$ eV? How about for an insulator with a gap $E_g = 5$ eV.

10.4 Band gap

For bulk, direct gap, semiconductors it is often common to plot α^2 versus E (α is the absorption coefficient of the semiconductor which is proportional to its joint density of states). In this manner, the bulk band gap E_g of the system can be estimated. Show why this is the case.

10.5 Excess energy

Show for a photon with energy $E_{h\nu}$ ($E_{h\nu} > E_g$) absorbed by a semiconductor, that the excess energy $E_{h\nu} - E_g$ is partitioned among the electron and hole as follows. For the electron,

$$\Delta E_e = (E_{h\nu} - E_g)\left(\frac{1}{1 + m_e/m_h}\right),$$

and for the hole,

$$\Delta E_h = (E_{h\nu} - E_g)\left(\frac{1}{1 + m_h/m_e}\right).$$

10.6 Kronig–Penney, model: δ-function barriers

Starting from the following expression derived for the finite width barrier problem,

$$\frac{\beta^2 - \alpha^2}{2\alpha\beta} \sinh \beta b \sin \alpha a + \cosh \beta b \cos \alpha a = \cos k(a + b),$$

with $\alpha = \sqrt{2m_0 E/\hbar^2}$ and $\beta = \sqrt{2m_0(V - E)/\hbar^2}$, derive the results of the δ-function barrier approximation.

10.7 Tight binding approximation

Show that the wavefunction used in the tight binding approximation,

$$\psi(r + a) = e^{ika}\psi(r),$$

satisfies Bloch's theorem.

10.8 Tight binding approximation

Consider a Taylor expansion of the band energy about an extremum in k-space, i.e.,

$$E(k) = E_0 + \left.\frac{dE}{dk}\right|_{k=k_0} (k - k_0) + \frac{1}{2}\left.\frac{d^2E}{dk^2}\right|_{k=k_0} (k - k_0)^2 + \cdots.$$

For small values of $k - k_0$, the higher-order terms can be ignored. Confirm that the band energy obtained from the one-dimensional tight binding model exhibits an extremum about $k = 0$ and yields the correct parabolic E versus k energy dependence.

10.9 Tight binding approximation: SC lattice

Show from the tight binding approximation that the energy of an electron in a solid with a simple cubic unit cell can be expressed as

$$E(k) \simeq E_a - A - 2B(\cos k_x a + \cos k_y a + \cos k_z a),$$

where a is the lattice constant. Then show that near $k_x = k_y = k_z = 0$,

$$E(k) \simeq E_a - A - 6B + Ba^2k^2,$$

with $k^2 = k_x^2 + k_y^2 + k_z^2$.

10.10 Tight binding approximation: BCC lattice

Using the tight binding approximation, show that the energy of an electron in a BCC lattice is given by

$$E = E_a - A - 8B \cos\frac{k_x a}{2} \cos\frac{k_y a}{2} \cos\frac{k_z a}{2}.$$

Hint: There are eight nearest neighbors in the BCC lattice. It might also help to consider a change in the origin of the unit cell coordinate system.

10.11 Tight binding approximation: FCC lattice

Using the tight binding approximation, show that the energy of an electron in a FCC lattice is given by

$$E = E_a - A - 4B\left(\cos\frac{k_x a}{2}\cos\frac{k_y a}{2} + \cos\frac{k_x a}{2}\cos\frac{k_z a}{2} + \cos\frac{k_y a}{2}\cos\frac{k_z a}{2}\right).$$

Hint: There are 12 nearest-neighbor atoms in the FCC lattice.

10.12 Tight binding approximation: effective mass

Consider the band energy obtained from the one-dimensional tight binding approximation. Show that within the first Brillouin zone, the tight binding model readily predicts a positive effective mass near the bottom of a band and a negative effective mass near the top. Plot the general shape of the effective mass relationship as a function of k.

10.13 Tight binding approximation: effective mass

Using the one-dimensional tight binding approximation band energy, find the value of B in eV that leads to the appropriate value for the electron effective mass in GaAs ($m_e = 0.067m_0$). The lattice constant of GaAs is $a = 5.65$ Å.

10.14 Tight binding approximation: effective mass

In general, within a rectangular three-dimensional lattice with distances a, b, and c between nearest-neighbor atoms along the x, y, and z directions, the electron energy can be written as

$$E(k_x, k_y, k_z) = E_a - A - 2B_x \cos k_x a - 2B_y \cos k_y b - 2B_z \cos k_z c.$$

From this, one sees that, in general, effective masses will differ depending on direction. Thus, the reciprocal of the effective mass is actually a tensor quantity. To illustrate, the effective mass in the x direction is

$$m_x = \frac{\hbar^2}{\partial^2 E/\partial k_x^2},$$

while it is

$$m_y = \frac{\hbar^2}{\partial^2 E/\partial k_y^2}$$

in the y direction and

$$m_z = \frac{\hbar^2}{\partial^2 E/\partial k_z^2}$$

in the z direction. There are also other terms such as

$$m_{xy} = \frac{\hbar^2}{\partial^2 E/\partial k_x \partial k_y}.$$

For illustration purposes, find m_x, m_y, and m_z. Show also that in a rectangular lattice,

$$\frac{1}{m_{xy}} = \frac{1}{m_{yx}} = \frac{1}{m_{xz}} = \frac{1}{m_{zx}} = \frac{1}{m_{yz}} = \frac{1}{m_{zy}}.$$

10.15 Tight binding approximation: effective mass

Given the effective mass tensor

$$\frac{1}{m_{\text{eff}}} = \begin{pmatrix} \alpha_{xx} & 0 & \alpha_{xz} \\ 0 & \alpha_{yy} & 0 \\ \alpha_{xz} & 0 & \alpha_{zz} \end{pmatrix},$$

find the associated energy $E(k_x, k_y, k_z)$.

10.8 REFERENCES

Kronig R de L, Penney WG (1931) Quantum mechanics of electrons in crystal lattices. *Proc. R. Soc. Lond. A* **130**, 499.

Wolfe JC (1978) Summary of the Kronig–Penney electron. *Am. J. Phys.* **46**, 1012.

10.9 FURTHER READING

General

Ashcroft NW, Mermin ND (1976) *Solid State Physics.* Brooks/Cole, Belmont, CA.

Kittel C (2005) *Introduction to Solid State Physics*, 8th edn. Wiley, New York.

Singleton J (2001) *Band Theory and Electronic Properties of Solids.* Oxford University Press, Oxford, UK.

Linear Algebra

Arfken GB, Weber HJ (2005) *Mathematical Methods for Physicists*, 6th edn. Academic Press, San Diego.

Kreyszig E (2011) *Advanced Engineering Mathematics*, 10th edn. Wiley, New York.

McQuarrie DA (2003) *Mathematical Methods for Scientists and Engineers*, University Science Books, Sausalito, CA.

Rectangular Barrier Kronig–Penney Model

Brennan KF (1999) *The Physics of Semiconductors: With Applications to Optoelectronic Devices.* Cambridge University Press, Cambridge, UK.

McQuarrie DA (1996) The Kronig–Penney model: a single lecture illustrating the band structure of solids. *Chem. Ed.* **1**, 1.

Delta-Function Barrier Kronig–Penney Model

Dominguez-Adame F (1987) Relativistic and nonrelativistic Kronig–Penney models. *Am. J. Phys.* **55**, 1003.

Wisey ND, Goodman RR (1967) Optical absorption in a Kronig–Penney semiconductor. *Am. J. Phys.* **35**, 35.

Tight Binding and Nearly Free Electron Model

Davies JH (1998) *The Physics of Low-Dimensional Semiconductors: An Introduction.* Cambridge University Press, Cambridge, UK.

Rosencher E, Vinter B (2002) *Optoelectronics.* Cambridge University Press, Cambridge, UK.

Other Band Model and $k \cdot p$ References

Chuang SL (1995) *Physics of Optoelectronic Devices.* Wiley, New York.

Harrison P (2010) *Quantum Wells, Wires and Dots*, 3rd edn. Wiley, Hoboken, NJ.

Wolfe CM, Holonyak N Jr, Stillman GE (1989) *Physical Properties of Semiconductors.* Prentice Hall, Englewood Cliffs, NJ.

Time-Dependent Perturbation Theory

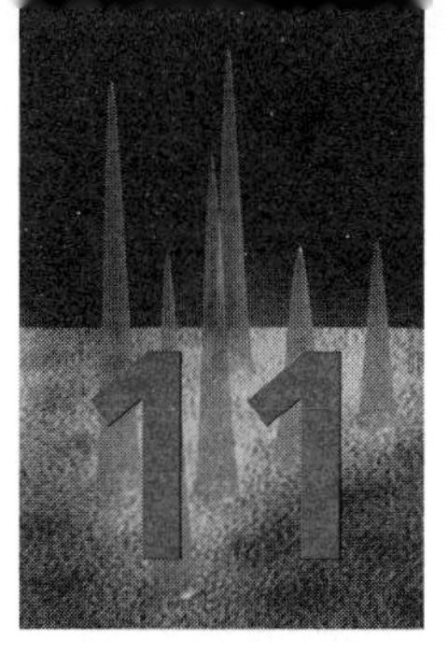

11.1 INTRODUCTION

When treating the optical properties of nanostructures, one often deals with the absorption and emission of light. We can model these processes using results from our density-of-states calculation in Chapter 9 and what is called time-dependent perturbation theory. Specifically, it will be shown in this chapter that the transition rate between electronic states is given by Fermi's (second) Golden Rule, often called the "Golden Rule" for simplicity. It is a quantum mechanical version of a classical rate.

Specifically, for a given initial state (i) and a final state (f), we find that the transition rate is given by

$$R_{i\to f} = \frac{2\pi}{\hbar}|M|^2\rho(E), \tag{11.1}$$

where $|M|$ is a matrix element with constants, linking the initial and final states, and $\rho(E)$ is the density of available final states. Note that this is where our previous density-of-states calculation (Chapter 9) begins to pay dividends.

In what follows, we will work in what is referred to as the Schrödinger representation of quantum mechanics. For future reference, there are three "pictures" of quantum mechanics: the Schrödinger representation, the Heisenberg representation, and the interaction representation. When doing time-dependent perturbation theory, the interaction representation is generally preferred. However, we obtain the same results working in the Schrödinger representation, which we have been using all along. The interested reader may consult more detailed quantum mechanics texts such as those in Further Reading for additional information about these other representations.

11.1.1 Preliminaries

The Time-Dependent Schrödinger Equation

Recall from Chapter 6 that the time-dependent Schrödinger equation is written as

$$i\hbar\frac{\partial}{\partial t}\psi(\mathbf{r},t) = \hat{H}\psi(\mathbf{r},t), \tag{11.2}$$

where $\hat{H}$ is the Hamiltonian,

$$\hat{H} = -\frac{\hbar^2}{2m}\nabla^2 + V$$

and the potential V (for now) is time-independent. To solve the Schrödinger equation, we look for separable solutions of the form

$$\psi(\mathbf{r}, t) = \psi(\mathbf{r}) f(t) \equiv \psi f.$$

The last equivalence simply makes our notation simpler in what follows. We replace this into the above time-dependent Schrödinger equation and simplify to obtain

$$\begin{aligned} i\hbar \frac{\partial}{\partial t}(\psi f) &= \hat{H} \psi f \\ &= \left(-\frac{\hbar^2}{2m} \nabla^2 + V \right) \psi f. \end{aligned}$$

When the t versus $\mathbf{r}$ dependences of ψ and f are considered, we obtain

$$\psi i\hbar \frac{df}{dt} = f \left(\frac{-\hbar^2}{2m} \right) \nabla^2 \psi + V \psi f.$$

Dividing by ψf then gives

$$i\hbar \frac{1}{f} \frac{df}{dt} = \frac{1}{\psi} \left(-\frac{\hbar^2}{2m} \nabla^2 \psi + V \psi \right).$$

Note that the left-hand side of this expression depends only on time while the right-hand side depends only on position. This is true for any value of t or position. As a consequence, both sides must equal a constant. We call it E in anticipation of the fact that it is our particle's energy.

Evaluating the Left-Hand Side

We have

$$i\hbar \frac{1}{f} \frac{df}{dt} = E,$$

which we rearrange to

$$\frac{df}{f} = -\frac{iE}{\hbar} dt.$$

This is then integrated to obtain

$$f(t) = Ce^{-iEt/\hbar},$$

where C is a constant.

Evaluating the Right-Hand Side

We have

$$\frac{1}{\psi} \left(-\frac{\hbar^2}{2m} \nabla^2 \psi + V \psi \right) = E,$$

which is rearranged to

$$-\frac{\hbar^2}{2m} \nabla^2 \psi + V \psi = E \psi$$

and is the usual form of the Schrödinger equation we have solved many times in the past (i.e., in Chapters 7, 8, and 10), yielding $\psi(\mathbf{r})$ and E.

When everything is put together, the general form of the total wavefunction is

$$\psi(\mathbf{r}, t) = \psi(\mathbf{r})e^{-iEt/\hbar}$$

and is referred to as a stationary state since its probability density, $\psi^*\psi$, is time-independent. The reader may verify this as an exercise.

11.2 TIME-DEPENDENT PERTURBATION THEORY

Now, the basic idea behind time-dependent perturbation theory is relatively straightforward. First, if a time-dependent perturbation $V(t)$ (later, we will denote this by $\hat{H}^{(1)}(t)$) is absent, the eigenfunctions and eigenvalues of the particle are obtained by solving the standard time-independent Schrödinger equation

$$\hat{H}^{(0)}\psi^{(0)} = E^{(0)}\psi^{(0)}.$$

Note the slight notational change. The use of the superscript (0) refers to the unperturbed zeroth-order Hamiltonian wavefunctions and energies. We saw this earlier in Chapter 6.

Next, a general concept in quantum mechanics is that any state of a particle being modeled can be expressed as a linear combination of these zeroth-order states, known as basis states (Chapter 6). We denote this as follows:

$$\psi(\mathbf{r}, t = 0) = \sum_n c_n e^{-iE_n^{(0)}t/\hbar}\psi_n^{(0)}(\mathbf{r}),$$

where c_n are constant weights, reflecting the contribution of a given basis state to the total wavefunction.

In the absence of a perturbation, we have for all times the same c_n such that:

$$\psi(\mathbf{r}, t > 0) = \sum_n c_n e^{-iE_n^{(0)}t/\hbar}\psi_n^{(0)}(\mathbf{r}).$$

However, in the presence of a perturbation and, in this case a time-dependent one, the above linear combination is no longer a valid solution to the Schrödinger equation. We can still express the solution as a linear combination of zeroth-order basis states, but now the coefficients become time-dependent. We denote this by

$$\begin{aligned}\psi(\mathbf{r}, t) &= \sum_n c_n(t)e^{-iE_n^{(0)}t/\hbar}\psi_n^{(0)}(\mathbf{r}) \\ &= \sum_n c_n(t)\psi_n^{(0)}(\mathbf{r}, t).\end{aligned}$$

This wavefunction is then inserted into the general time-dependent Schrödinger equation

$$i\hbar\frac{\partial\psi(\mathbf{r}, t)}{\partial t} = \hat{H}\psi(\mathbf{r}, t),$$

where the Hamiltonian is written as a zeroth-order (unperturbed) term $\hat{H}^{(0)}$, which can contain a time-independent potential $V(\mathbf{r})$, plus a second term $\hat{H}^{(1)}$, which contains the time-dependent perturbation:

$$\hat{H} = \hat{H}^{(0)} + \hat{H}^{(1)}(t).$$

We then have

$$i\hbar\frac{\partial\psi(\mathbf{r},t)}{\partial t}=(\hat{H}^{(0)}+\hat{H}^{(1)})\psi(\mathbf{r},t),$$

$$i\hbar\frac{\partial}{\partial t}\left[\sum_n c_n(t)\psi_n^{(0)}(\mathbf{r},t)\right]=(\hat{H}^{(0)}+\hat{H}^{(1)})\sum_n c_n(t)\psi_n^{(0)}(\mathbf{r},t),$$

whereupon the left-hand side of this equation can be expanded using the product rule for differentiation, leading to

$$i\hbar\sum_n c_n(t)\frac{\partial\psi_n^{(0)}(\mathbf{r},t)}{\partial t}+i\hbar\sum_n\frac{dc_n(t)}{dt}\psi_n^{(0)}(\mathbf{r},t)$$
$$=\hat{H}^{(0)}\sum_n c_n(t)\psi_n^{(0)}(\mathbf{r},t)+\hat{H}^{(1)}\sum_n c_n(t)\psi_n^{(0)}(\mathbf{r},t).$$

It turns out that the first terms on the left- and right-hand sides cancel. To illustrate, let $n=1$. We then have (considering only these two terms)

$$i\hbar c_1(t)\frac{\partial\psi_1^{(0)}(\mathbf{r},t)}{\partial t}=c_1(t)\hat{H}^{(0)}\psi_1^{(0)}(\mathbf{r},t),$$

$$i\hbar\frac{\partial\psi_1^{(0)}(\mathbf{r},t)}{\partial t}=\hat{H}^{(0)}\psi_1^{(0)}(\mathbf{r},t),$$

which is just the time-dependent Schrödinger equation for an eigenstate of the system. This same equality holds for all other n. As a consequence, we are left with

$$i\hbar\sum_n\frac{dc_n(t)}{dt}\psi_n^{(0)}(\mathbf{r},t)=\hat{H}^{(1)}\sum_n c_n(t)\psi_n^{(0)}(\mathbf{r},t).$$

At this point, we wish to solve the equation for the time-dependent coefficients of the states. Say we choose a state k (not to be confused with the wavevector discussed in earlier chapters). We first multiply both sides of the equation by $\psi_k^{(0)*}(\mathbf{r})$ and integrate over all space. To simplify the notation, the integral is left out of the following equations. Recall also from our introduction to quantum mechanics in Chapter 6 that all states of the same basis where $n\neq k$ are orthogonal. As a consequence, we find that

$$i\hbar\sum_n\psi_k^{(0)*}(\mathbf{r})\frac{dc_n(t)}{dt}\psi_n^{(0)}(\mathbf{r},t)=\psi_k^{(0)*}\hat{H}^{(1)}\sum_n c_n(t)\psi_n^{(0)}(\mathbf{r},t).$$

When all terms are consolidated and the order of operation on the right-hand side is maintained, we obtain

$$i\hbar\sum_n\frac{dc_n(t)}{dt}\psi_k^{(0)*}(\mathbf{r})\psi_n^{(0)}(\mathbf{r},t)=\sum_n c_n(t)\psi_k^{(0)*}\hat{H}^{(1)}\psi_n^{(0)}(\mathbf{r},t).$$

Expanding $\psi_n^{(0)}(\mathbf{r},t)$ then gives

$$i\hbar\sum_n\frac{dc_n(t)}{dt}\psi_k^{(0)*}(\mathbf{r})\psi_n^{(0)}(\mathbf{r})e^{-iE_n^{(0)}t/\hbar}=\sum_n c_n(t)\psi_k^{(0)*}(\mathbf{r})\hat{H}^{(1)}\psi_n^{(0)}(\mathbf{r})e^{-iE_n^{(0)}t/\hbar}.$$

On the left-hand side, owing to orthogonality, the only nonzero term that remains is the one where $n=k$. We thus have

$$i\hbar\frac{dc_k(t)}{dt}\psi_k^{(0)*}(\mathbf{r})\psi_k^{(0)}(\mathbf{r})e^{-iE_k^{(0)}t/\hbar}=\sum_n c_n(t)\psi_k^{(0)*}(\mathbf{r})\hat{H}^{(1)}\psi_n^{(0)}(\mathbf{r})e^{-iE_n^{(0)}t/\hbar}.$$

Next, if $\psi_k^{(0)}$ is normalized, $\psi_k^{(0)*}\psi_k^{(0)} = 1$, since we have implicitly integrated over all space. This leaves

$$i\hbar\frac{dc_k(t)}{dt}e^{-iE_k^{(0)}t/\hbar} = \sum_n c_n(t)\psi_k^{(0)*}(\mathbf{r})\hat{H}^{(1)}\psi_n^{(0)}(\mathbf{r})e^{-iE_n^{(0)}t/\hbar}.$$

Multiplying both sides by $e^{iE_n^{(0)}t/\hbar}$ and consolidating terms then results in

$$i\hbar\frac{dc_k(t)}{dt} = \sum_n c_n(t)\psi_k^{(0)*}(\mathbf{r})\hat{H}^{(1)}\psi_n^{(0)}(\mathbf{r})e^{i(E_k^{(0)}-E_n^{(0)})t/\hbar},$$

whereupon letting $\omega_{kn} = (E_k^{(0)} - E_n^{(0)})/\hbar$ yields

$$i\hbar\frac{dc_k(t)}{dt} = \sum_n c_n(t)\psi_k^{(0)*}(\mathbf{r})\hat{H}^{(1)}\psi_n^{(0)}(\mathbf{r})e^{i\omega_{kn}t}.$$

Finally, switching to bra–ket notation gives

$$i\hbar\frac{dc_k(t)}{dt} = \sum_n c_n(t)\langle k|\hat{H}^{(1)}|n\rangle e^{i\omega_{kn}t}. \tag{11.3}$$

Since there are many n for every state k of interest, Equation 11.3 is basically a system of simultaneous linear differential equations. It is therefore best to use matrix notation at this point. We write

$$i\hbar\frac{d}{dt}\begin{pmatrix} c_1 \\ c_2 \\ c_3 \\ \cdot \\ \cdot \\ \cdot \\ c_k \end{pmatrix} = \begin{pmatrix} V_{11} & V_{12}e^{i\omega_{12}t} & V_{13}e^{i\omega_{13}t} & \cdots \\ V_{21}e^{-i\omega_{12}t} & V_{22} & V_{23}e^{i\omega_{23}t} & \cdots \\ V_{31}e^{-i\omega_{13}t} & V_{32}e^{-i\omega_{23}t} & V_{33} & \cdots \\ \cdot & \cdot & \cdot & \cdot \\ \cdot & \cdot & \cdot & \cdot \\ \cdot & \cdot & \cdot & \cdot \end{pmatrix}\begin{pmatrix} c_1 \\ c_2 \\ c_3 \\ \cdot \\ \cdot \\ \cdot \\ c_k \end{pmatrix},$$

where

$$\begin{aligned} V_{11} &= \langle 1|\hat{H}^{(1)}|1\rangle, \\ V_{22} &= \langle 2|\hat{H}^{(1)}|2\rangle, \\ V_{33} &= \langle 3|\hat{H}^{(1)}|3\rangle, \\ &\text{etc.} \end{aligned}$$

and

$$\begin{aligned} V_{12} &= \langle 1|\hat{H}^{(1)}|2\rangle, \\ V_{21} &= \langle 2|\hat{H}^{(1)}|1\rangle, \\ V_{13} &= \langle 1|\hat{H}^{(1)}|3\rangle, \\ V_{31} &= \langle 3|\hat{H}^{(1)}|1\rangle, \\ &\text{etc.} \end{aligned}$$

In all cases, subscripts refer to row/column notation. For example, V_{23} refers to $k = 2$ and $n = 3$.

Now what? This matrix equation is difficult to solve in a brute force fashion. To tame it, we make some assumptions about the size of the perturbation and, in turn, what $c_1, c_2, c_3, \ldots, c_k$ are.

To illustrate, suppose the system starts out in state 1. Then

$$c_1(t=0) = 1,$$
$$c_2(t=0) = 0,$$
$$c_3(t=0) = 0,$$
$$\text{etc.}$$

If there are no perturbations, the system stays in state 1 forever:

$$c_1(t) = 1,$$
$$c_2(t) = 0,$$
$$c_3(t) = 0,$$
$$\text{etc.}$$

If the perturbation is weak (a relative term, but implicit is that the magnitude of the perturbation is much smaller than the energy difference between states of interest—we have seen this assumption in earlier chapters), then for the most part one has

$$c_1(t) \simeq 1,$$
$$c_2(t) \simeq 0,$$
$$c_3(t) \simeq 0,$$
$$\text{etc.}$$

Now to get the first-order approximation for $c_k(t)$, we insert these values into our matrix to obtain

$$i\hbar\frac{dc_k(t)}{dt} = c_1\langle k|\hat{H}^{(1)}|1\rangle e^{i\omega_{k1}t},$$

where $c_1(t) \simeq 1$. All other terms are effectively zero, since $c_2(t) \simeq 0$, $c_3(t) \simeq 0$, $c_4(t) \simeq 0$, etc. Alternatively, we write

$$\frac{dc_k(t)}{dt} = \frac{1}{i\hbar}\langle k|\hat{H}^{(1)}|1\rangle e^{i\omega_{k1}t},$$

which then gets integrated to

$$c_k(t) = \frac{1}{i\hbar}\int_0^t \langle k|\hat{H}^{(1)}|1\rangle e^{i\omega_{k1}t'}\,dt' + \text{constant}.$$

The trailing constant of integration is simply $c_k(t=0) = \delta_{k1}$.

Our final expression for the desired time-dependent coefficient is then

$$c_k(t) = \delta_{k1} + \frac{1}{i\hbar}\int_0^t \langle k|\hat{H}^{(1)}|1\rangle e^{i\omega_{k1}t'}\,dt'. \tag{11.4}$$

The reason we care about this time-dependent weight is that the probability of being in a particular state (in this case, k) is the square of the weight:

$$P_k = |c_k(t)|^2. \tag{11.5}$$

More specifically,

$$P_k = \frac{1}{\hbar^2}\left|\int_0^t \langle k|\hat{H}^{(1)}|1\rangle e^{i\omega_{k1}t'}\,dt'\right|^2 \tag{11.6}$$

is the probability of being in state k, having started out in state 1. Along the same lines, the transition rate R into the final state is the time derivative of P_k:

$$R = \frac{dP_k}{dt}. \tag{11.7}$$

Thus, from Equations 11.5–11.7, we can find both the transition probability and the transition rate from some initial state into a final state. These values are, in turn, related to the absorption and emission efficiencies of a material. Recall that in Chapter 5 we introduced the absorption coefficient and absorption cross section as measures of how well a system absorbs light. We will return to this later in more detail when we evaluate actual transition probabilities and rates for various low-dimensional systems in Chapter 12.

11.3 EXAMPLE: A TWO-LEVEL SYSTEM

Let us illustrate the above concepts with a concrete example. This will be the case of a two-level system where the transition occurs as a result of a sinusoidal perturbation (i.e., light) between a ground state and an excited state. The situation is illustrated in **Figure 11.1**. We will work in the Schrödinger representation, although this could just as well be done in the interaction representation.

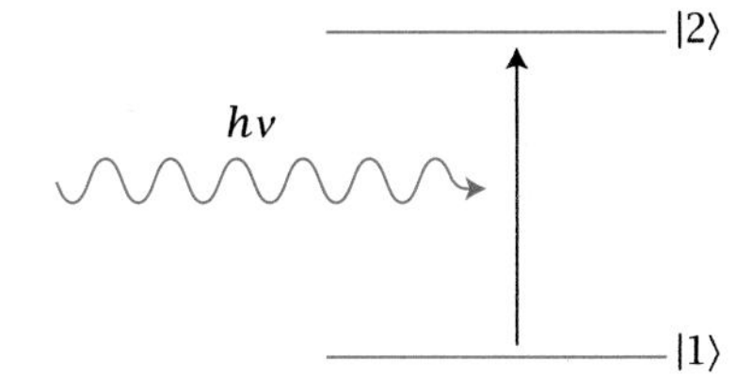

Figure 11.1 Illustration of a two-level system.

Recall that our time-dependent Schrödinger equation is

$$i\hbar\frac{\partial}{\partial t}\psi(\mathbf{r},t) = \hat{H}\psi(\mathbf{r},t),$$

where the Hamiltonian is $\hat{H} = \hat{H}^{(0)} + \hat{H}^{(1)}(t)$. Thus,

$$i\hbar\frac{\partial}{\partial t}\psi(\mathbf{r},t) = (\hat{H}^{(0)} + \hat{H}^{(1)})\psi(\mathbf{r},t).$$

Recall also that a general solution to this time-dependent Schrödinger equation is a linear combination of unperturbed zeroth-order wavefunctions, provided that we have time-dependent coefficients. For the case where we only consider two states ($|1\rangle$ and $|2\rangle$), we have

$$\psi(\mathbf{r},t) = c_1(t)\psi_1^{(0)}(\mathbf{r},t) + c_2(t)\psi_2^{(0)}(\mathbf{r},t),$$

with

$$\psi_1^{(0)}(\mathbf{r},t) = \psi_1^{(0)}(\mathbf{r})e^{-iE_1^{(0)}t/\hbar}$$
$$\psi_2^{(0)}(\mathbf{r},t) = \psi_2^{(0)}(\mathbf{r})e^{-iE_2^{(0)}t/\hbar}.$$

Let us now be more specific about what $\hat{H}^{(1)}(t)$ looks like. Again, the superscript (0) refers to the system's unperturbed energies and wavefunctions. For the case here, consider the interaction between the two-level system and an incident excitation having an electric field of the form

$$\mathbf{E} = E_0\mathbf{e}\cos\omega t, \tag{11.8}$$

where E_0 is the amplitude of the field, ω is its frequency, and $\mathbf{e}$ is a unit polarization vector. The interaction between the initial state and the final state can then be written as

$$\hat{H}^{(1)} = -\hat{\mu}\cdot\mathbf{E}, \tag{11.9}$$

where $\hat{\mu}$ is the system's dipole moment operator and the dot product is between $\hat{\mu} = (\mu_x, \mu_y, \mu_z)$ and $\mathbf{E} = (E_x, E_y, E_z)$. Note that there exists an alternative but equivalent expression for $\hat{H}^{(1)}$ in terms of the vector potential $\mathbf{A}$ of the incident light. We will see this in more detail in Chapter 12 as well as in the appendix to this chapter. For now, though, let us continue with the above approach. For convenience, we choose a polarization for the light, say along the z direction of a Cartesian coordinate system, such that $\mathbf{e} = (0, 0, 1)$. As a consequence,

$$\hat{H}^{(1)} = -\mu_z E_0 \cos \omega t. \tag{11.10}$$

At this point, let us rederive our expression for the time-dependent coefficient associated with the final state, since this is ultimately related to the transition probability and the transition rate (Equations 11.5 and 11.7). We introduce our expression for the wavefunction into the time-dependent Schrödinger equation to get

$$i\hbar \frac{\partial}{\partial t}\left[c_1(t)\psi_1^{(0)}(\mathbf{r}, t) + c_2(t)\psi_2^{(0)}(\mathbf{r}, t)\right] = (\hat{H}^{(0)} + \hat{H}^{(1)})\left[c_1(t)\psi_1^{(0)}(\mathbf{r}, t) + c_2(t)\psi_2^{(0)}(\mathbf{r}, t)\right].$$

The product rule of differentiation is then used to simplify the left-hand side. This gives

$$\begin{aligned} i\hbar\left[c_1(t)\frac{\partial\psi_1^{(0)}(\mathbf{r}, t)}{\partial t} + \psi_1^{(0)}(\mathbf{r}, t)\frac{dc_1(t)}{dt} + c_2(t)\frac{\partial\psi_2^{(0)}(\mathbf{r}, t)}{\partial t} + \psi_2^{(0)}(\mathbf{r}, t)\frac{dc_2(t)}{dt}\right] \\ = \hat{H}^{(0)}\left[c_1(t)\psi_1^{(0)}(\mathbf{r}, t) + c_2(t)\psi_2^{(0)}(\mathbf{r}, t)\right] \\ +\hat{H}^{(1)}\left[c_1(t)\psi_1^{(0)}(\mathbf{r}, t) + c_2(t)\psi_2^{(0)}(\mathbf{r}, t)\right], \end{aligned}$$

which then rearranges to

$$\begin{aligned} i\hbar\left[c_1(t)\frac{\partial\psi_1^{(0)}(\mathbf{r}, t)}{\partial t} + c_2(t)\frac{\partial\psi_2^{(0)}(\mathbf{r}, t)}{\partial t}\right] + i\hbar\left[\psi_1^{(0)}(\mathbf{r}, t)\frac{dc_1(t)}{dt} + \psi_2^{(0)}(\mathbf{r}, t)\frac{dc_2(t)}{dt}\right] \\ = \hat{H}^{(0)}\left[c_1(t)\psi_1^{(0)}(\mathbf{r}, t) + c_2(t)\psi_2^{(0)}(\mathbf{r}, t)\right] \\ +\hat{H}^{(1)}\left[c_1(t)\psi_1^{(0)}(\mathbf{r}, t) + c_2(t)\psi_2^{(0)}(\mathbf{r}, t)\right]. \end{aligned}$$

We have previously seen that the first terms on each side of the equation cancel, since

$$\begin{aligned} i\hbar\frac{\partial\psi_1^{(0)}(\mathbf{r}, t)}{\partial t} &= \hat{H}^{(0)}\psi_1^{(0)}(\mathbf{r}, t), \\ i\hbar\frac{\partial\psi_2^{(0)}(\mathbf{r}, t)}{\partial t} &= \hat{H}^{(0)}\psi_2^{(0)}(\mathbf{r}, t). \end{aligned}$$

As a consequence, we are left with the following equation:

$$\begin{aligned} i\hbar\psi_1^{(0)}(\mathbf{r}, t)\frac{dc_1(t)}{dt} + i\hbar\psi_2^{(0)}(\mathbf{r}, t)\frac{dc_2(t)}{dt} = c_1(t)\hat{H}^{(1)}\psi_1^{(0)}(\mathbf{r}, t) \\ + c_2(t)\hat{H}^{(1)}\psi_2^{(0)}(\mathbf{r}, t), \end{aligned}$$

which we solve for the coefficient $c_2(t)$.

To do this, we multiply/integrate both sides by $\psi_2^{(0)*}(\mathbf{r})$ (the integral is omitted for notational convenience):

$$i\hbar\psi_2^{(0)*}(\mathbf{r})\psi_1^{(0)}(\mathbf{r},t)\frac{dc_1(t)}{dt} + i\hbar\psi_2^{(0)*}(\mathbf{r})\psi_2^{(0)}(\mathbf{r},t)\frac{dc_2(t)}{dt}$$
$$= c_1(t)\psi_2^{(0)*}(\mathbf{r})\hat{H}^{(1)}\psi_1^{(0)}(\mathbf{r},t) + c_2(t)\psi_2^{(0)*}(\mathbf{r})\hat{H}^{(1)}\psi_2^{(0)}(\mathbf{r},t).$$

When $\psi_1^{(0)}(\mathbf{r},t)$ and $\psi_2^{(0)}(\mathbf{r},t)$ are expanded to reveal their explicit time dependences, we obtain

$$i\hbar\psi_2^{(0)*}(\mathbf{r})\psi_1^{(0)}(\mathbf{r})e^{-iE_1^{(0)}t/\hbar}\frac{dc_1(t)}{dt} + i\hbar\psi_2^{(0)*}(\mathbf{r})\psi_2^{(0)}(\mathbf{r})e^{-iE_2^{(0)}t/\hbar}\frac{dc_2(t)}{dt}$$
$$= c_1(t)\psi_2^{(0)*}(\mathbf{r})\hat{H}^{(1)}\psi_1^{(0)}(\mathbf{r})e^{-iE_1^{(0)}t/\hbar} + c_2(t)\psi_2^{(0)*}(\mathbf{r})\hat{H}^{(1)}\psi_2^{(0)}(\mathbf{r})e^{-iE_2^{(0)}t/\hbar}.$$

To eliminate the first term on the left-hand side, we now recall the orthogonality between basis functions and their normalization when integrated over all space (i.e., $\int \psi_2^{(0)*}(\mathbf{r})\psi_2^{(0)}(\mathbf{r})\,d^3r = 1$). We are left with

$$i\hbar e^{-iE_2^{(0)}t/\hbar}\frac{dc_2(t)}{dt} = c_1(t)\psi_2^{(0)*}(\mathbf{r})\hat{H}^{(1)}\psi_1^{(0)}(\mathbf{r})e^{-iE_1^{(0)}t/\hbar}$$
$$+ c_2(t)\psi_2^{(0)*}(\mathbf{r})\hat{H}^{(1)}\psi_2^{(0)}(\mathbf{r})e^{-iE_2^{(0)}t/\hbar}.$$

At this point, for convenience, let us switch to bra–ket notation to simplify the appearance of the expression's right-hand side:

$$i\hbar e^{-iE_2^{(0)}t/\hbar}\frac{dc_2(t)}{dt} = c_1(t)\langle 2|\hat{H}^{(1)}|1\rangle e^{-iE_1^{(0)}t/\hbar} + c_2(t)\langle 2|\hat{H}^{(1)}|2\rangle e^{-iE_2^{(0)}t/\hbar}.$$

To further simplify things, we invoke the assumptions inherent to time-dependent perturbation theory. Namely, since the perturbation is small,

$$c_1(t) \simeq 1,$$
$$c_2(t) \simeq 0.$$

This gives

$$i\hbar e^{-iE_2^{(0)}t/\hbar}\frac{dc_2(t)}{dt} = \langle 2|\hat{H}^{(1)}|1\rangle e^{-iE_1^{(0)}t/\hbar},$$

whereupon multiplying both sides by $e^{iE_2^{(0)}t/\hbar}$ and dividing by $i\hbar$ gives

$$\frac{dc_2(t)}{dt} = \frac{1}{i\hbar}\langle 2|\hat{H}^{(1)}|1\rangle e^{i(E_2^{(0)}-E_1^{(0)})t/\hbar}.$$

At this point, we introduce the expression $\hat{H}^{(1)}(t) = -\mu_z E_0\cos\omega t$. But rather than use it directly, let us employ the Euler relation to obtain

$$\hat{H}^{(1)}(t) = -\mu_z E_0\left(\frac{e^{i\omega t}+e^{-i\omega t}}{2}\right).$$

When this is introduced into the above expression,

$$\frac{dc_2(t)}{dt} = -\frac{1}{2i\hbar}\langle 2|\mu_z E_0(e^{i\omega t}+e^{-i\omega t})|1\rangle e^{i(E_2^{(0)}-E_1^{(0)})t/\hbar}.$$

We can remove all terms except μ_z from the bra–ket. This is because E_0 is a constant and $e^{i\omega t}$ and $e^{-i\omega t}$ do not have any spatial dependences (i.e., they do not depend on z). The μ_z dipole moment term, however, cannot be removed, because it implicitly contains a z dependence (namely, $\mu_z = qz$). This leaves us with

$$\frac{dc_2(t)}{dt} = -\frac{E_0}{2i\hbar}\langle 2|\mu_z|1\rangle(e^{i\omega t} + e^{-i\omega t})e^{i(E_2^{(0)}-E_1^{(0)})t/\hbar},$$

with $\langle 2|\mu_z|1\rangle$ called the *transition dipole moment.*

Let us continue consolidating terms by making the substitution $\omega = E/\hbar$ that comes from $E = \hbar\omega$:

$$\frac{dc_2(t)}{dt} = -\frac{E_0}{2i\hbar}\langle 2|\mu_z|1\rangle(e^{i[E+(E_2^{(0)}-E_1^{(0)})]t/\hbar} + e^{-i[E-(E_2^{(0)}-E_1^{(0)})]t/\hbar}).$$

This integrates to yield

$$c_2(t) = -\frac{E_0}{2i\hbar}\langle 2|\mu_z|1\rangle \int_0^t e^{i[E+(E_2^{(0)}-E_1^{(0)})]t'/\hbar} + e^{-i[E-(E_2^{(0)}-E_1^{(0)})]t'/\hbar}\,dt' + \text{constant}.$$

The constant of integration is simply $c_2(t=0) = \delta_{kn}$, and since at $t = 0$, $k \neq n$, we have $\delta_{kn} = 0$. Therefore,

$$c_2(t) = -\frac{E_0}{2i\hbar}\langle 2|\mu_z|1\rangle \int_0^t e^{i[E+(E_2^{(0)}-E_1^{(0)})]t'/\hbar} + e^{-i[E-(E_2^{(0)}-E_1^{(0)})]t'/\hbar}\,dt',$$

which subsequently integrates to

$$c_2(t) = -\frac{E_0}{2i\hbar}\langle 2|\mu_z|1\rangle \left\{\frac{\hbar(e^{i[E+(E_2^{(0)}-E_1^{(0)})]t/\hbar} - 1)}{i[E + (E_2^{(0)} - E_1^{(0)})]} - \frac{\hbar(e^{-i[E-(E_2^{(0)}-E_1^{(0)})]t/\hbar} - 1)}{i[E - (E_2^{(0)} - E_1^{(0)})]}\right\}.$$

At this point, another approximation is made. We have two terms in the curly brackets. The one with $[E - (E_2^{(0)} - E_1^{(0)})]$ in the denominator is called the resonant term. The other, with $[E + (E_2^{(0)} - E_1^{(0)})]$, is called the nonresonant term. This terminology arises because when the incident light energy E equals the energy difference between the two states, $E = E_2^{(0)} - E_1^{(0)}$, the former term dominates and one is said to be "on resonance." Since we are primarily interested in the transition between the two states, we ignore the nonresonant term, leaving us with

$$c_2(t) \simeq -\frac{E_0\langle 2|\mu_z|1\rangle}{2}\,\frac{e^{-i[E-(E_2^{(0)}-E_1^{(0)})]t/\hbar} - 1}{E - (E_2^{(0)} - E_1^{(0)})}.$$

Note that the approximation we have just made is called the *rotating wave approximation.*

We continue simplifying the expression. Namely, since

$$c_2(t) \simeq -\frac{E_0\langle 2|\mu_z|1\rangle}{2} e^{-i[E-(E_2^{(0)}-E_1^{(0)})]t/2\hbar} \times \frac{e^{-i[E-(E_2^{(0)}-E_1^{(0)})]t/2\hbar} - e^{i[E-(E_2^{(0)}-E_1^{(0)})]t/2\hbar}}{E - (E_2^{(0)} - E_1^{(0)})},$$

we invoke the Euler relation to obtain our desired final expression:

$$c_2(t) \simeq iE_0\langle 2|\mu_z|1\rangle e^{-i[E-(E_2^{(0)}-E_1^{(0)})]t/2\hbar}\,\frac{\sin\left[\dfrac{E - (E_2^{(0)} - E_1^{(0)})}{2\hbar}t\right]}{E - (E_2^{(0)} - E_1^{(0)})}. \tag{11.11}$$

Equation 11.11 describes the time-dependent coefficient associated with the final state. From this, the probability of finding the system in

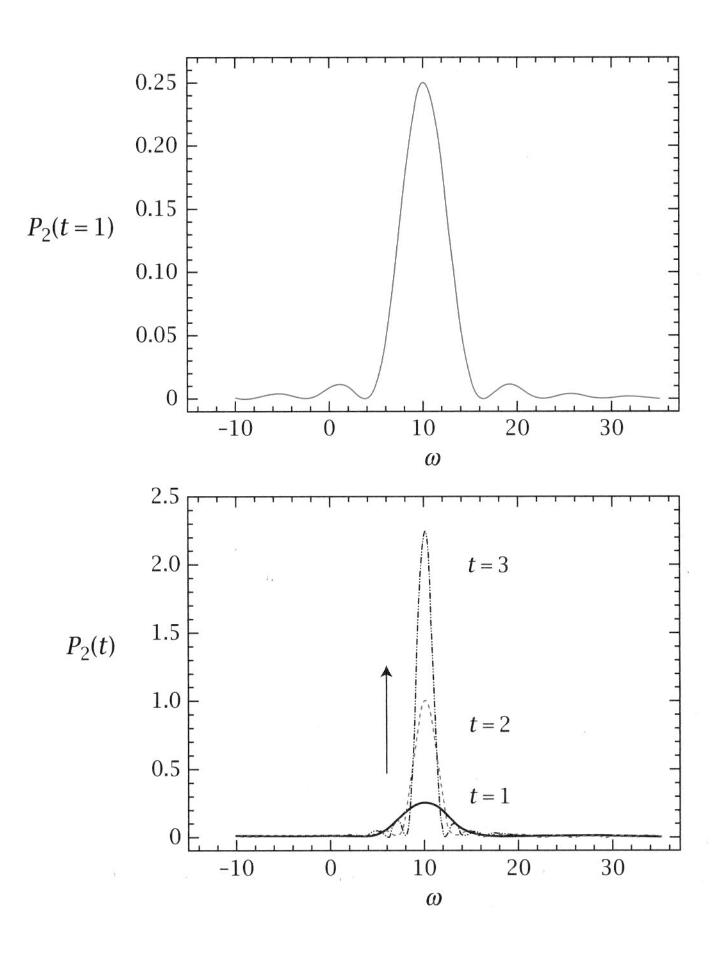

Figure 11.2 Illustration of the transition probability for a two-level system. The bottom plot shows the time dependence of the probability. Arbitrary units are used throughout.

$|2\rangle$ is simply the square of this value, $P_2(t) = |c_2(t)|^2$ (Equation 11.5):

$$P_2(E, t) \simeq E_0^2 |\langle 2|\mu_z|1\rangle|^2 \frac{\sin^2\left[\frac{(E - E_{21})t}{2\hbar}\right]}{(E - E_{21})^2}, \tag{11.12}$$

where $E_{21} = E_2^{(0)} - E_1^{(0)}$. Alternatively, one sometimes sees the probability written in terms of frequency as opposed to energy. Since $E = \hbar\omega$, Equation 11.12 can also be written as

$$P_2(\omega, t) \simeq \frac{E_0^2}{\hbar^2} |\langle 2|\mu_z|1\rangle|^2 \frac{\sin^2\left[\frac{1}{2}(\omega - \omega_{21})t\right]}{(\omega - \omega_{21})^2}, \tag{11.13}$$

where $\omega_{21} = \omega_2^{(0)} - \omega_1^{(0)}$. Both Equations 11.12 and 11.13 have the functional form of a sinc function. A plot of P_k is shown in **Figure 11.2**. Note that it peaks strongly about a resonance frequency ($\omega = 10$ in **Figure 11.2**) and increases sharply as a function of time. In fact, as $t \to \infty$, we obtain a delta function:

$$\delta(\alpha) = \lim_{t\to\infty} \frac{\sin^2 \alpha t}{\pi\alpha^2 t}. \tag{11.14}$$

11.4 RATES

Next, let us find the transition rate into a final state or, more generally, into a group of final states. The latter case is where the density of states matters. For a two-level system, one has the following time-dependent rate, which is the time derivative of P_k above: $R = dP(t)/dt$.

We begin with Equation 11.13:

$$P_2(\omega, t) \simeq \frac{E_0^2}{\hbar^2}|\langle 2|\mu_z|1\rangle|^2 \frac{\sin^2\left[\frac{1}{2}(\omega - \omega_{21})t\right]}{(\omega - \omega_{21})^2}.$$

For notational convenience, let $\alpha = \frac{1}{2}(\omega - \omega_{21})$ so that $2\alpha = \omega - \omega_{21}$. We then have

$$P_2(t) \simeq \frac{E_0^2}{\hbar^2}|\langle 2|\mu_z|1\rangle|^2 \frac{\sin^2 \alpha t}{4\alpha^2}.$$

The associated transition rate is $R_2(t) = dP_2(t)/dt$, giving

$$R_2(t) = \frac{E_0^2}{4\alpha\hbar^2}|\langle 2|\mu_z|1\rangle|^2 2 \sin \alpha t \cos \alpha t.$$

Since $\frac{1}{2}\sin 2\alpha t = \sin \alpha t \cos \alpha t$, we obtain

$$R_2(t) = \frac{E_0^2}{4\alpha\hbar^2}|\langle 2|\mu_z|1\rangle|^2 \sin 2\alpha t,$$

whereupon replacing frequencies back into the expression yields

$$R_2(t) = \frac{E_0^2}{2\hbar^2}|\langle 2|\mu_z|1\rangle|^2 \frac{\sin[(\omega - \omega_{21})t]}{\omega - \omega_{21}} \tag{11.15}$$

as our desired two-level system transition rate. Note that it oscillates as a function of time.

Alternatively, we can re-express the transition probability in the limit where $t \to \infty$. This is because, as can be seen from the graphs of $P_2(t)$ in **Figure 11.2**, the central peak becomes narrower and taller at longer time and ultimately approaches a delta function. Thus, in the above case, we have

$$P_2(t) \simeq \frac{E_0^2}{\hbar^2}|\langle 2|\mu_z|1\rangle|^2 \frac{\sin^2 \alpha t}{4\alpha^2}.$$

This rearranges to

$$P_2(t) \simeq \frac{\pi E_0^2}{4\hbar^2}|\langle 2|\mu_z|1\rangle|^2 \frac{\sin^2 \alpha t}{\pi\alpha^2 t} t,$$

and as $t \to \infty$ one finds that

$$P_2(t) = \frac{\pi E_0^2}{4\hbar^2}|\langle 2|\mu_z|1\rangle|^2 \delta(\alpha) t.$$

The long-time transition probability for a two-level system is therefore

$$\lim_{t\to\infty} P_2(t) = \frac{\pi E_0^2}{4\hbar^2}|\langle 2|\mu_z|1\rangle|^2 \delta\left(\frac{\omega - \omega_{21}}{2}\right) t.$$

Furthermore, since

$$\delta(cx) = \frac{1}{|c|}\delta(x), \tag{11.16}$$

we may also express our result as

$$\lim_{t\to\infty} P_2(t) = \frac{\pi E_0^2}{2\hbar^2}|\langle 2|\mu_z|1\rangle|^2 \delta(\omega - \omega_{21}) t. \tag{11.17}$$

Equation 11.17 can alternatively be written in terms of energy by employing the additional relationship

$$\int \delta(\omega - \omega_0)\, d\omega = \int \delta(E - E_0)\, dE,$$

to find

$$\delta(\omega - \omega_0) = \delta(E - E_0)\frac{dE}{d\omega},$$

where $E = \hbar\omega$ and $dE/d\omega = \hbar$. This shows that

$$\delta(\omega - \omega_0) = \hbar\delta(E - E_0) \tag{11.18}$$

and allows us to obtain

$$\lim_{t\to\infty} P_2(t) = \frac{\pi E_0^2}{2\hbar}|\langle 2|\mu_z|1\rangle|^2\delta(E - E_{21})t. \tag{11.19}$$

Finally, the associated transition rate $R_2 = dP_2(t)/dt$, giving

$$R_{2,t\to\infty} = \frac{\pi E_0^2}{2\hbar}|\langle 2|\mu_z|1\rangle|^2\delta(E - E_{21}). \tag{11.20}$$

We see that it is time-independent and is the quantum mechanical analog of a classical rate. Note that these expressions have assumed a particular polarization of the incident light.

11.4.1 More Than One Final State

Now, rather than a two-level system, what about the case where many possible final states exist? This situation could arise when dealing with a spectrally broad laser exciting our sample instead of a perfectly monochromatic (i.e., one-color) source. In the previous two-level system limit, we found that

$$P_k(\omega, t) \simeq \frac{E_0^2}{\hbar^2}|\langle k|\mu_z|n\rangle|^2\frac{\sin^2\left[\frac{1}{2}(\omega - \omega_{kn})t\right]}{(\omega - \omega_{kn})^2},$$

where the notation has been generalized, using k to denote the final state and n to denote the initial state. This time, however, since we want to consider the transition into a group of final states, we integrate $P_k(\omega, t)$ over this spread of final states characterized by a density $\rho(E_k)$ or $\rho(\omega_k)$ with units of number per unit energy per unit volume or number per unit frequency per unit volume. We therefore have

$$P_{\text{tot}} = \int_{-\infty}^{\infty} P_k(\omega, t)\rho(\omega_k)\, d\omega_k.$$

As $t \to \infty$, the sinc function in $P_k(\omega, t)$ becomes a delta function (Equation 11.14). Thus, rearranging things to match this form yields

$$\begin{aligned}
P_{\text{tot}} &= \frac{E_0^2}{\hbar^2}|\langle k|\mu_z|n\rangle|^2\int_{-\infty}^{\infty}\frac{\sin^2\left[\frac{1}{2}(\omega - \omega_{kn})t\right]}{(\omega - \omega_{kn})^2}\rho(\omega_k)\, d\omega_k \\
&= \frac{E_0^2}{\hbar^2}|\langle k|\mu_z|n\rangle|^2\int_{-\infty}^{\infty}\frac{\sin^2 \alpha t}{4\alpha^2}\rho(\omega_k)\, d\omega_k \\
&= \frac{E_0^2}{\hbar^2}\left(\frac{\pi t}{4}\right)|\langle k|\mu_z|n\rangle|^2\int_{-\infty}^{\infty}\frac{\sin^2 \alpha t}{\pi\alpha^2 t}\rho(\omega_k)\, d\omega_k,
\end{aligned}$$

where $\alpha = \frac{1}{2}(\omega - \omega_{kn})$. Then, as $t \to \infty$, we find

$$P_{\text{tot}} = \frac{E_0^2}{\hbar^2}\left(\frac{\pi t}{4}\right)|\langle k|\mu_z|n\rangle|^2 \int_{-\infty}^{\infty} \delta\left(\frac{\omega - \omega_{kn}}{2}\right)\rho(\omega_k)\, d\omega_k$$

and, since $\delta(cx) = |c|^{-1}\delta(x)$, we also have

$$P_{\text{tot}} = \frac{E_0^2}{\hbar^2}\left(\frac{\pi t}{2}\right)|\langle k|\mu_z|n\rangle|^2 \int_{-\infty}^{\infty} \delta(\omega - \omega_{kn})\rho(\omega_k)\, d\omega_k,$$

leading to

$$P_{\text{tot}} = \frac{\pi E_0^2}{2\hbar^2}\rho(\omega + \omega_n)|\langle k|\mu_z|n\rangle|^2 t. \tag{11.21}$$

Equation 11.21 can be expressed in terms of energy by recalling that $\rho(\omega + \omega_n) = \hbar\rho(E + E_n)$:

$$P_{\text{tot}} = \frac{\pi E_0^2}{2\hbar}\rho(E + E_n)|\langle k|\mu_z|n\rangle|^2 t. \tag{11.22}$$

Finally, the desired transition rate into a group of final states is $R_{\text{tot}} = dP_{\text{tot}}/dt$:

$$R_{\text{tot}} = \frac{\pi E_0^2}{2\hbar}\rho(E + E_n)|\langle k|\mu_z|n\rangle|^2, \tag{11.23}$$

which is time-independent and is analogous to a classical rate.

11.5 SUMMARY

In this chapter, we have found the transition probability and transition rate of a two-level system when excited with light. We have then generalized these expressions to account for a distribution of final states. This has allowed us to invoke our previous density-of-states calculation, first seen in Chapter 9. In Chapter 12, we will use these results to evaluate interband transitions between the valence and conduction bands of bulk as well as low-dimensional semiconductors. This will, in turn, allow us to obtain expressions for their absorption coefficients.

11.6 APPENDIX

Finally, note that other texts sometimes have different looking expressions for the above two-level system transition probabilities and rates. In fact, Equation 11.1 is an example of this. Notice the $2\pi/\hbar$ prefactor, which differs from the $\pi/2\hbar$ prefactor in the expressions above. This is because authors often start with an alternative form of the interaction Hamiltonian $\hat{H}^{(1)}$ when deriving P_{tot} and R_{tot}. Since these differences are bound to cause confusion, we present an abbreviated derivation of those alternative transition probabilities and rates in this section, *identical* to the ones above.

Starting with a generalized version of Equation 11.6,

$$P_k = \frac{1}{\hbar^2}\left|\int_0^t \langle k|\hat{H}^{(1)}|n\rangle e^{i\omega_{kn}t'}\, dt'\right|^2,$$

we use the following (equivalent) interaction Hamiltonian

$$\hat{H}^{(1)} = -\frac{q}{m_0}\mathbf{A}\cdot\hat{\mathbf{p}},$$

where $\mathbf{A}$ is called the vector potential and $\hat{\mathbf{p}}$ is the quantum mechanical momentum operator. The reader may find a more thorough discussion of this in Chapter 12. For now, simply take this for granted. Next, $\mathbf{A}$ has the following harmonic form:

$$\mathbf{A} = A_0 \mathbf{e} \sin(\mathbf{k}\cdot\mathbf{r} - \omega t),$$

where $\mathbf{e}$ is a unit polarization vector. As a consequence,

$$\mathbf{A}\cdot\hat{\mathbf{p}} = A_0 \sin(\mathbf{k}\cdot\mathbf{r} - \omega t)(\mathbf{e}\cdot\hat{\mathbf{p}}).$$

The resulting interaction Hamiltonian is then

$$\hat{H}^{(1)} = -\frac{q}{m_0} A_0 \sin(\mathbf{k}\cdot\mathbf{r} - \omega t)(\mathbf{e}\cdot\hat{\mathbf{p}}).$$

In complex form, it can be written as

$$\hat{H}^{(1)} = -\frac{q}{m_0} A_0 \left[\frac{e^{i(\mathbf{k}\cdot\mathbf{r}-\omega t)} - e^{-i(\mathbf{k}\cdot\mathbf{r}-\omega t)}}{2i}\right](\mathbf{e}\cdot\hat{\mathbf{p}}).$$

We now replace this into our transition probability. This gives

$$\begin{aligned} P_k &= \frac{1}{\hbar^2}\left| -\frac{qA_0}{m_0}\int_0^t \langle k| \left(\frac{e^{i(\mathbf{k}\cdot\mathbf{r}-\omega t')} - e^{-i(\mathbf{k}\cdot\mathbf{r}-\omega t')}}{2i}\right)(\mathbf{e}\cdot\hat{\mathbf{p}})|n\rangle e^{i\omega_{kn}t'}\, dt'\right|^2 \\ &= \frac{1}{\hbar^2}\left(\frac{qA_0}{2m_0}\right)^2 \left|\int_0^t \langle k|(e^{i\mathbf{k}\cdot\mathbf{r}} e^{-i(\omega-\omega_{kn})t'} - e^{-i\mathbf{k}\cdot\mathbf{r}} e^{i(\omega+\omega_{kn})t'})(\mathbf{e}\cdot\hat{\mathbf{p}})|n\rangle\, dt'\right|^2 . \end{aligned}$$

Looking ahead, the first (resonant) term leads to absorption while the second leads to stimulated emission. As a consequence, we keep only the first term. This simplification is called the rotating wave approximation, as seen earlier. We therefore have

$$\begin{aligned} P_k &\simeq \frac{1}{\hbar^2}\left(\frac{qA_0}{2m_0}\right)^2 \left|\int_0^t \langle k|e^{i\mathbf{k}\cdot\mathbf{r}} e^{-i(\omega-\omega_{kn})t'}(\mathbf{e}\cdot\hat{\mathbf{p}})|n\rangle\, dt'\right|^2 \\ &\simeq \frac{1}{\hbar^2}\left(\frac{qA_0}{2m_0}\right)^2 \left|\int_0^t \langle k|e^{i\mathbf{k}\cdot\mathbf{r}}(\mathbf{e}\cdot\hat{\mathbf{p}})|n\rangle e^{-i(\omega-\omega_{kn})t'}\, dt'\right|^2 . \end{aligned}$$

At this point, the electric dipole approximation, $e^{i\mathbf{k}\cdot\mathbf{r}} \simeq 1$, is invoked. Its justification is provided in Chapter 12. We thus have

$$P_k \simeq \frac{1}{\hbar^2}\left(\frac{qA_0}{2m_0}\right)^2 \left|\int_0^t \langle k|\mathbf{e}\cdot\hat{\mathbf{p}}|n\rangle e^{-i(\omega-\omega_{kn})t'}\, dt'\right|^2 .$$

The bra–ket term can now come out of the time integral, giving

$$P_k \simeq \frac{1}{\hbar^2}\left(\frac{qA_0}{2m_0}\right)^2 |\langle k|\mathbf{e}\cdot\hat{\mathbf{p}}|n\rangle|^2 \left|\int_0^t e^{-i(\omega-\omega_{kn})t'}\, dt'\right|^2 ,$$

whereupon the time integral yields

$$\begin{aligned} \int_0^t e^{-i(\omega-\omega_{kn})t'}\, dt' &= \frac{e^{-i(\omega-\omega_{kn})t} - 1}{-i(\omega-\omega_{kn})} \\ &= \frac{2}{\omega-\omega_{kn}} e^{-i(\omega-\omega_{kn})t/2} \frac{-e^{-i(\omega-\omega_{kn})t/2} + e^{i(\omega-\omega_{kn})t/2}}{2i} \\ &= \frac{2}{\omega-\omega_{kn}} e^{-i(\omega-\omega_{kn})t/2} \sin\left[\tfrac{1}{2}(\omega-\omega_{kn})t\right]. \end{aligned}$$

Replacing this result back into P_k then gives

$$P_k \simeq \frac{1}{\hbar^2}\left(\frac{qA_0}{2m_0}\right)^2 |\langle k|\mathbf{e}\cdot\hat{\mathbf{p}}|n\rangle|^2 \left|\frac{2}{\omega-\omega_{kn}} e^{-i(\omega-\omega_{kn})t/2}\sin\left[\tfrac{1}{2}(\omega-\omega_{kn})t\right]\right|^2,$$

which simplifies to

$$P_k \simeq \frac{1}{\hbar^2}\left(\frac{qA_0}{m_0}\right)^2 |\langle k|\mathbf{e}\cdot\hat{\mathbf{p}}|n\rangle|^2 \frac{\sin^2\left[\frac{1}{2}(\omega-\omega_{kn})t\right]}{(\omega-\omega_{kn})^2} \quad (11.24)$$

and is the desired two-level system transition probability, starting from the alternative interaction Hamiltonian $\hat{H}^{(1)}$. In terms of energy, one writes

$$P_k \simeq \left(\frac{qA_0}{m_0}\right)^2 |\langle k|\mathbf{e}\cdot\hat{\mathbf{p}}|n\rangle|^2 \frac{\sin^2\left(\frac{E-E_{kn}}{2\hbar}t\right)}{(E-E_{kn})^2}, \quad (11.25)$$

since $E=\hbar\omega$. Note that both Equations 11.24 and 11.25 are general expressions, which do not assume a polarization for the incident light. We can, however, choose to consider z-polarized light as before. In this case, $\mathbf{e}\cdot\hat{\mathbf{p}} = p_z$ in Equations 11.24 and 11.25, with $\hat{p}_z = -i\hbar\,\partial/\partial z$ the z quantum mechanical momentum operator.

11.6.1 Associated Rates

Next, to obtain the associated transition rate, we begin with Equation 11.25 (alternatively with Equation 11.24):

$$P_k \simeq \left(\frac{qA_0}{m_0}\right)^2 |\langle k|\mathbf{e}\cdot\hat{\mathbf{p}}|n\rangle|^2 \frac{\sin^2\left(\frac{E-E_{kn}}{2\hbar}t\right)}{(E-E_{kn})^2}$$

and let $\alpha = (E-E_{kn})/2\hbar$ so that

$$P_k \simeq \left(\frac{qA_0}{m_0}\right)^2 |\langle k|\mathbf{e}\cdot\hat{\mathbf{p}}|n\rangle|^2 \frac{\sin^2\alpha t}{4\hbar^2\alpha^2}.$$

The transition rate is then the time derivative of this expression:

$$\begin{aligned} R_k = \frac{dP_k}{dt} &\simeq \left(\frac{qA_0}{m_0}\right)^2 \frac{|\langle k|\mathbf{e}\cdot\hat{\mathbf{p}}|n\rangle|^2}{4\hbar^2\alpha^2}\frac{d\sin^2\alpha t}{dt} \\ &\simeq \left(\frac{qA_0}{m_0}\right)^2 \frac{|\langle k|\mathbf{e}\cdot\hat{\mathbf{p}}|n\rangle|^2}{4\hbar^2\alpha^2} 2\alpha\sin\alpha t\cos\alpha t \\ &\simeq \left(\frac{qA_0}{m_0}\right)^2 \frac{|\langle k|\mathbf{e}\cdot\hat{\mathbf{p}}|n\rangle|^2}{4\hbar^2\alpha}\sin 2\alpha t. \end{aligned}$$

Replacing an actual value for α then yields our desired two-level system transition rate

$$R_k \simeq \left(\frac{qA_0}{m_0}\right)^2 \frac{|\langle k|\mathbf{e}\cdot\hat{\mathbf{p}}|n\rangle|^2}{2\hbar} \frac{\sin\left(\frac{E-E_{kn}}{\hbar}t\right)}{E-E_{kn}}. \quad (11.26)$$

Alternatively, we can re-express the rate in the limit where $t\to\infty$. Thus, starting with

$$P_k \simeq \left(\frac{qA_0}{m_0}\right)^2 |\langle k|\mathbf{e}\cdot\hat{\mathbf{p}}|n\rangle|^2 \frac{\sin^2\alpha t}{4\hbar^2\alpha^2},$$

we recognize that (Equation 11.14)

$$\lim_{t\to\infty}\frac{\sin^2\alpha t}{\pi\alpha^2 t}=\delta(\alpha)$$

and therefore that

$$P_k \simeq \left(\frac{qA_0}{m_0}\right)^2 |\langle k|\mathbf{e}\cdot\hat{\mathbf{p}}|n\rangle|^2\left(\frac{\pi t}{4\hbar^2}\right)\left(\frac{\sin^2\alpha t}{\pi\alpha^2 t}\right).$$

The infinite-time limit then gives

$$\begin{aligned}P_{k,t\to\infty} &= \left(\frac{qA_0}{m_0}\right)^2 |\langle k|\mathbf{e}\cdot\hat{\mathbf{p}}|n\rangle|^2\left(\frac{\pi t}{4\hbar^2}\right)\delta(\alpha)\\ &= \left(\frac{qA_0}{m_0}\right)^2 |\langle k|\mathbf{e}\cdot\hat{\mathbf{p}}|n\rangle|^2\left(\frac{\pi t}{4\hbar^2}\right)\delta\left(\frac{E-E_{kn}}{2\hbar}\right),\end{aligned}$$

and since

$$\delta(cx)=\frac{1}{|c|}\delta(x),$$

we find

$$P_{k,t\to\infty}=\frac{\pi}{2\hbar}\left(\frac{qA_0}{m_0}\right)^2|\langle k|\mathbf{e}\cdot\hat{\mathbf{p}}|n\rangle|^2\delta(E-E_{kn})t \qquad (11.27)$$

as the desired transition probability.

Finally, $R_k = dP_{k,t\to\infty}/dt$ gives

$$R_k=\frac{2\pi}{\hbar}\left(\frac{qA_0}{2m_0}\right)^2|\langle k|\mathbf{e}\cdot\hat{\mathbf{p}}|n\rangle|^2\delta(E-E_{kn}), \qquad (11.28)$$

which is our desired rate. Notice the factor of $2\pi/\hbar$ that emerges.

11.7 THOUGHT PROBLEMS

11.1 Interaction Hamiltonian

When deriving the time-dependent transition probability and transition rate for a two-level system, an interaction Hamiltonian using $\cos\omega t$ was assumed. Show that the same results can be found by assuming an interaction Hamiltonian proportional to $\sin\omega t$.

11.2 Transition probability

Plot the effective transition probability

$$P_2(\omega)\propto\frac{\sin^2\left[\frac{1}{2}(\omega-\omega_{21})t\right]}{(\omega-\omega_{21})^2}$$

using $\omega_{21}=50$ in arbitrary frequency units and times that take values of $t = 1, 1.5, 2, 3$. In addition, plot the associated (effective) transition rate as a function of the same parameters.

11.3 Transition probability

Show using l'Hôpital's rule that for monochromatic light, on resonance with the transition of a two-level system, the transition probability grows quadratically with time. Start with

$$P_k=\frac{E_0^2|\langle k|\mu_z|n\rangle|^2}{\hbar^2}\frac{\sin^2\left[\frac{1}{2}(\omega-\omega_{kn})t\right]}{(\omega-\omega_{kn})^2}$$

and plot a few points when $\omega=\omega_{kn}$, comparing these values with $f(t)=t^2$.

11.4 Transition probability

For a given system, find the transition probability to an excited state $|k\rangle$ at long times when the interaction Hamiltonian has the form

$$\hat{H}^{(1)}=H_0(x)(1-e^{-t/\tau}).$$

In this expression, τ reflects the timescale for turning on the perturbation. To simplify things, assume very long times and the case where $\omega_{kl}\gg 1/\tau$.

11.5 Transition probability

Model the absorption of a two-level system to first order when subjected to a perturbation of the form

$$\hat{H}^{(1)} = -\mu_z E_0 e^{\eta t} \sin \omega t.$$

η is a small positive number that can be used to turn on the perturbation slowly rather than instantaneously. Provide expressions for $C_2(t)$, $P_2(t)$, and $R_2(t)$. At the end, let η go to zero. Hint: Let $t_0 \to -\infty$ to simplify things.

11.6 Transition probability

Consider the one-dimensional harmonic oscillator problem discussed previously in Chapter 8. The wavefunctions of the ground and first two excited states are

$$|0\rangle = \left(\frac{\gamma}{\pi}\right)^{1/4} e^{-\gamma x^2/2},$$

$$|1\rangle = \left(\frac{4\gamma^3}{\pi}\right)^{1/4} x e^{-\gamma x^2/2},$$

$$|2\rangle = \left(\frac{\gamma}{4\pi}\right)^{1/4} (2\gamma x^2 - 1) e^{-\gamma x^2/2},$$

where $\gamma = m\omega/\hbar$. Assume that a potential pulse is applied to the system with an interaction Hamiltonian of the form

$$\hat{H}^{(1)}(t) = \begin{cases} 0 & (t < 0) \\ qE_0 \hat{x} e^{-\beta t} & (t \geq 0). \end{cases}$$

Find the probability of finding the system (a) in the first and (b) in the second excited state at long times (i.e., as $t \to \infty$).

11.7 Transition probability

Consider the hydrogen atom problem first seen in Chapter 6 (Problems 6.1–6.4 and 6.22). A potential pulse is applied to the system with an interaction potential having the form

$$\hat{H}^{(1)} = \begin{cases} 0 & (t < 0), \\ qE_0 \hat{z} e^{-\beta t} & (t \geq 0). \end{cases}$$

Determine the probability of finding the electron in (a) the 2s and (b) the 2p ($m = 0$) excited state. The ground and first two excited states of interest have the following wavefunctions:

$$\psi_{1s} = \frac{1}{\sqrt{\pi}} \left(\frac{1}{a_0}\right)^{3/2} e^{-r/a_0}$$

$$\psi_{2s} = \frac{1}{\sqrt{32\pi}} \left(\frac{1}{a_0}\right)^{3/2} \left(2 - \frac{r}{a_0}\right) e^{-r/2a_0},$$

$$\psi_{2p,m=0} = \frac{1}{\sqrt{32\pi}} \left(\frac{1}{a_0}\right)^{3/2} \left(\frac{r}{a_0}\right) e^{-r/2a_0} \cos\theta.$$

To simplify things, consider the long-time limit where $t \to \infty$.

11.8 Transition probability

A particle is in the ground state of an infinite one-dimensional box with walls at $x = 0$ and $x = a$. At $t > 0$, it is subjected to a perturbation of the form

$$\hat{H}^{(1)}(t) = x^2 e^{-t/\tau}.$$

Calculate the probability of finding the particle in its first excited state at $t > 0$ to first order.

Hints: As $t \to \infty$,

$$P_2 = \left(\frac{16a^2}{9\pi^2\hbar}\right)^2 \left(\frac{9\pi^4\hbar^2}{4m^2a^4} + \frac{1}{\tau^2}\right)^{-1}.$$

In addition, from a table of integrals,

$$\int \sin ax \sin bx \, dx = \frac{\sin(a-b)x}{2(a-b)} - \frac{\sin(a+b)x}{2(a+b)}.$$

11.9 Transition probability

Consider the transition $|1\rangle \to |k\rangle$ where the perturbation is a square-like pulse and the interaction Hamiltonian is

$$\hat{H}^{(1)} = \begin{cases} V_0 \dfrac{\cos \omega t}{\sqrt{t_1 - t_0}} & (t_0 \leq t \leq t_1), \\ 0 & (t < t_0;\, t > t_1). \end{cases}$$

(a) Calculate the associated transition probability $P_k(\omega)$ under-near-resonance conditions (i.e., $\omega \simeq \omega_{k1}$).

(b) Plot $P_k(\omega)$.

(c) Evaluate the peak transition probability on resonance (Hint: use l'Hôpital's rule) and describe how the peak probability varies with pulse width, $t_1 - t_0$.

(d) Finally, instead of a two-level system, consider a near-continuum of final states with a density of states $\rho(E_k)$, where $E_k = \hbar\omega_{k1}$. Find the associated transition probability. There are a number ways to approach this last part. Is there any dependence of the transition probability with pulse width?

11.10 Gaussian integral

In preparation for the problems below, show that the following Gaussian integral is true:

$$\int_{-\infty}^{\infty} e^{-ax^2} \, dx = \sqrt{\frac{\pi}{a}}.$$

This relationship occurs quite frequently and is useful to know.

Hint: If

$$I = \int_{-\infty}^{\infty} e^{-ax^2} \, dx,$$

then

$$I^2 = \int_{-\infty}^{\infty} e^{-ax^2} \, dx \int_{-\infty}^{\infty} e^{-ay^2} \, dy.$$

Finally, switch to polar coordinates.

11.11 Gaussian Fourier transform

In preparation for the problems below, consider the following Gaussian function:

$$\phi(t) = A e^{-a(t-t_0)^2/4},$$

where both A and a are constants. Find its Fourier transform

$$\phi(\omega) = \frac{1}{\sqrt{2\pi}} \int_{-\infty}^{\infty} \phi(t) e^{i\omega t}\, dt.$$

Hint: Complete the square.

11.12 Transition probability

Consider a system where a Gaussian pulse electric field is turned on. The interaction Hamiltonian has the form

$$\hat{H}^{(1)} = H_0(x) e^{-t^2/\tau^2}.$$

For simplicity, assume limits of $-\infty$ and ∞ in any subsequent integrals. Evaluate the probability of finding the system in an excited state k.

11.13 Transition probability

Consider the following $|1\rangle \to |k\rangle$ transition where the perturbation is a Gaussian pulse and the interaction Hamiltonian is

$$\hat{H}^{(1)} = \begin{cases} V_0 \dfrac{\cos \omega t}{\sqrt{\tau}} e^{-(t-t_0)^2/\tau^2} & (t_0 \le t \le t_1), \\ 0 & (t < t_0;\, t > t_1). \end{cases}$$

(a) Calculate the associated transition probability $P_k(\omega)$ under near-resonance conditions (i.e., $\omega \simeq \omega_{k1}$).

(b) Plot $P_k(\omega)$.

(c) Evaluate the peak transition probability on resonance and describe how the peak probability varies with the pulse width τ.

(d) Finally, instead of a two-level system, consider a near-continuum of final states with a density of states $\rho(E_k)$, where $E_k = \hbar\omega_{k1}$. Find the associated transition probability. Is there any dependence of the transition probability with pulse width?

11.14 Instantaneous perturbation

We have seen that time-dependent perturbation theory assumes that the magnitude of the perturbation is small. We can also consider how quickly or how slowly the perturbation is turned on. For simplicity, consider the case where the perturbation is turned on instantaneously. In this case, the problem involves evaluating a change of basis. Namely, to consider the likelihood of being in a given final state of the new basis, one needs to evaluate the overlap integral between the initial state (expressed in the old basis) and the final state in the new basis (i.e., $P = |\langle k_{\text{new basis}} | 1_{\text{old basis}}\rangle|^2$).

Consider the particle in a one-dimensional box with walls at 0 and a. It is initially in its ground state. Now assume that a perturbation is applied, instantaneously doubling the width of the box. Find the likelihood of finding the particle in its ground and first excited states when in the new box.

11.15 Instantaneous perturbation

A particle is initially in the ground state of a one-dimensional harmonic oscillator potential where $V(x) = \frac{1}{2} k_{\text{Hooke}} x^2$. Assume that the system's spring constant is suddenly doubled owing to an unspecified perturbation. Estimate the likelihood of finding the particle in the ground state of the new potential.

11.16 Instantaneous perturbation

A particle is initially in the ground state of a two-dimensional harmonic oscillator potential where

$$V(x, y) = \tfrac{1}{2} k_{\text{Hooke}} (x^2 + y^2).$$

(a) If the spring constant k_{Hooke} is suddenly doubled, calculate the probability of finding the particle in the ground state of the new potential.

(b) Calculate the probability of finding it in the first and second excited states.

11.17 Oscillator strength

Show that the oscillator strength f_{kn} of a transition, defined as

$$f_{kn} = \frac{2m\omega}{\hbar} |\langle k|z|n\rangle|^2$$

(note the z polarization of the excitation), can be derived from the ratio of the quantum mechanical expectation value for a transition dipole (obtained using time-dependent perturbation theory) with the dipole moment found from a classical oscillator picture. We alluded to such classical models in Chapter 3. Namely, we found that the following second-order differential equation described the time-dependent displacement of an electron under illumination:

$$m\frac{d^2 x(t)}{dt^2} + \Gamma \frac{dx(t)}{dt} + m\omega_{kn}^2 x(t) = -qE(t).$$

In this expression, $x(t)$ is the electron's displacement, m is its mass, ω_{kn} is the oscillator's characteristic resonance frequency, and Γ is a phenomenological damping constant.

To do this problem, first assume $\Gamma = 0$ for an ideal oscillator that experiences no damping. Next, assume $E = E_0 \cos \omega t$ for the electric field of the incident light. Then solve for the time-dependent dipole $\mu(t) = qx(t)$ to obtain

$$\mu(t) = \frac{\dfrac{q^2 E_0}{m} \cos \omega t}{\omega_{kn}^2 - \omega^2}.$$

At this point, use time-dependent perturbation theory to find the time-dependent weight $c_k(t)$, leading to the following wavefunction for a given state $|n\rangle$, under excitation:

$$|\psi(t)\rangle = c_n(t)|n\rangle e^{-i\omega_n t} + \sum_k c_k(t) e^{-i\omega_k t}|k\rangle.$$

To simplify things, assume only one other state, k, in the system, leading to

$$|\psi(t)\rangle = c_n(t)|n\rangle e^{-i\omega_n t} + c_k(t) e^{-i\omega_k t}|k\rangle,$$

where $c_n(t) \simeq 1$. Then find the expectation value of the dipole via

$$\langle \mu_{kn}\rangle = \langle \psi(t)|\mu_z|\psi(t)\rangle,$$

which results in

$$\langle \mu_{kn}\rangle = \frac{2q^2 E_0 \omega_{kn} |\langle k|z|n\rangle|^2 \cos \omega t}{\hbar(\omega_{kn}^2 - \omega^2)}.$$

The ratio of $\langle \mu_{kn}\rangle$ to $\mu(t)$ is then the desired oscillator strength.

11.8 FURTHER READING

Green NJB (1998) *Quantum Mechanics 2: The Toolkit.* Oxford University Press, Oxford, UK.

Griffiths DJ (2004) *Introduction to Quantum Mechanics*, 2nd edn. Prentice Hall, Upper Saddle River, NJ.

Liboff RL (2003) *Introductory Quantum Mechanics*, 4th edn. Addison Wesley, San Francisco, CA.

McQuarrie DA (2008) *Quantum Chemistry*, 2nd edn. University Science Books, Mill Valley, CA.

Schatz GC, Ratner MA (2002) *Quantum Mechanics in Chemistry.* Dover Publications, New York.

Zettili N (2009) *Quantum Mechanics, Concepts and Applications*, 2nd edn. Wiley, Chichester, UK.

Interband Transitions

12.1 INTRODUCTION

In this section, we wish to model the absorption of light by semiconductors. This will be accomplished using results from Chapter 11. Namely, Fermi's Golden Rule will be used to describe the rate of an optical transition, in order to evaluate the absorption coefficient of a system.

Recall from time-dependent perturbation theory that on applying a time-dependent perturbation, eigenstates of the system in the absence of the perturbation are no longer valid solutions to Schrödinger's equation. Instead, general solutions can be described by linear combinations of these original eigenstates. The relative contributions of each are given by time-dependent weights denoted by $c_k(t)$, with k an index identifying a specific state.

In particular, we found that $c_k(t)$ could be expressed as

$$c_k(t) = \frac{1}{i\hbar}\int_0^t \langle k|\hat{H}^{(1)}|n\rangle e^{i\omega_{kn}t'}\,dt',$$

which is a generalized version of Equation 11.4, where n denotes the index of the initial state, k is the index of the final state, and ω_{kn} is the frequency difference between the two. $\hat{H}^{(1)}$ is our time-dependent perturbation, which can be expressed in various ways as will be seen shortly. Keep in mind that light is our implicit time-dependent harmonic perturbation, underlying $\hat{H}^{(1)}$.

Now, the probability of being in a final state k is simply the square of the time-dependent weight:

$$P_k = |c_k(t)|^2$$

(Equation 11.5). As a consequence, the following transition probability exists between an initial state and a final state (Equation 11.6):

$$P_k = \frac{1}{\hbar^2}\left|\int_0^t \langle k|\hat{H}^{(1)}|n\rangle e^{i\omega_{kn}t'}\,dt'\right|^2.$$

In what follows, we shall derive more specific expressions for P_k and its associated transition rate for a bulk semiconductor by introducing appropriate wavefunctions for $|k\rangle$ and $|n\rangle$. Ultimately, though, what we want are expressions for P_k and the transition rate for the low-dimensional systems we have discussed earlier. We then end the chapter with a more detailed comparison of these theoretical transitions with those seen experimentally in the case of a model quantum dot system, namely CdSe nanocrystals.

12.2 PRELIMINARIES: $-\hat{\mu}\cdot\mathbf{E}$ VERSUS $-(q/m_0)\mathbf{A}\cdot\hat{\mathbf{p}}$

Before proceeding, we need some preliminaries. We saw in Chapter 11 that the interaction Hamiltonian $\hat{H}^{(1)}$ between $|k\rangle$ and $|n\rangle$ could be written as (Equation 11.9)

$$\hat{H}^{(1)} = -\hat{\mu} \cdot \mathbf{E}.$$

In this expression, $\hat{\mu}$ is the dipole moment operator and $\mathbf{E}$ is the electric field of the incident light. We also saw in Chapter 11 that $E \propto \cos \omega t$ (Equation 11.8).

It is important to note that one alternatively sees written

$$\hat{H}^{(1)} = -\frac{q}{m_0}(\mathbf{A} \cdot \hat{\mathbf{p}}), \tag{12.1}$$

where $\mathbf{A}$ is the vector potential of the incident light and $\hat{\mathbf{p}}$ is the quantum mechanical momentum operator. Although different in appearance, both expressions ultimately lead to the same result. Because one often sees both expressions in the literature, we shall show their equivalence in what follows.

The origin of this second version of the interaction Hamiltonian stems from the classical Hamiltonian, resulting from the Lorentz force on a charged particle (i.e., the force on a charged particle in the presence of an electric and magnetic field). This Hamiltonian has the approximate form

$$\hat{H} \simeq -\frac{\hbar^2}{2m_0}\nabla^2 + V - \frac{q}{m_0}(\mathbf{A} \cdot \hat{\mathbf{p}}),$$

where a small second-order term has been dropped.

12.2.1 The Vector Potential A

Next, the vector potential of the incident light is related to its electric and magnetic fields through

$$\mathbf{E} = -\nabla\phi - \frac{\partial \mathbf{A}}{\partial t}, \tag{12.2}$$

$$\mathbf{B} = \nabla \times \mathbf{A}. \tag{12.3}$$

Both expressions are obtained from Maxwell's equations. In Equation 12.2, ϕ is a scalar potential associated with the electromagnetic field of the incident light.

In principle, knowing ϕ and $\mathbf{A}$ allows us to evaluate $\mathbf{E}$ and $\mathbf{B}$. However, they are not unique expressions since many combinations of ϕ and $\mathbf{A}$ lead to $\mathbf{E}$ and $\mathbf{B}$. This is what is meant by having multiple gauges. It is therefore customary to impose constraints on ϕ or $\mathbf{A}$ to add specificity. One common constraint is to let the divergence of the vector potential be zero:

$$\nabla \cdot \mathbf{A} = 0.$$

This is called the Coulomb gauge, and with this, one finds in the absence of free charges that

$$\mathbf{E} = -\frac{\partial \mathbf{A}}{\partial t},$$

$$\mathbf{B} = \nabla \times \mathbf{A}.$$

Thus, for a monochromatic plane wave where

$$\mathbf{E} = E_0\mathbf{e}\cos(\mathbf{k} \cdot \mathbf{r} - \omega t),$$

$$\mathbf{B} = \frac{\mathbf{k}}{\omega} \times \mathbf{E} = \frac{\mathbf{k}}{\omega} \times \mathbf{e}E_0\cos(\mathbf{k} \cdot \mathbf{r} - \omega t),$$

with $\mathbf{e}$ a unit polarization vector, E_0 the amplitude of the electric field, and $\mathbf{k}$ the wavevector pointing in the direction of light propagation, we have

$$\mathbf{A} = \frac{E_0}{\omega}\mathbf{e}\sin(\mathbf{k}\cdot\mathbf{r} - \omega t). \tag{12.4}$$

Alternatively,

$$\mathbf{A} = A_0\mathbf{e}\sin(\mathbf{k}\cdot\mathbf{r} - \omega t). \tag{12.5}$$

Note that the amplitude of the vector potential is related to that of the electric field by

$$A_0 = \frac{E_0}{\omega}. \tag{12.6}$$

It can be shown, although for brevity, this is not done here, that **A** and **E** are parallel to each other and are orthogonal to the magnetic field; i.e.,

$$\mathbf{E} \parallel \mathbf{A},$$
$$\mathbf{E} \perp \mathbf{B},$$
$$\mathbf{A} \perp \mathbf{B}.$$

Furthermore, **E**, **A**, and **B** are all orthogonal to **k**:

$$\mathbf{E} \perp \mathbf{k},$$
$$\mathbf{B} \perp \mathbf{k},$$
$$\mathbf{A} \perp \mathbf{k}.$$

One therefore has the situation depicted in **Figure 12.1**.

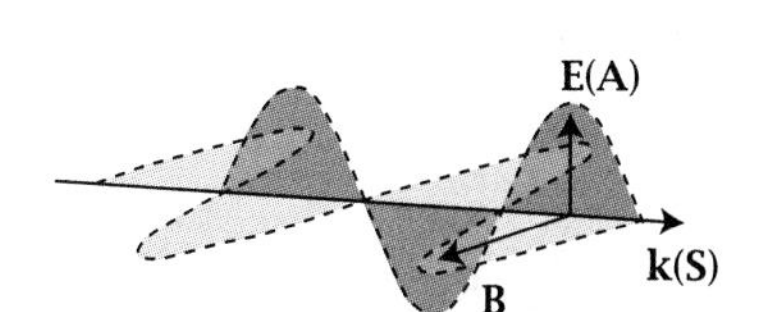

Figure 12.1 Illustration showing the relative orientations of the electric field **E**, magnetic field **B**, and associated vector potential **A** of an electromagnetic wave. Also shown is the Poynting vector **S** (introduced in Section 12.8.1), as well as their relationship to the wavevector **k** pointing in the direction of wave propagation.

12.3 BACK TO TRANSITION PROBABILITIES

At this point, starting with the interaction Hamiltonian

$$\hat{H}^{(1)} = -\frac{q}{m_0}\mathbf{A}\cdot\hat{\mathbf{p}},$$

where we have just found that

$$\mathbf{A} = A_0\mathbf{e}\sin(\mathbf{k}\cdot\mathbf{r} - \omega t),$$

we use the Euler relation to write

$$\hat{H}^{(1)} = -\frac{q}{m_0}A_0\left[\frac{e^{i(\mathbf{k}\cdot\mathbf{r}-\omega t)} - e^{-i(\mathbf{k}\cdot\mathbf{r}-\omega t)}}{2i}\right](\mathbf{e}\cdot\hat{\mathbf{p}}).$$

When this is substituted into our generic expression for the transition probability, we obtain

$$\begin{aligned}
P_k &= \frac{1}{\hbar^2}\left|\int_0^t \langle k|\hat{H}^{(1)}|n\rangle e^{i\omega_{kn}t'}dt'\right|^2 \\
&= \frac{1}{\hbar^2}\left(\frac{qA_0}{2m_0}\right)^2\left|\int_0^t \langle k|(e^{i(\mathbf{k}\cdot\mathbf{r}-\omega t')} - e^{-i(\mathbf{k}\cdot\mathbf{r}-\omega t')})(\mathbf{e}\cdot\hat{\mathbf{p}})|n\rangle e^{i\omega_{kn}t'}dt'\right|^2 \\
&= \frac{1}{\hbar^2}\left(\frac{qA_0}{2m_0}\right)^2\left|\int_0^t \langle k|(e^{i\mathbf{k}\cdot\mathbf{r}}e^{-i(\omega-\omega_{kn})t'} - e^{-i\mathbf{k}\cdot\mathbf{r}}e^{i(\omega+\omega_{kn})t'})(\mathbf{e}\cdot\hat{\mathbf{p}})|n\rangle dt'\right|^2.
\end{aligned}$$

Before going on, we caution the reader to make the distinction between the wavevector **k** and the final state $|k\rangle$. Notice now the two terms sandwiched by the bra and the ket. They contain $e^{-i(\omega-\omega_{kn})t'}$ and $e^{i(\omega+\omega_{kn})t'}$. The second, when integrated, leads to a nonresonant term, related to stimulated emission (Chapter 11). Since we are only interested in the former resonant term, which describes the absorption of light, we ignore the latter term and are left with

$$P_k \simeq \frac{1}{\hbar^2}\left(\frac{qA_0}{2m_0}\right)^2 \left|\int_0^t \langle k|e^{i\mathbf{k}\cdot\mathbf{r}} e^{-i(\omega-\omega_{kn})t'}(\mathbf{e}\cdot\hat{\mathbf{p}})|n\rangle dt'\right|^2 .$$

Recall that this procedure was called the rotating wave approximation. Removing $e^{-i(\omega-\omega_{kn})t'}$ from the bra–ket then yields

$$P_k \simeq \frac{1}{\hbar^2}\left(\frac{qA_0}{2m_0}\right)^2 \left|\int_0^t \langle k|e^{i\mathbf{k}\cdot\mathbf{r}}(\mathbf{e}\cdot\hat{\mathbf{p}})|n\rangle e^{-i(\omega-\omega_{kn})t'} dt'\right|^2 , \tag{12.7}$$

which is our generic transition probability, since we have not specified what our initial $|n\rangle$ and final $|k\rangle$ actually look like.

12.4 BULK SEMICONDUCTOR

Let us now introduce the bulk Bloch wavefunctions for $|k\rangle$ and $|n\rangle$. They are normalized forms that yield an electron density of one electron per state in the crystal's total volume (V_{tot}). Later, when discussing low-dimensional materials we shall use different wavefunctions, specific to wells, wires, and dots.

The general form of the bulk wavefunction is

$$\psi_{\text{tot}} = \frac{1}{\sqrt{V_{\text{tot}}}} e^{i\mathbf{k}\cdot\mathbf{r}} u(\mathbf{r}) = \frac{1}{\sqrt{NV_{\text{unit}}}} e^{i\mathbf{k}\cdot\mathbf{r}} u(\mathbf{r}), \tag{12.8}$$

where $u(\mathbf{r})$ represents the unit-cell part of the total Bloch wavefunction and $e^{i\mathbf{k}\cdot\mathbf{r}}$ is the associated plane-wave term, normalized by $1/\sqrt{V_{\text{tot}}}$. N represents the total number of unit cells in the finite crystal and V_{unit} is the unit-cell volume. Note that the unit-cell functions $u_n(\mathbf{r})$ and $u_k(\mathbf{r})$ are normalized such that

$$\langle u_i|u_i\rangle = \int_{V_{\text{unit}}} |u_i(\mathbf{r})|^2\, d^3r = V_{\text{unit}}. \tag{12.9}$$

In turn, the total Bloch wavefunction is normalized:

$$\langle \psi_{\text{tot}}|\psi_{\text{tot}}\rangle = \int_{V_{\text{tot}}=NV_{\text{unit}}} |\psi_{\text{tot}}|^2\, d^3r = 1. \tag{12.10}$$

We can illustrate this explicitly. Beginning with

$$\begin{aligned}\int_{V_{\text{tot}}} |\psi_{\text{tot}}|^2\, d^3r &= \int_{NV_{\text{unit}}} \left[\frac{1}{\sqrt{NV_{\text{unit}}}} e^{-i\mathbf{k}_i\cdot\mathbf{r}} u_i^*(\mathbf{r})\right]\left[\frac{1}{\sqrt{NV_{\text{unit}}}} e^{i\mathbf{k}_i\cdot\mathbf{r}} u_i(\mathbf{r})\right] d^3r\\ &= \frac{1}{NV_{\text{unit}}}\int_{NV_{\text{unit}}} |u_i(\mathbf{r})|^2\, d^3r,\end{aligned}$$

we approximate the integral as a sum of N equivalent unit-cell integrals to find

$$\begin{aligned}\int_{V_{\text{tot}}} |\psi_{\text{tot}}|^2 \, d^3r &\simeq \frac{1}{N}\sum_N \frac{1}{V_{\text{unit}}}\int_{V_{\text{unit}}} |u_i(\mathbf{r})|^2 \, d^3r \\ &\simeq \frac{1}{N}\sum_N \frac{V_{\text{unit}}}{V_{\text{unit}}} = \frac{1}{N}\sum_N 1 \\ &\simeq \frac{N}{N} = 1.\end{aligned}$$

This illustrates that the overall bulk wavefunction is normalized.

Returning to P_k, the initial and final states of the transition are therefore

$$\begin{aligned}\psi_n = |n\rangle &= \frac{1}{\sqrt{V_{\text{tot}}}} e^{i\mathbf{k}_n\cdot\mathbf{r}} u_n(\mathbf{r}) = \frac{1}{\sqrt{NV_{\text{unit}}}} e^{i\mathbf{k}_n\cdot\mathbf{r}} u_n(\mathbf{r}), \\ \psi_k = |k\rangle &= \frac{1}{\sqrt{V_{\text{tot}}}} e^{i\mathbf{k}_k\cdot\mathbf{r}} u_k(\mathbf{r}) = \frac{1}{\sqrt{NV_{\text{unit}}}} e^{i\mathbf{k}_k\cdot\mathbf{r}} u_k(\mathbf{r}).\end{aligned}$$

The first represents the hole left behind in the valence band, while the second represents the electron left behind in the conduction band.

Because of the translational symmetry of bulk Bloch wavefunctions, P_k can be written so that the wavevector $\mathbf{k}$ belongs exclusively to the first Brillouin zone. We therefore have

$$P_k \simeq \frac{1}{\hbar^2}\left(\frac{qA_0}{2m_0}\right)^2 \left|\int_0^t \langle k|e^{i\mathbf{k}\cdot\mathbf{r}}(\mathbf{e}\cdot\hat{\mathbf{p}})|n\rangle e^{-i(\omega-\omega_{kn})t'} dt'\right|^2,$$

where we can write the bra–ket term equivalently as

$$\begin{aligned}\langle k|e^{i\mathbf{k}\cdot\mathbf{r}}(\mathbf{e}\cdot\hat{\mathbf{p}})|n\rangle &= \int_{V_{\text{tot}}} \left[\frac{1}{\sqrt{NV_{\text{unit}}}} e^{-i\mathbf{k}_k\cdot\mathbf{r}} u_k^*(\mathbf{r})\right] e^{i\mathbf{k}\cdot\mathbf{r}}(\mathbf{e}\cdot\hat{\mathbf{p}}) \\ &\quad \times \left[\frac{1}{\sqrt{NV_{\text{unit}}}} e^{i\mathbf{k}_n\cdot\mathbf{r}} u_n(\mathbf{r})\right] d^3r \\ &= \frac{1}{NV_{\text{unit}}}\int_{V_{\text{tot}}} e^{i(\mathbf{k}-\mathbf{k}_k)\cdot\mathbf{r}} u_k^*(\mathbf{r})(\mathbf{e}\cdot\hat{\mathbf{p}})u_n(\mathbf{r})e^{i\mathbf{k}_n\cdot\mathbf{r}} d^3r.\end{aligned}$$

To continue evaluating the integral, let us make an important approximation. Namely, we recognize that the length scales over which $u_n(\mathbf{r})$ and $u_k(\mathbf{r})$ vary are significantly different from the length scales over which the plane-wave terms $e^{i\mathbf{k}_k\cdot\mathbf{r}}$, $e^{i\mathbf{k}_n\cdot\mathbf{r}}$, and $e^{i\mathbf{k}\cdot\mathbf{r}}$ fluctuate. The former oscillate quickly on the atomic scale, whereas the latter plane-wave terms are, relatively speaking, nearly constant. This is illustrated schematically in Figure 12.2. As a consequence, these plane-wave functions can be removed from the integral, and we can simply consider a sum of N unit-cell integrals:

$$\langle k|e^{i\mathbf{k}\cdot\mathbf{r}}(\mathbf{e}\cdot\hat{\mathbf{p}})|n\rangle \simeq \frac{1}{N}\sum_N \frac{e^{i[\mathbf{k}+(\mathbf{k}_n-\mathbf{k}_k)]\cdot\mathbf{r}}}{V_{\text{unit}}}\int_{V_{\text{unit}}} u_k^*(\mathbf{r})(\mathbf{e}\cdot\hat{\mathbf{p}})u_n(\mathbf{r})\, d^3r.$$

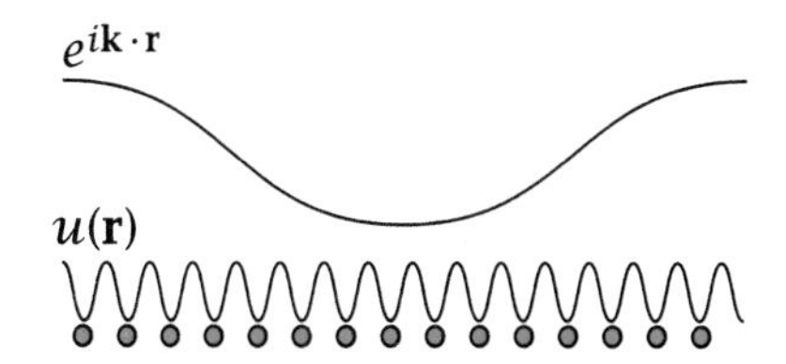

Figure 12.2 Illustration of the long-wavelength or electric dipole approximation.

Next, because of momentum conservation during transitions between the initial and final state, we have

$$\begin{aligned}\hbar\mathbf{k}_k &= \hbar\mathbf{k}_n + \hbar\mathbf{k}, \\ \mathbf{k}_k &= \mathbf{k}_n + \mathbf{k}.\end{aligned}$$

Thus,

$$\begin{aligned}
\langle k|e^{i\mathbf{k}\cdot\mathbf{r}}(\mathbf{e}\cdot\hat{\mathbf{p}})|n\rangle &\simeq \frac{1}{N}\sum_{N}\frac{1}{V_{\text{unit}}}\int_{V_{\text{unit}}} u_k^*(\mathbf{r})(\mathbf{e}\cdot\hat{\mathbf{p}})u_n(\mathbf{r})\,d^3r \\
&\simeq \frac{1}{N}\left[\frac{N}{V_{\text{unit}}}\int_{V_{\text{unit}}} u_k^*(\mathbf{r})(\mathbf{e}\cdot\hat{\mathbf{p}})u_n(\mathbf{r})\,d^3r\right] \\
&\simeq \frac{1}{V_{\text{unit}}}\int_{V_{\text{unit}}} u_k^*(\mathbf{r})(\mathbf{e}\cdot\hat{\mathbf{p}})u_n(\mathbf{r})\,d^3r \\
&\simeq \langle u_k(\mathbf{r})|\mathbf{e}\cdot\hat{\mathbf{p}}|u_n(\mathbf{r})\rangle,
\end{aligned}$$

where

$$\langle u_k(\mathbf{r})|\mathbf{e}\cdot\hat{\mathbf{p}}|u_n(\mathbf{r})\rangle = \frac{1}{V_{\text{unit}}}\int_{V_{\text{unit}}} u_k^*(\mathbf{r})(\mathbf{e}\cdot\hat{\mathbf{p}})u_n(\mathbf{r})\,d^3r. \tag{12.11}$$

When this is replaced into our expression for the transition probability, we get

$$P_k \simeq \frac{1}{\hbar^2}\left(\frac{qA_0}{2m_0}\right)^2\left|\int_0^t \langle u_k(\mathbf{r})|\mathbf{e}\cdot\hat{\mathbf{p}}|u_n(\mathbf{r})\rangle\, e^{-i(\omega-\omega_{kn})t'}\,dt'\right|^2.$$

At this point, the bra–ket term comes out of the integral since it does not depend on time. This yields

$$P_k \simeq \frac{1}{\hbar^2}\left(\frac{qA_0}{2m_0}\right)^2 |\langle u_k(\mathbf{r})|\mathbf{e}\cdot\hat{\mathbf{p}}|u_n(\mathbf{r})\rangle|^2\left|\int_0^t e^{-i(\omega-\omega_{kn})t'}\,dt'\right|^2$$

and the last integral evaluates to

$$\int_0^t e^{-i(\omega-\omega_{kn})t'}\,dt' = \frac{e^{-i(\omega-\omega_{kn})t}-1}{-i(\omega-\omega_{kn})},$$

or

$$\begin{aligned}
&= \frac{2}{\omega-\omega_{kn}}e^{-i(\omega-\omega_{kn})t/2}\left[\frac{e^{i(\omega-\omega_{kn})t/2}-e^{-i(\omega-\omega_{kn})t/2}}{2i}\right] \\
&= \frac{2}{\omega-\omega_{kn}}e^{-i(\omega-\omega_{kn})t/2}\sin\left(\frac{\omega-\omega_{kn}}{2}t\right).
\end{aligned}$$

We then have

$$\begin{aligned}
P_k \simeq{}& \frac{1}{\hbar^2}\left(\frac{qA_0}{2m_0}\right)^2 |\langle u_k(\mathbf{r})|\mathbf{e}\cdot\hat{\mathbf{p}}|u_n(\mathbf{r})\rangle|^2 \\
&\times\left|\frac{2}{\omega-\omega_{kn}}e^{-i(\omega-\omega_{kn})t/2}\sin\left(\frac{\omega-\omega_{kn}}{2}t\right)\right|^2,
\end{aligned}$$

which ultimately reduces to our desired transition probability, expressed in terms of frequency:

$$P_k(\omega) \simeq \frac{1}{\hbar^2}\left(\frac{qA_0}{m_0}\right)^2 |\langle u_k(\mathbf{r})|\mathbf{e}\cdot\hat{\mathbf{p}}|u_n(\mathbf{r})\rangle|^2\,\frac{\sin^2\left(\frac{\omega-\omega_{kn}}{2}t\right)}{(\omega-\omega_{kn})^2}. \tag{12.12}$$

We can leave P_k written this way. However, we can also convert it into an expression with energies using $E = \hbar\omega$:

$$P_k(E) \simeq \left(\frac{qA_0}{m_0}\right)^2 |\langle u_k(\mathbf{r})|\mathbf{e}\cdot\hat{\mathbf{p}}|u_n(\mathbf{r})\rangle|^2\,\frac{\sin^2\left(\frac{E-E_{kn}}{2\hbar}t\right)}{(E-E_{kn})^2}. \tag{12.13}$$

Both Equations 12.12 and 12.13 are general and do not assume any specific polarization of the incident light.

Polarization-Specific Expressions

We can be more specific by assuming that $\mathbf{e}$ points along a given direction, say along the z direction of a Cartesian coordinate system, that is, $\mathbf{e} = (0, 0, 1)$. Then $\mathbf{e}\cdot\hat{\mathbf{p}} = (0, 0, 1)\cdot(\hat{p}_x, \hat{p}_y, \hat{p}_z) = \hat{p}_z = -i\hbar\,\partial/\partial z$, which is the z projection of the quantum mechanical momentum operator. The following polarization-specific transition probabilities result:

$$P_k(\omega) \simeq \frac{1}{\hbar^2}\left(\frac{qA_0}{m_0}\right)^2 \left|\langle u_k(\mathbf{r})|\hat{p}_z|u_n(\mathbf{r})\rangle\right|^2 \frac{\sin^2\left(\frac{\omega-\omega_{kn}}{2}t\right)}{(\omega-\omega_{kn})^2} \tag{12.14}$$

and

$$P_k(E) \simeq \left(\frac{qA_0}{m_0}\right)^2 \left|\langle u_k(\mathbf{r})|\hat{p}_z|u_n(\mathbf{r})\rangle\right|^2 \frac{\sin^2\left(\frac{E-E_{kn}}{2\hbar}t\right)}{(E-E_{kn})^2}. \tag{12.15}$$

Note that these transition probabilities are analogous to Equations 11.13 and 11.12, which were derived starting from the alternative form of the interaction Hamiltonian, $\hat{H}^{(1)} = -\hat{\mu}\cdot\mathbf{E}$. In this regard, generalized versions of Equations 11.13 and 11.12 are

$$P_k(\omega) \simeq \frac{E_0^2}{\hbar^2}\left|\langle f|\mu_z|i\rangle\right|^2 \frac{\sin^2\left(\frac{\omega-\omega_{fi}}{2}t\right)}{(\omega-\omega_{fi})^2} \tag{12.16}$$

and

$$P_k(E) \simeq E_0^2\left|\langle f|\mu_z|i\rangle\right|^2 \frac{\sin^2\left(\frac{E-E_{fi}}{2\hbar}t\right)}{(E-E_{fi})^2}, \tag{12.17}$$

whereupon we will now demonstrate the equivalence of Equations 12.14 and 12.16, as well as Equations 12.15 and 12.17.

12.5 EQUIVALENCE OF $\hat{H}^{(1)} = -(q/m_0)\mathbf{A}\cdot\hat{\mathbf{p}}$ AND $\hat{H}^{(1)} = -\hat{\mu}\cdot\mathbf{E}$

The resulting transition probabilities (Equations 12.14 and 12.15) should look familiar to what we saw earlier in Chapter 11 when first discussing Fermi's Golden Rule. The only real difference, apart from the prefactors, is the bra–ket term. Whereas we previously had the square of the transition dipole moment $\left|\langle f|\mu_z|i\rangle\right|^2$, we now have $\left|\langle f|\hat{p}_z|i\rangle\right|^2$. Here i and f simply represent generic initial and final states. In what follows, we show that these two expressions for P_k, obtained from two different interaction Hamiltonians

$$\hat{H}^{(1)} = -\frac{q}{m_0}\mathbf{A}\cdot\hat{\mathbf{p}}$$

and

$$\hat{H}^{(1)} = -\hat{\mu}\cdot\mathbf{E},$$

yield the same result. This was briefly alluded to in the Appendix to Chapter 11.

To begin, the P_k expressions, Equations 12.15 and 12.17, coincide if we recognize that

$$\hat{p}_z = \frac{im_0}{\hbar}[\hat{H}^{(0)}, \hat{z}] = \frac{im_0}{\hbar}(\hat{H}^{(0)}z - z\hat{H}^{(0)}). \tag{12.18}$$

Equation 12.18, $\hat{H}^{(0)}$ is the system's original unperturbed Hamiltonian, where $\hat{z} = z$ is the z-position operator (**Table 6.1**). For simplicity, we will assume the specific case of z-polarized light.

This expression can be justified by simply revisiting some of the earlier commutator relationships provided in Chapter 6. Specifically, starting with

$$\begin{aligned}\hat{p}_z &= \frac{im_0}{\hbar}[\hat{H}^{(0)}, \hat{z}] \\ &= \frac{i}{2\hbar}[\hat{p}_z^2, \hat{z}],\end{aligned}$$

we invoke the commutator relationship (**Table 6.2**)

$$[\hat{A}\hat{B}, \hat{C}] = \hat{A}[\hat{B}, \hat{C}] + [\hat{A}, \hat{C}]\hat{B},$$

to obtain

$$[\hat{p}_z^2, \hat{z}] = \hat{p}_z[\hat{p}_z, \hat{z}] + [\hat{p}_z, \hat{z}]\hat{p}_z.$$

Next, recall that

$$[\hat{p}_z, \hat{z}] = -i\hbar,$$

which can be seen by operating on a generic placeholder wavefunction:

$$\begin{aligned}[\hat{p}_z, \hat{z}]\psi &= (\hat{p}_z z - z\hat{p})\psi \\ &= -i\hbar\frac{\partial}{\partial z}(z\psi) - z\left(-i\hbar\frac{\partial\psi}{\partial z}\right) \\ &= -i\hbar\left(z\frac{\partial\psi}{\partial z} + \psi\right) + i\hbar z\frac{\partial\psi}{\partial z} \\ &= -i\hbar\psi.\end{aligned}$$

As a consequence,

$$\begin{aligned}[\hat{p}_z^2, \hat{z}] &= \hat{p}_z(-i\hbar) + (-i\hbar)\hat{p}_z, \\ &= -2i\hbar\hat{p}_z\end{aligned}$$

and, when reintroduced into our original expression, we obtain

$$\hat{p}_z = \frac{i}{2\hbar}[\hat{p}_z^2, \hat{z}] = \hat{p}_z.$$

This shows that the invoked expression (Equation 12.18) makes sense.

Now, given

$$\hat{p}_z = \frac{im_0}{\hbar}[\hat{H}^{(0)}, \hat{z}],$$

we insert this into $\langle f|\hat{p}_z|i\rangle$ to see what happens. We find

$$\begin{aligned}\langle f|\hat{p}_z|i\rangle &= \langle f|\frac{im_0}{\hbar}[\hat{H}^{(0)},\hat{z}]|i\rangle \\ &= \langle f|\frac{im_0}{\hbar}(\hat{H}^{(0)}z - z\hat{H}^{(0)})|i\rangle \\ &= \frac{im_0}{\hbar}(\langle f|\hat{H}^{(0)}z|i\rangle - \langle f|z\hat{H}^{(0)}|i\rangle),\end{aligned}$$

which is equivalent to

$$\begin{aligned}\langle f|\hat{p}_z|i\rangle &= \frac{im_0}{\hbar}\left(\langle f|E_f z|i\rangle - \langle f|zE_i|i\rangle\right) \\ &= \frac{im_0}{\hbar}\langle f|z|i\rangle(E_f - E_i).\end{aligned}$$

In the resulting expression, E_i and E_f are the energies of the initial and final states. Since $E = \hbar\omega$, we also find that, in terms of frequency,

$$\begin{aligned}\langle f|\hat{p}_z|i\rangle &= im_0\langle f|z|i\rangle(\omega_f - \omega_i) \\ &= im_0\langle f|z|i\rangle\omega_{fi},\end{aligned}$$

where $\omega_{fi} = \omega_f - \omega_i$.

Inserting this into $P_k(E)$ (Equation 12.15) then gives

$$\begin{aligned}P_k(E) &\simeq \left(\frac{qA_0}{m_0}\right)^2 |\langle f|\hat{p}_z|i\rangle|^2 \frac{\sin^2\left(\frac{E-E_{fi}}{2\hbar}t\right)}{(E-E_{fi})^2} \\ &\simeq \left(\frac{qA_0}{m_0}\right)^2 |im_0\langle f|z|i\rangle\omega_{fi}|^2 \frac{\sin^2\left(\frac{E-E_{fi}}{2\hbar}t\right)}{(E-E_{fi})^2} \\ &\simeq A_0^2\omega_{fi}^2 |\langle f|qz|i\rangle|^2 \frac{\sin^2\left(\frac{E-E_{fi}}{2\hbar}t\right)}{(E-E_{fi})^2}\end{aligned}$$

and, since $\mu_z = qz$, we also find that

$$P_k(E) \simeq A_0^2\omega_{fi}^2 |\langle f|\mu_z|i\rangle|^2 \frac{\sin^2\left(\frac{E-E_{fi}}{2\hbar}t\right)}{(E-E_{fi})^2}.$$

Finally, recall that $A_0 = E_0/\omega$ ($\omega = \omega_{fi}$, Equation 12.6), to obtain

$$P_k(E) \simeq E_0^2 |\langle f|\mu_z|i\rangle|^2 \frac{\sin^2\left(\frac{E-E_{fi}}{2\hbar}t\right)}{(E-E_{fi})^2},$$

which is identical to Equation 12.17. The same procedure can be used to show that Equation 12.14 is equivalent to Equation 12.16. This is left as an exercise for the reader.

12.6 MULTIPLE STATES

We now have the transition probability of a two-level system. In addition, we have seen that, irrespective of whether one starts with $\hat{H}^{(1)} = -\hat{\mu}\cdot\mathbf{E}$ or $\hat{H}^{(1)} = -(q/m_0)\mathbf{A}\cdot\hat{\mathbf{p}}$, the final results are the same. Let us now consider a more realistic scenario. Instead of an effective two-level system, what if there are multiple transitions possible? We then need to consider the total transition probability by integrating P_k over all available states involved. This means evaluating

$$P_{\text{tot}} = \int_{-\infty}^{\infty} \rho_{\text{energy}}(E_{kn}) P_k(E)\, dE_{kn},$$

where $\rho_{\text{energy}}(E_{kn})$ is the density of states available for the transition within the energy range of interest. In semiconductors, we have seen that transitions occur from the valence band to the conduction band. Thus, instead of ρ_{energy}, we use the joint density of states, $\rho_{\text{joint}}(E_{kn})$, first introduced in Chapter 9.

For continuity, let us also start with $P_k(E)$ from above, where $f \to |u_k(\mathbf{r})\rangle$ and $i \to |u_n(\mathbf{r})\rangle$. Recall that this is a polarization-specific transition probability. More general expressions can be obtained using polarization-nonspecific transition probabilities. We leave this to the reader. We thus find that

$$\begin{aligned} P_{\text{tot}} &\simeq \int_{-\infty}^{\infty} \rho_{\text{joint}}(E_{kn}) E_0^2 \left|\langle u_k(\mathbf{r})|\mu_z|u_n(\mathbf{r})\rangle\right|^2 \frac{\sin^2\left(\frac{E-E_{kn}}{2\hbar}t\right)}{(E-E_{kn})^2}\, dE_{kn} \\ &\simeq E_0^2 \left|\langle u_k(\mathbf{r})|\mu_z|u_n(\mathbf{r})\rangle\right|^2 \int_{-\infty}^{\infty} \rho_{\text{joint}}(E_{kn}) \frac{\sin^2\left(\frac{E-E_{kn}}{2\hbar}t\right)}{(E-E_{kn})^2}\, dE_{kn}, \end{aligned}$$

where the bulk joint density of states found previously is (Equation 9.35)

$$\rho_{\text{joint}}(E_{kn}) = \frac{1}{2\pi^2}\left(\frac{2\mu}{\hbar^2}\right)^{3/2} \sqrt{E_{kn} - E_g}.$$

In ρ_{joint}, μ is the electron–hole reduced mass and E_g is the bulk band gap.

As $t \to \infty$, the last term in P_{tot} gets taller and narrower. It ultimately becomes the Dirac delta function (**Figure 11.2**). In this regard, we have previously seen that (Equation 11.14)

$$\lim_{t\to\infty} \frac{\sin^2 \alpha t}{\pi \alpha^2 t} = \delta(\alpha).$$

As a consequence, on re-expressing terms, we find

$$P_{\text{tot}} \simeq E_0^2 \left|\langle u_k(\mathbf{r})|\mu_z|u_n(\mathbf{r})\rangle\right|^2 \int_{-\infty}^{\infty} \rho_{\text{joint}}(E_{kn}) \left(\frac{\pi t}{4\hbar^2}\right) \frac{\sin^2 \alpha t}{\pi \alpha^2 t}\, dE_{kn},$$

where $\alpha = (E - E_{kn})/2\hbar$. Then, as $t \to \infty$

$$P_{\text{tot},t\to\infty} \simeq E_0^2 \left|\langle u_k(\mathbf{r})|\mu_z|u_n(\mathbf{r})\rangle\right|^2 \int_{-\infty}^{\infty} \rho_{\text{joint}}(E_{kn}) \left(\frac{\pi t}{4\hbar^2}\right) \delta\left(\frac{E-E_{kn}}{2\hbar}\right) dE_{kn}.$$

Alternatively, since $\delta(cx) = |c|^{-1}\delta(x)$,

$$P_{\text{tot},t\to\infty} \simeq E_0^2 \left|\langle u_k(\mathbf{r})|\mu_z|u_n(\mathbf{r})\rangle\right|^2 \int_{-\infty}^{\infty} \rho_{\text{joint}}(E_{kn}) \left(\frac{\pi t}{2\hbar}\right) \delta(E - E_{kn})\, dE_{kn}.$$

The integral is now easy to evaluate, and yields

$$P_{\text{tot},t\to\infty} \simeq \frac{\pi}{2\hbar} E_0^2 \left|\langle u_k(\mathbf{r})|\mu_z|u_n(\mathbf{r})\rangle\right|^2 \rho_{\text{joint}}(E = E_{kn})t.$$

When the full expression for ρ_{joint} is introduced, we find that

$$P_{\text{tot},t\to\infty} \simeq \frac{\pi}{2\hbar} E_0^2 \left|\langle u_k(\mathbf{r})|\mu_z|u_n(\mathbf{r})\rangle\right|^2 \left[\frac{1}{2\pi^2}\left(\frac{2\mu}{\hbar^2}\right)^{3/2}\sqrt{E - E_g}\right] t. \tag{12.19}$$

This is our bulk transition probability per unit volume into a group of final states. An alternative but equivalent expression is obtained by integrating Equation 12.15 over all energies, considering the joint density of final states. The reader may verify the following result:

$$P_{\text{tot},t\to\infty} \simeq \frac{\pi}{2\hbar}\left(\frac{qA_0}{m_0}\right)^2 \left|\langle u_k(\mathbf{r})|\hat{p}_z|u_n(\mathbf{r})\rangle\right|^2 \left[\frac{1}{2\pi^2}\left(\frac{2\mu}{\hbar^2}\right)^{3/2}\sqrt{E - E_g}\right] t. \tag{12.20}$$

12.7 FERMI'S GOLDEN RULE AND THE ASSOCIATED TRANSITION RATE

Finally, from Fermi's Golden Rule, we can find the associated transition rate by taking the time derivative of $P_{\text{tot},t\to\infty}$:

$$R_{\text{tot}} = \frac{dP_{\text{tot},t\to\infty}}{dt}.$$

For the two equivalent versions of the transition probability just derived, we find

$$R_{\text{tot}} \simeq \frac{\pi}{2\hbar} E_0^2 \left|\langle u_k(\mathbf{r})|\mu_z|u_n(\mathbf{r})\rangle\right|^2 \left[\frac{1}{2\pi^2}\left(\frac{2\mu}{\hbar^2}\right)^{3/2}\sqrt{E - E_g}\right] \tag{12.21}$$

and

$$R_{\text{tot}} \simeq \frac{\pi}{2\hbar}\left(\frac{qA_0}{m_0}\right)^2 \left|\langle u_k(\mathbf{r})|\hat{p}_z|u_n(\mathbf{r})\rangle\right|^2 \left[\frac{1}{2\pi^2}\left(\frac{2\mu}{\hbar^2}\right)^{3/2}\sqrt{E - E_g}\right]. \tag{12.22}$$

Either represents the desired transition rate per unit volume in the bulk solid.

12.8 ABSORPTION COEFFICIENT α

We now close the loop by connecting R_{tot} to the bulk absorption coefficient α, first discussed in Chapter 5. Recall that it has typical units of inverse length (cm^{-1}) and represents the absorption efficiency of the system. Recall also that for nanowires and quantum dots the absorption cross-section σ, with units of cm^2, was a preferred measure of the system's absorption efficiency.

Before connecting R_{tot} to α, we first need a more detailed description of the intensity of the incident light (or, alternatively, the irradiance), I (with units of $\text{W}\,\text{cm}^{-2}$). To do this, we introduce the Poynting vector.

12.8.1 Poynting Vector

The Poynting vector **S** "points" in the direction of propagation of the incident light. It reflects the direction of energy flow, since its magnitude is the power per unit area crossing a surface normal to the propagation direction. Formally, the Poynting vector is defined as

$$\mathbf{S} = \frac{1}{\mu\mu_0}(\mathbf{E} \times \mathbf{B}), \tag{12.23}$$

where **E** and **B** are the electric and magnetic fields of the incident radiation and μ (μ_0) is the relative permeability (permeability of free space) (note that μ should not be confused with a reduced mass). The cross product also means that **S** is oriented normally to both **E** and **B**, as shown in **Figure 12.1**.

To be more specific, let us assume that the incident light propagates along the z direction of a Cartesian coordinate system. In our case, the electric field is oriented along the x axis while the magnetic field points along the y axis. Both **E** and **B** take the form of traveling waves:

$$\mathbf{E} = (E_0 e^{i(kz-\omega t)}, 0, 0)$$
$$\mathbf{B} = (0, B_0 e^{i(kz-\omega t)}, 0).$$

We now evaluate **S**. We have (with $\mathbf{e}_x$, $\mathbf{e}_y$, and $\mathbf{e}_z$ unit vectors in the x, y, and z directions)

$$\begin{aligned}\mathbf{E} \times \mathbf{B} &= \begin{vmatrix} \mathbf{e}_x & \mathbf{e}_y & \mathbf{e}_z \\ E_0 e^{i(kz-\omega t)} & 0 & 0 \\ 0 & B_0 e^{i(kz-\omega t)} & 0 \end{vmatrix} \\ &= \mathbf{e}_x(0) - \mathbf{e}_y(0) + \mathbf{e}_z(E_0 e^{i(kz-\omega t)} B_0 e^{i(kz-\omega t)}) \\ &= \mathbf{e}_z E_0 B_0 e^{2i(kz-\omega t)},\end{aligned}$$

which yields

$$\mathbf{S} = \left(0, 0, \frac{E_0 B_0}{\mu\mu_0} e^{2i(kz-\omega t)}\right)$$

with

$$S_z = \frac{E_0 B_0}{\mu\mu_0} e^{2i(kz-\omega t)}.$$

Note that if the reader is unclear on how to obtain the cross product, this can easily be looked up in most elementary linear algebra texts.

At this point, the resulting expression is simplified using an additional relationship between E_0 and B_0. Recall from Maxwell's equations that

$$\nabla \times \mathbf{E} = -\frac{\partial \mathbf{B}}{\partial t}. \tag{12.24}$$

The cross product is then

$$\begin{aligned}\nabla \times \mathbf{E} &= \begin{vmatrix} \mathbf{e}_x & \mathbf{e}_y & \mathbf{e}_z \\ \dfrac{\partial}{\partial x} & \dfrac{\partial}{\partial y} & \dfrac{\partial}{\partial z} \\ E_0 e^{i(kz-\omega t)} & 0 & 0 \end{vmatrix} \\ &= \mathbf{e}_x(0) + \mathbf{e}_y(ikE_0 e^{i(kz-\omega t)}) + \mathbf{e}_z(0),\end{aligned}$$

and yields

$$\nabla \times \mathbf{E} = (0, ikE_0 e^{i(kz-\omega t)}, 0).$$

At the same time, the right-hand side of Equation 12.24 gives

$$-\frac{\partial \mathbf{B}}{\partial t} = i\omega B_0 e^{i(kz-\omega t)} \mathbf{e}_y$$

or

$$-\frac{\partial \mathbf{B}}{\partial t} = (0, i\omega B_0 e^{i(kz-\omega t)}, 0).$$

As a consequence,

$$ikE_0 e^{i(kz-\omega t)} = i\omega B_0 e^{i(kz-\omega t)}$$

$$kE_0 = \omega B_0$$

and

$$\frac{|E_0|}{|B_0|} = \frac{\omega}{k}.$$

Since $\omega = kc$, we also find that

$$\frac{|E_0|}{|B_0|} = c. \tag{12.25}$$

Note that if not operating in vacuum, the speed of light changes with the medium's refractive index n_m, yielding

$$\frac{|E_0|}{|B_0|} = \frac{c}{n_m}. \tag{12.26}$$

We now use Equation 12.26 in our Poynting vector expression to obtain

$$S_z = \frac{n_m}{c\mu\mu_0} e^{2i(kz-\omega t)} E_0^2.$$

Finally, the light intensity (or irradiance) is the time average of the Poynting vector:

$$I = \langle S \rangle = \frac{n_m}{2c\mu\mu_0} E_0^2. \tag{12.27}$$

The factor of $\frac{1}{2}$ comes from averaging an oscillating function with limits 0 and 1. Alternatively, since $c = 1/\sqrt{\mu_0\epsilon_0}$ we find that $c\mu_0 = 1/c\epsilon_0$, which, in turn, gives

$$I = \frac{n_m c\epsilon_0}{2\mu} E_0^2.$$

If dealing with nonmagnetic materials, $\mu \sim 1$, and thus

$$I = \frac{n_m c\epsilon_0}{2} E_0^2. \tag{12.28}$$

In terms of the vector potential, where $E_0 = \omega A_0$ (Equation 12.6), an equivalent expression is

$$I = \frac{n_m c\epsilon_0 \omega^2}{2} A_0^2, \tag{12.29}$$

which we will use in our subsequent evaluation of the bulk absorption coefficient.

At this point, the number of photons, n_p, incident per unit area per unit time is

$$n_p = \frac{I}{\hbar\omega} \tag{12.30}$$

with $\hbar\omega$ the photon energy (in units of J) and I the incident light intensity (in units of $\mathrm{W\,cm^{-2}}$). The reader may verify Equation 12.30 through a unit analysis. The desired absorption coefficient is then

$$\alpha = \frac{R_{\text{tot}}}{n_p}, \tag{12.31}$$

where R_{tot} is the transition rate per unit volume derived earlier. Note that Equation 12.31 can also be verified using a unit analysis. Finally, combining Equations 12.30 and 12.31 gives

$$\alpha = \frac{R_{\text{tot}}\hbar\omega}{I}$$

with $I = \frac{1}{2} n_m c\epsilon_0\omega^2 A_0^2$. We thus find that

$$\alpha = \frac{2\hbar\omega}{n_m c\epsilon_0\omega^2 A_0^2} R_{\text{tot}}, \tag{12.32}$$

with R_{tot} from Equation 12.22:

$$R_{\text{tot}} = \frac{\pi}{2\hbar}\left(\frac{qA_0}{m_0}\right)^2 |\langle u_k(\mathbf{r})|\hat{p}_z|u_n(\mathbf{r})\rangle|^2 \left[\frac{1}{2\pi^2}\left(\frac{2\mu}{\hbar^2}\right)^{3/2}\sqrt{E-E_g}\right],$$

giving

$$\alpha = \frac{\pi q^2}{n_m c\epsilon_0 m_0^2\omega} |\langle u_k(\mathbf{r})|\hat{p}_z|u_n(\mathbf{r})\rangle|^2 \left[\frac{1}{2\pi^2}\left(\frac{2\mu}{\hbar^2}\right)^{3/2}\sqrt{E-E_g}\right]. \tag{12.33}$$

Often there is an additional change of notation with the introduction of the following variable:

$$E_p = \frac{2|\langle u_k(\mathbf{r})|\hat{p}_z|u_n(\mathbf{r})\rangle|^2}{m_0}. \tag{12.34}$$

It typically takes values between 20 and 25 eV for many semiconductors and can be found using the Kane model of bandstructures (Kane 1957). We did not discuss the Kane model in Chapter 10; however, the interested reader may consult other more advanced texts for an in-depth description of the model, for example Chuang (1995) and Rosencher and Vinter (2000).

We now have

$$\alpha = \frac{\pi\hbar q^2}{2n_m c\epsilon_0 m_0}\left(\frac{E_p}{\hbar\omega}\right)\left[\frac{1}{2\pi^2}\left(\frac{2\mu}{\hbar^2}\right)^{3/2}\sqrt{E-E_g}\right] \tag{12.35}$$

as the bulk semiconductor absorption coefficient. It is proportional to $\sqrt{E-E_g}$, showing that the overall shape of the bulk joint density of states is preserved in the system's linear absorption, as first suggested in Chapter 9.

Finally, an equivalent absorption cross section σ is (effectively) found by taking α and multiplying it by the nanostructure volume V_{tot}:

$$\sigma = \alpha V_{\text{tot}}. \tag{12.36}$$

We saw this earlier in Chapter 5 when discussing nanowire and quantum dot absorption cross sections.

12.9 TRANSITIONS IN LOW-DIMENSIONAL SEMICONDUCTORS

Let us now use our knowledge about the quantum mechanics of bulk interband transitions to describe the linear absorption of low-dimensional systems such as quantum wells, quantum wires, and quantum dots. We also seek to obtain selection rules that dictate whether certain transitions are allowed or not. To do this, we must first introduce the effective mass approximation (EMA), which is sometimes called the envelope function approximation. This is because we will need more representative descriptions of what the initial ($|n\rangle$) and final ($|k\rangle$) states of our low-dimensional systems look like.

12.9.1 The Effective Mass Approximation

We begin with the time-independent Schrödinger equation from Chapter 7:

$$\hat{H}\psi = E\psi.$$

Recall that ψ is the wavefunction describing our carrier, E is its energy, and $\hat{H}$ is the Hamiltonian operator

$$\hat{H} = -\frac{\hbar^2}{2m_0}\nabla^2 + V.$$

In this expression, m_0 is the free electron mass and V represents *all* of the potential energy terms that the particle experiences.

Specifically, the carrier (e.g., an electron) in a crystal does not experience a constant or a null potential. Rather it experiences a periodic potential due to the regular crystalline arrangement of atoms. Thus, as seen earlier in our discussion of bands (Chapter 10), we are ultimately dealing with the following equation:

$$\left[-\frac{\hbar^2}{2m_0}\nabla^2 + V_{\text{periodic}}(\mathbf{r})\right]\psi(\mathbf{r}) = E\psi(\mathbf{r}),$$

where V_{periodic} is the periodic potential due to the lattice. Recall that solutions to this equation are Bloch wavefunctions of the form

$$\psi(\mathbf{r}) = u_{sk}(\mathbf{r})e^{i\mathbf{k}\cdot\mathbf{r}}$$

with u_{sk} a dual-index unit-cell periodic function, having the same periodicity as the underlying lattice (s is an index that represents a given band).

The above (complicated) Schrödinger equation therefore becomes

$$\hat{H}^{(0)}\psi(\mathbf{r}) = E\psi(\mathbf{r}),$$

with $\hat{H}^{(0)} = -(\hbar^2/2m_0)\nabla^2 + V_{\text{periodic}}(\mathbf{r})$ an effective zeroth-order Hamiltonian that incorporates the periodic potential.

Now, in general, we could have other potentials present, V_{other}. For example, they may arise from an applied electric or magnetic field or even impurity centers (see Luttinger and Kohn 1955). As a consequence, our general Schrödinger equation is

$$(\hat{H}^{(0)} + V_{\text{other}})\psi(\mathbf{r}) = E\psi(\mathbf{r}), \tag{12.37}$$

where we realize that the solutions $\psi(\mathbf{r})$ can be described through linear combinations of our original Bloch wavefunctions, denoted in what follows by ϕ. In fact, we write $\psi(\mathbf{r})$ in one dimension as

$$\psi(x) = \sum_s \frac{1}{2\pi} \int_{-\pi/a}^{\pi/a} \tilde{\chi}_s(k)\phi_{s,k}(x)\, dk,$$

where $\phi_{s,k}(x) = u_{sk}(x)e^{ikx}$, or in three dimensions as

$$\psi(\mathbf{r}) = \sum_s \frac{1}{(2\pi)^3} \int_{BZ} \tilde{\chi}_s(\mathbf{k})\phi_{s,k}(\mathbf{r})\, d^3k,$$

where in either case $\tilde{\chi}_s(k)(\tilde{\chi}_s(\mathbf{k}))$ are expansion coefficients and in the latter case the integral is over the first Brillouin zone. Furthermore, in either case, the sum accounts for possible contributions from multiple bands and the integral over the first Brillouin zone accounts for all k contributions within a given band. This derivation follows the work of Davies (2000).

To simplify things, assume now that only a single band contributes significantly to the wavefunction and that we consider only a small region of k-space near $k = 0$. As a consequence, we drop the sum over s and have $u_{sk}(x) \to u_0(x)$, where $u_0(x)$ is essentially independent of k in this region. We therefore have

$$\psi(x) \simeq \frac{1}{2\pi} \int_{-\pi/a}^{\pi/a} \tilde{\chi}(k)\phi_k(x)\, dk \qquad (12.38)$$

$$\simeq \frac{1}{2\pi} \int_{-\pi/a}^{\pi/a} \tilde{\chi}(k)[u_0(x)e^{ikx}]\, dk$$

$$\simeq \frac{u_0(x)}{2\pi} \int_{-\pi/a}^{\pi/a} \tilde{\chi}(k)e^{ikx}\, dk = u_0(x)\chi(x). \qquad (12.39)$$

Note that the integral is in the form of an inverse Fourier transform and we refer to $\chi(x)$ as an "envelope" function, with the implicit assumption that $\chi(x)$ varies slowly with x.

Returning to the Schrödinger equation written in one dimension, we then have

$$(\hat{H}^{(0)} + V_{\text{other}})\psi(x) = E\psi(x),$$

which we will simplify in what follows using our new wavefunction (Equation 12.38). First,

$$\hat{H}^{(0)}\psi(x) + V_{\text{other}}\psi(x) = E\psi(x),$$

and, focusing on the first term, we have

$$\hat{H}^{(0)}\psi(x) \simeq \hat{H}^{(0)}\frac{1}{2\pi} \int_{-\pi/a}^{\pi/a} \tilde{\chi}(k)\phi_k(x)\, dk$$

$$\simeq \frac{1}{2\pi} \int_{-\pi/a}^{\pi/a} \tilde{\chi}(k)\hat{H}^{(0)}\phi_k(x)\, dk.$$

Recall that $\phi_k(x)$ are eigenfunctions of the effective zeroth-order Hamiltonian; i.e., $\hat{H}^{(0)}\phi_k(x) = E\phi_k(x)$. Thus,

$$\hat{H}^{(0)}\psi(x) \simeq \frac{1}{2\pi} \int_{-\pi/a}^{\pi/a} \tilde{\chi}(k)E\phi_k(x)\, dk$$

$$\simeq \frac{1}{2\pi} \int_{-\pi/a}^{\pi/a} \tilde{\chi}(k)E[u_0(x)e^{ikx}]\, dk$$

$$\simeq \frac{u_0(x)}{2\pi} \int_{-\pi/a}^{\pi/a} \tilde{\chi}(k)Ee^{ikx}\, dk.$$

At this point, E is expanded in terms of a power series in k,

$$E = \sum_m a_m k^m,$$

with a_m a weight and m an index. Thus,

$$\begin{aligned}\hat{H}^{(0)}\psi(x) &\simeq \frac{u_0(x)}{2\pi}\int_{-\pi/a}^{\pi/a} \tilde{\chi}(k)\left[\sum_m a_m k^m\right] e^{ikx}\, dk \\ &\simeq u_0(x)\sum_m a_m \frac{1}{2\pi}\int_{-\pi/a}^{\pi/a}\left[k^m \tilde{\chi}(k)\right] e^{ikx}\, dk.\end{aligned}$$

The result is the inverse Fourier transform of the function $k^m\tilde{\chi}(k)$ and one can show that it equals $(-i\, d/dx)^m\, \chi(x)$, i.e.,

$$F^{-1}\left(k^m\tilde{\chi}(k)\right) = \left(-i\frac{d}{dx}\right)^m \chi(x). \tag{12.40}$$

Equation 12.40 can be obtained by considering the forward Fourier transform of a first derivative and generalizing it to higher-order derivatives. We illustrate this in the interlude below. The reader may skip ahead if desired.

Interlude

To justify Equation 12.40, the forward Fourier transform of a derivative (ignoring constants) is

$$\int\left[\frac{df(x)}{dx}\right] e^{-ikx}\, dx.$$

Upon integrating by parts and assuming that $f(x) \to 0$ as $x \to \infty$ and $x \to -\infty$, we find

$$\int\left[\frac{df(x)}{dx}\right] e^{-ikx}\, dx = ik\int f(x)e^{-ikx}\, dx$$

or that

$$F(f'(x)) = ikF(f(x)) = ik\tilde{f}(k). \tag{12.41}$$

One can then see that the inverse Fourier transform of the function $k\tilde{f}(k)$ just yields

$$F^{-1}(k\tilde{f}(k)) = -if'(x) = -i\frac{df(x)}{dx}.$$

Next, for higher-order derivatives, we start with Equation 12.41, $F(f'(x)) = ikF(f(x))$ and apply it multiple times. For a second derivative, we find

$$F(f''(x)) = ikF(f'(x)),$$

where we have just seen that $F(f'(x)) = ik\tilde{f}(k)$. As a consequence,

$$F(f''(x)) = (ik)^2\tilde{f}(k)$$

and, by further extending this, the following general result is obtained:

$$F(f^m(x)) = (ik)^m\tilde{f}(k). \tag{12.42}$$

Finally, reversing the last expression yields our desired relationship. Namely,

$$F(f^m(x)) = (ik)^m\tilde{f}(k),$$
$$\int \frac{d^m}{dx^m} f(x)e^{-ikx}\,dx = (ik)^m\tilde{f}(k),$$
$$\int \left[-i^m \frac{d^m}{dx^m} f(x)\right] e^{-ikx}\,dx = [k^m\tilde{f}(k)]$$

reveals that the inverse Fourier transform of the function $k^m\tilde{f}(k)$ is

$$F^{-1}(k^m\tilde{f}(k)) = \left(-i\frac{d}{dx}\right)^m f(x), \tag{12.43}$$

which is Equation 12.40 when f is replaced with χ.

Back to Where We Left Off

Returning to where we left off, we see that

$$\begin{aligned}\hat{H}^{(0)}\psi(x) &\simeq u_0(x)\sum_m a_m \frac{1}{2\pi}\int_{-\pi/a}^{\pi/a}[k^m\tilde{\chi}(k)]e^{ikx}\,dk \\ &\simeq u_0(x)\sum_m a_m F^{-1}[k^m\tilde{\chi}(k)] \\ &\simeq u_0(x)\sum_m a_m\left(-i\frac{d}{dx}\right)^m \chi(x).\end{aligned}$$

Recall also that $E(k)$ has been expanded in terms of a power series in k (i.e., $E(k) = \sum_m a_m k^m$). As a consequence, we recognize that

$$\hat{H}^{(0)}\psi(x) \simeq u_0(x)E\left(-i\frac{d}{dx}\right)\chi(x),$$

with E written as a function of $-i\,d/dx$ given that $k = -i\,d/dx$.

On replacing everything back into our one-dimensional Schrödinger equation, we find

$$\begin{aligned}(\hat{H}^{(0)} + V_{\text{other}})\psi(x) &= E\psi(x), \\ \hat{H}^{(0)}\psi(x) + V_{\text{other}}\psi(x) &= E\psi(x), \\ u_0(x)E\left(-i\frac{d}{dx}\right)\chi(x) + V_{\text{other}}u_0(x)\chi(x) &= Eu_0(x)\chi(x).\end{aligned}$$

Dropping the common $u_0(x)$ term then gives

$$E\left(-i\frac{d}{dx}\right)\chi(x) + V_{\text{other}}\chi(x) = E\chi(x).$$

Finally, if we invoke a parabolic relationship between E and k, namely, $E = \hbar^2k^2/2m_{\text{eff}}$ (valid near the bottom of the conduction band for the electron or the top of the valence band for the hole) along with m_{eff} (an effective electron or hole mass in the crystal), we find on replacing k with $-i\,d/dx$ that

$$\begin{aligned}E(k) &= \frac{\hbar^2k^2}{2m_{\text{eff}}} \\ &= -\frac{\hbar^2}{2m_{\text{eff}}}\frac{d^2}{dx^2}.\end{aligned}$$

This leads to

$$-\frac{\hbar^2}{2m_{\text{eff}}}\frac{d^2}{dx^2}\chi(x) + V_{\text{other}}\chi(x) = E\chi(x),$$

which can be written more compactly as

$$\left(-\frac{\hbar^2}{2m_{\text{eff}}}\frac{d^2}{dx^2}+V_{\text{other}}\right)\chi(x)=E\chi(x). \tag{12.44}$$

Equation 12.44 is essentially the Schrödinger equation for a free carrier with an effective mass m_{eff} (previously we used m_e for the electron and m_h for the hole in Chapters 9 and 10). More generally, in three dimensions, we write the above effective mass Schrödinger equation as

$$\left(-\frac{\hbar^2}{2m_{\text{eff}}}\nabla^2+V_{\text{other}}\right)\chi(\mathbf{r})=E\chi(\mathbf{r}). \tag{12.45}$$

In either case, the complexity of the periodic potential is now hidden within m_{eff}, greatly simplifying things. Note that what we solve for now is the envelope part of the total Bloch wavefunction $\chi(\mathbf{r})$, as well as E, the particle's energy. We alluded to this earlier in Chapter 7 when we solved several representative model systems for quantum wells, quantum wires, and quantum dots. The total carrier wavefunction, however, is still the product of $\chi(\mathbf{r})$ and a unit-cell term:

$$\psi_{\text{tot}}(\mathbf{r})=u_k(\mathbf{r})\chi(\mathbf{r}).$$

Note that the unit-cell function $u_k(\mathbf{r})$ is assumed to be common for all systems of interest and possesses the symmetry of the material's conduction band or valence band, depending on whether one is describing the electron or the hole. This is an important assumption that must be kept in mind.

12.9.2 Quantum Well

With this established, let us now evaluate the transition probability and transition rate for a quantum well. Before beginning, a few additional things need to be said. Specifically, for this example let us assume that the quantum well is oriented in such a way so that the single degree of confinement occurs along the z direction of a Cartesian coordinate system. The two degrees of freedom for the carrier then occur within the (x, y) plane. This is illustrated in **Figure 12.3**.

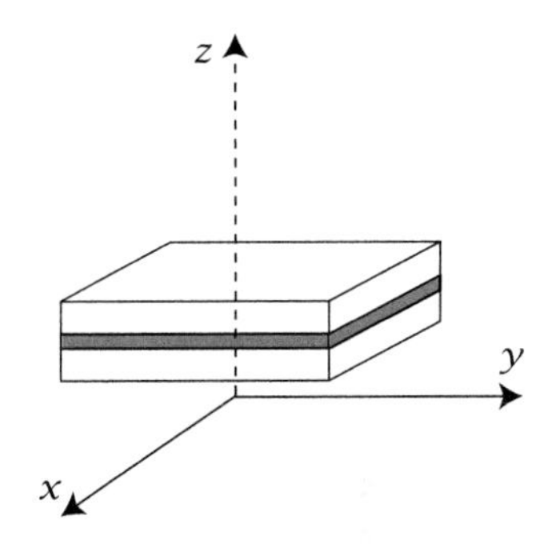

Figure 12.3 Illustration of the orientation of a model quantum well in a Cartesian coordinate system. The shaded region represents the quantum well.

Next, a general form for the quantum well carrier wavefunction is

$$\psi_{\text{tot}}=\left[\frac{1}{\sqrt{A}}\phi(z)e^{i\mathbf{k}\cdot\mathbf{r}_{xy}}\right]u(\mathbf{r}_{xy},z), \tag{12.46}$$

where the term in square brackets is simply the envelope function $\chi(\mathbf{r}_{xy}, z)$. Namely, in the expression first seen in Chapter 7, A is the well's area in the (x, y) plane and $\phi(z)$ is the confined direction's contribution to the full envelope function. We thus showed that

$$\chi(\mathbf{r}_{xy},z)=\frac{1}{\sqrt{A}}\sqrt{\frac{2}{a}}\sin\left(\frac{n_z\pi}{a}z\right)e^{i\mathbf{k}\cdot\mathbf{r}_{xy}}$$

(Equation 7.15). The latter $u(\mathbf{r}_{xy}, z)$ term is the unit-cell part of the total wavefunction. Recall that we implicitly assume that the unit-cell terms are the same as in the bulk solid. In all cases, the notation $\mathbf{r}_{xy}$ implies a coordinate within the (x, y) plane.

There are additional things to note. Namely, both the envelope and unit-cell parts of the wavefunction are normalized as follows:

$$\langle u(\mathbf{r}_{xy},z)|u(\mathbf{r}_{xy},z)\rangle=\int_{V_{\text{unit}}}u^*(\mathbf{r}_{xy},z)u(\mathbf{r}_{xy},z)\,d^2r\,dz=V_{\text{unit}}, \tag{12.47}$$

along with

$$\langle \chi(\mathbf{r}_{xy}, z) | \chi(\mathbf{r}_{xy}, z) \rangle = \int_{V_{tot}} \chi^*(\mathbf{r}_{xy}, z) \chi(\mathbf{r}_{xy}, z) \, d^2r \, dz = 1. \quad (12.48)$$

Specifically, the unit-cell function is normalized to the volume of the unit cell, while the envelope part is normalized to one. In addition, within the envelope function, the plane-wave term is normalized by $1/\sqrt{A}$, while $\phi(z)$ separately satisfies

$$\langle \phi(z) | \phi(z) \rangle = \int_{L_z} \phi^*(z) \phi(z) \, dz = 1, \quad (12.49)$$

with L_z the width of the well. In turn, the total wavefunction (Equation 12.46) is normalized such that ψ_{tot} represents one carrier per state in the well.

This can be demonstrated explicitly. We have

$$\begin{aligned} \langle \psi_{tot} | \psi_{tot} \rangle &= \int_{V_{tot}} \psi^*_{tot} \psi_{tot} \, d^2r \, dz \\ &= \int_{V_{tot}} \left[\frac{1}{\sqrt{A}} \phi(z) e^{i\mathbf{k} \cdot \mathbf{r}_{xy}} \right]^* u^*(\mathbf{r}_{xy}, z) \left[\frac{1}{\sqrt{A}} \phi(z) e^{i\mathbf{k} \cdot \mathbf{r}_{xy}} \right] u(\mathbf{r}_{xy}, z) \, d^2r \, dz \\ &= \frac{1}{A} \int_{V_{tot}} \phi^*(z) \phi(z) u^*(\mathbf{r}_{xy}, z) u(\mathbf{r}_{xy}, z) \, d^2r \, dz. \end{aligned}$$

To continue, let us make an approximation seen earlier. Namely, assume that $\phi(z)$ varies over much longer distances than $u(\mathbf{r}_{xy}, z)$, which varies on the atomic scale. As a consequence, let us write the integral differently as follows. Namely, we consider the following sum over N unit-cell integrals due to this separation of length scales over which $\phi(z)$ and $u(\mathbf{r}_{xy}, z)$ vary:

$$\begin{aligned} \langle \psi_{tot} | \psi_{tot} \rangle &\simeq \frac{1}{A} \sum_N \phi^*(z) \phi(z) \int_{V_{unit}} u^*(\mathbf{r}_{xy}, z) u(\mathbf{r}_{xy}, z) \, d^2r \, dz \\ &\simeq \frac{1}{A} \sum_N \phi^*(z) \phi(z) V_{unit}. \end{aligned}$$

Provided that the sum is large (i.e., there are many unit cells), we can return to an integral over the total quantum well volume:

$$\langle \psi_{tot} | \psi_{tot} \rangle \simeq \frac{1}{A} \int_{V_{tot}} \phi^*(z) \phi(z) \, d^2r \, dz.$$

Note that V_{unit} has been recast as $d^2r \, dz$, the unit volume element. We then have

$$\begin{aligned} \langle \psi_{tot} | \psi_{tot} \rangle &\simeq \frac{1}{A} \int_A d^2r \int_{L_z} \phi^*(z) \phi(z) \, dz \\ &\simeq \frac{A}{A} \int_{L_z} \phi^*(z) \phi(z) \, dz \\ &\simeq \int_{L_z} \phi^*(z) \phi(z) \, dz = 1, \end{aligned}$$

which shows that the total carrier wavefunction is normalized.

Transition Probability

At this point, to evaluate the quantum well transition probability, we start with Equation 12.7,

$$P_k \simeq \frac{1}{\hbar^2}\left(\frac{qA_0}{2m_0}\right)^2 \left|\int_0^t \langle k|e^{i\mathbf{k}\cdot\mathbf{r}}(\mathbf{e}\cdot\hat{\mathbf{p}})|n\rangle e^{-i(\omega-\omega_{kn})t'}dt'\right|^2$$

and introduce the quantum well wavefunctions for $|k\rangle$ and $|n\rangle$. Namely,

$$|k\rangle = \left[\frac{1}{\sqrt{A}}\phi_k(z)e^{i\mathbf{k}_k\cdot\mathbf{r}_{xy}}\right]u_k(\mathbf{r}_{xy}, z)$$

represents the conduction band electron, while

$$|n\rangle = \left[\frac{1}{\sqrt{A}}\phi_n(z)e^{i\mathbf{k}_n\cdot\mathbf{r}_{xy}}\right]u_n(\mathbf{r}_{xy}, z)$$

represents the valence band hole. Both are inserted into P_k. To simplify the result, evaluate $\langle k|e^{i\mathbf{k}\cdot\mathbf{r}}(\mathbf{e}\cdot\hat{\mathbf{p}})|n\rangle$ as follows:

$$\begin{aligned}\langle k|e^{i\mathbf{k}\cdot\mathbf{r}}(\mathbf{e}\cdot\hat{\mathbf{p}})|n\rangle &= \int_{V_{\text{tot}}}\left[\frac{1}{\sqrt{A}}\phi_k(z)e^{i\mathbf{k}_k\cdot\mathbf{r}_{xy}}\right]^* u_k^*(\mathbf{r}_{xy}, z)\left(e^{i\mathbf{k}\cdot\mathbf{r}}(\mathbf{e}\cdot\hat{\mathbf{p}})\right)\\ &\quad\times\left[\frac{1}{\sqrt{A}}\phi_n(z)e^{i\mathbf{k}_n\cdot\mathbf{r}_{xy}}\right]u_n(\mathbf{r}_{xy}, z)\,d^2r\,dz\\ &= \frac{1}{A}\int_{V_{\text{tot}}}\phi_k^*(z)e^{-i\mathbf{k}_k\cdot\mathbf{r}_{xy}}u_k^*(\mathbf{r}_{xy}, z)\left(e^{i\mathbf{k}\cdot\mathbf{r}}(\mathbf{e}\cdot\hat{\mathbf{p}})\right)\\ &\quad\times\phi_n(z)e^{i\mathbf{k}_n\cdot\mathbf{r}_{xy}}u_n(\mathbf{r}_{xy}, z)\,d^2r\,dz.\end{aligned}$$

We now invoke the assumption that all of the plane-wave and envelope function terms vary over much longer length scales than the unit-cell functions. As a consequence, we approximate the integral as a sum over N unit-cell integrals:

$$\begin{aligned}\langle k|e^{i\mathbf{k}\cdot\mathbf{r}}(\mathbf{e}\cdot\hat{\mathbf{p}})|n\rangle &\simeq \frac{1}{A}\sum_N \phi_k^*(z)\phi_n(z)e^{i\mathbf{k}\cdot\mathbf{r}}e^{i(\mathbf{k}_n-\mathbf{k}_k)\cdot\mathbf{r}_{xy}}\\ &\quad\times\int_{V_{\text{unit}}}u_k^*(\mathbf{r}_{xy}, z)(\mathbf{e}\cdot\hat{\mathbf{p}})u_n(\mathbf{r}_{xy}, z)\,d^2r\,dz.\end{aligned}$$

We can then express this as

$$\begin{aligned}\langle k|e^{i\mathbf{k}\cdot\mathbf{r}}(\mathbf{e}\cdot\hat{\mathbf{p}})|n\rangle &\simeq \frac{1}{A}\sum_N \phi_k^*(z)\phi_n(z)(e^{ik_z z}e^{i\mathbf{k}_{xy}\cdot\mathbf{r}_{xy}})e^{i(\mathbf{k}_n-\mathbf{k}_k)\cdot\mathbf{r}_{xy}}\\ &\quad\times\int_{V_{\text{unit}}}u_k^*(\mathbf{r}_{xy}, z)(\mathbf{e}\cdot\hat{\mathbf{p}})u_n(\mathbf{r}_{xy}, z)d^2r\,dz.\end{aligned}$$

What results is

$$\begin{aligned}\langle k|e^{i\mathbf{k}\cdot\mathbf{r}}(\mathbf{e}\cdot\hat{\mathbf{p}})|n\rangle &\simeq \frac{1}{A}\sum_N \phi_k^*(z)\phi_n(z)e^{ik_x z}e^{i[\mathbf{k}_{xy}+(\mathbf{k}_n-\mathbf{k}_k)]\cdot\mathbf{r}_{xy}}\\ &\quad\times\int_{V_{\text{unit}}}u_k^*(\mathbf{r}_{xy}, z)(\mathbf{e}\cdot\hat{\mathbf{p}})u_n(\mathbf{r}_{xy}, z)d^2r\,dz,\end{aligned}$$

where, by momentum conservation,

$$\begin{aligned}\hbar\mathbf{k}_k &= \hbar\mathbf{k}_n + \hbar\mathbf{k}_{xy},\\ \mathbf{k}_k &= \mathbf{k}_n + \mathbf{k}_{xy}.\end{aligned}$$

As a consequence,

$$\langle k|e^{i\mathbf{k}\cdot\mathbf{r}}(\mathbf{e}\cdot\hat{\mathbf{p}})|n\rangle \simeq \frac{1}{A}\sum_N \phi_k^*(z)\phi_n(z)e^{ik_z z} \times \int_{V_{\text{unit}}} u_k^*(\mathbf{r}_{xy},z)(\mathbf{e}\cdot\hat{\mathbf{p}})u_n(\mathbf{r}_{xy},z)d^2r\,dz,$$

and since the magnitude of $k_z \propto 1/\lambda$, with λ the wavelength of light as well as z of molecular dimensions, $e^{ik_z z} \simeq 1$. The latter approximation is the long-wavelength approximation we have seen previously. We are left with

$$\begin{aligned}
\langle k|e^{i\mathbf{k}\cdot\mathbf{r}}(\mathbf{e}\cdot\hat{\mathbf{p}})|n\rangle &\simeq \frac{1}{A}\sum_N \phi_k^*(z)\phi_n(z)\int_{V_{\text{unit}}} u_k^*(\mathbf{r}_{xy},z)(\mathbf{e}\cdot\hat{\mathbf{p}})u_n(\mathbf{r}_{xy},z)\,d^2r\,dz \\
&\simeq \frac{1}{A}\sum_N \phi_k^*(z)\phi_n(z)\frac{V_{\text{unit}}}{V_{\text{unit}}}\int_{V_{\text{unit}}} u_k^*(\mathbf{r}_{xy},z)(\mathbf{e}\cdot\hat{\mathbf{p}})u_n(\mathbf{r}_{xy},z)d^2r\,dz \\
&\simeq \frac{1}{A}\sum_N \phi_k^*(z)\phi_n(z)\left\langle u_k(\mathbf{r}_{xy},z)\middle|\mathbf{e}\cdot\hat{\mathbf{p}}\middle|u_n(\mathbf{r}_{xy},z)\right\rangle V_{\text{unit}},
\end{aligned}$$

where

$$\left\langle u_k(\mathbf{r}_{xy},z)\middle|\mathbf{e}\cdot\hat{\mathbf{p}}\middle|u_n(\mathbf{r}_{xy},z)\right\rangle = \frac{1}{V_{\text{unit}}}\int_{V_{\text{unit}}} u_k^*(\mathbf{r}_{xy},z)(\mathbf{e}\cdot\hat{\mathbf{p}})u_n(\mathbf{r}_{xy},z)\,d^2r\,dz. \tag{12.50}$$

Finally, provided sufficient unit cells, we revert back to integral form over the total quantum well volume:

$$\begin{aligned}
\langle k|e^{i\mathbf{k}\cdot\mathbf{r}}(\mathbf{e}\cdot\hat{\mathbf{p}})|n\rangle &\simeq \frac{1}{A}\int_{V_{\text{tot}}} \phi_k^*(z)\phi_n(z)\left\langle u_k(\mathbf{r}_{xy},z)\middle|\mathbf{e}\cdot\hat{\mathbf{p}}\middle|u_n(\mathbf{r}_{xy},z)\right\rangle d^2r\,dz \\
&\simeq \frac{1}{A}\left\langle u_k(\mathbf{r}_{xy},z)\middle|\mathbf{e}\cdot\hat{\mathbf{p}}\middle|u_n(\mathbf{r}_{xy},z)\right\rangle \int_{V_{\text{tot}}} \phi_k^*(z)\phi_n(z)\,d^2r\,dz \\
&\simeq \frac{1}{A}\left\langle u_k(\mathbf{r}_{xy},z)\middle|\mathbf{e}\cdot\hat{\mathbf{p}}\middle|u_n(\mathbf{r}_{xy},z)\right\rangle \int_A d^2r\int_{L_z} \phi_k^*(z)\,\phi_n(z)\,dz \\
&\simeq \frac{A}{A}\left\langle u_k(\mathbf{r}_{xy},z)\middle|\mathbf{e}\cdot\hat{\mathbf{p}}\middle|u_n(\mathbf{r}_{xy},z)\right\rangle \int_{L_z} \phi_k^*(z)\phi_n(z)\,dz \\
&\simeq \left\langle u_k(\mathbf{r}_{xy},z)\middle|\mathbf{e}\cdot\hat{\mathbf{p}}\middle|u_n(\mathbf{r}_{xy},z)\right\rangle \left\langle \phi_k(z)\middle|\phi_n(z)\right\rangle.
\end{aligned}$$

This allows us to conclude that

$$\langle k|e^{i\mathbf{k}\cdot\mathbf{r}}(\mathbf{e}\cdot\hat{\mathbf{p}})|n\rangle \simeq \left\langle u_k(\mathbf{r}_{xy},z)\middle|\mathbf{e}\cdot\hat{\mathbf{p}}\middle|u_n(\mathbf{r}_{xy},z)\right\rangle \left\langle \phi_k(z)\middle|\phi_n(z)\right\rangle.$$

When this is inserted into P_k, we find

$$\begin{aligned}
P_k &\simeq \frac{1}{\hbar^2}\left(\frac{qA_0}{2m_0}\right)^2 \left|\int_0^t \left\langle u_k(\mathbf{r}_{xy},z)\middle|\mathbf{e}\cdot\hat{\mathbf{p}}\middle|u_n(\mathbf{r}_{xy},z)\right\rangle \left\langle \phi_k(z)\middle|\phi_n(z)\right\rangle \times e^{-i(\omega-\omega_{kn})t'}dt'\right|^2 \\
&\simeq \frac{1}{\hbar^2}\left(\frac{qA_0}{2m_0}\right)^2 \left|\left\langle u_k(\mathbf{r}_{xy},z)\middle|\mathbf{e}\cdot\hat{\mathbf{p}}\middle|u_n(\mathbf{r}_{xy},z)\right\rangle\right|^2 \left|\left\langle \phi_k(z)\middle|\phi_n(z)\right\rangle\right|^2 \times \left|\int_0^t e^{-i(\omega-\omega_{kn})t'}\,dt'\right|^2.
\end{aligned}$$

As seen earlier, this integral term evaluates to

$$\frac{4}{(\omega-\omega_{kn})^2}\sin^2\left(\frac{\omega-\omega_{kn}}{2}t\right).$$

The desired transition probability is therefore

$$P_k(\omega) \simeq \frac{1}{\hbar^2}\left(\frac{qA_0}{m_0}\right)^2 \left|\langle u_k(\mathbf{r}_{xy}, z)\left|\mathbf{e}\cdot\hat{\mathbf{p}}\right|u_n(\mathbf{r}_{xy}, z)\rangle\right|^2 \times \left|\langle\phi_k(z)\left|\phi_n(z)\rangle\right|^2 \frac{\sin^2\left(\frac{\omega-\omega_{kn}}{2}t\right)}{(\omega-\omega_{kn})^2} \qquad (12.51)$$

and looks essentially like Equation 12.12 except with an additional envelope function overlap integral (i.e., $|\langle\phi_k(z)|\phi_n(z)\rangle|^2$). The result may also be written in terms of energy using $E = \hbar\omega$:

$$P_k(E) \simeq \left(\frac{qA_0}{m_0}\right)^2 \left|\langle u_k(\mathbf{r}_{xy}, z)\left|\mathbf{e}\cdot\hat{\mathbf{p}}\right|u_n(\mathbf{r}_{xy}, z)\rangle\right|^2 \times \left|\langle\phi_k(z)\left|\phi_n(z)\rangle\right|^2 \frac{\sin^2\left(\frac{E-E_{kn}}{2\hbar}t\right)}{(E-E_{kn})^2}. \qquad (12.52)$$

These expressions are general apart from the fact that we have assumed an orientation to the quantum well, as shown in **Figure 12.3**.

Multiple States

Next, if there are multiple possible transitions, we must integrate P_k over the energy range of interest by invoking the quantum well joint density of states. In this case, we use ρ_{joint} found earlier in Chapter 9. Thus, starting with

$$P_{\text{tot}} = \int_{-\infty}^{\infty} \rho_{\text{joint}}(E_{kn}) P_k(E)\, dE_{kn},$$

we find, on substituting $P_k(E)$ from above, that

$$P_{\text{tot}} \simeq \left(\frac{qA_0}{m_0}\right)^2 \left|\langle u_k(\mathbf{r}_{xy}, z)\left|\mathbf{e}\cdot\hat{\mathbf{p}}\right|u_n(\mathbf{r}_{xy}, z)\rangle\right|^2 \left|\langle\phi_k(z)\left|\phi_n(z)\rangle\right|^2 \times \int_{-\infty}^{\infty} \rho_{\text{joint}}(E_{kn}) \frac{\sin^2\left(\frac{E-E_{kn}}{2\hbar}t\right)}{(E-E_{kn})^2}\, dE_{kn}.$$

Note that the sinc-like function within the integral becomes taller and narrower with time and ultimately becomes a delta function (**Figure 11.2**). Namely (Equation 11.14),

$$\lim_{t\to\infty} \frac{\sin^2 \alpha t}{\pi\alpha^2 t} = \delta(\alpha).$$

We therefore rearrange the last term of the integral to obtain

$$P_{\text{tot}} \simeq \left(\frac{qA_0}{m_0}\right)^2 \left|\langle u_k(\mathbf{r}_{xy}, z)\left|\mathbf{e}\cdot\hat{\mathbf{p}}\right|u_n(\mathbf{r}_{xy}, z)\rangle\right|^2 \left|\langle\phi_k(z)\left|\phi_n(z)\rangle\right|^2 \times \int_{-\infty}^{\infty} \rho_{\text{joint}}(E_{kn}) \left(\frac{\pi t}{4\hbar^2}\right) \frac{\sin^2 \alpha t}{\pi\alpha^2 t}\, dE_{kn}$$

with $\alpha = (E - E_{kn})/2\hbar$. Then

$$P_{\text{tot},t\to\infty} \simeq \left(\frac{qA_0}{m_0}\right)^2 \left|\langle u_k(\mathbf{r}_{xy}, z)\left|\mathbf{e}\cdot\hat{\mathbf{p}}\right|u_n(\mathbf{r}_{xy}, z)\rangle\right|^2 \left|\langle\phi_k(z)\left|\phi_n(z)\rangle\right|^2 \times \int_{-\infty}^{\infty} \rho_{\text{joint}}(E_{kn}) \left(\frac{\pi t}{4\hbar^2}\right) \delta\left(\frac{E-E_{kn}}{2\hbar}\right) dE_{kn},$$

and, since $\delta(cx) = |c|^{-1}\delta(x)$,

$$P_{\text{tot},t\to\infty} \simeq \left(\frac{qA_0}{m_0}\right)^2 |\langle u_k(\mathbf{r}_{xy}, z)|\mathbf{e}\cdot\hat{\mathbf{p}}|u_n(\mathbf{r}_{xy}, z)\rangle|^2 |\langle\phi_k(z)|\phi_n(z)\rangle|^2$$
$$\times \int_{-\infty}^{\infty} \rho_{\text{joint}}(E_{kn}) \left(\frac{\pi t}{2\hbar}\right) \delta(E - E_{kn})\, dE_{kn}.$$

This leads to

$$P_{\text{tot},t\to\infty} \simeq \frac{\pi}{2\hbar}\left(\frac{qA_0}{m_0}\right)^2 |\langle u_k(\mathbf{r}_{xy}, z)|\mathbf{e}\cdot\hat{\mathbf{p}}|u_n(\mathbf{r}_{xy}, z)\rangle|^2 |\langle\phi_k(z)|\phi_n(z)\rangle|^2$$
$$\times \rho_{\text{joint}}(E = E_{kn})t.$$

Finally, inserting $\rho_{\text{joint}} = \mu/\pi\hbar^2$ for a single quantum well subband (Equation 9.36) leads to the total quantum well transition probability

$$P_{\text{tot},t\to\infty} \simeq \frac{\pi}{2\hbar}\left(\frac{qA_0}{m_0}\right)^2 |\langle u_k(\mathbf{r}_{xy}, z)|\mathbf{e}\cdot\hat{\mathbf{p}}|u_n(\mathbf{r}_{xy}, z)\rangle|^2$$
$$\times |\langle\phi_k(z)|\phi_n(z)\rangle|^2 \left(\frac{\mu}{\pi\hbar^2}\right) t. \tag{12.53}$$

The associated transition rate (number per unit area per unit time) is $R_{\text{tot}} = dP_{\text{tot},t\to\infty}/dt$ and results in

$$R_{\text{tot}} \simeq \frac{\pi}{2\hbar}\left(\frac{qA_0}{m_0}\right)^2 |\langle u_k(\mathbf{r}_{xy}, z)|\mathbf{e}\cdot\hat{\mathbf{p}}|u_n(\mathbf{r}_{xy}, z)\rangle|^2$$
$$\times |\langle\phi_k(z)|\phi_n(z)\rangle|^2 \left(\frac{\mu}{\pi\hbar^2}\right).$$

Alternatively, a transition rate of number per unit volume per unit time is

$$R_{\text{tot}} \simeq \frac{\pi}{2\hbar}\left(\frac{qA_0}{m_0}\right)^2 \left(\frac{1}{L_z}\right) |\langle u_k(\mathbf{r}_{xy}, z)|\mathbf{e}\cdot\hat{\mathbf{p}}|u_n(\mathbf{r}_{xy}, z)\rangle|^2$$
$$\times |\langle\phi_k(z)|\phi_n(z)\rangle|^2 \left(\frac{\mu}{\pi\hbar^2}\right). \tag{12.54}$$

Note that apart from assuming an orientation to the quantum well, both $P_{\text{tot},t\to\infty}$ and R_{tot} are generic expressions.

Absorption Coefficient

We can now find the associated quantum well absorption coefficient. Namely, from Equation 12.32, we have

$$\alpha = \frac{2\hbar\omega}{n_m c\epsilon_0\omega^2 A_0^2} R_{\text{tot}},$$

which leads to

$$\alpha = \frac{\pi q^2}{n_m c\epsilon_0\omega m_0^2 L_z} |\langle u_k(\mathbf{r}_{xy}, z)|\mathbf{e}\cdot\hat{\mathbf{p}}|u_n(\mathbf{r}_{xy}, z)\rangle|^2$$
$$\times |\langle\phi_k(z)|\phi_n(z)\rangle|^2 \left(\frac{\mu}{\pi\hbar^2}\right). \tag{12.55}$$

As alluded to earlier in Chapter 9, the appearance of the original quantum well joint density of states is preserved in the system's absorption spectrum.

Selection Rules

Finally, note that P_{tot} and R_{tot} contain within them the following two terms:

$$\left|\langle u_k(\mathbf{r}_{xy}, z)\middle|\mathbf{e}\cdot\hat{\mathbf{p}}\middle|u_n(\mathbf{r}_{xy}, z)\rangle\right|^2$$

and

$$\left|\langle\phi_k(z)\middle|\phi_n(z)\rangle\right|^2.$$

Both determine selection rules for quantum well optical transitions. The latter is easier to discuss, because we have previously calculated this overlap integral in Chapter 7. Recall for our model quantum well that if the confined contribution to the total envelope function (Equation 7.15) is

$$\phi(z) = \sqrt{\frac{2}{L_z}}\sin\frac{n_z\pi z}{L_z},$$

then the overlap integral requires that the quantum number n_z not change during the transition. Furthermore, we found from the transverse plane-wave contributions to the total envelope function that the associated transverse wavevectors could not change either. Hence $|\langle\phi_k(z)|\phi_n(z)\rangle|^2$ yields

$$\Delta n_z = 0 \qquad (12.56)$$

and

$$\Delta k = 0 \qquad (12.57)$$

as quantum well optical transition selection rules.

On top of this, we have selection rules dictated by the term $|\langle u_k(\mathbf{r}_{xy}, z)|\mathbf{e}\cdot\hat{\mathbf{p}}|u_n(\mathbf{r}_{xy}, z)\rangle|^2$. This second term is sensitive to the polarization of the incoming light and to the symmetry of the unit-cell functions. Thus, to provide more insight into these selection rules, we make some assumptions and simplifications. First, we assume that in our semiconductor, the conduction band is made of a linear combination of s atomic orbitals (Chapter 10). It therefore possesses $|s\rangle$-like symmetry and hence is an even function. We denote this by

$$|u_k(\mathbf{r}_{xy}, z)\rangle \rightarrow |s\rangle\uparrow,$$

where $\uparrow$ represents spin up.

Next, the semiconductor's valence band is made up of a linear combination of atomic p orbitals. Thus, $|u_n(\mathbf{r}_{xy}, z)\rangle$ has $|p\rangle$-like symmetry and is an odd function. There are a number of ways to describe this. For example, let us assume that

$$|u_n(\mathbf{r}_{xy}, z)\rangle = \frac{1}{\sqrt{2}}(|X\rangle + i|Y\rangle)\uparrow,$$

where $|X\rangle$ and $|Y\rangle$ denote odd functions that can be operated on by the momentum operators $\hat{p}_x$ and $\hat{p}_y$ and $\uparrow$ again means spin up. Note that the coordinate system of these functions does not necessarily correspond to that of the quantum well (**Figure 12.3**). In fact, it is referenced to the unit-cell coordinate system (Chapter 2). For simplicity in what follows, let us assume that the two coincide. More comprehensive lists of labels for the conduction and valence band Bloch functions, typically seen in the literature, follow in **Tables 12.1** and **12.2**.

Table 12.1 Conduction band Bloch functions and their symmetry representations

State	Label
$\left|\frac{1}{2},\frac{1}{2}\right\rangle$	$|s\rangle\uparrow$
$\left|\frac{1}{2},-\frac{1}{2}\right\rangle$	$|s\rangle\downarrow$

Table 12.2 Valence band Bloch functions and their symmetry representations

State	Label
$\|\frac{3}{2},\frac{3}{2}\rangle$	$\frac{1}{\sqrt{2}}(X+iY)\uparrow$
$\|\frac{3}{2},-\frac{3}{2}\rangle$	$\frac{i}{\sqrt{2}}(X-iY)\downarrow$
$\|\frac{3}{2},\frac{1}{2}\rangle$	$\frac{i}{\sqrt{6}}[(X+iY)\downarrow-2Z\uparrow]$
$\|\frac{3}{2},-\frac{1}{2}\rangle$	$\frac{1}{\sqrt{6}}[(X-iY)\uparrow+2Z\downarrow]$
$\|\frac{1}{2},\frac{1}{2}\rangle$	$\frac{1}{\sqrt{3}}[(X+iY)\downarrow+Z\uparrow]$
$\|\frac{1}{2},-\frac{1}{2}\rangle$	$\frac{i}{\sqrt{3}}[-(X-iY)\uparrow+Z\downarrow]$

We therefore have the following matrix element to consider:

$$\langle u_k(\mathbf{r}_{xy},z)|\mathbf{e}\cdot\hat{\mathbf{p}}|u_n(\mathbf{r}_{xy},z)\rangle = \langle s|\mathbf{e}\cdot\hat{\mathbf{p}}\left[\frac{1}{\sqrt{2}}(|X\rangle + i|Y\rangle)\right].$$

The spin-up labels have been dropped for convenience. However, recall that both point in the same direction to indicate the lack of spin flips during optical transitions.

Case 1: *z*-polarized light Assume that the incoming light is z-polarized (i.e., $\mathbf{e}=\mathbf{e}_z=(0,0,1)$). This leads to $\mathbf{e}\cdot\hat{\mathbf{p}}=\hat{p}_z$. The matrix element is then

$$\langle s|\hat{p}_z\left[\frac{1}{\sqrt{2}}(|X\rangle + i|Y\rangle)\right]$$

and evaluates to zero since $\hat{p}_z$ does not operate on either $|X\rangle$ or $|Y\rangle$. We conclude that z-polarized light does not induce any optical transitions in our particular quantum well.

Case 2: *x*-polarized light Next, assume that the incoming light is x-polarized (i.e., $\mathbf{e}=\mathbf{e}_x=(1,0,0)$). Thus, $\mathbf{e}\cdot\hat{\mathbf{p}}=\hat{p}_x$ and the matrix element is

$$\langle s|\hat{p}_x\left[\frac{1}{\sqrt{2}}(|X\rangle + i|Y\rangle)\right] = \frac{1}{\sqrt{2}}\langle s|\hat{p}_x|X\rangle.$$

We see that once $\hat{p}_x$ operates on $|X\rangle$ it yields an even function because what results is the derivative of the original (odd) function. The integral of an (overall) even function about a symmetric interval is nonzero. Hence, $\langle u_k(\mathbf{r}_{xy},z)|\hat{p}_x|u_n(\mathbf{r}_{xy},z)\rangle \neq 0$ and we see that x-polarized light induces a transition in the quantum well.

Case 3: *y*-polarized light Finally, assume that the incoming light is y-polarized (i.e., $\mathbf{e}=\mathbf{e}_y=(0,1,0)$). Then $\mathbf{e}\cdot\hat{\mathbf{p}}=\hat{p}_y$ and the matrix element is

$$\langle s|\hat{p}_y\left[\frac{1}{\sqrt{2}}(|X\rangle + i|Y\rangle)\right] = \frac{i}{\sqrt{2}}\langle s|\hat{p}_y|Y\rangle.$$

Note that $\hat{p}_y$ operating on $|Y\rangle$ yields an even function. The integral of an (overall) even function is then nonzero, $\langle u_k(\mathbf{r}_{xy},z)|\hat{p}_y|u_n(\mathbf{r}_{xy},z)\rangle \neq 0$. We therefore see that y-polarized light induces a transition in our quantum well.

To summarize, for the three cases considered, light polarized in the (x,y) plane induces optical transitions. However, light polarized in the z direction does not. These rules then add to the $\Delta n_z=0$ and $\Delta k=0$ selection rules, which arise from the envelope part of the total wavefunction.

12.9.3 Nanowire Transitions

In a similar way, we can calculate the transition probability and transition rate for a nanowire. The procedure is nearly identical to what we just did for the quantum well and the reader may therefore skip ahead if desired.

We begin with the nanowire carrier wavefunction. It takes the general form

$$\psi_{\mathrm{tot}} = \left[\frac{1}{\sqrt{L_z}}\phi(x,y)e^{ikz}\right]u(x,y,z), \tag{12.58}$$

where the term in square brackets is the envelope part of the total wavefunction $\chi(x, y, z)$ first seen in Chapter 7 and $u(x, y, z)$ is the unit-cell part. Note that we have implicitly assumed an orientation to the wire. Namely, we have aligned it with its long axis along the z direction of a Cartesian coordinate system. The confined directions of the wire are then along the remaining x and y directions. This is illustrated in **Figure 12.4**.

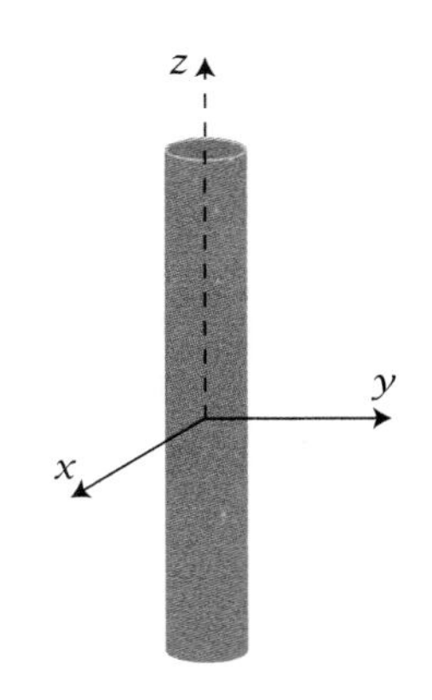

Figure 12.4 Illustration of the orientation of a model nanowire in a Cartesian coordinate system.

Both the envelope and unit-cell parts of the total wavefunction are normalized independently. Specifically,

$$\langle u(x, y, z)|u(x, y, z)\rangle = \int_{V_{\text{unit}}} u^*(x, y, z)u(x, y, z)\, dx\, dy\, dz = V_{\text{unit}} \tag{12.59}$$

and

$$\langle \chi(x, y, z)|\chi(x, y, z)\rangle = \int_{V_{\text{tot}}} \chi^*(x, y, z)\chi(x, y, z)\, dx\, dy\, dz = 1. \tag{12.60}$$

Note further that within the envelope function the plane-wave term is normalized by $1/\sqrt{L_z}$, where L_z is the wire's length. Likewise, $\phi(x, y)$ satisfies

$$\langle \phi(x, y)|\phi(x, y)\rangle = \int_A \phi^*(x, y)\phi(x, y)\, dx\, dy = 1. \tag{12.61}$$

Thus, the total nanowire wavefunction is normalized. The reader may verify this as follows:

$$\begin{aligned}
\langle \psi_{\text{tot}}|\psi_{\text{tot}}\rangle &= \int_{V_{\text{tot}}} \psi^*_{\text{tot}}\psi_{\text{tot}}\, dx\, dy\, dz \\
&= \int_{V_{\text{tot}}} \left[\frac{1}{\sqrt{L_z}}\phi(x, y)e^{ikz}\right]^* u^*(x, y, z) \left[\frac{1}{\sqrt{L_z}}\phi(x, y)e^{ikz}\right] \\
&\quad \times u(x, y, z)\, dx\, dy\, dz \\
&= \frac{1}{L_z}\int_{V_{\text{tot}}} \phi^*(x, y)\phi(x, y)u^*(x, y, z)u(x, y, z)\, dx\, dy\, dz.
\end{aligned}$$

We now make the usual assumption that $\phi(x, y)$ varies on a much longer length scale than $u(x, y, z)$. As a consequence, we can revert to a sum over N unit cell integrals:

$$\begin{aligned}
\langle \psi_{\text{tot}}|\psi_{\text{tot}}\rangle &\simeq \frac{1}{L_z}\sum_N \phi^*(x, y)\phi(x, y)\int_{V_{\text{unit}}} u^*(x, y, z)u(x, y, z)\, dx\, dy\, dz \\
&\simeq \frac{1}{L_z}\sum_N \phi^*(x, y)\phi(x, y)V_{\text{unit}}.
\end{aligned}$$

Next, if the sum is sufficiently large, we revert to integral form over the total nanowire volume to obtain

$$\langle \psi_{\text{tot}}|\psi_{\text{tot}}\rangle \simeq \frac{1}{L_z}\int_{V_{\text{tot}}} \phi^*(x, y)\phi(x, y)\, dx\, dy\, dz.$$

Note that V_{unit} has been recast as $dx\, dy\, dz$ above. We continue simplifying,

$$\begin{aligned}
\langle \psi_{\text{tot}}|\psi_{\text{tot}}\rangle &\simeq \frac{1}{L_z}\int_{L_z} dz \int_A \phi^*(x, y)\phi(x, y)\, dx\, dy \\
&\simeq \frac{L_z}{L_z}\int_A \phi^*(x, y)\phi(x, y)\, dx\, dy = 1
\end{aligned}$$

to show that the total wavefunction is normalized.

Transition Probability

Equation 12.7 is now used to evaluate the nanowire's transition probability:

$$P_k \simeq \frac{1}{\hbar^2}\left(\frac{qA_0}{2m_0}\right)^2 \left|\int_0^t \langle k|e^{i\mathbf{k}\cdot\mathbf{r}}(\mathbf{e}\cdot\hat{\mathbf{p}})|n\rangle e^{-i(\omega-\omega_{kn})t'}\,dt'\right|^2 .$$

We introduce the wavefunction

$$|k\rangle = \left[\frac{1}{\sqrt{L_z}}\phi_k(x,y)e^{ik_k z}\right]u_k(x,y,z)$$

for the electron in the conduction band and

$$|n\rangle = \left[\frac{1}{\sqrt{L_z}}\phi_n(x,y)e^{ik_n z}\right]u_n(x,y,z)$$

for the hole in the valence band. When inserted into P_k, we will need to evaluate $\langle k|e^{i\mathbf{k}\cdot\mathbf{r}}(\mathbf{e}\cdot\hat{\mathbf{p}})|n\rangle$. Let us preempt this by seeing what it looks like:

$$\begin{aligned}\langle k|e^{i\mathbf{k}\cdot\mathbf{r}}(\mathbf{e}\cdot\hat{\mathbf{p}})|n\rangle &= \int_{V_{\mathrm{tot}}}\left[\frac{1}{\sqrt{L_z}}\phi_k(x,y)e^{ik_k z}\right]^* u_k^*(x,y,z)e^{i\mathbf{k}\cdot\mathbf{r}}(\mathbf{e}\cdot\hat{\mathbf{p}})\\ &\quad\times\left[\frac{1}{\sqrt{L_z}}\phi_n(x,y)e^{ik_n z}\right]u_n(x,y,z)\,dx\,dy\,dz\\ &= \frac{1}{L_z}\int_{V_{\mathrm{tot}}}\phi_k^*(x,y)e^{-ik_k z}u_k^*(x,y,z)e^{i\mathbf{k}\cdot\mathbf{r}}(\mathbf{e}\cdot\hat{\mathbf{p}})\phi_n(x,y)\\ &\quad\times e^{ik_n z}u_n(x,y,z)\,dx\,dy\,dz.\end{aligned}$$

To simplify this, let us invoke the assumption that all of the plane-wave and envelope terms vary over much longer length scales than the unit-cell functions. As a consequence, we approximate the integral as a sum of N unit-cell integrals:

$$\begin{aligned}\langle k|e^{i\mathbf{k}\cdot\mathbf{r}}(\mathbf{e}\cdot\hat{\mathbf{p}})|n\rangle &\simeq \frac{1}{L_z}\sum_N \phi_k^*(x,y)\phi_n(x,y)e^{i\mathbf{k}\cdot\mathbf{r}}e^{i(k_n-k_k)z}\\ &\quad\times\int_{V_{\mathrm{unit}}} u_k^*(x,y,z)(\mathbf{e}\cdot\hat{\mathbf{p}})u_n(x,y,z)\,dx\,dy\,dz.\end{aligned}$$

In turn, this can be written as

$$\begin{aligned}\langle k|e^{i\mathbf{k}\cdot\mathbf{r}}(\mathbf{e}\cdot\hat{\mathbf{p}})|n\rangle &\simeq \frac{1}{L_z}\sum_N \phi_k^*(x,y)\phi_n(x,y)(e^{ik_z z}e^{i\mathbf{k}_{xy}\cdot\mathbf{r}_{xy}})e^{i(k_n-k_k)z}\\ &\quad\times\int_{V_{\mathrm{unit}}} u_k^*(x,y,z)(\mathbf{e}\cdot\hat{\mathbf{p}})u_n(x,y,z)\,dx\,dy\,dz.\\ &\simeq \frac{1}{L_z}\sum_N \phi_k^*(x,y)\phi_n(x,y)e^{i\mathbf{k}_{xy}\cdot\mathbf{r}_{xy}}e^{i[k_z+(k_n-k_k)]z}\\ &\quad\times\int_{V_{\mathrm{unit}}} u_k^*(x,y,z)(\mathbf{e}\cdot\hat{\mathbf{p}})u_n(x,y,z)\,dx\,dy\,dz.\end{aligned}$$

Then, due to momentum conservation,

$$\begin{aligned}\hbar k_k &= \hbar k_n + \hbar k_z,\\ k_k &= k_n + k_z,\end{aligned}$$

we have

$$\langle k|e^{i\mathbf{k}\cdot\mathbf{r}}(\mathbf{e}\cdot\hat{\mathbf{p}})|n\rangle \simeq \frac{1}{L_z}\sum_N \phi_k^*(x,y)\phi_n(x,y)e^{i\mathbf{k}_{xy}\cdot\mathbf{r}_{xy}} \times \int_{V_{\text{unit}}} u_k^*(x,y,z)(\mathbf{e}\cdot\hat{\mathbf{p}})u_n(x,y,z)\,dx\,dy\,dz.$$

At this point, taking the long-wavelength approximation with $|\mathbf{k}_{xy}| \propto 1/\lambda$ and $\mathbf{r}_{xy}$ of molecular dimensions, $e^{i\mathbf{k}_{xy}\cdot\mathbf{r}_{xy}} \simeq 1$. Thus, we are left with

$$\begin{aligned}
\langle k|e^{i\mathbf{k}\cdot\mathbf{r}}(\mathbf{e}\cdot\hat{\mathbf{p}})|n\rangle &\simeq \frac{1}{L_z}\sum_N \phi_k^*(x,y)\phi_n(x,y) \times \int_{V_{\text{unit}}} u_k^*(x,y,z)(\mathbf{e}\cdot\hat{\mathbf{p}})u_n(x,y,z)\,dx\,dy\,dz \\
&\simeq \frac{1}{L_z}\sum_N \phi_k^*(x,y)\phi_n(x,y)\frac{V_{\text{unit}}}{V_{\text{unit}}} \times \int_{V_{\text{unit}}} u_k^*(x,y,z)(\mathbf{e}\cdot\hat{\mathbf{p}})u_n(x,y,z)\,dx\,dy\,dz \\
&\simeq \frac{1}{L_z}\sum_N \phi_k^*(x,y)\phi_n(x,y)\left\langle u_k(x,y,z)\middle|\mathbf{e}\cdot\hat{\mathbf{p}}\middle|u_n(x,y,z)\right\rangle V_{\text{unit}},
\end{aligned}$$

where

$$\left\langle u_k(x,y,z)\middle|\mathbf{e}\cdot\hat{\mathbf{p}}\middle|u_n(x,y,z)\right\rangle = \frac{1}{V_{\text{unit}}}\int_{V_{\text{unit}}} u_k^*(x,y,z)(\mathbf{e}\cdot\hat{\mathbf{p}})u_n(x,y,z)\,dx\,dy\,dz. \tag{12.62}$$

Provided there are a sufficient number of unit cells, we revert back to integral form ($V_{\text{unit}} \to dx\,dy\,dz$)

$$\begin{aligned}
\langle k|e^{i\mathbf{k}\cdot\mathbf{r}}(\mathbf{e}\cdot\hat{\mathbf{p}})|n\rangle &\simeq \frac{1}{L_z}\int_{V_{\text{tot}}} \phi_k^*(x,y)\phi_n(x,y)\left\langle u_k(x,y,z)\middle|\mathbf{e}\cdot\hat{\mathbf{p}}\middle|u_n(x,y,z)\right\rangle dx\,dy\,dz \\
&\simeq \frac{1}{L_z}\left\langle u_k(x,y,z)\middle|\mathbf{e}\cdot\hat{\mathbf{p}}\middle|u_n(x,y,z)\right\rangle \int_{V_{\text{tot}}} \phi_k^*(x,y)\phi_n(x,y)\,dx\,dy\,dz \\
&\simeq \frac{1}{L_z}\left\langle u_k(x,y,z)\middle|\mathbf{e}\cdot\hat{\mathbf{p}}\middle|u_n(x,y,z)\right\rangle \int_{L_z} dz \int_A \phi_k^*(x,y)\phi_n(x,y)\,dx\,dy \\
&\simeq \frac{L_z}{L_z}\left\langle u_k(x,y,z)\middle|\mathbf{e}\cdot\hat{\mathbf{p}}\middle|u_n(x,y,z)\right\rangle \int_A \phi_k^*(x,y)\phi_n(x,y)\,dx\,dy, \\
&\simeq \left\langle u_k(x,y,z)\middle|\mathbf{e}\cdot\hat{\mathbf{p}}\middle|u_n(x,y,z)\right\rangle \left\langle \phi_k(x,y)\middle|\phi_n(x,y)\right\rangle
\end{aligned}$$

to find

$$\langle k|e^{i\mathbf{k}\cdot\mathbf{r}}(\mathbf{e}\cdot\hat{\mathbf{p}})|n\rangle \simeq \left\langle u_k(x,y,z)\middle|\mathbf{e}\cdot\hat{\mathbf{p}}\middle|u_n(x,y,z)\right\rangle \left\langle \phi_k(x,y)\middle|\phi_n(x,y)\right\rangle.$$

When this is inserted into P_k, we obtain

$$P_k \simeq \frac{1}{\hbar^2}\left(\frac{qA_0}{2m_0}\right)^2 \left|\int_0^t \left\langle u_k(x,y,z)\middle|\mathbf{e}\cdot\hat{\mathbf{p}}\middle|u_n(x,y,z)\right\rangle \times \left\langle \phi_k(x,y)\middle|\phi_n(x,y)\right\rangle e^{-i(\omega-\omega_{kn})t'}\,dt'\right|^2,$$

and the reader may verify that this simplifies to

$$P_k(\omega) \simeq \frac{1}{\hbar^2}\left(\frac{qA_0}{m_0}\right)^2 \left|\langle u_k(x,y,z)|\mathbf{e}\cdot\hat{\mathbf{p}}|u_n(x,y,z)\rangle\right|^2 \times \left|\langle \phi_k(x,y)|\phi_n(x,y)\rangle\right|^2 \frac{\sin^2\left(\frac{\omega-\omega_{kn}}{2}t\right)}{(\omega-\omega_{kn})^2}. \tag{12.63}$$

Alternatively, in terms of energy,

$$P_k(E) \simeq \left(\frac{qA_0}{m_0}\right)^2 \left|\langle u_k(x,y,z)|\mathbf{e}\cdot\hat{\mathbf{p}}|u_n(x,y,z)\rangle\right|^2 \times \left|\langle \phi_k(x,y)|\phi_n(x,y)\rangle\right|^2 \frac{\sin^2\left(\frac{E-E_{kn}}{2\hbar}t\right)}{(E-E_{kn})^2}. \tag{12.64}$$

Multiple States

As before, to account for a distribution of possible transitions, we integrate either Equation 12.63 or Equation 12.64 over the range of possible energies, taking into account the nanowire joint density of states found in Chapter 9:

$$P_{\text{tot}}(E_{kn}) = \int_{-\infty}^{\infty} \rho_{\text{joint}}(E_{kn}) P_k(E)\, dE_{kn}.$$

Using Equation 12.64, we find that

$$P_{\text{tot}} \simeq \left(\frac{qA_0}{m_0}\right)^2 \left|\langle u_k(x,y,z)|\mathbf{e}\cdot\hat{\mathbf{p}}|u_n(x,y,z)\rangle\right|^2 \left|\langle \phi_k(x,y)|\phi_n(x,y)\rangle\right|^2 \times \int_{-\infty}^{\infty} \rho_{\text{joint}}(E_{kn}) \frac{\sin^2\left(\frac{E-E_{kn}}{2\hbar}t\right)}{(E-E_{kn})^2} dE_{kn},$$

where the sinc-like function in the integral becomes a delta function as $t \to \infty$. Namely (Equation 11.14)

$$\lim_{t\to\infty} \frac{\sin^2 \alpha t}{\pi\alpha^2 t} = \delta(\alpha).$$

Rearranging the last term in our expression to resemble this then gives

$$P_{\text{tot}} \simeq \left(\frac{qA_0}{m_0}\right)^2 \left|\langle u_k(x,y,z)|\mathbf{e}\cdot\hat{\mathbf{p}}|u_n(x,y,z)\rangle\right|^2 \left|\langle \phi_k(x,y)|\phi_n(x,y)\rangle\right|^2 \times \int_{-\infty}^{\infty} \rho_{\text{joint}}(E_{kn}) \left(\frac{\pi t}{4\hbar^2}\right) \frac{\sin^2 \alpha t}{\pi\alpha^2 t} dE_{kn},$$

where $\alpha = (E - E_{kn})/2\hbar$. From this,

$$P_{\text{tot},t\to\infty} \simeq \left(\frac{qA_0}{m_0}\right)^2 \left|\langle u_k(x,y,z)|\mathbf{e}\cdot\hat{\mathbf{p}}|u_n(x,y,z)\rangle\right|^2 \left|\langle \phi_k(x,y)|\phi_n(x,y)\rangle\right|^2 \times \int_{-\infty}^{\infty} \rho_{\text{joint}}(E_{kn}) \left(\frac{\pi t}{4\hbar^2}\right) \delta\left(\frac{E-E_{kn}}{2\hbar}\right) dE_{kn}$$

and recalling that $\delta(cx) = |c|^{-1}\delta(x)$ gives

$$P_{\text{tot},t\to\infty} \simeq \left(\frac{qA_0}{m_0}\right)^2 \left|\langle u_k(x,y,z)|\mathbf{e}\cdot\hat{\mathbf{p}}|u_n(x,y,z)\rangle\right|^2 \left|\langle \phi_k(x,y)|\phi_n(x,y)\rangle\right|^2 \times \int_{-\infty}^{\infty} \rho_{\text{joint}}(E_{kn}) \left(\frac{\pi t}{2\hbar}\right) \delta(E-E_{kn})\, dE_{kn},$$

from which

$$P_{\text{tot},t\to\infty} \simeq \frac{\pi}{2\hbar}\left(\frac{qA_0}{m_0}\right)^2 \left|\langle u_k(x,y,z)|\mathbf{e}\cdot\hat{\mathbf{p}}|u_n(x,y,z)\rangle\right|^2 \left|\langle\phi_k(x,y)|\phi_n(x,y)\rangle\right|^2 \times \rho_{\text{joint}}(E=E_{kn})t.$$

If we insert the nanowire joint density of states (Equation 9.38),

$$\rho_{\text{joint}}(E_{kn}) = \frac{1}{\pi}\left(\frac{2\mu}{\hbar^2}\right)^{1/2} \frac{1}{\sqrt{E_{kn} - E_{n_{x,e},n_{y,e},n_{x,h},n_{y,h}}}}$$

(recall that this is for one subband and that the indices $n_{x,e}$, $n_{y,e}$, $n_{x,h}$, $n_{y,h}$ represent quantum numbers describing the radial confinement energy), we obtain

$$P_{\text{tot},t\to\infty} \simeq \frac{\pi}{2\hbar}\left(\frac{qA_0}{m_0}\right)^2 \left|\langle u_k(x,y,z)|\mathbf{e}\cdot\hat{\mathbf{p}}|u_n(x,y,z)\rangle\right|^2 \times \left|\langle\phi_k(x,y)|\phi_n(x,y)\rangle\right|^2 \left[\frac{1}{\pi}\left(\frac{2\mu}{\hbar^2}\right)^{1/2}\frac{1}{\sqrt{E-E'}}\right]t,$$

with $E' = E_{n_{x,e},n_{y,e},n_{x,h},n_{y,h}}$. The associated transition rate with units of number per unit length per unit time is then $R_{\text{tot}} = dP_{\text{tot},t\to\infty}/dt$:

$$R_{\text{tot}} \simeq \frac{\pi}{2\hbar}\left(\frac{qA_0}{m_0}\right)^2 \left|\langle u_k(x,y,z)|\mathbf{e}\cdot\hat{\mathbf{p}}|u_n(x,y,z)\rangle\right|^2 \times \left|\langle\phi_k(x,y)|\phi_n(x,y)\rangle\right|^2 \left[\frac{1}{\pi}\left(\frac{2\mu}{\hbar^2}\right)^{1/2}\frac{1}{\sqrt{E-E'}}\right].$$

Alternatively, the transition rate with units of number per unit volume per unit time is

$$R_{\text{tot}} \simeq \frac{\pi}{2\hbar}\left(\frac{qA_0}{m_0}\right)^2 \left(\frac{1}{A}\right) \left|\langle u_k(x,y,z)|\mathbf{e}\cdot\hat{\mathbf{p}}|u_n(x,y,z)\rangle\right|^2 \times \left|\langle\phi_k(x,y)|\phi_n(x,y)\rangle\right|^2 \left[\frac{1}{\pi}\left(\frac{2\mu}{\hbar^2}\right)^{1/2}\frac{1}{\sqrt{E-E'}}\right],$$

where A is the cross-sectional area of the wire. Note that in real systems, the singularity that occurs when $E = E'$ is often washed out by Coulomb interactions between the electron and hole, as these are just idealized expressions. These coulomb interactions are complex since in many cases dielectric contrast effects are prominent. However, a discussion of these effects is beyond the scope of this text.

Absorption Coefficient

Finally, as with the quantum well and bulk examples, we can find the nanowire's associated absorption coefficient. Namely, the incident light intensity is (Equation 12.29)

$$I = \frac{n_m c\epsilon_0\omega^2}{2}A_0^2,$$

while the number of incident photons per unit area per unit time is (Equation 12.30)

$$n_p = \frac{I}{\hbar\omega}.$$

The wire's absorption cross-section is then (Equation 12.36)

$$\sigma = \alpha V_{\text{tot}}$$

with

$$\alpha = \frac{R_{\text{tot}}}{n_p} = \frac{2\hbar\omega}{n_m c \epsilon_0 \omega^2 A_0^2} R_{\text{tot}}$$

(Equation 12.32). As a consequence,

$$\sigma = \frac{\pi q^2 L_z}{n_m c \epsilon_0 \omega m_0^2} \left|\langle u_k(x,y,z)|\mathbf{e}\cdot\hat{\mathbf{p}}|u_n(x,y,z)\rangle\right|^2 \times \left|\langle \phi_k(x,y)|\phi_n(x,y)\rangle\right|^2 \left[\frac{1}{\pi}\left(\frac{2\mu}{\hbar^2}\right)^{1/2} \frac{1}{\sqrt{E-E'}}\right] \tag{12.65}$$

is the desired nanowire absorption cross section. Notice that the $E^{-1/2}$ dependence of the joint density of states is preserved in the wire's linear absorption.

Selection Rules

Finally, $P_{\text{tot},t\to\infty}$ and R_{tot} contain the following two terms:

$$\left|\langle u_k(x,y,z)|\mathbf{e}\cdot\hat{\mathbf{p}}|u_n(x,y,z)\rangle\right|^2$$

and

$$\left|\langle \phi_k(x,y)|\phi_n(x,y)\rangle\right|^2.$$

Both determine selection rules for nanowire optical transitions. As before, the latter is easier to discuss because we have previously calculated this overlap integral in Chapter 7. Recall for our model cylindrical nanowire (Equation 7.34) that if

$$\phi(x,y) = \phi(r,\theta) = NJ_m\left(\frac{\alpha_{m,n}}{a} r\right) e^{im\theta}$$

represents the confined contribution to the envelope function with N a normalization constant, $J_m(\alpha_{m,n} r/a)$ regular Bessel functions, a the radius of the nanowire, $\alpha_{m,n}$ the root of the Bessel function, m an angular quantum number, and n a radial quantum number, the following selection rules arise:

$$\Delta n = 0, \tag{12.66}$$

$$\Delta m = 0. \tag{12.67}$$

Namely, the radial quantum number should not change during the transition, nor should the angular quantum number. On top of this, there is an additional requirement

$$\Delta n_z = 0 \tag{12.68}$$

that stems from the plane-wave contribution to Equation 7.34. This means that the longitudinal quantum number should not change either. The statement is identical to saying that $\Delta k = 0$ since $\Delta k = (2\pi/L_z)\Delta n_z$.

Finally, we have additional selection rules that arise from the former $|\langle u_k(x,y,z)|\mathbf{e}\cdot\hat{\mathbf{p}}|u_n(x,y,z)\rangle|^2$ term. These selection rules depend on the polarization of the incoming light and on the symmetry of the unit-cell functions. Since we do not assume differences in these functions between bulk materials, wells, wires, and dots, the reader may refer to our earlier discussion on quantum well polarization selection rules to see what these terms look like.

12.9.4 Quantum Dot Transitions

In this last section, we calculate the transition probability and transition rate for a quantum dot. We start with the quantum dot wavefunction

$$\psi_{\text{tot}} = \chi(r,\theta,\phi)u(r,\theta,\phi), \tag{12.69}$$

where $\chi(r,\theta,\phi)$ is the envelope part of the total wavefunction and $u(r,\theta,\phi)$ is its unit-cell term. Both functions are normalized independently, with

$$\langle u(r,\theta,\phi)|u(r,\theta,\phi)\rangle = \int_{V_{\text{unit}}} u^*(r,\theta,\phi)u(r,\theta,\phi)\,d^3r = V_{\text{unit}} \tag{12.70}$$

and

$$\langle \chi(r,\theta,\phi)|\chi(r,\theta,\phi)\rangle = \int_{V_{\text{tot}}} \chi^*(r,\theta,\phi)\chi(r,\theta,\phi)\,d^3r = 1. \tag{12.71}$$

The total wavefunction is then normalized, as the reader can verify. Specifically,

$$\langle \psi_{\text{tot}}|\psi_{\text{tot}}\rangle = \int_{V_{\text{tot}}} \chi^*(r,\theta,\phi)u^*(r,\theta,\phi)\chi(r,\theta,\phi)u(r,\theta,\phi)\,d^3r,$$

whereupon we now make the common assumption that $\chi(r,\theta,\phi)$ and $u(r,\theta,\phi)$ vary over very different length scales, with the former changing over much larger distances. As a consequence, $\chi(r,\theta,\phi)$ is more or less constant across the unit cell and we can rewrite the above integral as a sum over N unit-cell integrals:

$$\begin{aligned}\langle \psi_{\text{tot}}|\psi_{\text{tot}}\rangle &\simeq \sum_N \chi^*(r,\theta,\phi)\chi(r,\theta,\phi)\int_{V_{\text{unit}}} u^*(r,\theta,\phi)u(r,\theta,\phi)\,d^3r \\ &\simeq \sum_N \chi^*(r,\theta,\phi)\chi(r,\theta,\phi)V_{\text{unit}}.\end{aligned}$$

Provided there are a sufficient number of unit cells (clearly an approximation), we can then re-express this as an integral over the entire quantum dot volume:

$$\langle \psi_{\text{tot}}|\psi_{\text{tot}}\rangle \simeq \int_{V_{\text{tot}}} \chi^*(r,\theta,\phi)\chi(r,\theta,\phi)\,d^3r = 1.$$

This shows that the total wavefunction is normalized. Note that V_{unit} was recast as d^3r above.

Transition Probability

Let us now evaluate the quantum dot transition probability using Equation 12.7:

$$P_k \simeq \frac{1}{\hbar^2}\left(\frac{qA_0}{2m_0}\right)^2 \left|\int_0^t \langle k|e^{i\mathbf{k}\cdot\mathbf{r}}(\mathbf{e}\cdot\hat{\mathbf{p}})|n\rangle e^{-i(\omega-\omega_{kn})t'}\,dt'\right|^2$$

along with our quantum dot wavefunctions for $|k\rangle$ and $|n\rangle$. For the electron in the conduction band, we write

$$|k\rangle = \chi_k(r,\theta,\phi)u_k(r,\theta,\phi),$$

while for the hole in the valence band, we write

$$|n\rangle = \chi_n(r,\theta,\phi)u_n(r,\theta,\phi).$$

We then evaluate the matrix element

$$\begin{aligned}\langle k|e^{i\mathbf{k}\cdot\mathbf{r}}(\mathbf{e}\cdot\hat{\mathbf{p}})|n\rangle = \int_{V_{\text{tot}}} &[\chi_k(r,\theta,\phi)u_k(r,\theta,\phi)]^*\, e^{i\mathbf{k}\cdot\mathbf{r}}(\mathbf{e}\cdot\hat{\mathbf{p}}) \\ &\times [\chi_n(r,\theta,\phi)u_n(r,\theta,\phi)]\,d^3r\end{aligned}$$

$$= \int_{V_{\text{tot}}} \chi_k^*(r,\theta,\phi) u_k^*(r,\theta,\phi) e^{i\mathbf{k}\cdot\mathbf{r}} (\mathbf{e}\cdot\hat{\mathbf{p}}) \chi_n(r,\theta,\phi) \times u_n(r,\theta,\phi) d^3r,$$

where, to simplify things, we shall assume that both the envelope functions and $e^{i\mathbf{k}\cdot\mathbf{r}}$ vary over much longer length scales than $u_n(r,\theta,\phi)$ or $u_k(r,\theta,\phi)$. The integral is then expressed as a sum over N unit-cell integrals:

$$\langle k|e^{i\mathbf{k}\cdot\mathbf{r}}(\mathbf{e}\cdot\hat{\mathbf{p}})|n\rangle \simeq \sum_N \chi_k^*(r,\theta,\phi)\chi_n(r,\theta,\phi)e^{i\mathbf{k}\cdot\mathbf{r}} \times \int_{V_{\text{unit}}} u_k^*(r,\theta,\phi)(\mathbf{e}\cdot\hat{\mathbf{p}})u_n(r,\theta,\phi)\, d^3r.$$

At this point, invoking the long-wavelength or electric dipole approximation, $e^{i\mathbf{k}\cdot\mathbf{r}} \simeq 1$, gives

$$\begin{aligned}
\langle k|e^{i\mathbf{k}\cdot\mathbf{r}}(\mathbf{e}\cdot\hat{\mathbf{p}})|n\rangle &\simeq \sum_N \chi_k^*(r,\theta,\phi)\chi_n(r,\theta,\phi) \int_{V_{\text{unit}}} u_k^*(r,\theta,\phi)(\mathbf{e}\cdot\hat{\mathbf{p}})u_n(r,\theta,\phi)\, d^3r \\
&\simeq \sum_N \chi_k^*(r,\theta,\phi)\chi_n(r,\theta,\phi)\frac{V_{\text{unit}}}{V_{\text{unit}}} \\
&\quad \times \int_{V_{\text{unit}}} u_k^*(r,\theta,\phi)(\mathbf{e}\cdot\hat{\mathbf{p}})u_n(r,\theta,\phi)\, d^3r \\
&\simeq \sum_N \chi_k^*(r,\theta,\phi)\chi_n(r,\theta,\phi)\big\langle u_k(r,\theta,\phi)\big|\mathbf{e}\cdot\hat{\mathbf{p}}\big|u_n(r,\theta,\phi)\big\rangle\, V_{\text{unit}},
\end{aligned}$$

with

$$\big\langle u_k(r,\theta,\phi)\big|\mathbf{e}\cdot\hat{\mathbf{p}}\big|u_n(r,\theta,\phi)\big\rangle = \frac{1}{V_{\text{unit}}}\int_{V_{\text{unit}}} u_k^*(r,\theta,\phi)(\mathbf{e}\cdot\hat{\mathbf{p}})u_n(r,\theta,\phi)\, d^3r. \tag{12.72}$$

Assuming a large enough sum, we revert back to integral form to obtain

$$\begin{aligned}
\langle k|e^{i\mathbf{k}\cdot\mathbf{r}}(\mathbf{e}\cdot\hat{\mathbf{p}})|n\rangle &\simeq \int_{V_{\text{tot}}} \chi_k^*(r,\theta,\phi)\chi_n(r,\theta,\phi)\big\langle u_k(r,\theta,\phi)\big|\mathbf{e}\cdot\hat{\mathbf{p}}\big|u_n(r,\theta,\phi)\big\rangle\, d^3r \\
&\simeq \big\langle u_k(r,\theta,\phi)\big|\mathbf{e}\cdot\hat{\mathbf{p}}\big|u_n(r,\theta,\phi)\big\rangle \int_{V_{\text{tot}}} \chi_k^*(r,\theta,\phi)\chi_n(r,\theta,\phi)\, d^3r \\
&\simeq \big\langle u_k(r,\theta,\phi)\big|\mathbf{e}\cdot\hat{\mathbf{p}}\big|u_n(r,\theta,\phi)\big\rangle\big\langle \chi_k(r,\theta,\phi)\big|\chi_n(r,\theta,\phi)\big\rangle.
\end{aligned}$$

When this result is introduced into P_k, we find that

$$\begin{aligned}
P_k &\simeq \frac{1}{\hbar^2}\left(\frac{qA_0}{2m_0}\right)^2 \bigg|\int_0^t \big\langle u_k(r,\theta,\phi)\big|\mathbf{e}\cdot\hat{\mathbf{p}}\big|u_n(r,\theta,\phi)\big\rangle \\
&\quad \times \big\langle \chi_k(r,\theta,\phi)\big|\chi_n(r,\theta,\phi)\big\rangle e^{-i(\omega-\omega_{kn})t'}\, dt'\bigg|^2 \\
&\simeq \frac{1}{\hbar^2}\left(\frac{qA_0}{2m_0}\right)^2 \big|\big\langle u_k(r,\theta,\phi)\big|\mathbf{e}\cdot\hat{\mathbf{p}}\big|u_n(r,\theta,\phi)\big\rangle\big|^2 \big|\big\langle \chi_k(r,\theta,\phi)\big|\chi_n(r,\theta,\phi)\big\rangle\big|^2 \\
&\quad \times \left|\int_0^t e^{-i(\omega-\omega_{kn})t'}\, dt'\right|^2.
\end{aligned}$$

This integral has been evaluated before and ultimately yields the total quantum dot transition probability

$$P_k(\omega) \simeq \frac{1}{\hbar^2}\left(\frac{qA_0}{m_0}\right)^2 \left|\langle u_k(r,\theta,\phi)|\mathbf{e}\cdot\hat{\mathbf{p}}|u_n(r,\theta,\phi)\rangle\right|^2 \times \left|\langle \chi_k(r,\theta,\phi)|\chi_n(r,\theta,\phi)\rangle\right|^2 \frac{\sin^2\left(\frac{\omega-\omega_{kn}}{2}t\right)}{(\omega-\omega_{kn})^2}. \quad (12.73)$$

In terms of energy, an equivalent expression is

$$P_k(E) \simeq \left(\frac{qA_0}{m_0}\right)^2 \left|\langle u_k(r,\theta,\phi)|\mathbf{e}\cdot\hat{\mathbf{p}}|u_n(r,\theta,\phi)\rangle\right|^2 \times \left|\langle \chi_k(r,\theta,\phi)|\chi_n(r,\theta,\phi)\rangle\right|^2 \frac{\sin^2\left(\frac{E-E_{kn}}{2\hbar}t\right)}{(E-E_{kn})^2}. \quad (12.74)$$

Multiple States

Finally, if there are multiple possible transitions, we need to integrate P_k by considering the density of states. However, from Chapter 9, the quantum dot joint density of states is simply a series of delta functions. As a consequence, we have

$$P_{\text{tot}} = \int_{-\infty}^{\infty} \rho_{\text{joint}}(E_{kn})P_k(E)\, dE_{kn},$$

where $\rho_{\text{joint}}(E_{kn}) = 2\delta(E_{kn} - E_{n_{x,e},n_{y,e},n_{z,e},n_{x,h},n_{y,h},n_{z,h}}) = 2\delta(E_{kn} - E')$ (Equation 9.40) for a given transition. This gives, using $P_k(E)$,

$$P_{\text{tot}} \simeq \int_{-\infty}^{\infty} 2\delta(E_{kn}-E')\left(\frac{qA_0}{m_0}\right)^2 \left|\langle u_k(r,\theta,\phi)|\mathbf{e}\cdot\hat{\mathbf{p}}|u_n(r,\theta,\phi)\rangle\right|^2 \times \left|\langle \chi_k(r,\theta,\phi)|\chi_n(r,\theta,\phi)\rangle\right|^2 \frac{\sin^2\left(\frac{E-E_{kn}}{2\hbar}t\right)}{(E-E_{kn})^2} dE_{kn}$$

$$\simeq 2\left(\frac{qA_0}{m_0}\right)^2 \left|\langle u_k(r,\theta,\phi)|\mathbf{e}\cdot\hat{\mathbf{p}}|u_n(r,\theta,\phi)\rangle\right|^2 \left|\langle \chi_k(r,\theta,\phi)|\chi_n(r,\theta,\phi)\rangle\right|^2 \times \frac{\sin^2\left(\frac{E-E'}{2\hbar}t\right)}{(E-E')^2}.$$

Then, as $t \to \infty$, we know that (Equation 11.14)

$$\lim_{t\to\infty} \frac{\sin^2 \alpha t}{\pi\alpha^2 t} = \delta(\alpha),$$

where $\alpha = (E-E')/2\hbar$. Rearranging things yields

$$P_{\text{tot},t\to\infty} \simeq 2\left(\frac{qA_0}{m_0}\right)^2 \left|\langle u_k(r,\theta,\phi)|\mathbf{e}\cdot\hat{\mathbf{p}}|u_n(r,\theta,\phi)\rangle\right|^2 \times \left|\langle \chi_k(r,\theta,\phi)|\chi_n(r,\theta,\phi)\rangle\right|^2 \left(\frac{\pi t}{4\hbar^2}\right)\delta\left(\frac{E-E'}{2\hbar}\right)$$

and, since $\delta(cx) = |c|^{-1}\delta(x)$,

$$P_{\text{tot},t\to\infty} \simeq \frac{\pi}{\hbar}\left(\frac{qA_0}{m_0}\right)^2 \left|\langle u_k(r,\theta,\phi)|\mathbf{e}\cdot\hat{\mathbf{p}}|u_n(r,\theta,\phi)\rangle\right|^2 \times \left|\langle \chi_k(r,\theta,\phi)|\chi_n(r,\theta,\phi)\rangle\right|^2 \delta(E-E')t. \quad (12.75)$$

This is the total quantum dot transition probability.

Finally, the transition rate expressed with units of number per unit time is simply $R_{\text{tot}} = dP_{\text{tot},t\to\infty}/dt$, giving

$$R_{\text{tot}} \simeq \frac{\pi}{\hbar}\left(\frac{qA_0}{m_0}\right)^2 \left|\langle u_k(r,\theta,\phi)|\mathbf{e}\cdot\hat{\mathbf{p}}|u_n(r,\theta,\phi)\rangle\right|^2 \times \left|\langle \chi_k(r,\theta,\phi)|\chi_n(r,\theta,\phi)\rangle\right|^2 \delta\left(E-E'\right).$$

Alternatively, the rate expressed with units of number per unit volume per unit time is

$$R_{\text{tot}} \simeq \frac{\pi}{\hbar}\left(\frac{qA_0}{m_0}\right)^2\left(\frac{1}{V_{\text{tot}}}\right) \left|\langle u_k(r,\theta,\phi)|\mathbf{e}\cdot\hat{\mathbf{p}}|u_n(r,\theta,\phi)\rangle\right|^2 \times \left|\langle \chi_k(r,\theta,\phi)|\chi_n(r,\theta,\phi)\rangle\right|^2 \delta(E-E'). \quad (12.76)$$

We will use this version of R_{tot} below.

Absorption Coefficient

As with nanowires, we can go on to find an expression for the quantum dot absorption cross section. Namely, since

$$\sigma = \alpha V_{\text{tot}},$$

where (Equation 12.32)

$$\alpha = \frac{2\hbar\omega}{n_m c\epsilon_0\omega^2 A_0^2} R_{\text{tot}},$$

we find that

$$\sigma = \frac{2\pi q^2}{n_m c\epsilon_0\omega m_0^2} \left|\langle u_k(r,\theta,\phi)|\mathbf{e}\cdot\hat{\mathbf{p}}|u_n(r,\theta,\phi)\rangle\right|^2 \times \left|\langle \chi_k(r,\theta,\phi)|\chi_n(r,\theta,\phi)\rangle\right|^2 \delta\left(E-E'\right). \quad (12.77)$$

This cross section reflects the efficiency of the transition and also shows that the discreteness of the quantum dot's joint density of states is preserved in its linear absorption.

Selection Rules

Now, as before, we see that both $P_{\text{tot},t\to\infty}$ and R_{tot} depend on two matrix elements

$$\left|\langle u_k(r,\theta,\phi)|\mathbf{e}\cdot\hat{\mathbf{p}}|u_n(r,\theta,\phi)\rangle\right|^2$$

and

$$\left|\langle \chi_k(r,\theta,\phi)|\chi_n(r,\theta,\phi)\rangle\right|^2.$$

Both determine the selection rules for allowed quantum dot transitions. As in earlier examples, the latter is easier to discuss, because we have already calculated this overlap integral in Chapter 7. Recall that for our model (spherical) quantum dot, we had (Equation 7.42)

$$\chi(r,\theta,\phi) = Nj_l\left(\frac{\alpha_{l,n}}{a}r\right) Y_l^m(\theta,\phi),$$

where N is a normalization constant, $j_l(\alpha_{l,n}r/a)$ are spherical Bessel functions, l is the order of the Bessel function, n is the index of the root (also called a radial quantum number), a is the radius of the sphere,

$Y_l^m(\theta,\phi)$ are spherical harmonics, and m is a momentum quantum number.

From our earlier calculations, we found the following selection rules:

$$\Delta n = 0, \tag{12.78}$$

$$\Delta l = 0, \tag{12.79}$$

$$\Delta m = 0. \tag{12.80}$$

Thus, the radial quantum number should not change during transitions, nor should the momentum quantum number. Transitions also only occur between states having the same symmetry (i.e., only S to S, P to P, etc. transitions are allowed).

For convenience, we label these quantum dot transitions

$$n_h l_h n_e l_e,$$

with the hole quantum numbers/labels written first. As a consequence, one labels the lowest energy transition in this model $1S_h1S_e$, where S_h (S_e) represents $l_h = 0$ ($l_e = 0$). A higher-energy transition would be $2S_h2S_e$. In practice, things become more complicated due to additional corrections in the semiconductor valence band, as we shall see shortly.

Finally, like the bulk, quantum well, and nanowire examples before, we have additional selection rules that arise from $|\langle u_k(r,\theta,\phi)|\mathbf{e}\cdot\hat{\mathbf{p}}|u_n(r,\theta,\phi)\rangle|^2$. These selection rules depend on the polarization of the incoming light and the symmetry of the unit-cell functions. Since we do not assume differences in unit-cell functions between bulk and low-dimensional systems, the reader can refer to our earlier discussion about quantum well polarization selection rules to see what these terms look like.

12.9.5 Towards a Comparison with Experimental Data

Symmetry of the CdSe Bloch Terms

Now, because we will eventually compare the model developed here with experimental data on CdSe quantum dots, let us be more specific about both the notation and energies of $u_k(r,\theta,\phi)$ and $u_n(r,\theta,\phi)$. In particular, note that the conduction band of CdSe arises from a linear combination of Cd 5s atomic orbitals while the valence band arises from a linear combination of Se 4p atomic orbitals.

This can be seen through the Aufbau principle. For cadmium, the atomic number is $Z = 48$. Its atomic configuration is therefore

$$\mathrm{Cd{:}1s^2 2s^2 2p^6 3s^2 3p^6 4s^2 3d^{10} 4p^6 5s^2 4d^{10}}.$$

The outermost orbital is 5s. Likewise, for selenium, the atomic number is $Z = 34$, with a corresponding atomic configuration of

$$\mathrm{Se{:}1s^2 2s^2 2p^6 3s^2 3p^6 4s^2 3d^{10} 4p^4}.$$

The outermost orbital is 4p. We thus find that $u_k(r,\theta,\phi)$ has s-like symmetry while $u_n(r,\theta,\phi)$ has p-like symmetry.

Bloch Function Angular Momentum Labels

To continue, let us revisit the addition of angular momenta. What we care about is the angular momentum descriptions of $u_k(r,\theta,\phi)$ and $u_n(r,\theta,\phi)$. The reason for this is that we will eventually change notation since unit-cell function labels in the literature are often denoted differently. Thus, to establish this notation, we need further information about the addition of angular momenta and a better description of the CdSe valence band.

First, recall that if we have angular momenta J_1 and J_2, their sum yields

$$J_{tot} = J_1 + J_2, J_1 + J_2 - 1, \ldots, |J_1 - J_2|,$$

with momentum projections

$$m_j = J_{tot}, J_{tot} - 1, \ldots, -J_{tot}.$$

More about the addition of angular momenta can be found in elementary quantum mechanics texts.

Next, for atomic orbitals, we consider the addition of orbital (l) and spin (s) angular momenta. We thus deal with

$$J = l + s, \tag{12.81}$$

where J represents the carrier's total angular momentum. As a consequence, for the electron in the CdSe conduction band, $l_e = 0$ (s orbital) and $s_e = \frac{1}{2}$ (spin), giving

$$J_e = \tfrac{1}{2}.$$

Its associated momentum projections along the z axis of a unit-cell coordinate system are $\pm\frac{1}{2}$. Likewise, for the valence band hole, $l_h = 1$ (p orbital) and $s_h = \frac{1}{2}$ (spin). We therefore have

$$J_h = \tfrac{3}{2}, \tfrac{1}{2}$$

with momentum projections $\pm\frac{3}{2}, \pm\frac{1}{2}$, and $\pm\frac{1}{2}$. Again, these are projections along the z axis of a unit-cell coordinate system.

As a consequence, we can alternatively denote $u_k(r,\theta,\phi)$ and $u_n(r,\theta,\phi)$ in bra–ket fashion using their total angular momentum and associated projections as indices (i.e., as $|J, m_j\rangle$). For the conduction band, we therefore find

$$u_k(r,\theta,\phi) \equiv |\tfrac{1}{2},\tfrac{1}{2}\rangle \quad \text{and} \quad |\tfrac{1}{2},-\tfrac{1}{2}\rangle.$$

The function is twofold-degenerate. Likewise, for the valence band, we have

$$u_n(r,\theta,\phi) \equiv |\tfrac{3}{2},\tfrac{3}{2}\rangle, |\tfrac{3}{2},\tfrac{1}{2}\rangle, |\tfrac{3}{2},-\tfrac{1}{2}\rangle, |\tfrac{3}{2},-\tfrac{3}{2}\rangle, |\tfrac{1}{2},\tfrac{1}{2}\rangle, |\tfrac{1}{2},-\tfrac{1}{2}\rangle$$

and is sixfold-degenerate. In either case, the angular momentum descriptions have the symmetry-equivalent functions listed in **Tables 12.1** and **12.2**.

At this point, it is apparent that the lowest-energy quantum dot optical transitions occur between states in the valence band having $u_n(r,\theta,\phi) = |\frac{3}{2},\frac{3}{2}\rangle, |\frac{3}{2},\frac{1}{2}\rangle, |\frac{3}{2},-\frac{1}{2}\rangle, |\frac{3}{2},-\frac{3}{2}\rangle, |\frac{1}{2},\frac{1}{2}\rangle$, or $|\frac{1}{2},-\frac{1}{2}\rangle$ and states in the conduction band having $u_k(r,\theta,\phi) = |\frac{1}{2},\frac{1}{2}\rangle$ or $|\frac{1}{2},-\frac{1}{2}\rangle$. As an example, a transition occurs between the final state

$$\begin{aligned}\psi_e = |k\rangle &= \chi(r,\theta,\phi)u_k(r,\theta,\phi) \\ &= \left[Nj_0\left(\frac{\alpha_{01}}{a}r\right)Y_0^0(\theta,\phi)\right]u_k(r,\theta,\phi) \\ &= \left[\frac{1}{\sqrt{2\pi a}}\frac{\sin(\pi r/a)}{r}\right]|\tfrac{1}{2},\tfrac{1}{2}\rangle\end{aligned}$$

and the initial state

$$\psi_h = |n\rangle = \left[\frac{1}{\sqrt{2\pi a}}\frac{\sin(\pi r/a)}{r}\right]|\tfrac{3}{2},\tfrac{3}{2}\rangle.$$

In both $|k\rangle$ and $|n\rangle$, Equation 7.43 has been used to represent the lowest-energy envelope function contribution to the total wavefunction. Furthermore, note that the transition occurs due to a change in angular momentum of -1 in the unit-cell terms, stemming from the absorption of a photon (i.e., going from $|\frac{3}{2},\frac{3}{2}\rangle$ to $|\frac{1}{2},\frac{1}{2}\rangle$ implies a change of the momentum projection by -1). While we will not go into this in any great detail, the transition arises from the absorption of circularly polarized light.

In parallel, we have an energetically degenerate transition between

$$\psi_e = |k\rangle = \left[\frac{1}{\sqrt{2\pi a}}\frac{\sin(\pi r/a)}{r}\right]|\tfrac{1}{2},-\tfrac{1}{2}\rangle$$

and

$$\psi_h = |n\rangle = \left[\frac{1}{\sqrt{2\pi a}}\frac{\sin(\pi r/a)}{r}\right]|\tfrac{3}{2},-\tfrac{3}{2}\rangle.$$

due to the absorption of a photon, which changes the angular momentum by +1. This again entails circularly polarized light. Other similarly degenerate transitions involve the following final/initial states:

$$\psi_e = |k\rangle = \left[\frac{1}{\sqrt{2\pi a}}\frac{\sin(\pi r/a)}{r}\right]|\tfrac{1}{2},\tfrac{1}{2}\rangle$$

and

$$\psi_h = |n\rangle = \left[\frac{1}{\sqrt{2\pi a}}\frac{\sin(\pi r/a)}{r}\right]|\tfrac{3}{2},\tfrac{1}{2}\rangle,$$

as well as

$$\psi_e = |k\rangle = \left[\frac{1}{\sqrt{2\pi a}}\frac{\sin(\pi r/a)}{r}\right]|\tfrac{1}{2},-\tfrac{1}{2}\rangle$$

and

$$\psi_h = |n\rangle = \left[\frac{1}{\sqrt{2\pi a}}\frac{\sin(\pi r/a)}{r}\right]|\tfrac{3}{2},-\tfrac{1}{2}\rangle.$$

Readers may verify this for themselves. In either of these two latter cases, there is no change in the angular momentum projection. Unlike the previous examples, these transitions therefore involve linearly polarized light.

Within our current CdSe picture, all of these degenerate transitions are labeled

$$1S_h 1S_e.$$

Higher-energy transitions arise exclusively due to changes in the envelope part of the total wavefunction. To illustrate, the second lowest-energy transition involves the next smallest spherical Bessel root (α_{11}), and, as a consequence, the envelope functions for both the electron and hole change to

$$\left[N j_1\left(\frac{\alpha_{11}}{a}r\right)Y_1^{1,0,-1}(\theta,\phi)\right],$$

while the unit-cell terms remain the same. These transitions are labeled $1P_h 1P_e$.

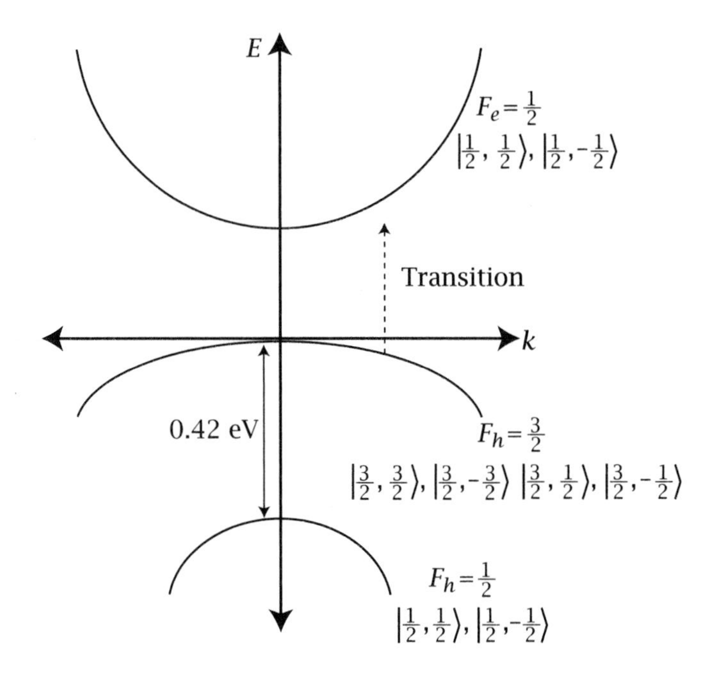

Figure 12.5 CdSe band structure with spin–orbit coupling lifting the degeneracy of hole states with $F_h = \frac{3}{2}$ and $F_h = \frac{1}{2}$. What results is a fourfold-degenerate band and a twofold-degenerate band at higher energies. The lowest-energy transition is denoted by the dashed arrow.

Additional Complications, a Three-Valence-Band Model

Unfortunately, the valence band of CdSe is complicated by spin–orbit coupling, which lifts the degeneracy of $u_n(r,\theta,\phi)$. States with $J_h = \frac{3}{2}$ (i.e., $|\frac{3}{2},\frac{3}{2}\rangle$, $|\frac{3}{2},\frac{1}{2}\rangle$, $|\frac{3}{2},-\frac{1}{2}\rangle$, and $|\frac{3}{2},-\frac{3}{2}\rangle$) and $J_h = \frac{1}{2}$ (i.e., $|\frac{1}{2},\frac{1}{2}\rangle$ and $|\frac{1}{2},-\frac{1}{2}\rangle$) therefore differ in energy, with the twofold-degenerate $F_h = \frac{1}{2}$ states being 0.42 eV higher in energy. These $u_k(r,\theta,\phi)$ and $u_n(r,\theta,\phi)$ terms are summarized in **Figure 12.5**, where one sees that the degeneracy between $J_h = \frac{3}{2}$ and $J_h = \frac{1}{2}$ has been lifted due to spin–orbit coupling. Even more complications exist. Namely, crystal field splitting exists in wurtzite CdSe. This further lifts the fourfold degeneracy of the lowest $J_h = \frac{3}{2}$ band, with the magnitude of the splitting being 25 meV.

The CdSe valence band therefore consists of three bands. The lowest is called the heavy hole (HH) band or alternatively, the "A" band. Slightly higher in energy is the light hole (LH) band, sometimes called the "B" band. Finally, there is the split-off hole (SOH) band, sometimes referred to as the "C" band. The angular momentum labels for their unit-cell functions are then

$$\begin{aligned}\text{HH:}&\quad |\tfrac{3}{2},\tfrac{3}{2}\rangle, |\tfrac{3}{2},-\tfrac{3}{2}\rangle\\ \text{LH:}&\quad |\tfrac{3}{2},\tfrac{1}{2}\rangle, |\tfrac{3}{2},-\tfrac{1}{2}\rangle\\ \text{SOH:}&\quad |\tfrac{1}{2},\tfrac{1}{2}\rangle, |\tfrac{1}{2},-\tfrac{1}{2}\rangle.\end{aligned}$$

For the conduction band, we still have

$$\text{CB:}\quad |\tfrac{1}{2},\tfrac{1}{2}\rangle, |\tfrac{1}{2},-\tfrac{1}{2}\rangle.$$

All are shown schematically in **Figure 12.6**. Let us assume for now that the valence bands are all independent. As a consequence, each possesses its own progression of excited-state transitions with the conduction band. Furthermore, their energies are just offset from each other by the energy differences between the HH, LH, and SOH bands.

In this revised CdSe picture with three independent valence bands, the lowest-energy transitions therefore occur between the HH band and the conduction band. This is followed by transitions between the LH

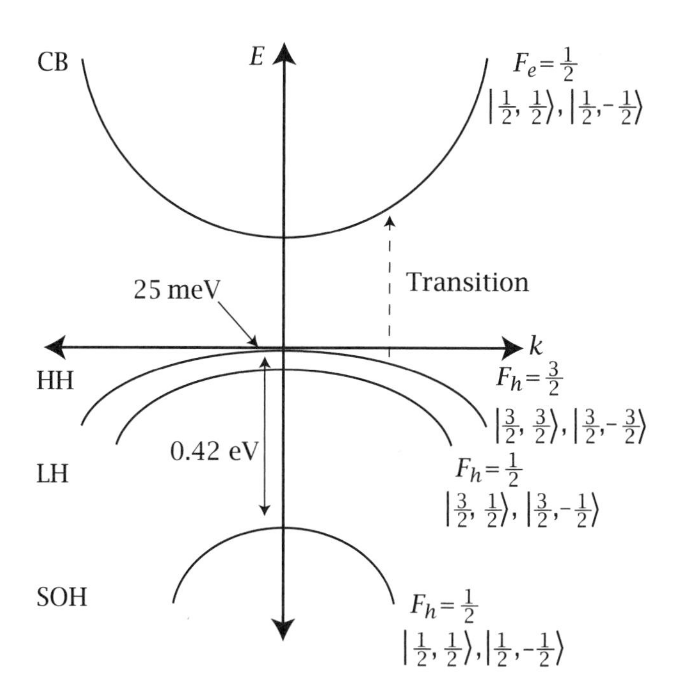

Figure 12.6 CdSe band structure including the effects of spin–orbit coupling and crystal field splitting in the valence band. What results are three twofold-degenerate bands called the heavy hole, light hole, and split off hole bands. The lowest-energy transition is denoted by the dashed arrow.

Table 12.3 Four smallest spherical Bessel function roots, associated spherical Bessel functions, and symmetry labels

	First-lowest root	Second-lowest root	Third-lowest root	Fourth-lowest root
$\alpha_{l,n}$	$\alpha_{01} \simeq 3.14$	$\alpha_{11} \simeq 4.49$	$\alpha_{21} \simeq 5.76$	$\alpha_{02} \simeq 6.28$
Spherical Bessel	$j_0(x)$	$j_1(x)$	$j_2(x)$	$j_0(x)$
Envelope function symmetry	$l = 0$ (S)	$l = 1$ (P)	$l = 2$ (D)	$l = 0$ (S)

band and the conduction band and, in turn, by higher-energy transitions between the SOH band and the conduction band. We again label these transitions using the notation

$$n_h l_{J_h} n_e l_{J_e},$$

where J_e and J_h are the total electron and hole unit cell angular momentum quantum numbers.

In what follows, let us calculate an excited-state progression using this model. Information that we need, namely, the four lowest spherical Bessel function roots, the associated spherical Bessel functions, and their angular momentum symmetry labels, are listed in **Table 12.3**.

Since the lowest-energy transitions occur from the heavy hole band to the conduction band, as illustrated in **Figure 12.7**, a final (initial) wavefunction for the electron (hole) could be

$$\psi_e = |k\rangle = \left[N j_0\left(\frac{\alpha_{01}}{a} r\right) Y_0^0(\theta, \phi)\right] |\tfrac{1}{2}, \tfrac{1}{2}\rangle,$$
$$\psi_h = |n\rangle = \left[N j_0\left(\frac{\alpha_{01}}{a} r\right) Y_0^0(\theta, \phi)\right] |\tfrac{3}{2}, \tfrac{3}{2}\rangle.$$

Note that the transition is twofold-degenerate. As a consequence, a complementary transition between an electron (hole) state, having the same envelope function but with $|\frac{1}{2}, -\frac{1}{2}\rangle$ ($|\frac{3}{2}, -\frac{3}{2}\rangle$) as its unit-cell term, can also be written. This is left to the reader as an exercise. Both transitions

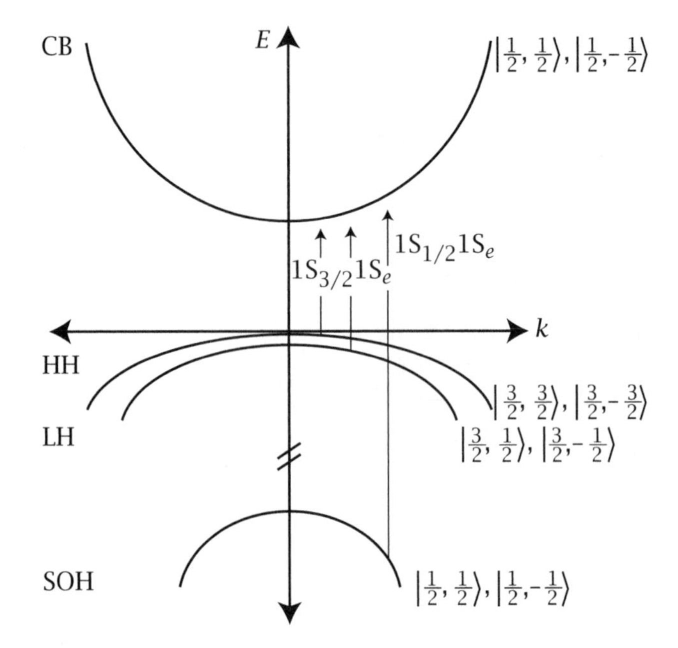

Figure 12.7 Interband transition labels of a refined CdSe quantum dot model, involving three independent valence bands.

involve circularly polarized light, since the momentum changes by ± 1. Either transition is labeled $1S_{3/2}1S_e$, where it is understood that the subscript e refers to $\frac{1}{2}$.

The next lowest-energy transitions occur from the light hole band to the conduction band, as illustrated in **Figure 12.7**. Recall that they are only slightly higher in energy (25 meV) than the earlier HH transitions. As before, the final (initial) electron (hole) wavefunction can be written as

$$\psi_e = |k\rangle = \left[Nj_0\left(\frac{\alpha_{01}}{a}r\right) Y_0^0(\theta,\phi)\right] |\tfrac{1}{2},\tfrac{1}{2}\rangle,$$
$$\psi_h = |n\rangle = \left[Nj_0\left(\frac{\alpha_{01}}{a}r\right) Y_0^0(\theta,\phi)\right] |\tfrac{3}{2},\tfrac{1}{2}\rangle.$$

The transition is degenerate, and a second pair of final/initial states with identical envelope functions but with $|\frac{1}{2},-\frac{1}{2}\rangle$ ($|\frac{3}{2},-\frac{1}{2}\rangle$) as the electron (hole) unit-cell contribution can be written. These transitions are also labeled $1S_{3/2}1S_e$, although it should be understood that they originate from the LH band. Such transitions also involve linearly polarized light, since there is no change in the unit-cell function momentum projections.

For the next transition, we must decide whether it involves a 1S-to-1S SOH-to-conduction band transition or whether it is a 1P–1P transition, involving the HH band. In a $a = 2.5$ nm radius quantum dot, the 1P-to-1P (HH) transition energy exceeds 0.5 eV added to E_g. Since the spin–orbit splitting is 0.42 eV, this suggests that the 1S-to-1S SOH transition is the next lowest-energy transition. This is illustrated in **Figure 12.7**. In this case, the final (initial) electron (hole) wavefunction is

$$\psi_e = |k\rangle = \left[Nj_0\left(\frac{\alpha_{01}}{a}r\right) Y_0^0(\theta,\phi)\right] |\tfrac{1}{2},\tfrac{1}{2}\rangle$$
$$\psi_h = |n\rangle = \left[Nj_0\left(\frac{\alpha_{01}}{a}r\right) Y_0^0(\theta,\phi)\right] |\tfrac{1}{2},\tfrac{1}{2}\rangle.$$

As with the earlier HH and LH transitions, a complementary degenerate transition involving a final (initial) electron (hole) state with the same envelope function but with $|\frac{1}{2},-\frac{1}{2}\rangle$ ($|\frac{1}{2},-\frac{1}{2}\rangle$) as the unit-cell term exists. Verification of this is left as an exercise for the reader. We label such transitions $1S_{1/2}1S_e$ and they involve the absorption of linearly polarized light.

At this point, the next higher-energy transition is the 1P-to-1P HH transition with a final (initial) electron (hole) wavefunction of

$$\psi_e = |k\rangle = \left[Nj_1\left(\frac{\alpha_{11}}{a}r\right)Y_1^{1,0,-1}(\theta,\phi)\right]\left|\tfrac{1}{2},\tfrac{1}{2}\right\rangle$$

$$\psi_h = |n\rangle = \left[Nj_1\left(\frac{\alpha_{11}}{a}r\right)Y_1^{1,0,-1}(\theta,\phi)\right]\left|\tfrac{3}{2},\tfrac{3}{2}\right\rangle.$$

An accompanying energetically degenerate transition, involving an electron (hole) wavefunction with the same envelope function but with $\left|\frac{1}{2},-\frac{1}{2}\right\rangle$ ($\left|\frac{3}{2},-\frac{3}{2}\right\rangle$) as the electron (hole) unit-cell term, also exists. Both transitions are labeled $1P_{3/2}1P_e$ and involve the absorption of circularly polarized light.

We can go on in this fashion to assign other higher-energy transitions. What results though is the following excited-state progression for our simple three-valence-band CdSe model. From lower to higher energies, we have

$1S_{3/2}1S_e$ (HH)
$1S_{3/2}1S_e$ (LH)
$1S_{1/2}1S_e$ (SOH)
$1P_{3/2}1P_e$ (HH)
$1P_{3/2}1P_e$ (LH)
and so on

S–D and Valence Band Mixing

Now, even this more refined model misses a few important things. Namely, it does not explain the presence of avoided crossings between states seen in actual experimental data. As a consequence, to account for such observations, we need additional modifications to the model. The resulting theory goes beyond the scope of this text. However, the reader can see more about it in the references. In what follows, we simply highlight some of its key results.

In particular, the quantum numbers J_e and J_h seen earlier are actually not good quantum numbers of the system. Instead, a better quantum number turns out to be

$$F = L + J, \tag{12.82}$$

where L is the angular momentum quantum number of the envelope function and J is the unit-cell momentum contribution seen earlier. For the electron, we then have

$$F_e = L_e + J_e,$$

while for the hole

$$F_h = L_h + J_h.$$

This means that the various valence band states are actually coupled together. Thus, a given hole unit-cell function consists of a linear combination of heavy hole, light hole, and split-off hole band terms. Furthermore, a similar mixing of envelope functions occurs. This is referred to as S–D mixing since envelope functions with L and $L+2$ momenta mix. Thus, a nominally "S" state has "D" character to it, and vice versa.

Representative hole wavefunctions for even-symmetry states can therefore be written as follows. We shall not derive these functions but instead refer the interested reader to the Further Reading for more details. We find

$$\begin{aligned}
|\tfrac{3}{2},+\tfrac{3}{2}\rangle^{+} &= \left(R_{h2}(r)Y_0^0+\sqrt{\tfrac{1}{5}}R_{h1}(r)Y_2^0\right)|\tfrac{3}{2},\tfrac{3}{2}\rangle+\left(2\sqrt{\tfrac{1}{10}}R_{h1}(r)Y_2^2\right)|\tfrac{3}{2},-\tfrac{1}{2}\rangle\\
&\quad+\left(-2\sqrt{\tfrac{1}{10}}R_{h1}(r)Y_2^1\right)|\tfrac{3}{2},\tfrac{1}{2}\rangle+\left(-\sqrt{\tfrac{1}{5}}R_s(r)Y_2^1\right)|\tfrac{1}{2},\tfrac{1}{2}\rangle\\
&\quad+\left(2\sqrt{\tfrac{1}{5}}R_s(r)Y_2^2\right)|\tfrac{1}{2},-\tfrac{1}{2}\rangle,\\
|\tfrac{3}{2},-\tfrac{3}{2}\rangle^{+} &= \left(R_{h2}(r)Y_0^0+\sqrt{\tfrac{1}{5}}R_{h1}(r)Y_2^0\right)|\tfrac{3}{2},-\tfrac{3}{2}\rangle+\left(2\sqrt{\tfrac{1}{10}}R_{h1}(r)Y_2^{-2}\right)|\tfrac{3}{2},\tfrac{1}{2}\rangle\\
&\quad+\left(-2\sqrt{\tfrac{1}{10}}R_{h1}(r)Y_2^{-1}\right)|\tfrac{3}{2},-\tfrac{1}{2}\rangle+\left(-2\sqrt{\tfrac{1}{5}}R_s(r)Y_2^{-2}\right)|\tfrac{1}{2},\tfrac{1}{2}\rangle\\
&\quad+\left(\sqrt{\tfrac{1}{5}}R_s(r)Y_2^{-1}\right)|\tfrac{1}{2},-\tfrac{1}{2}\rangle,\\
|\tfrac{3}{2},+\tfrac{1}{2}\rangle^{+} &= \left(\sqrt{\tfrac{2}{5}}R_{h1}(r)Y_2^{-1}\right)|\tfrac{3}{2},\tfrac{3}{2}\rangle+\left(\sqrt{\tfrac{2}{5}}R_{h1}(r)Y_2^2\right)|\tfrac{3}{2},-\tfrac{3}{2}\rangle\\
&\quad+\left(R_{h2}(r)Y_0^0\right)|\tfrac{3}{2},\tfrac{1}{2}\rangle+\left(-\sqrt{\tfrac{1}{5}}R_{h1}(r)Y_2^0\right)|\tfrac{3}{2},\tfrac{1}{2}\rangle\\
&\quad+\left(-\sqrt{\tfrac{2}{5}}R_s(r)Y_2^0\right)|\tfrac{1}{2},\tfrac{1}{2}\rangle+\left(\sqrt{\tfrac{3}{5}}R_s(r)Y_2^1\right)|\tfrac{1}{2},-\tfrac{1}{2}\rangle,\\
|\tfrac{3}{2},-\tfrac{1}{2}\rangle^{+} &= \left(\sqrt{\tfrac{2}{5}}R_{h1}(r)Y_2^{-2}\right)|\tfrac{3}{2},\tfrac{3}{2}\rangle+\left(\sqrt{\tfrac{2}{5}}R_{h1}(r)Y_2^1\right)|\tfrac{3}{2},-\tfrac{3}{2}\rangle\\
&\quad+\left(R_{h2}(r)Y_0^0-\sqrt{\tfrac{1}{5}}R_{h1}(r)Y_2^0\right)|\tfrac{3}{2},-\tfrac{1}{2}\rangle+\left(\sqrt{\tfrac{2}{5}}R_s(r)Y_2^0\right)|\tfrac{1}{2},-\tfrac{1}{2}\rangle\\
&\quad+\left(-\sqrt{\tfrac{3}{5}}R_s(r)Y_2^{-1}\right)|\tfrac{1}{2},\tfrac{1}{2}\rangle,\\
|\tfrac{1}{2},+\tfrac{1}{2}\rangle^{+} &= \sqrt{\tfrac{2}{5}}R_{h1}(r)Y_2^2|\tfrac{3}{2},-\tfrac{3}{2}\rangle-\sqrt{\tfrac{1}{10}}R_{h1}(r)Y_2^{-1}|\tfrac{3}{2},\tfrac{3}{2}\rangle\\
&\quad-\sqrt{\tfrac{3}{10}}R_{h1}(r)Y_2^1|\tfrac{3}{2},-\tfrac{1}{2}\rangle+\sqrt{\tfrac{1}{5}}R_{h1}(r)Y_2^0|\tfrac{3}{2},\tfrac{1}{2}\rangle+R_s(r)Y_0^0|\tfrac{1}{2},\tfrac{1}{2}\rangle,\\
|\tfrac{1}{2},-\tfrac{1}{2}\rangle^{+} &= \sqrt{\tfrac{1}{10}}R_{h1}(r)Y_2^1|\tfrac{3}{2},-\tfrac{3}{2}\rangle-\sqrt{\tfrac{2}{5}}R_{h1}(r)Y_2^{-2}|\tfrac{3}{2},\tfrac{3}{2}\rangle\\
&\quad+\sqrt{\tfrac{3}{10}}R_{h1}(r)Y_2^{-1}|\tfrac{3}{2},\tfrac{1}{2}\rangle-\sqrt{\tfrac{1}{5}}R_{h1}(r)Y_2^0|\tfrac{3}{2},-\tfrac{1}{2}\rangle+R_s(r)Y_0^0|\tfrac{1}{2},-\tfrac{1}{2}\rangle.
\end{aligned}$$

For convenience, the (θ,ϕ) dependences of the spherical harmonics have been omitted to make the notation more compact. Furthermore, the radial functions $R_{h1}(r)$, $R_{h2}(r)$, and $R_s(r)$ simply represent linear combinations of spherical Bessel functions. We shall not go into what they actually look like. Again, the interested reader may consult the Further Reading for more details. Notice that the bra–ket angular momentum labels ($|\tfrac{3}{2},-\tfrac{3}{2}\rangle^{+}$, for example) now simply represent hole wavefunctions. The important thing to notice is that they contain linear combinations of S and D envelope functions. Furthermore, they contain unit-cell terms from the HH, LH, and SOH bands. The electron wavefunctions remain more or less unchanged in this model.

Finally, since only F and its momentum projections are good quantum numbers, the selection rules we saw earlier break down. Specifically, the $\Delta n = 0$ selection rule found earlier does not hold anymore due to the mixed character of resulting hole wavefunctions. As a consequence, transitions such as $2S_{3/2}1S_e$ exist. In fact, if we calculate the energy of such a mixed transition, we find that it is the second lowest-energy transition in our new excited-state progression. Furthermore, the $\Delta l = 0$ selection rule seen earlier also breaks down. This leaves open the possibility of S-to-P transitions. While we shall not derive it, one finds the following excited-state progression in our quantum dot. From lower to higher energies, we have

$1S_{3/2}1S_e$

$2S_{3/2}1S_e$

$1S_{1/2}1S_e$

$1P_{3/2}1P_e$

$2S_{1/2}1S_e$

and so on

12.9.6 Photoluminescence Excitation Spectroscopy (PLE)

At this point, we introduce the experimental data by first describing the technique used to obtain it. In this regard, even the best colloidal quantum dot samples today exhibit residual size distributions. As a consequence, one seeks to reduce the inhomogeneous broadening of optical transitions by performing so-called size-selective spectroscopies. One of these techniques is based on monitoring the emission intensity of an ensemble as the excitation frequency is scanned to the blue (i.e., to higher energies) of the monitored (emission) wavelength. The logic behind this approach is that, since the emission intensity is proportional to the amount of light absorbed, one can map out a sample's absorption spectrum indirectly.

The size selectivity of the technique arises from the fact that larger quantum dots in the ensemble emit predominantly on the red edge of the spectrum while smaller ones emit towards the blue. This arises due to the carrier confinement effects we have discussed previously. Thus, by monitoring either the blue or red edge of the ensemble emission, the inhomogeneous broadening of the absorption is suppressed and, in turn, yields a line-narrowed spectrum. Usually, these experiments are carried out at low temperatures using a commercial fluorimeter with double monochromators and a Xenon arc lamp.

Figure 12.8 illustrates representative PLE data from an ensemble of CdSe quantum dots (Norris 1995; Norris and Bawendi 1996). One sees clear features in the PLE-extracted spectrum relative to the original linear absorption profile. This PLE data can also be compared with **Figure 1.3**, which shows the linear absorption of a small-size series of analogous CdSe quantum dots. In all cases, it is evident that the PLE technique removes much of the inhomogeneous broadening present in the sample.

The red edge of the PLE data is simply cut off because of the finite Stokes shift between the absorption and emission. Discrimination between the emission and excitation therefore becomes a problem in this region, resulting in excitation bleed-through (i.e., one starts to see the excitation light and not the emission). The blue edge is usually limited by the output of the arc lamp, since its power density falls off towards higher energies.

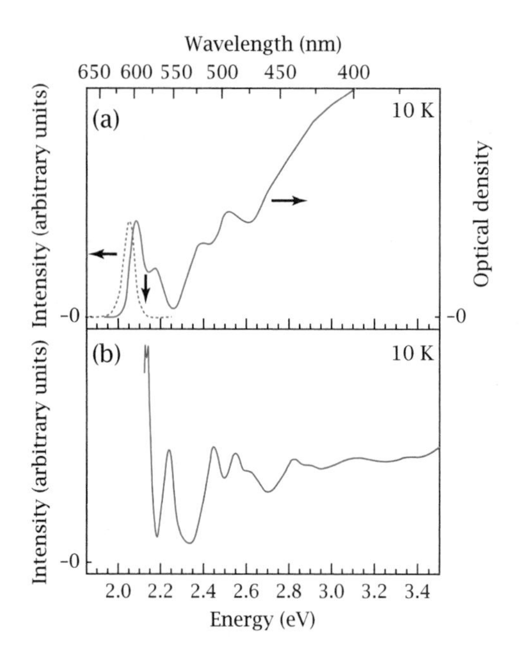

Figure 12.8 Comparison between (a) the CdSe quantum dot linear absorption spectrum and (b) the corresponding PLE spectrum. Figure used with permission from Norris and Bawendi (1996).

In **Figure 12.9**, the extracted PLE spectrum is fit with a sum of Gaussians. Each represents a separate quantum dot transition. The features are labeled alphabetically in order of increasing energy. A cubic background function is also used to improve the quality of the fit. With this procedure, up to 10 excited states in the quantum dot linear absorption spectrum have been extracted.

The PLE experiment is then repeated for different quantum dot sizes, enabling the size dependences of these 10 excited states to be mapped out. This is shown in **Figure 12.10**, where the excited-state energies of states (a)–(j) are plotted relative to the energy of (a). The reason for this plotting scheme is that the energy of the lowest excited state is better defined than the actual mean physical size of a given ensemble.

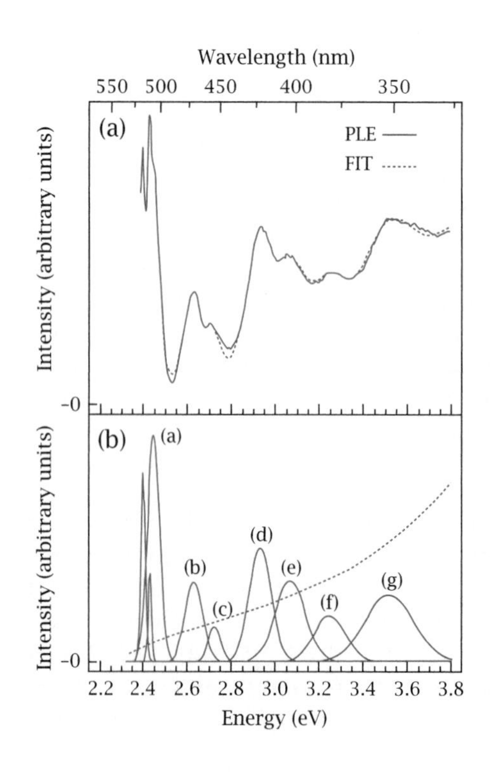

Figure 12.9 PLE data from an ensemble of colloidal CdSe quantum dots. (a) Extracted data and corresponding fit to a sum of Gaussians. (b) Extracted Gaussian lineshapes, each representing a separate quantum dot optical transition. A cubic background function has been used to improve the fit. Figure used with permission from Norris and Bawendi (1996).

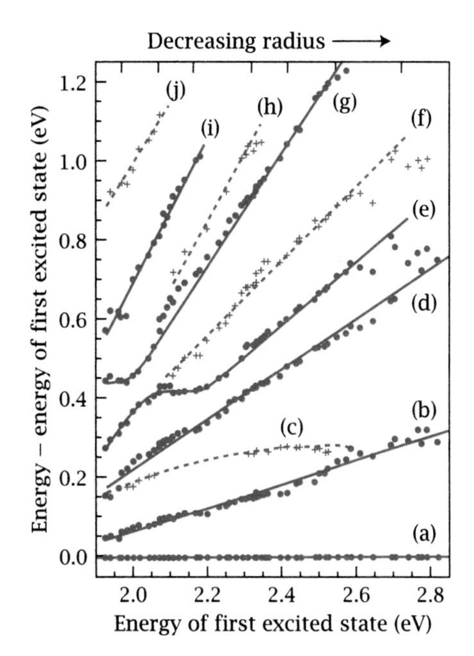

Figure 12.10 Summary of the size-dependent excited state progression in colloidal CdSe quantum dots. Strong transitions are denoted with solid circles. Weaker transitions are denoted with crosses. Lines are guides to the eye. Figure used with permission from Norris and Bawendi (1996).

In the figure, strong transitions are denoted with solid circles while weaker transitions are denoted by crosses. Lines are drawn through all experimental points to guide the eye.

Figures 12.11 and **12.12** show the same data, but this time with assignments for the various transitions provided. Also shown is a fit using the more complicated quantum dot model mentioned at the end of the last section. We see that the lowest-energy transition is $1S_{3/2}1S_e$, as expected. The next higher-energy transition is $2S_{3/2}1S_e$. This is followed by $1S_{1/2}1S_e$ and $1P_{3/2}1P_e$. If we summarize the five lowest-energy transitions, we have the following excited-state progression:

$$1S_{3/2}1S_e$$
$$2S_{3/2}1S_e$$

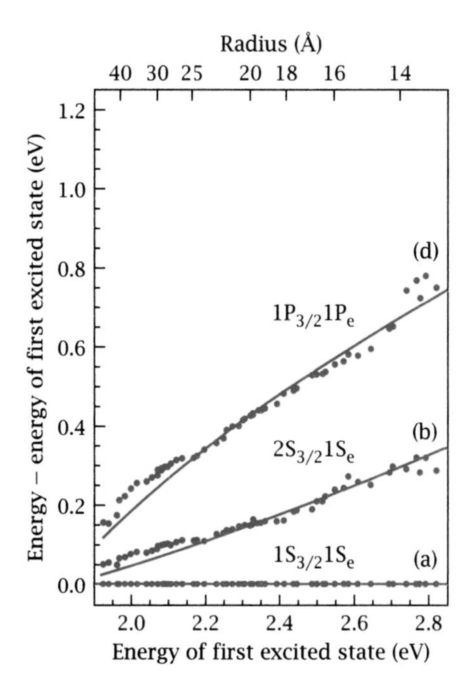

Figure 12.11 Extracted PLE data for three excited states with their corresponding assignments. Solid lines denote theoretical predictions from a more involved CdSe quantum dot model that accounts for valence band and envelope function mixing. Figure used with permission from Norris and Bawendi (1996).

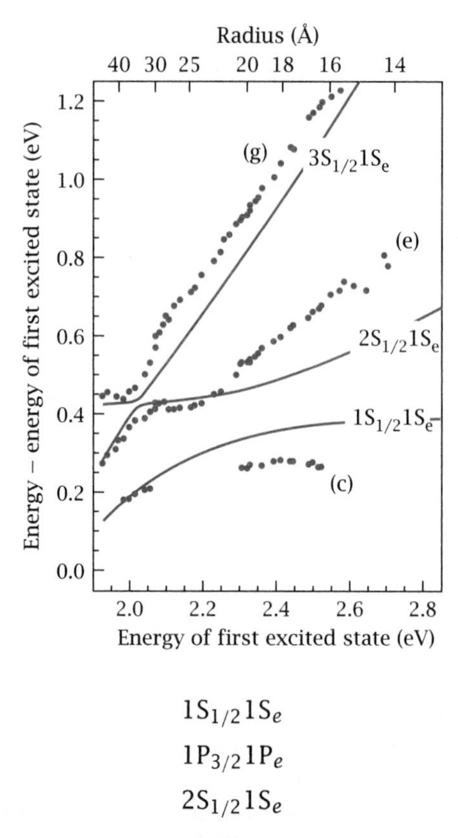

Figure 12.12 Extracted PLE data for three additional excited states with their corresponding assignments. This data set highlights the avoided transitions between levels, not explained by a basic particle-in-a-sphere quantum dot transition model. Solid lines are theoretical predictions from the more complicated valence band and envelope function mixing model. Figure used with permission from Norris and Bawendi (1996).

$$1S_{1/2}1S_e$$
$$1P_{3/2}1P_e$$
$$2S_{1/2}1S_e$$

Notice that it qualitatively resembles the progression predicted earlier using our simple quantum dot model with three independent valence bands. However, it is also clear that the data is in much better agreement with the refined model, which accounts for valence band as well as envelope function mixing. The solid lines in **Figures 12.11** and **12.12** are theoretical fits that demonstrate the theory's success. More importantly, the avoided crossings in **Figure 12.12** are well reproduced by the model. This illustrates the power of the effective mass approximation in describing the optical properties of nanostructures.

12.9.7 Fine Structure

Finally, there is one last complication in CdSe that we briefly touch on. It is not central to our understanding the experimental results above. However, it did prove instrumental in resolving some of the more unusual properties about the absorption, emission, and excited-state lifetime of colloidal CdSe quantum dots. Namely, it turns out that the electron–hole exchange interaction is enhanced in quantum dots. As a consequence, F_e and F_h are no longer good quantum numbers and one must actually consider their sum

$$N = F_e + F_h \tag{12.83}$$

as the good quantum number in the system. Given that we have previously seen that $F_e = \frac{1}{2}$ and $F_h = \frac{3}{2}, \frac{1}{2}$, their mutual sum (for $F_e = \frac{1}{2}$ and $F_h = \frac{3}{2}$) gives

$$N = 2, 1,$$

with corresponding momentum projections of $N_m = \pm 2, \pm 1, 0$ and $N_m = \pm 1, 0$.

As a consequence, five twofold-degenerate states result when the exchange interaction, spin–orbit coupling, crystal field splitting, and

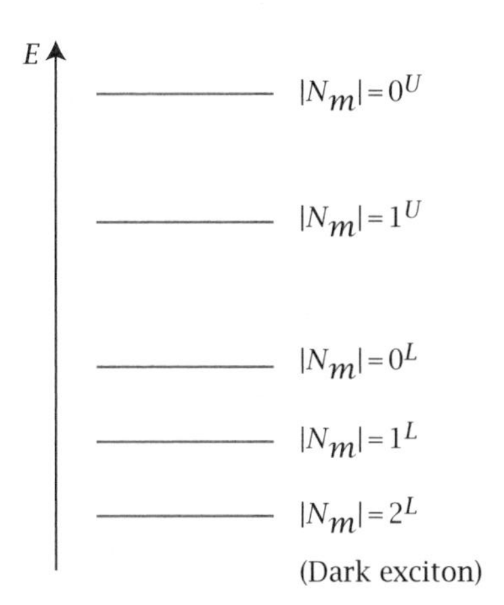

Figure 12.13 Size-dependent CdSe band edge fine structure with a low-lying dark exciton.

even the prolate shape of actual colloidal quantum dots are considered in tandem. In the literature, these states are referred to as band edge fine structure states and are ordered energetically as shown in **Figure 12.13**.

The superscripts L and U refer to lower and upper respectively. Note that the lowest energy state is referred to as a "dark" exciton, given its triplet-like nature. The presence of exciton fine structure, as well as the existence of a triplet-like dark exciton, then explains many of the unusual features of the CdSe quantum dot emission, for example, its long microsecond lifetimes at low temperatures as well as its size-dependent "resonant" and "nonresonant" Stokes shifts. The reader may consult the Further Reading for additional details.

12.10 SUMMARY

With this, we complete our journey begun in Chapter 6, with the purpose of understanding interband transitions as well as the size-dependent optical properties of low-dimensional materials. We have emphasized CdSe quantum dots as a model system since they have been extensively studied and allow comparisons with simple model predictions. We now move on to describing the synthesis (Chapter 13), characterization (Chapter 14), and applications (Chapter 15) of these low-dimensional materials.

12.11 THOUGHT PROBLEMS

12.1 Fourier transforms

We utilized the concept of Fourier transforms when discussing the effective mass approximation. To continue illustrating their use, show that the Fourier transform of an exponential decay

$$f(t) = \begin{cases} 0 & (t < 0) \\ e^{-at} & (t \geq 0) \end{cases}$$

yields a Lorentzian lineshape in frequency. This is commonly seen in spectroscopy where states exhibiting exponential relaxation processes possess transitions with Lorentzian lineshapes. Recall that

$$F(\omega) = \frac{1}{\sqrt{2\pi}} \int_{-\infty}^{\infty} f(t) e^{-i\omega t}\, dt$$

and consider only the real part of the resulting function, i.e., $\mathrm{Re}(F(\omega))$.

12.2 Fourier transforms

Show that the Fourier transform of a Gaussian function

$$f(t) = e^{-at^2},$$

with a a constant, yields another Gaussian function in frequency. Again, recall that

$$F(\omega) = \frac{1}{\sqrt{2\pi}} \int_{-\infty}^{\infty} f(t) e^{-i\omega t}\, dt.$$

12.3 Absorption

In the main text, we found an expression for the absorption coefficient of a semiconductor at 0 K, $\alpha(0\,\mathrm{K})$. However, at finite temperatures, the absorption coefficient can be expressed as

$$\alpha(T) = \alpha(0\,\mathrm{K})[1 - f_e(E_e) - f_h(E_h)],$$

where $f_e(E_e)$ and $f_h(E_h)$ are distribution functions for electrons and holes. First show that $\alpha(T) = \alpha(0\,\mathrm{K})$ at $T = 0\,\mathrm{K}$. Then show that the above finite-temperature expression can essentially be found from the net absorption rate of a system where both the absorption and stimulated emission rates are considered in tandem, giving a net absorption rate proportional to the absorption coefficient. Note that the absorption and stimulated emission rates have the same prefactor, which is essentially the Einstein B coefficient first seen in Chapter 5.

12.4 Absorption coefficient

The absorption coefficient of a bulk semiconductor was found to be (Equation 12.35)

$$\alpha(E_{kn}) = \frac{\pi \hbar q^2}{2 n_m c \epsilon_0 m_0}\left(\frac{E_p}{\hbar\omega}\right)\left[\frac{1}{2\pi^2}\left(\frac{2\mu}{\hbar^2}\right)^{3/2}\sqrt{E_{kn} - E_g}\right].$$

It basically reduces to

$$\alpha(E_{kn}) = K\sqrt{E_{kn} - E_g}.$$

For a transition 0.1 eV above the band gap of CdSe, find α, assuming the following parameters

$$m_e = 0.13 m_0,$$
$$m_h = 0.45 m_0,$$
$$n_m = 2.75,$$
$$E_p = 25\,\mathrm{eV},$$
$$E_g = 1.74\,\mathrm{eV}\ \text{at}\ 300\,\mathrm{K}.$$

Compare the result with the experimental absorption coefficients first seen in Chapter 5.

12.5 Emission

Consider the emission of a semiconductor in the limit of thermal equilibrium. The emission rate R_{em} is proportional to the product of the electron and hole occupation factors $f_e(E)$ and $f_h(E)$. Show that in the limit of thermal equilibrium where E_F is common to both electrons and holes that

$$R_{\mathrm{em}}(h\nu) \propto C f_e(E) f_h(E) = C e^{-h\nu/kT},$$

with C a constant of proportionality. Make simplifications as needed.

12.12 REFERENCES

Chuang SL (1995) *Physics of Optoelectronic Devices.* Wiley, New York.

Davies JH (2000) *The Physics of Low Dimensional Semiconductors.* Cambridge University Press, Cambridge, UK.

Kane EO (1957) Band structure of indium antimonide. *J. Phys. Chem. Sol.* **1**, 249.

Luttinger JM, Kohn W (1955) Motion of electrons and holes in perturbed periodic fields. *Phys. Rev.* **97**, 869.

Norris DJ (1995) Measurement and assignment of the size-dependent optical spectrum in cadmium selenide (CdSe) quantum dots. PhD Thesis, MIT.

Norris DJ, Bawendi MG (1996) Measurement and assignment of the size-dependent optical spectrum in CdSe quantum dots. *Phys. Rev. B* **53**, 16338.

Rosencher E, Vinter B (2002) *Optoelectronics.* Cambridge University Press, Cambridge, UK.

12.13 FURTHER READING

SD Mixing and Valence Band Mixing

Ekimov AI, Hache F, Schanne-Klein MC, et al. (1993) Absorption and intensity-dependent photoluminescence measurements on CdSe quantum dots: assignment of the first electronic transitions. *J. Opt. Soc. Am. B* **10**, 100.

Grigoryan GB, Kazaryan EM, Efros AL, Yazeva TV (1990) Quantized holes and the absorption edge in spherical semiconductor microcrystals with a complex valence band structure. *Sov. Phys. Solid State* **32**, 1031.

Klimov VI (ed.) (2004) *Semiconductor and Metal Nanocrystals. Synthesis and Electronic and Optical Properties.* Marcel Dekker, New York.

Band-Edge Fine Structure

Efros AL, Rosen M (2000) The electronic structure of semiconductor nanocrystals. *Annu. Rev. Mater. Sci.* **30**, 475.

Efros AL, Rosen M, Kuno M, et al. (1996) Band-edge exciton in quantum dots of semiconductors with a degenerate valence band: dark and bright exciton states. *Phys. Rev. B* **54**, 4843.

Kuno M (1998) Band edge spectroscopy of CdSe quantum dots. PhD Thesis, MIT.

Kuno M, Lee JK, Dabbousi BO, et al. (1997) The band edge luminescence of surface modified CdSe nanocrystallites: probing the luminescing state. *J. Chem. Phys.* **106**, 9869.

Synthesis

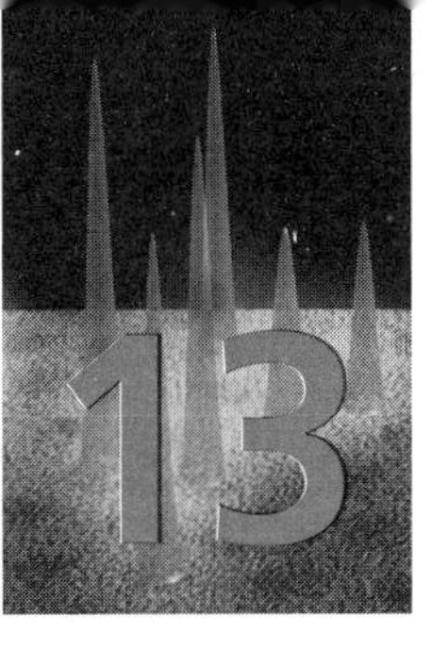

13.1 INTRODUCTION

In this chapter, we describe the synthesis of nanoscale materials. Our emphasis will be on nanocrystals and nanowires as opposed to quantum wells, since the latter have been well established. Furthermore, we focus on colloidal syntheses of these zero- and one-dimensional systems. Where possible, simple recipes for making them will be provided to illustrate how these preparations are conducted in practice.

We begin with colloidal metal nanoparticles and semiconductor quantum dots, since they were some of the first nanostructures to be made, dating from Faraday's time and even earlier. We conclude with nanowires, since only recently have high-quality one-dimensional materials been made abundantly. Our discussion is followed by a description of common characterization techniques in Chapter 14 and an overview of nanocrystal/nanowire applications in Chapter 15.

13.2 MOLECULAR BEAM EPITAXY (MBE)

The first technique we describe is a high-tech bottom-up synthesis of quantum wells and quantum dots, referred to as molecular beam epitaxy (MBE). The basic idea behind the approach is fairly straightforward. However, in practice, its realization is more involved and expensive.

MBE uses an ultrahigh-vacuum chamber into which a substrate is loaded onto a sample holder. Precursors of desired elements are introduced into heated crucibles. The sources are outfitted with computer-controlled shutters. When opened, beams of atoms emerge into the main chamber, directed towards the substrate. Under the low pressures of the experiment, the atoms have long mean free paths. This allows them to reach the substrate without experiencing any collisions with other species in the chamber. Next, by controlling the crucible temperature as well as the sequence and timing of the opening and closing of the shutters, uniform films of semiconductors can be deposited. Because of the low overall flux of atoms, atomically precise film thicknesses are possible. What results are thin epitaxial films, which often form the basis of quantum wells.

In the case of quantum dots, conditions exist under which small semiconductor islands can be grown. Namely, the idea is to deposit semiconductors having different lattice constants on top of one another. In practice, a layer of one material is first grown over the substrate. Afterwards, the system is abruptly switched to the growth of a second material. After several layers of the new, lattice-mismatched, semiconductor have been deposited, the strain at the interface between the two materials changes the mode of growth from within the plane to out of the plane (called Stranski–Krastanov growth). As a consequence, small islands form that are referred to as quantum dots. To distinguish them from the chemist's colloidal quantum dots, which we will see below, they are often referred to as epitaxial or self-assembled quantum dots. The MBE approach has been used to make a variety of quantum dot systems. More information about them can be found in Bimberg et al. (2001).

13.3 COLLOIDAL GROWTH OF NANOCRYSTALS

Colloidal growth is a chemist's approach for making nanostructures. However, many variations exist. Thus, no blanket description perfectly encompasses all of them. Generally speaking, though, colloidal syntheses are characterized by the use of a precipitation approach to make nanocrystals. This effectively entails three stages of growth: nucleation, normal growth, and competitive growth. We will see examples of each throughout the course of the following discussion.

We begin with a short description of metallic nanoparticles. This will be followed by a much longer historical overview of an important model system, CdSe quantum dots. This is because their development over the last 20 years illustrates many of the classic aspects of colloidal growth as well as the strategies that have been developed to obtain high-quality samples with narrow size distributions.

13.3.1 Metal Nanoparticles

The history of colloidal metal nanoparticles is very old. It is now well established that stained glass windows in medieval cathedrals contain metal nanostructures. In fact, many red glasses take their hue from the gold nanoparticles embedded within them. This makes the glass makers and painters of the time some of the earliest nanotechnologists that we know of. Schalm et al. (2009) provide an interesting overview of the different elements and compositions used to make stained glass. Fredrickx et al. (2002) have also conducted a transmission electron microscopy (TEM: Chapter 14) analysis of gold nanoparticles in a red glass fragment found in the excavated workshop of Johann Kunckel (1610–1705), a well known 17th century glass maker.

The production of metal nanoparticles, however, has earlier origins. To illustrate, the Roman Lycurgus cup from the 4th century AD is a well preserved example of a Roman "cage" cup. It depicts the mythological story of King Lycurgus' persecution of Dionysos and his subsequent death (Freestone et al. 2007). Today, the Lycurgus cup can be found in the British Museum in London.

Apart from its historical significance, what makes the cup special is its unique colors that change depending on viewing angle. Namely, by directly looking at the cup's transmitted light, it appears red. However, if observed at an angle, it appears green. This multicolor phenomenon arises because of the different frequencies of light scattered and transmitted by the glass and ultimately stems from the presence of metal nanoparticles embedded within it. Barber and Freestone (1990) carried out a TEM study of a small fragment of the cup and unambiguously showed the inclusion of small 50–100 nm diameter gold/silver nanoparticles. These two historical examples—medieval stained glasses and the Lycurgus cup—illustrate that humans have been making, living with, and using metal nanoparticles for nearly 2000 years.

The modern age of scientific inquiry begins with Faraday's work on gold nanoparticles. Intrigued by the different colors exhibited by gold sols, Faraday, among other things, proceeded to develop rational syntheses for these particles. A perusal of his original paper shows that the approaches he developed often entail reducing gold chloride in various ways (Faraday 1857). Note that Faraday was not specific in the nature of the gold chloride used and, in fact, his starting gold solutions were obtained by dissolving bulk gold with aqua regia (a mixture of nitric acid and hydrochloric acid). As a consequence, more likely than not, chloroaurate ($AuCl_4^-$) was his starting gold ionic species.

Through his work, Faraday found that he could reduce gold chloride with phosphorus dissolved in various solvents. Other reducing agents successfully used include ether, tartaric acid, sugar, glycerol, and even an ox's gut! Faraday also discovered that heating his gold solutions reduced them. He further noted subsequent particle growth upon prolonged heating. This is described in his paper as the transition of an initial ruby-colored solution into a blue-colored one. It is clear, however, that Faraday struggled with determining the actual cause of this color change. Tweney (2006) provides an interesting account of recent attempts to follow in Faraday's scientific footsteps.

Today, there are two general syntheses for gold nanoparticles. The first is called the Turkevich approach and was reported in 1951 (Turkevich et al. 1951). The second is the Brust approach, which was developed later in 1994 (Brust et al. 1994). The primary difference between the two is the medium in which the gold nanoparticles are synthesized. The Turkevich synthesis is an aqueous preparation while the Brust approach is conducted in an organic solvent.

In either case, a gold precursor, chloroauric acid ($HAuCl_4$), is dissolved in an appropriate solvent. On dissolution, $AuCl_4^-$ ions are present, with gold in the Au(III) oxidation state. The addition of a reducing agent causes their reduction to Au(0). An effective supersaturation of the solution then occurs, leading to the formation of small Au(0) clusters. Subsequent addition of atoms to their surfaces produces gold nanoparticles. Provided below are two simple recipes that illustrate these preparations.

EXAMPLE 13.1

The Turkevich Approach to Making Gold Nanoparticles To make gold nanoparticles using the Turkevich approach, a gold stock solution is prepared. Namely, 0.155 g chloroauric acid is added to 91 mL deionized water. Note that the gold salt is hygroscopic and picks up water on exposure to air. It should therefore be weighed out quickly and stored in a dessicator. Furthermore, any metal spatulas used should be Teflon-taped to prevent them from corroding on contact with the salt. To prepare the citrate solution, 0.25 g sodium citrate is added to 50 mL deionized water.

Next, to carry out the actual gold nanoparticle synthesis, 1 mL of the chloroauric acid stock is added to 18 mL deionized water. The solution is then brought to a vigorous boil on a hotplate. At this point, 1 mL of the sodium citrate stock is added. This initiates the reduction process, whereupon the reaction mixture turns color and eventually takes a purple hue due to the plasmon resonance of the resulting particles. (The plasmon resonance of Au nanoparticles will be described in more detail in Chapter 15.) When the solution exhibits a dark purple color (admittedly this is a little subjective), it is removed from the hotplate and is allowed to cool. If desired, the solution can be diluted.

The resulting gold nanoparticles have negatively charged citrate groups on their surface. This, in turn, prevents them from agglomerating and falling out of solution. Citrate therefore reduces the gold ions at the outset of the reaction and also electrostatically passivates the surfaces of resulting nanoparticles. Electron micrographs of Turkevich-synthesized gold nanoparticles have previously been shown (**Figures 4.11** and **4.12**) and more about their characterization will be described in Chapter 14.

The Brust Approach to Making Gold Nanoparticles Alternatively, to prepare gold nanoparticles using the Brust approach, the following solutions are prepared. A solution of tetra-*n*-octylammonium bromide (TOAB) is made by adding 0.55 g to 20 mL toluene. Next, a gold solution is prepared by adding 0.09 g chloroauric acid to 7.5 mL deionized water.

The two solutions are then combined. The resulting mixture is stirred vigorously, since TOAB extracts gold ions from the aqueous phase and brings them into the organic phase. As a consequence, after a short period of time, the orange color of the gold ions transfers into the toluene phase, leaving behind a clear aqueous layer.

At this point, a solution consisting of 0.095 g sodium borohydride in 6.25 mL deionized water is prepared. It is then slowly added to the reaction mixture (under vigorous stirring) over the course of approximately 30 minutes. The borohydride reduces the gold ions much like the citrate in the Turkevich approach. This initiates nanoparticle nucleation and subsequent growth. The resulting mixture is then left stirring for another 10–20 minutes to completely reduce the gold.

The toluene phase, containing the resulting nanoparticles, is subsequently isolated using a separatory funnel. Approximately 3 mL trioctylphosphine (TOP) is then added to allow the phosphine to datively bind to the surfaces of the nanoparticles. This sterically stabilizes them and keeps them from aggregating as well as falling out of solution. Note that the original Brust preparation uses a thiol. However, the use here of a phosphine does not lead to any major differences.

The TOP-passivated particles are subsequently isolated by precipitating them from solution. This is done by adding a nonsolvent such as methanol. The resulting suspension is then centrifuged to recover the nanocrystals. If desired, the recovered material can be resuspended in fresh toluene. More about precipitating nanoparticles from solution using a nonsolvent will be discussed shortly within the context of colloidal semiconductor quantum dots.

EXAMPLE 13.2

13.4 SEMICONDUCTOR NANOCRYSTALS

13.4.1 Colloidal CdSe Quantum Dots

We now describe the historical development of colloidal cadmium selenide (CdSe) quantum dots as a representative (model) semiconductor system. The reason for this is that there have been over 20 years of work on this material. It was the first to yield nanocrystal ensembles, exhibiting narrow size distributions, large yields, a high degree of crystallinity, and good optical properties (as characterized by structure in the linear absorption, relatively large emission quantum yields, and the absence of defect emission). Furthermore, the historical development of CdSe illustrates many of the strategies used today in the solution phase growth of other nanoscale systems that possess more complicated compositions or morphologies.

13.4.2 Glasses

Before beginning our historical overview, we mention that some of the original CdSe quantum dots were made using supersaturated metal and chalcogen solutions in molten glasses. As a consequence, this approach complements the solution-based syntheses that we will see below.

In practice, quantum dot growth in glasses consists of two stages. First, molten glasses are doped with metal salts of a desired semiconductor at levels of a few weight percent. The melt is then rapidly quenched, yielding seed nuclei of the desired semiconductor. The resulting material is then processed using a secondary heat treatment with temperatures ranging from 400 to 1000 °C. This enables the resulting seed particles to grow. By varying the temperature as well as duration of the secondary heat treatment, their mean size can be varied. More about the growth of quantum dots in glasses can be found in Gaponenko (1999).

Note that colored glass filters sold by Schott, Hoya and Corning are produced in a similar manner. As a consequence, they contain small cadmium chalcogenide nanocrystals. This, in turn, makes them some of the first commercially successful quantum dot products. More recent examples will be seen in Chapter 15.

13.4.3 Arrested Precipitation

Around the same time, other solution chemistries were being developed to make nanoscale cadmium chalcogenides. Many early nanoparticle syntheses employ a strategy referred to as arrested precipitation to "arrest" or halt semiconductor growth soon after nucleation. This is because simple thermodynamic considerations show that the metal and chalcogen precursors react readily to yield bulk materials when unhindered. As a consequence, a means of controlling (or halting) the growth kinetics is needed in order to isolate nanoscale bulk fragments.

One of the earliest reports in this area describes the production of cadmium sulfide (CdS) nanocrystals by rapidly introducing chalcogen precursors into a metal salt solution at low temperature (Rossetti et al. 1985). Following nucleation, subsequent growth of the nucleated particles is suppressed because of the low reaction temperatures maintained. In this manner, researchers found that they could isolate small, metastable, CdS particles. Note that, although the growth of CdS precedes that of CdSe, the synthesis of the latter is identical, with the only difference being the chalcogen precursor employed.

Two other reports around the same time describe the use of polymers to arrest CdS growth (Kalyanasundaram et al. 1981; Rossetti and Brus 1982). Namely, on introducing metal salt solutions into a separate aqueous chalcogen solution containing a polymer, nucleation occurs. However, subsequent growth is retarded because the polymer encapsulates the resulting particles, preventing them from growing uncontrollably towards the bulk. As a consequence, the polymer arrests growth and stabilizes the kinetic as opposed to the thermodynamically

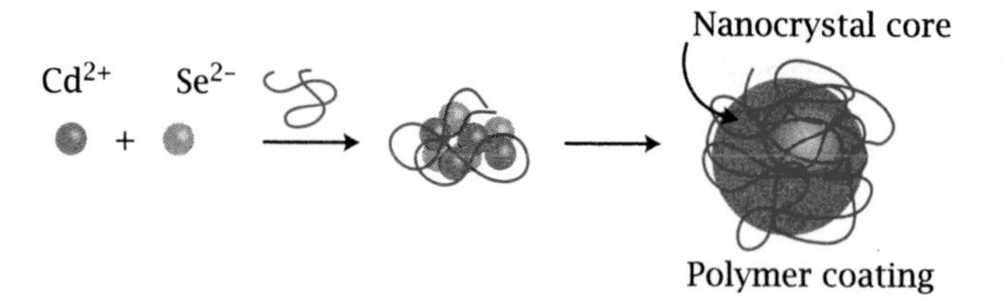

Figure 13.1 Illustration of arrested precipitation using polymers.

favored (bulk) product. Other similar approaches, using restricted environments such as zeolites, lipid membranes, and micelles, were also developed around this time. We will see examples of the latter in what follows. The arrested precipitation procedure is outlined in **Figure 13.1**.

Key Advance

- The ability to arrest precipitation using surfactants or restricted environments.

As a brief historical note, early arrested precipitation syntheses of colloidal CdS and CdSe quantum dots likely originated in the 19th century. In this regard, cadmium chalcogenides are well-known pigments in cadmium "yellows", "oranges," and "reds". These paints were favored by many artists of the time, such as Matisse, Monet, and Van Gogh, because of their brilliant colors and their reputed stability (Perkowitz 1998). The introduction of selenium into these mixtures also extended the range of available colors from light yellow to deep red.

As stated in a review about the development of cadmium pigments (Fiedler and Bayard 1986), "pale [yellow] shades could be prepared by partial precipitation from cold dilute solutions of cadmium slats or by rapid precipitation from acid solutions ... very light hues could readily be attained using cadmium nitrate and sodium sulfide or ammonium thiosulfate." This blue shift of the CdS absorption to higher energies likely stems from the confinement effects we have discussed earlier. Furthermore, many of the other color changes observed have probable origins in colloidal, size-dependent, light scattering effects. This phenomenon was also seen in Faraday's gold nanoparticle solutions, whose colors changed with size and shape. Light scattering will be discussed in Chapter 14 within the context of nanoparticle sizing. These observations therefore suggest that early 19th century pigment chemists were unknowingly making colloidal nanostructures much like the stained glass painters and artisans of medieval and Roman times.

13.4.4 Inverse Micelles

As part of learning to arrest precipitation, inverse micelles (alternatively called reverse micelles) were explored as a way to make nanocrystals. Motivating some of these early studies was interest in using CdS nanoparticles for solar photochemical applications (e.g., water splitting). In general, the inverse micelle approach entails using a surfactant that possesses a polar head group and an aliphatic tail to create nanometer-sized "beakers" for reactions. It was believed that these inverse micelles not only could suppress (uncontrolled) growth but could also restrict the size of resulting quantum dots to their physical dimensions. This would, in turn, allow one to obtain different sized particles in a controlled manner.

A common surfactant used in these preparations is sodium bis(2-ethylhexyl) sulfosuccinate, also referred to as Aerosol OT (AOT). Other surfactants such as sodium dodecyl sulfate (SDS) and cetyltrimethylammonium bromide (CTAB) are used as well. Because of their mixed affinities for polar or apolar media, when these materials are introduced into alkanes with small amounts of water (called "water-in-oil" emulsions), the surfactant molecules arrange themselves in a self segregating manner as illustrated in **Figure 13.2**. Namely, their polar head groups cluster together to enclose a hydrophilic interior, leaving behind their hydrophobic alkyl tails exposed to the solvent.

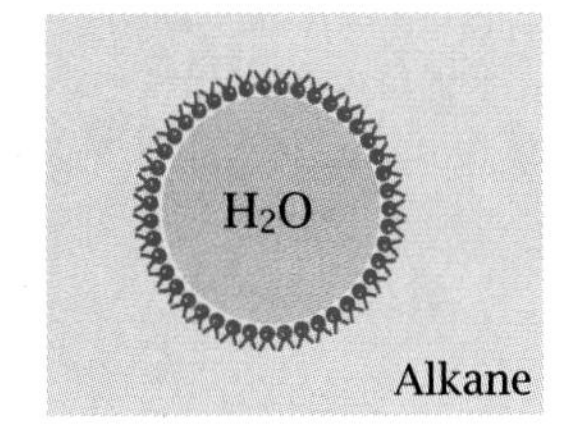

Figure 13.2 Schematic of a water-in-oil inverse micelle with a hydrophilic interior and a hydrophobic exterior.

The overall size of the hydrophilic interior is controlled by the ratio of water to surfactant, a parameter called the W value (sometimes written as ω). Specifically, the W value is

$$W = \frac{[H_2O]}{[AOT]} \tag{13.1}$$

and represents the molar ratio of water to surfactant. Note that W accounts for the amount of water deliberately added to the system. However, the surfactants mentioned above are often hygroscopic. As a consequence, they pick up water on their own. This additional water is often ignored in micelle preparations. What is important then is that as W increases, so too does the diameter of the resulting water pool. Empirically, as W increases from 1 to 50, the diameter of the corresponding water pool changes from about 0.6 nm to about 80 nm. **Table 13.1** shows a partial list of W values and associated water pool radii compiled from the literature.

Next, to produce nanocrystals using inverse micelles, water-soluble metal salts are encapsulated within the micelle's hydrophilic interior. A gas such as hydrogen sulfide or hydrogen selenide is then bubbled through the solution. The subsequent reaction between the metal and the chalcogen results in the formation of semiconductor particles. Alternatively, instead of gas-phase precursors, a second micellar solution, containing water-soluble chalcogen salts, can be mixed with the first to make quantum dots. This approach works because the micelles exchange contents as illustrated in **Figure 13.3**.

Although the outlined inverse micelle approach is facile and readily produces nanocrystals, a major drawback is the low degree of crystallinity exhibited by the resulting quantum dots. This is because they are produced at room temperature and thus possess many defects. Their optical properties, in turn, suffer. To illustrate, the emission quantum yield of micelle-derived nanocrystals is very low (<0.01). Furthermore, their emission is often dominated by defect-related contributions, called "deep-trap" emission.

There is, however, one important advantage of micelle-produced particles. Namely, their surfaces are accessible. As a consequence, this opens the door to the postsynthesis manipulation of nanocrystal ensembles. This would turn out to be instrumental in the subsequent development of higher-quality semiconductor quantum dots, as we shall see.

One of the first studies to illustrate this was a report showing the passivation of inverse micelle-generated nanocrystals with sodium hexametaphosphate (Fojtik et al. 1984). In the study, CdS quantum dots were isolated from solution by evaporating away the solvent. They could then be resuspended in fresh solvent without any apparent change to their mean size and/or size distribution. By contrast, identical particle

Table 13.1 Inverse micelle water pool radius as a function of the W value for the H_2O/AOT/cyclohexane system

W	Radius (nm)
1	0.6
2	0.8
3	1.0
4	1.2
5	1.4
6	1.5
7	1.7
8	1.9
10	2.1
15	2.7
20	3.3

Source: Maitra (1984).

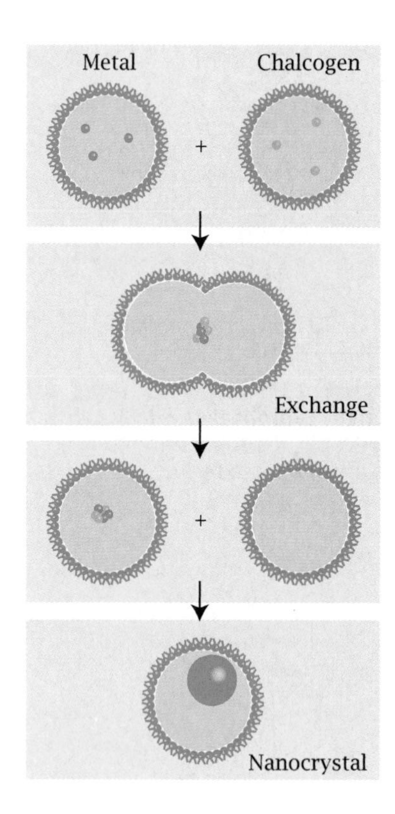

Figure 13.3 Use of inverse micelles and their ability to exchange contents to synthesize quantum dots.

isolation procedures using unpassivated nanocrystals inevitably led to their unwanted growth, a coarsening of their size distribution, and irreversible aggregation. The polymer capping was therefore an early example of redispersible nanocrystal "powders".

Continuing in this theme, it was soon discovered that the surfaces of CdSe nanocrystals could be covalently passivated with organoselenide groups (Steigerwald et al. 1988). In particular, phenyl(trimethylsilyl)selenium was added to a solution of inverse micelle-derived CdS nanocrystals, where subsequent cleavage of the Si–Se bond led to their phenyl passivation as depicted in **Figure 13.4**. In turn, the dots precipitated from solution. They could then be isolated as well as resuspended in different organic solvents without any apparent adverse effects.

> **Key Advance**
>
> - The ability to surface functionalize nanocrystals, leading to solubility changes and redispersible powders

13.4.5 Hybrid Preparations

The next advance in colloidal nanocrystal preparation was developed at Bell Laboratories in the late 1980s (Bawendi et al. 1990). Specifically, an approach was developed that combined the following advances made up to this point: the ability to make small seed particles by arrested precipitation in inverse micelles, the ability to subsequently passivate their surfaces with different organic ligands, and the ability to resuspend them in different solvents. The approach also introduced two additional advances, the first a concept discussed by LaMer for producing low-size-distribution (micrometer-sized) sulfur colloids and the second the use of high-boiling coordinating solvents.

Namely, LaMer showed that narrow-size-distribution micrometer-sized colloids could be obtained by inducing a discrete temporal nucleation event at the outset of a precipitation reaction (LaMer and Dinegar 1950). In this regard, precursors are rapidly added to a solution to dramatically increase an element's concentration. This is illustrated by the prenucleation stage in **Figure 13.5**. Once the concentration exceeds a critical nucleation threshold, small molecular clusters form. This is illustrated by the shaded nucleation stage in **Figure 13.5**. The element concentration then decreases due to cluster production and ultimately falls below the nucleation threshold. This halts any further nucleation. Subsequent particle growth then occurs through the addition of elements to the resulting nuclei in a controlled, uniform, manner. This is depicted by the third (normal) growth stage in **Figure 13.5**. LaMer's key point was that, provided a temporally narrow nucleation event, which yielded nuclei with low size polydispersities, subsequent growth

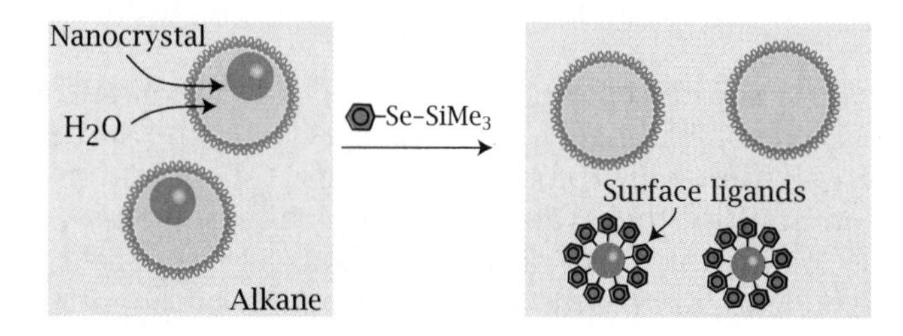

Figure 13.4 Illustration of the ligand passivation of inverse micelle-generated CdSe nanocrystals.

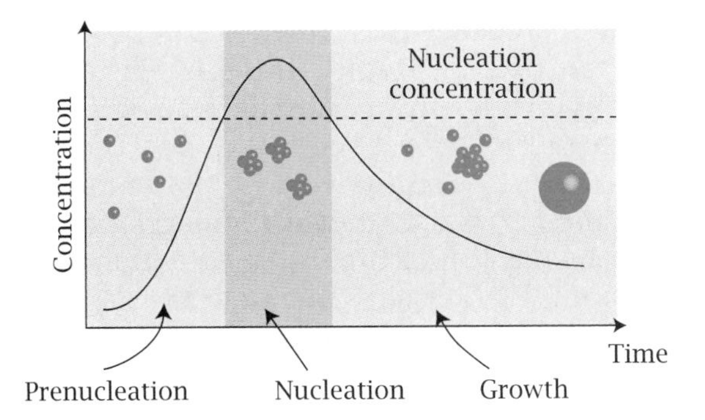

Figure 13.5 LaMer diagram illustrating the temporally discrete nucleation event, leading to the preparation of nearly monodisperse colloids.

over longer periods of time could preserve their original, (narrow) size distribution.

Within the context of the hybrid inverse micelle approach, this strategy was implemented using presynthesized CdSe particles as seed nuclei for the subsequent growth of nanocrystals. By physically separating particle nucleation from growth, a temporally discrete nucleation event could, in effect, be achieved.

In practice, the hybrid preparation entails using low-water-content inverse micelles ($W = 0$) to produce small CdSe seed nuclei. The resulting particles are then surface-functionalized by adding an excess of pyridine, which datively binds to their surfaces. As a consequence, they become insoluble in the aqueous phase of the micelle as well as in the alkane phase, causing them to fall out of solution. This allows them to be isolated and resuspended in a high-boiling organic solvent for subsequent growth. In the original preparation, a mixture of tributylphosphine oxide (TBPO) and tributylphosphine (TBP) was empirically found to be a good solvent. The hybrid preparation is outlined in **Figure 13.6**.

This brings us to the second major advance of the hybrid nanocrystal preparation. In this regard, surfactants play an important role in promoting the growth of high-quality nanocrystals. We have seen that at room temperature, ligands bind to nanoparticle surfaces to sterically stabilize them and to prevent them from agglomerating. The ligands also impart solubility to the nanocrystals, allowing them to be suspended in different solvents. They also slow down subsequent growth kinetics, as illustrated crudely by the polymer and inverse micelle

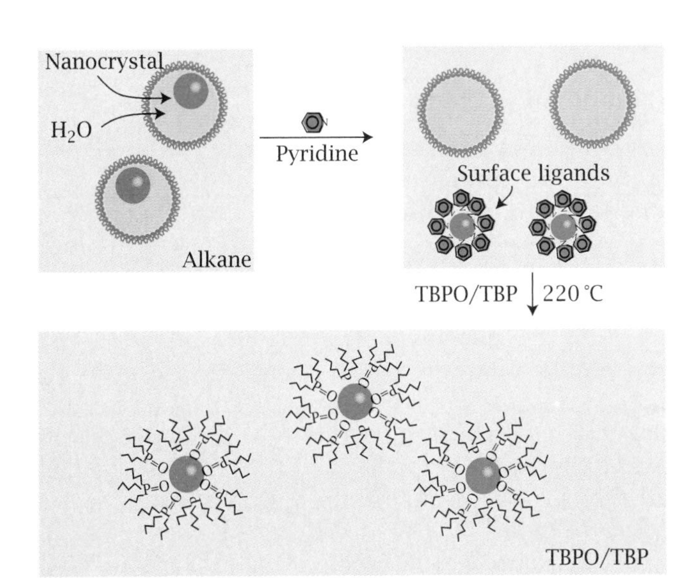

Figure 13.6 Hybrid nanocrystal synthesis. Seed particles are first made using inverse micelles. They are subsequently derivatized with pyridine and are isolated. The product is then heated in a mixture of tributylphosphine (TBP) and tributylphosphine oxide (TBPO) to induce growth.

arrested precipitation schemes described earlier. The new realization here was that certain ligands could also provide effective *electronic* stabilization of nanocrystal surfaces by helping to passivate dangling bonds, stemming from their abrupt termination.

As a consequence, growth in TBPO and TBP led to some of the first high-quality CdSe nanocrystals. Specifically, highly crystalline materials were obtained owing to the high growth temperatures that TBPO and TBP could sustain. Furthermore, the optical properties of resulting quantum dots were dramatically improved. As an example, obtained CdSe ensembles were some of the first to show structure in their linear absorption. They also exhibited near-band-edge emission with relatively large emission quantum yields (>0.1). Evidence of a phonon progression also existed in "line-narrowed" emission spectra. The most important achievement, though, was the ability to produce nanocrystal ensembles with narrow size distributions of the order of 10%. This was a significant improvement over the distributions displayed by other samples at the time.

To summarize, the hybrid approach incorporated the following developments:

- The ability to arrest precipitation and make small seed particles using inverse micelles.
- The ability to surface-functionalize them with organic ligands and to exploit their altered solubility properties to obtain redispersible powders.

It led to the following advances.

Key Advances

- The demonstration that a temporally discrete nucleation event (in this case seeded) could lead to improved size distributions.
- The demonstration that high-boiling coordinating solvents such as phosphines and phosphine oxides could control nanocrystal growth kinetics and could also electronically passivate their surfaces, leading to dramatic improvements in the optical properties of resulting particles.

13.4.6 Organometallic Preparations

The next major advance came in 1993 with the development of a complete "organometallic" preparation for CdSe nanocrystals, based on the pyrolysis of organometallic precursors in a coordinating solvent (Murray et al. 1993). Namely, metal and chalcogen precursors are injected at high temperatures into a suitable solvent to initiate CdSe nucleation *and* subsequent growth. Although foreshadowed by earlier studies, describing the pyrolysis of oligomeric cadmium chalcogenide complexes to obtain CdSe quantum dots (Brennan et al. 1989), this preparation encompasses all of the advances leading up to this point, including those of the hybrid preparation. It ultimately ushered in the modern age of nanocrystal growth.

In the original preparation, dimethylcadmium ($CdMe_2$) and trioctylphosphine selenide (TOPSe) were the metal and chalcogen precursors used. The coordinating solvent was trioctylphosphine oxide

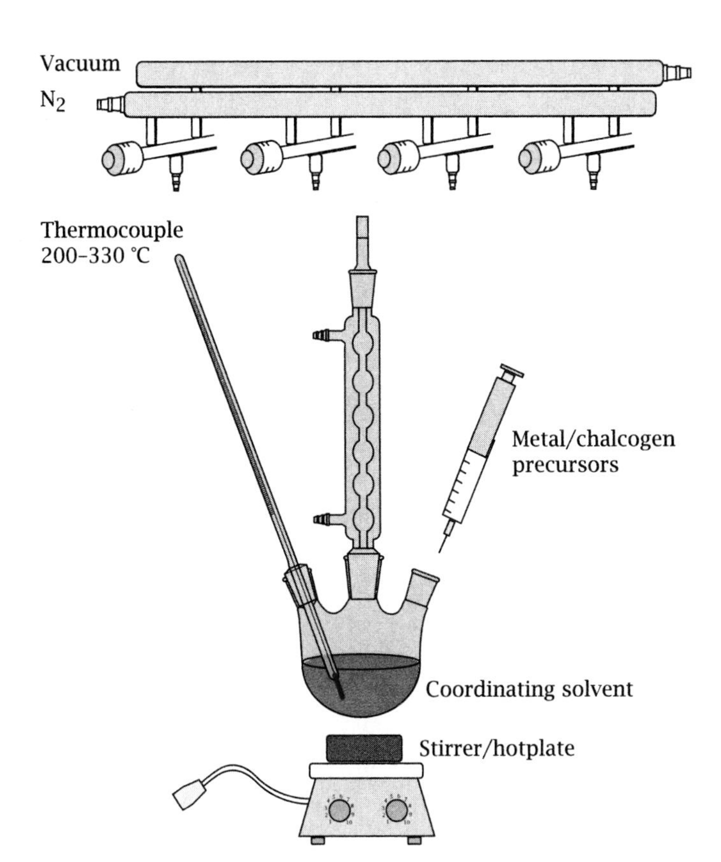

Figure 13.7 Apparatus used in the organometallic synthesis of colloidal CdSe quantum dots.

(TOPO), with its choice rationalized by the previous success of TBP and TBPO in the hybrid preparation. Furthermore, its high boiling point, above that of TBPO, could improve the crystallinity of resulting nanocrystals.

In practice, $CdMe_2$ and TOPSe are both diluted in trioctylphosphine (TOP) and are injected into heated TOPO as depicted in **Figure 13.7**. This leads to a dramatic drop in the reaction temperature because of the sizable volume of room-temperature injection solution introduced (up to 50% of the TOPO volume). The net result is a discrete temporal nucleation of CdSe seed particles, as achieved in the earlier seeded growth of CdSe nanocrystals. The dramatic temperature drop also prevents subsequent particle growth in a manner analogous to the rapid temperature quenching used to arrest and isolate early CdS nanocrystals.

Next, by slowly raising the reaction mixture's temperature to about 300 °C, particle growth is initiated. During the early stages of this process, any unreacted metal and chalcogen precursors are consumed to increase the mean particle size, as depicted in **Figure 13.8**. However, at later times, the depletion of these precursors eventually forces a transition into a competitive growth regime called Ostwald ripening.

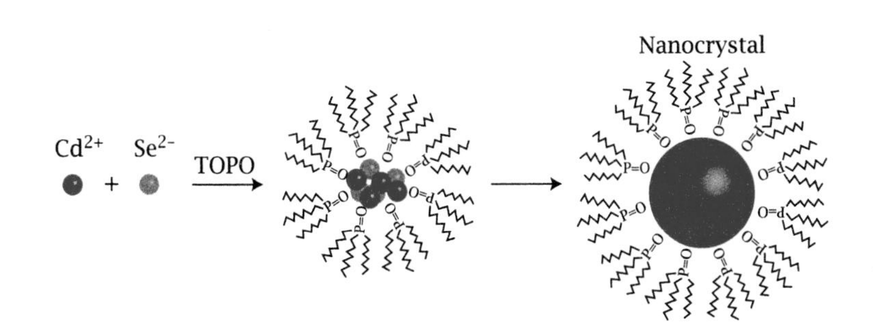

Figure 13.8 Schematic of the organometallic preparation of colloidal CdSe quantum dots. TOPO, trioctylphosphine oxide.

Table 13.2 Peak $1S_{3/2}1S_e$ absorption wavelength versus effective radii for colloidal CdSe quantum dots

Absorption wavelength (nm)	Size (Å)	Absorption wavelength (nm)	Size (Å)	Absorption wavelength (nm)	Size (Å)
460.00	10.41	524.00	16.54	588.00	24.26
462.00	10.79	526.00	16.76	590.00	24.51
464.00	11.13	528.00	16.98	592.00	24.78
466.00	11.43	530.00	17.21	594.00	25.05
468.00	11.71	532.00	17.44	596.00	25.33
470.00	11.96	534.00	17.68	598.00	25.63
472.00	12.19	536.00	17.92	600.00	25.94
474.00	12.40	538.00	18.16	602.00	26.26
476.00	12.60	540.00	18.40	604.00	26.61
478.00	12.78	542.00	18.65	606.00	26.98
480.00	12.94	544.00	18.90	608.00	27.37
482.00	13.10	546.00	19.15	610.00	27.79
484.00	13.25	548.00	19.40	612.00	28.25
486.00	13.40	550.00	19.65	614.00	28.74
488.00	13.54	552.00	19.90	616.00	29.26
490.00	13.68	554.00	20.15	618.00	29.83
492.00	13.82	556.00	20.39	620.00	30.45
494.00	13.95	558.00	20.64	622.00	31.12
496.00	14.09	560.00	20.89	624.00	31.84
498.00	14.24	562.00	21.13	626.00	32.62
500.00	14.38	564.00	21.38	628.00	33.47
502.00	14.53	566.00	21.62	630.00	34.39
504.00	14.69	568.00	21.86	632.00	35.39
506.00	14.84	570.00	22.10	634.00	36.47
508.00	15.01	572.00	22.33	636.00	37.64
510.00	15.18	574.00	22.57	638.00	38.91
512.00	15.36	576.00	22.81	640.00	40.27
514.00	15.54	578.00	23.04	642.00	41.75
516.00	15.73	580.00	23.28	644.00	43.35
518.00	15.92	582.00	23.52	646.00	45.07
520.00	16.12	584.00	23.76	648.00	46.92
522.00	16.33	586.00	24.01	650.00	48.91

Source: Kuno (1998).

This is a cannibalistic or competitive growth process that involves larger nanoparticles in solution growing at the expense of smaller ones. Physically, this means that the smaller particles dissolve to produce material for the larger ones. The ripening process is promoted by surface free energy differences that favor the stability of larger nanocrystals. While not a major issue in producing high-quality samples, subsequent improvements to the preparation would ultimately find ways to circumvent this latter growth stage.

In all cases, nanoparticle growth is monitored by taking small aliquots of the reaction mixture at different times. Then, by acquiring the absorption spectrum of the sample and comparing its $1S_{3/2}1S_e$ peak position (Chapter 12) with a known sizing curve (e.g., **Table 13.2**), the mean particle size can be determined. Alternatively, the size-dependent emission spectrum can be monitored and compared to an analogous sizing curve (e.g., **Table 13.3**). As an illustration, the size-dependent absorption and emission spectra of a CdSe ensemble obtained during growth are shown in **Figure 13.9**.

Table 13.3 Peak band-edge emission wavelength versus effective radii for colloidal CdSe quantum dots

Absorption wavelength (nm)	Size (A)	Absorption wavelength (nm)	Size (A)	Absorption wavelength (nm)	Size (Å)
516	14.52	566	20.01	616	27.74
518	14.59	568	20.25	618	28.08
520	14.70	570	20.49	620	28.44
522	14.86	572	20.75	622	28.81
524	15.04	574	21.00	624	29.20
526	15.26	576	21.27	626	29.62
528	15.49	578	21.55	628	30.08
530	15.73	580	21.83	630	30.58
532	15.99	582	22.12	632	31.14
534	16.24	584	22.42	634	31.77
536	16.50	586	22.73	636	32.50
538	16.76	588	23.05	638	33.33
540	17.01	590	23.37	640	34.29
542	17.26	592	23.69	642	35.40
544	17.51	594	24.03	644	36.69
546	17.75	596	24.36	646	38.19
548	17.98	598	24.70	648	39.93
550	18.21	600	25.03	650	41.95
552	18.44	602	25.37	652	44.29
554	18.66	604	25.71	654	47.00
556	18.88	606	26.05	656	50.12
558	19.11	608	26.39	658	53.72
560	19.33	610	26.72	660	57.84
562	19.55	612	27.06		
564	19.78	614	27.39		

Source: Kuno (1998).

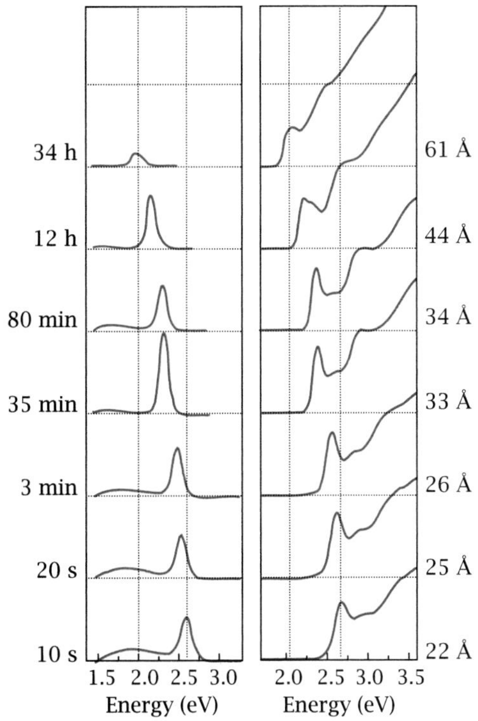

Figure 13.9 Evolution of CdSe quantum dot linear absorption (right) and emission (left) during growth. Associated mean nanocrystal diameters are listed. Used with permission from Mikulec (1999).

Table 13.4 Molar extinction coefficient versus effective diameter for colloidal CdSe quantum dots

Diameter (nm)	Extinction coefficient ($M^{-1}cm^{-1}$)	Diameter (nm)	Extinction coefficient ($M^{-1}cm^{-1}$)
1.5	5027.668800	3.8	279418.7500
1.6	22206.55000	3.9	286369.1510
1.7	38851.47660	4.0	292972.1862
1.8	54970.56120	4.1	299235.9681
1.9	70571.91620	4.2	305168.6091
2.0	85663.65420	4.3	310778.2220
2.1	100253.8878	4.4	316072.9192
2.2	114350.7296	4.5	321060.8133
2.3	127962.2921	4.6	325750.0169
2.4	141096.6878	4.7	330148.6424
2.5	153762.0293	4.8	334264.8026
2.6	165966.4292	4.9	338106.6098
2.7	177718.0000	5.0	341682.1767
2.8	189024.8543	5.1	344999.6158
2.9	199895.1046	5.2	348067.0397
3.0	210336.8635	5.3	350892.5610
3.1	220358.2435	5.4	353484.2921
3.2	229967.3572	5.5	355850.3456
3.3	239172.3172	5.6	357998.8342
3.4	247981.2359	5.7	359937.8703
3.5	256402.2260	5.8	361675.5665
3.6	264443.4000	5.9	363220.0353
3.7	272112.8705		

Source: Kuno (1998).

When a desired size is reached, the reaction is stopped by lowering the temperature of the reaction mixture below 200 °C. The particles are then isolated using a technique known as size-selective precipitation. This entails adding a polar nonsolvent to drive the particles out of solution. We will see more of this shortly.

Compilations of the absorption and emission curves needed to determine the particle size during the reaction are available, as are tables of molar extinction coefficients. An example is shown in **Table 13.4**. Similar values can be found in Yu et al. (2003), all of which involve extensive use of nanoparticle sizing techniques that we will see later in Chapter 14.

As with the hybrid preparation, highly crystalline nanocrystals are obtained. Their optical properties are also unparalleled when compared with those of other materials produced by similar means. Namely, there exists structure in the linear absorption. Furthermore, because different sized quantum dots can be produced, a size-dependent excited state progression towards higher energies is evident. All samples exhibit room temperature band-edge emission with quantum yields of the order of 10%. At low temperatures, these values increase to 90%. These advances allowed low-temperature photoluminescence excitation studies, which successfully assigned the different size-dependent optical transitions seen in the linear absorption. This was previously described in Chapter 12.

Key Advances

- The pyrolysis of organometallic precursors, involving a discrete temporal nucleation event, to obtain high-quality, narrow-size-distribution nanocrystal ensembles.
- The further use of TOPO as a high-boiling coordinating solvent to achieve effective electronic passivation of nanocrystal surface defects.

New Growth Paradigms

Although the organometallic synthesis described above enabled researchers to obtain some of the first high-quality nanocrystal samples, there were notable drawbacks. First and foremost was the batch-to-batch reproducibility of these preparations. In particular, the original synthesis of colloidal CdSe quantum dots employed technical-grade (90% purity) TOPO. As a consequence, the sample quality obtained could vary dramatically from one lot of TOPO to the next. Furthermore, using higher-purity (99%) TOPO generally led to uncontrolled growth, resulting in samples with poor size distributions. This suggested that impurities in the solvent also played a major role in dictating nanocrystal growth kinetics. Unfortunately, the identities of these impurities were not known at the time.

In addition, the original preparation used a highly pyrophoric metal precursor, dimethylcadmium. This presented not only fire hazards but also issues with its storage and handling. As a consequence, an inert atmosphere glovebox was needed to store/manipulate this substance. Scaling up reactions was also problematic given the precursor's exorbitant cost. Finally, although high-quality particles could be obtained, the production of larger nanocrystals was often difficult and could take days at high temperatures.

To circumvent these limitations, a number of subsequent studies aimed to replace both the solvent and the metal precursor in these preparations. An important breakthrough came with the realization that instead of dimethylcadmium more benign precursors such as cadmium oxide (CdO) or cadmium acetate ($CdAc_2$) could be used (Peng and Peng 2001). In particular, these studies showed that metal oxides and salts could be heated in the presence of metal-coordinating agents such as phosphonic acids to create new metal intermediates with decomposition kinetics, dictated by the ligand's stability constant. Early examples of these metal coordinating agents include hexylphosphonic acid (HPA) and tetradecylphosphonic acid (TDPA).

In tandem, it was realized that, by using ligand-coordinated metal precursors, the technical grade TOPO in the original preparation could be replaced with higher-purity 99% TOPO. In this respect, the TDPA or HPA acts as the "impurity," which helps control nanocrystal growth kinetics. Today, other metal-coordinating agents such as fatty acids are used in this way in conjunction with high-purity solvents, both coordinating and noncoordinating. As a consequence, the list of materials made using the above-mentioned approach has grown dramatically.

Key Advance

- The use of simple metal oxides and salts coordinated with ligands that possess defined stability constants in lieu of pyrophoric organometallic precursors such as dimethylcadmium.

Around this time, another major advance in colloidal quantum dot syntheses came with the introduction of multiple precursor injections following nucleation (Peng et al. 1998). This served to replenish material depleted during the initial stages of particle growth, and in this manner Ostwald ripening could be avoided. In turn, it was discovered that even better ensemble size distributions could be obtained, since the competitive growth in the later stages no longer occurred. Ensemble size distributions of about 5% could therefore be obtained without any additional post synthesis processing.

Key Advance

- The use of multiple injections, following nucleation and particle growth, to replenish depleted precursors in solution.

Today, nanocrystal preparations based on these concepts are widely used to make high-quality semiconductor quantum dots on industrially relevant scales.

EXAMPLE 13.3

Colloidal Synthesis of CdSe Quantum Dots The following illustrates a recipe to synthesize high-quality colloidal CdSe quantum dots. Note that many other procedures exist. However, all, including the one shown here—are variations on the same theme.

In a three-neck flask, 2–3 g 99% TOPO is mixed with 2 g dodecylamine (DDA), 0.3 g TDPA, and 0.05 g CdO. The solution is then dried and degassed under vacuum on a Schlenck line at 100 °C for 30 minutes. The reaction vessel is then backfilled with nitrogen and the temperature is raised to 330 °C. The color of the main reaction mixture begins to clear as the cadmium in solution coordinates with TDPA. In parallel, an injection solution consisting of 0.25 mL 1 M TOPSe and 4 mL trioctylphosphine (TOP) is prepared in a glovebox. Note that the glovebox is only used here to prevent TOP from oxidizing under ambient conditions.

When the reaction solution clears, the temperature is stabilized. The injection solution is then introduced quickly. This results in a gradual color change as the particles nucleate and begin to grow. Subsequent growth is monitored by periodically taking small aliquots of the mixture and measuring their absorption spectrum as outlined earlier. When a desired size is reached, the reaction is stopped by cooling the reaction vessel. Toluene is then added to prevent TOPO from solidifying. At this point, the resulting particles are precipitated from solution using a nonsolvent such as methanol. This purification protocol is described next.

13.4.7 Size-Selective Precipitation

An important postsynthesis processing technique developed for working with chemically synthesized nanomaterials is "size-selective precipitation." The name originates from its historical use to help reduce the size distribution of nanoparticle ensembles (Murray et al. 1993; Chemseddine and Weller 1993). Namely, early ensemble size distributions were on the order of 10%. To improve their size uniformity, it was found that by adding a nonsolvent, one could selectively isolate larger particles in the ensemble. As a consequence, the recovered material could be size-fractionated or "selected" to obtain samples exhibiting narrower size distributions.

However, size-selective precipitation is a much more useful and general technique by which to isolate chemically synthesized nanomaterials. Through this approach, it is possible to recover nanoparticle powders, which can subsequently be used in characterization techniques such as powder X-ray diffraction (Chapter 14). Furthermore, the extracted particles can be redispersed into other solvents, making possible the ligand exchange chemistries that we will see shortly.

In explaining size selective precipitation, we will assume that the nanocrystals have alkyl-tailed ligands on their surfaces. This makes them soluble in common organic solvents such as toluene but insoluble in polar media. The addition of a miscible alcohol such as methanol then reduces favorable solvent–surface-ligand interactions, leading to nanoparticle flocculation.

Classic calculations by Hamaker show that, to a first approximation, the van der Waals attraction between two equal-sized spheres increases with radius. Thus, if the polar solvent is added slowly and changes in solvent polarity are gradual, the largest nanoparticles in solution

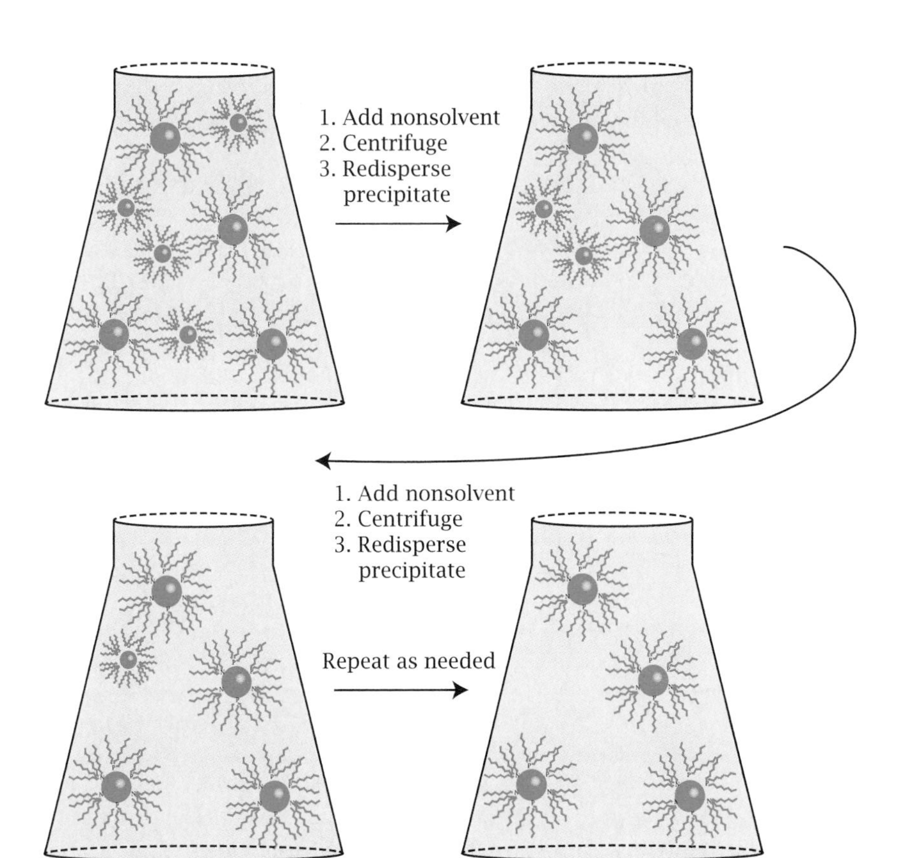

Figure 13.10 Overview of the steps involved in size-selective precipitation.

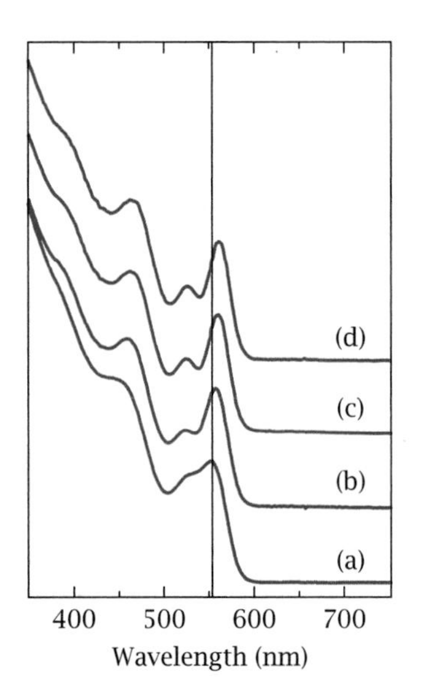

Figure 13.11 Effects of size-selective precipitation as seen in the optical spectra of an ensemble of colloidal CdSe quantum dots. (a) to (d) show spectra for an increasing number of size-selective precipitation steps. Used with permission from Mikulec (1999).

flocculate first. This leaves behind the smaller ones. Subsequent isolation of the precipitated material and resuspension in fresh solvent yields a size-fractionated ensemble. Then, by repeatedly applying this procedure, ensemble size distributions can be reduced to values of about 5%. **Figure 13.10** outlines the size-selective precipitation technique.

From an optical standpoint, the effects of size-selective precipitation are readily evident in the nanocrystal ensemble linear absorption. Namely, as the size distribution narrows, the width of the first $1S_{3/2}1S_e$ transition decreases. **Figure 13.11** also illustrates that the $2S_{3/2}1S_e$ transition becomes better defined, along with a corresponding sharpening of higher-energy transitions. In tandem, the band-edge emission linewidth narrows (not shown).

At the same time, if enough nonsolvent is added, *all* of the particles in solution flocculate. This indiscriminate precipitation is then a way to isolate ensembles of chemically synthesized nanoparticles. This will be useful for the ligand exchange procedures we will see next, as well as for subsequent characterization techniques to be introduced in Chapter 14.

Key Advance

- The ability to take advantage of ligand-imparted solubility differences to size-fractionate nanocrystal ensembles.

13.4.8 Ligand Exchange

We have seen that an important distinquishing feature of chemically synthesized nanomaterials is the ability to alter their surface passivation. This has enabled the development of surface functionalization chemistries, loosely referred to as "ligand exchange." Using these approaches, organic ligands on nanocrystal surfaces can be replaced with hydrophilic or hydrophobic ligands to make the particles soluble in polar or organic media.

Historically, we have seen that altering the surfaces of chemically synthesized nanocrystals allows them to be isolated. The particles are also stabilized against further growth, allowing them to be redispersed in different solvents without adverse effects on their size distribution. The appropriate choice of surface ligand also provides appropriate electronic passivation of surface defects, allowing dramatic improvements to nanocrystal optical properties.

Today, the general approach behind ligand exchange involves exposing functionalized nanoparticles to a large excess of a new ligand. Namely, unlike the static picture conveyed in **Figure 13.8**, the native surface ligands on the particle are actually in dynamic equilibrium with their surrounding solvent. They therefore come on and off the surface. The equilibrium position, however, is such that, on average, the particle remains passivated. On exposure to a large excess of a new ligand, the vastly increased number of new ligands near the particle surface eventually replace the old ligand. This, in turn, alters the nanocrystal surface chemistry.

In practice, the best way to achieve ligand exchange is to expose the nanoparticle precipitate to a new ligand in neat form (i.e., not diluted). Thus, if the ligand happens to be a low-melting solid, one melts it

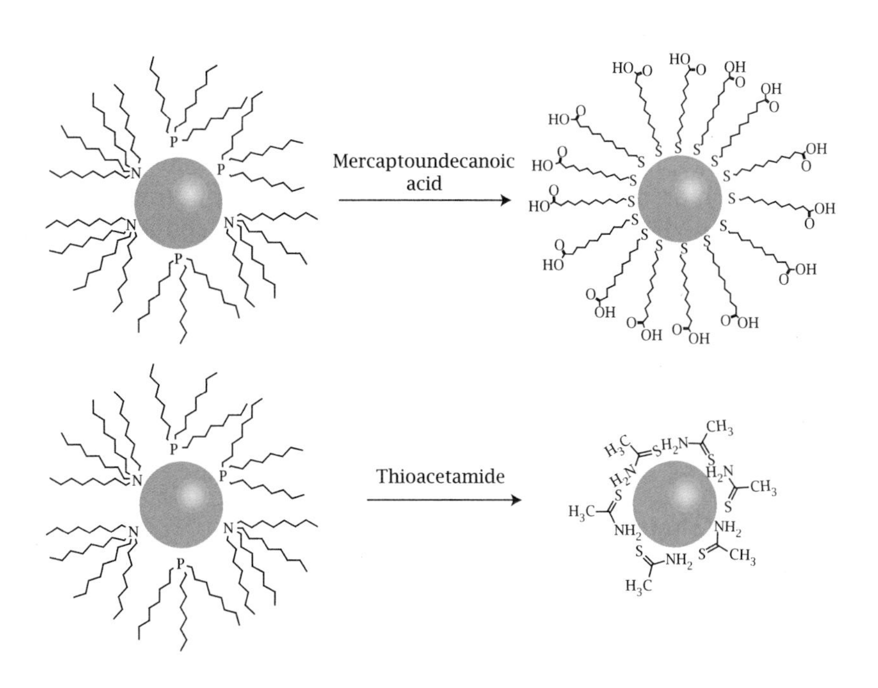

Figure 13.12 Illustration of ligand exchange with different molecules. The top shows ligand exchange with a thiol (mercaptoundecanoic acid) while the bottom shows exchange with thioacetamide.

and dissolves the nanocrystals within it. The resulting slurry is then kept heated to hasten the exchange. Alternatively, if the native state of the new ligand is a liquid, one disperses the nanoparticles within it and heats the resulting suspension. Occasionally, the nanoparticles are refluxed to improve the exchange efficiency.

In all cases, after prolonged heating, the particles are precipitated from solution using an appropriate nonsolvent. The material is then dispersed into a suitable solvent, whereupon the particles are used in applications or, if needed, precipitated again with a nonsolvent to "wash" them. This washing procedure is nothing more than a repetition of the previous size-selective precipitation step, using a very large excess of the nonsolvent to indiscriminately drive out all the nanoparticles from solution. Within the context of ligand exchange, these multiple washing steps are then used to remove as much excess ligand as possible from resulting samples. An illustration of ligand exchange with two different species is shown in **Figure 13.12**.

Key Advance

- The ability to alter the surface chemistry of quantum dots by exposing them to a large excess of a new ligand.

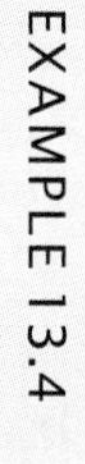

Ligand Exchange with Gold Nanoparticles As an example, the following ligand exchange demonstration can be conducted. This procedure alters the surface of gold nanoparticles made using the Brust approach. The initially hydrophobic particles can then be resuspended in water.

To carry out the exchange, a fraction of the initial gold nanoparticle solution (approximately one-quarter of the stock produced in Example 13.2) is added to a disposable 50 mL centrifuge tube. Then 0.1 g mercaptoundecanoic acid (MUA, **Figure 13.12**) is added and the mixture is shaken vigorously. The thiol head group of the MUA attaches to the gold nanoparticle surface. As a consequence, the solution turns cloudy. This stems from changes in the nanoparticle solubility. The resulting suspension is then centrifuged and a precipitate is recovered. To this, 2 mL tetrahydrofuran (THF) is added. At this point, a commercial 20 wt% solution of potassium t-butoxide in THF is added dropwise. As the t-butoxide is introduced, the nanoparticles fall out of solution. When complete, the suspension is centrifuged and a precipitate is recovered. Deionized water is then added to the recovered pellet to bring the particles into solution. This demonstrates the transfer of gold nanoparticles from an organic solvent into an aqueous environment.

13.4.9 Core/Shell Nanocrystals

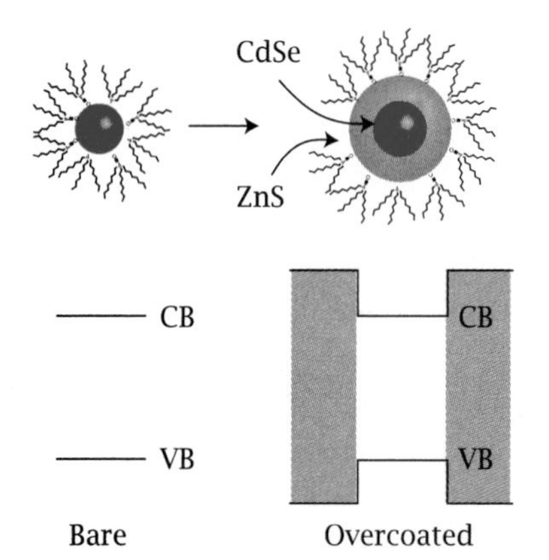

Figure 13.13 Illustration of the ZnS overcoating of CdSe quantum dots. The added potential barrier for both the electron and hole are depicted. The terms "conduction band" (CB) and "valence band" (VB) are used loosely here.

With the development of high-quality narrow-size-distribution colloidal quantum dots, increasing attention has been devoted to improving their optical and electrical properties. Of interest has been increasing their emission quantum yield, since many (as-made) quantum dot ensembles have values of about 10%. Although TOPO was key to achieving these respectable efficiencies, nonradiative relaxation still dominates their photoresponse. Furthermore, in smaller quantum dots, evidence of defect-related emission is sometimes apparent.

To address these deficiencies, better electronic passivation of CdSe nanocrystal surfaces was therefore sought by overcoating them with an inorganic shell. Materials were therefore chosen whose bulk band offsets could provide potential barriers for both the electron and the hole in the CdSe core. This is illustrated in **Figure 13.13**.

The first successful overcoating procedure was developed by Hines and Guyot-Sionnest in 1995 (Hines and Guyot-Sionnest 1995). They found that CdSe quantum dots could be overcoated with ZnS by slowly adding dimethylzinc ($ZnMe_2$) and hexamethyldisilathiane ($(TMS)_2S$) to a nanocrystal solution in TOPO. The underlying idea behind the slow addition was to prevent unwanted ZnS nucleation and to instead encourage its deposition onto the surface of existing particles. Furthermore, relatively low overcoating temperatures (200 °C) were used to prevent CdSe nanocrystals from undergoing any Ostwald ripening. Initial core size distributions could therefore be retained. Once the precursors had been added, the particles were subsequently isolated from solution using the same size-selective precipitation steps described earlier.

Emission quantum yields of these overcoated materials, in turn, ranged from 30% to 50%. Given these impressive numbers, other overcoating attempts were soon attempted (Dabbousi et al. 1997; Peng et al. 1997; Reiss et al. 2002). This led to ZnS-, ZnSe-, and CdS-coated CdSe nanocrystals with emission quantum yields exceeding 50%. Today, overcoated quantum dots with near-unity quantum yields exist, showing the beneficial aspects of passivating their surfaces with an inorganic shell.

Key Advance

- The passivation of quantum dot surface defects using an inorganic shell, leading to highly emissive core/shell structures.

EXAMPLE 13.5

Surface Overcoating of Colloidal CdSe Quantum Dots To illustrate the overcoating of CdSe nanocrystals, the following procedure can be attempted. Nanocrystals are first isolated from their growth mixture using methanol. The recovered material is then redispersed in a small amount of toluene. At this point, the mean nanocrystal size as well as ensemble concentration is determined using absorption/emission sizing tables as well as tables of size-dependent molar extinction coefficients (**Tables 13.2–13.4**).

The toluene suspension is subsequently added to a reaction mixture consisting of 5 g TOPO along with 2.5 g hexadecylamine (HDA). The mixture is dried and degassed under vacuum to drive off any toluene as well as any water picked up by the TOPO. Once this is complete, the temperature is raised to about 150 °C.

In tandem, an overcoating solution consisting of a small amount of $(TMS)_2S$ and diethylzinc ($ZnEt_2$) is prepared in an inert atmosphere glovebox. The amount of either compound is calculated beforehand to enable an appropriate number of ZnS "monolayers" to be deposited. This is where knowing the size and concentration of the dots becomes important. Both precursors are diluted in 5 mL TOP.

When the injection solution is ready, it is introduced into the reaction mixture using a syringe pump. The addition is done slowly over the course of 30 minutes to prevent any unwanted ZnS nucleation. The reaction can be monitored in real time using a hand-held ultraviolet lamp to observe increases in the nanocrystal's emission quantum yield.

Once the solution has been added, the reaction mixture is heated for some time to anneal the shell. The solution is subsequently cooled and toluene is added to prevent TOPO from solidifying. To isolate the particles, they are precipitated from solution by adding a nonsolvent such as methanol and are ultimately resuspended in toluene.

13.4.10 Shape Control, Anisotropic Nanocrystals

These days, there is also significant interest in controlling the morphology of nanocrystals beyond the spherical forms seen earlier. The impetus for this comes from the realization that a nanostructure's shape also has an impact on its optical and electrical response. Furthermore, an opportunity exists for learning how to control crystallization at the nano and even molecular levels.

Some of the earliest studies in this area therefore focused on producing CdSe nanorods (Peng et al. 1999). Originally, this was accomplished using an excess of phosphonic acids such as HPA and TDPA, which empirically promoted the asymmetric growth of CdSe. This observation was rationalized by the preferential binding of these ligands to the crystallographic faces of a wurztite CdSe nanocrystal but not to its Se-terminated face, oriented normally to the *c* axis. As a consequence, the existence of a reactive, unpassivated, face provided a crystal growth rate asymmetry that could be exploited to obtain nanorods, as opposed to quantum dots.

Around the same time, it was also realized that surfactants used in colloidal nanostructure growth could promote the formation of a given crystallographic phase. For example, certain surfactants were observed to promote the nucleation of zinc blende CdSe over its wurtzite form. This was, in turn, exploited in subsequent nanostructure shape control studies to obtain CdSe nanostructures possessing tetrapod and other more unusual morphologies. In the case of tetrapods, the existence of a zinc blende core leads to four equivalent {111} faces out of which arms can grow. This results in a characteristic "jumping jack" shape. More details about these and other shape control studies can be found in Manna et al. (2003).

Key Advance

- The ability to use ligands to alter the morphology of chemically synthesized nanostructures.

13.5 NANOWIRES

Along with the nonspherical quantum dot morphologies just described, an equally strong interest exists in developing one-dimensional materials. Much of this stems from the unique size-, shape- and dimensionality-dependent optical and electrical properties possessed by nanostructures, as seen throughout this text. Important differences between quantum dots and nanowires include their different density of states (Chapter 9), the intrinsic absorption and emission polarization anisotropies of wires, and their ability to transport charge efficiently along their length.

Perhaps the most popular approach today for producing high-quality semiconductor nanowires involves a technique referred to as vapor–liquid–solid growth (discussed below). Other growth approaches exist and we briefly touch on them. In the case of solution-based nanowire syntheses, the parallels with quantum dot growth are unavoidable. Namely, similar precipitation strategies are used to control nanowire growth, with a key distinguishing feature of these preparations being the use of surfactants to control nanostructure growth kinetics. Furthermore, the presence of surface ligands also enables the wires to be dispersed in different solvents. Finally, the accessibility of nanowire surfaces enables postsynthesis processing as well as potential surface functionalization chemistries.

13.5.1 Metal–Organic Chemical Vapor Deposition (MOCVD)

The first nanowire growth strategy we describe involves metal–organic chemical vapor deposition (MOCVD). This is an approach similar to MBE discussed earlier. However, the experimental apparatus differs greatly, since MOCVD is generally conducted using a quartz tube furnace. In this approach, a metal–organic compound such as trimethylaluminum, trimethylgallium, or trimethylindium is introduced through the vapor phase into a heated reactor. In the case of liquid organometallic precursors, these are brought into the reactor by bubbling an inert gas through them. At the elevated temperatures in the reactor, these precursors decompose to give the desired elements for the subsequent reactions described below.

13.5.2 Vapor–Liquid–Solid Growth (VLS)

Vapor–liquid–solid growth (VLS) is a technique that has been used to synthesize nanowires using small metal nanoparticle catalysts. The phenomenon was first discovered in the mid 1960s by Wagner and Ellis at Bell Laboratories (Wagner and Ellis 1964). They found that small droplets could be used to induce the asymmetric growth of semiconductors such as silicon from the thermolysis of gas-phase precursors. The droplets were said to "catalyze" crystal growth. Common seed particles include gold, whose choice was originally motivated by the large solubility of silicon in gold, as well as by the existence of a low-temperature eutectic. Note that because the resulting crystals had large micrometer-sized diameters they were (and still are) referred to as "whiskers" as opposed to nanowires.

Today's implementation of VLS entails the following. In all cases, small nanometer-sized metal droplets are used to induce nanowire growth. These metal nanoparticles are obtained in several ways. In one approach, a thin metal film is evaporated over a substrate where, on heating, it melts and balls up into tiny molten droplets. Owing to thermodynamic considerations, a lower size limit exists for the resulting particles. Typical droplet diameters are therefore of the order of hundreds of nanometers. This is potentially important, because a 1:1 correspondence exists between the catalyst nanoparticle size and that of the resulting wire. As a consequence, wires made using these catalyst particles have diameters of the order of 100 nm. They are thus not within their respective confinement regimes, as defined by the bulk exciton Bohr radius or the carrier's de Broglie wavelength (Chapter 3).

To circumvent this limitation, smaller metal droplets are synthesized using the colloidal procedures described above. As a consequence, nanoparticles with diameters ranging from several nanometers to tens of nanometers are possible. These particles are then isolated and reintroduced into the reactor by dropcasting or spin coating them onto an appropriate substrate.

In all cases, the molten droplets solubilize elements derived from vapor-phase precursors. These elements incorporate until saturation occurs, whereupon a nucleation event at the liquid droplet/solid substrate interface initiates nanowire growth. The catalyst particles then promote further nanowire growth by continuously adding elements to this interface. The VLS process is outlined schematically in **Figure 13.14**

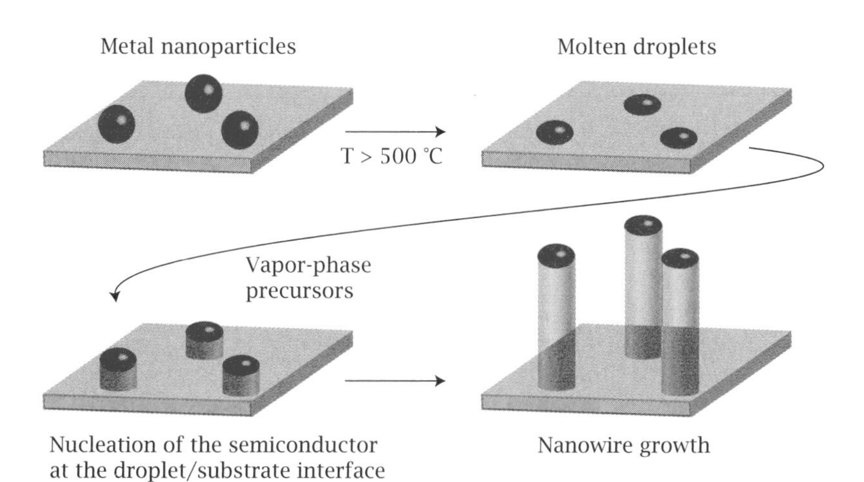

Figure 13.14 Illustration of the VLS nanowire growth procedure.

and a summary of the approach as well as its extensions can be found in Lu and Lieber (2006) and Wu et al. (2002).

13.5.3 Solution–Liquid–Solid Growth (SLS)

More recently, there have been advances in the solution-phase synthesis of nanowires. The general approach is referred to as solution–liquid–solid (SLS) growth and has a similar mechanism to the VLS procedure just described. The primary difference between the two is the lower temperatures used in SLS preparations. Namely, instead of growth temperatures between 500 and 1000 °C, values below 400 °C are used. This means that the employed metal catalyst particles must be made of elements that possess relatively low melting points. Examples include indium, tin, bismuth, and gallium.

The SLS growth mechanism proceeds as follows. First, solution-phase molecular precursors thermolyze to yield elements of the desired semiconductor. These elements are then solubilized within molten metal droplets. Upon supersaturation, a nucleation event at the liquid droplet/solution interface (as opposed to the liquid droplet/solid substrate interface in the VLS case) initiates nanowire growth. The continued addition of material to the nucleated seeds then promotes nanowire growth. In many cases, the diameter of the resulting wire is identical to that of the catalyst particle. The SLS growth mechanism is illustrated in **Figure 13.15**. More about it and its extensions can be found in Wang et al. (2006) and Kuno (2008).

Note that SLS growth has some unique features. First, despite its low temperatures, the resulting wires are crystalline. This has been shown in many of the TEM micrographs seen earlier. Examples include **Figures 4.2–4.5**. Yields can also be large. Furthermore, because the wires are grown in the presence of organic surfactants that bind datively or covalently to the surface, the resulting wires can be suspended in different solvents. Nanowire growth in the presence of surface binding agents also suppresses radial growth, leading to narrow wire diameters. In parallel, their surfaces are protected from oxidation. Subsequent surface functionalization chemistries are also possible using techniques analogous to those described earlier for quantum dots. Finally, SLS syntheses have empirically yielded some of the narrowest nanowires to date. As a consequence, wires with diameters within their respective confinement regimes are possible. This makes SLS-derived materials interesting from an optical standpoint.

In practice, SLS reactions are carried out in much the same way as colloidal quantum dot preparations. The reaction apparatus is identical to that shown in **Figure 13.7**. Furthermore, many of the same chemicals

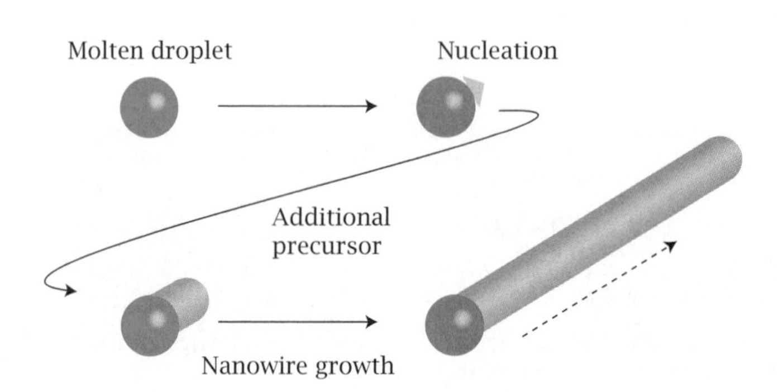

Figure 13.15 Illustration of the SLS nanowire growth procedure.

as well as methodologies are used. Specifically, a metal source such as CdO is mixed with a metal coordinating agent (e.g., a fatty acid), all within a high-boiling solvent such as TOPO. The reaction vessel is then dried and degassed under vacuum for a short period of time. Following this, the vessel is backfilled with an inert gas and the temperature is raised to values between 200 and 350 °C. A chalcogen precursor along with a bismuth catalyst is then introduced to initiate nanowire growth. The reaction is eventually stopped by lowering the temperature below 200 °C. At this point, the wires are processed in much the same way as colloidal quantum dots.

To illustrate the range of one-dimensional materials produced by SLS, **Table 13.5** lists recent nanowires that have been made this way. It shows the identity of the catalyst particles, the growth solvent, and the growth temperature. Resulting nanowire diameters are also stated so that they can be compared with the system's bulk exciton Bohr radius.

EXAMPLE 13.6

Solution-Based Growth of CdSe Nanowires To illustrate the SLS growth of CdSe nanowires, the following synthesis can be conducted. In an apparatus analogous to that shown in **Figure 13.7**, 0.025 g CdO, 0.23 mL octanoic acid, and 2.5 g 99% TOPO are mixed. The three-neck flask is then dried and degassed under vacuum at approximately 100 °C for 30 minutes. When this is complete, the vessel is back-filled with nitrogen and the temperature is raised to 320 °C. Above 250 °C, the solution becomes clear as the cadmium coordinates with octanoic acid. The temperature is then lowered to 250 °C. At this point, a premade injection solution, consisting of 25 μL 1 M TOPSe and 0.2 mL TOP, along with 12.5 μL 2 mM bismuth chloride ($BiCl_3$) solution in acetone, is introduced. On injection, a noticeable color change occurs, indicating nanowire growth. The reaction is then allowed to proceed for approximately 2 minutes, after which the reaction mixture is cooled. Toluene is added to prevent TOPO from solidifying and the resulting nanowires are centrifuged out of solution much like the size-selective precipitation steps carried out for quantum dots. The recovered precipitate is subsequently resuspended in fresh toluene.

13.5.4 Oriented Attachment

It should be noted that other solution-phase syntheses exist for making nanowires. One intriguing approach is referred to as oriented attachment. This is a technique whereby colloidal quantum dots are used to make nanowires by simply fusing them together along common crystallographic faces. The underlying mechanism stems from the existence of strong dipoles in unpassivated nanocrystals, resulting from their abrupt termination. Namely, in nanocrystals terminated in such a way as to possess a face with a partial negative charge and an opposing face with a partial positive charge, large dipoles exist. Magnitudes on the order of 50 D have been estimated. As a consequence, attractive dipole–dipole interactions allow multiple nanocrystals to come together to yield an extended wire-like structure. In turn, CdSe, CdTe, PbSe, and other nanowires have been produced this way. More information about oriented attachment can be found in Penn and Banfield (1998).

Table 13.5 Nanowires made using the SLS approach

Material	Nanoparticle catalyst	Solvent	Growth temperature (°C)	Diameter (nm)	Bulk exciton Bohr radius a_B (nm)
Ge	Bi	TOP	350–365	20–150	11.5
InP	Bi, In, Au	1,3-Diisopropylbenzene, TOPO, TOPO/TOP/HDA	203, 111–203, 320–340 240–300	3–11, 10–100, 10–100 4–12	10.0
InAs	Bi, Au	1,3-Diisopropylbenzene, TOPO/trioctylamine, trioctylamine, TOP/HDA	203, 320–340, 260–300	NA, 12–100, 5–13	50, 34
GaP	Bi	TOPO	320–340	20–100	—
GaAs	Bi, In	1,3-Diisopropylbenzene, trioctylamine, TOPO/TOP/ODE	203, 203, 320–340, 255	6–17, NA, 20–100, 4.5–9	12.4
$Al_xGa_{1-x}As$	Al/Ga	1,3-Diisopropylbenzene	203/700	20–200	—
ZnTe	Bi	TOPO/ODE	270–290	3.7–12	6.7
CdS	Bi	ODE/HDA	315	10–11	2.9
		TOPO	260–330	14	—
CdSe	Bi	TOPO	270–330, 240–300, 240–280, 7–12, 5–20, 4.7–15	5.6	—
CdTe	Bi	TOPO	275–330, 240–320	6–12, 5–20	7.5
PbS	Bi	TOPO/TOP	210–300	9–31	18
PbSe	Bi	Ph_2O/TOP	175	5–10	46

HDA, hexadecylamine; ODE, octadecene; Ph_2O, phenyl ether; TOP, trioctylphosphine; TOPO, trioctylphosphine oxide.

13.6 SUMMARY

With this, we conclude our brief introduction to the synthesis of nanoscale materials. We have seen that a precipitation approach can be used to make high-quality semiconductor quantum dots. Colloidal CdSe nanocrystals have been employed as a model system given the extensive work into their synthesis over the last two decades. In this regard, the historical development of CdSe highlights many of the ideas and concepts that underlie modern solution-phase nanostructure preparations. This has also enabled the subsequent development of solution-based nanowire syntheses. In what follows, we will see how these materials are characterized to determine critical parameters such as their size, size distribution, and chemical composition. This is followed by a discussion of their applications in Chapter 15.

13.7 THOUGHT PROBLEMS

13.1 History

Maya Blue is a pigment often found in ancient Mayan pottery and other artifacts. The source of its color, however, has been widely debated. More recently, its blue hue has been partially attributed to the inclusion of small metal nanoparticles. Read the article by Jose-Yacaman et al. (1996).

13.2 History

The chemistry of photographic emulsions owes a lot to the work of Mathew Carey Lea and George Eastman. For a biography of Lea, see Smith (1943). In the photographic process, silver halide particles are illuminated with light and are eventually reduced to elemental silver. In turn, resulting photographs contain colloidal silver particles and even nanoparticles. See, for example, Lam and Rossiter (1991).

13.3 History

Colloidal synthesis is an old area. In fact, researchers were making high quality micrometer-sized particles for many years before the advent of more recent colloidal nanostructures, described in the main text. Read, for example, the articles by Sugimoto (1987) and Matijevic (1993).

13.4 Magic numbers

Occasionally in the colloidal synthesis of nanocrystals, one encounters specimens that exhibit highly structured linear absorption spectra. Explain the concept of magic sizes or magic number nanocrystals and clusters. Why are such species stable? See Herron et al. (1993), Kasuya et al. (2004), and Ouyang et al. (2008), as well as references therein.

13.5 Soft templates

Explore the use of "soft templates" in making anisotropic nanostructures. Explain the concept. See, for example, Pileni (2003).

13.6 Templates

Explore the use of hard templates such as anodized aluminum oxide or nanochannel glass to make nanostructures such as nanowires. See, for example, Pearson and Tonucci (1995) and Yin et al. (2001).

13.7 Oriented attachment

Read about oriented attachment in Penn and Banfield (1998) and explain the concept as well as its application in making nanowires. See, for example, Tang et al. (2002).

13.8 Photolithography

Investigate the use of photolithography to make small structures. Find an example in the literature. Explain its advantages and limitations.

13.9 Electron beam lithography

Investigate the use of electron beam lithography to make small structures. Find an example in the literature. Explain its advantages and limitations.

13.10 Lithography

Explore other avenues for making nanostructures. One of them is called nanosphere lithography. Read the article by Hulteen and Van Duyne (1995).

13.11 Dip pen lithography

There are other ways to make nanostructures. Read about a technique called dip pen nanolithography, which operates in a similar manner to an old-fashioned quill pen, in Piner et al. (1999), Zhang et al. (2003), and Salaita et al. (2007).

13.12 Microcontact printing or stamping

Another way to make small structures involves a technique called microcontact printing. See the articles by Kumar and Whitesides (1993), Santhanam and Andres (2004), and Kraus et al. (2007), and explore how the process has been exploited in nanostructure device applications.

13.13 Biological syntheses

Along with researchers, nature has also found ways to make nanostructures. Read about biological syntheses of quantum dots and nanowires: for quantum dots, see Dameron et al. (1989), Sweeney et al. (2004), and Eisenstein (2005), and for nanowires, see Mao et al. (2003, 2004).

13.14 Magnetotactic bacteria

Read about magnetotactic bacteria that naturally contain magnetic particles (Blakemore 1975). Explain how these bacteria synthesize nanoparticles. Then ask yourself why they would want them in the first place.

13.15 Transformations

Explore the nanoscale Kirkendall effect. Read the articles by Buriak (2004) and Yin et al. (2004), and show how the phenomenon can be used to create other nanostructured morphologies.

13.16 Transformations

On a similar note, explore cation exchange in nanostructures. Read the articles by Son et al. (2004) and Robinson et al. (2007), and show how the effect has been used to make other nanosystems.

13.17 Surface chemistry

Describe the chemistry used to functionalize CdSe quantum dots in order to make them biocompatible. Are there common strategies for achieving this? See, for example, the two early studies in this area by Bruchez et al. (1998) and Chan and Nie (1998).

13.18 Doping

The introduction of magnetic or electronic impurities into nanostructures has been a long sought-after goal. However, efforts towards this in the past have often failed for a number of reasons. Read about them in Hanif et al. (2002), Norris et al. (2008), and Beaulac et al. (2008), and describe for yourself the potential uses of such doped nanostructures.

13.19 Nanotube purification

Carbon nanotubes come in many flavors, based on their chirality. As a consequence, some tubes are metallic while others are semiconducting. Unfortunately, the production of carbon nanotubes leads to a mixture of both types of materials. This has motivated many to find ways of purifying nanotube ensembles. Read about these approaches in Green and Hersam (2007) and Hersam (2008).

13.20 Assemblies

Nanocrystal superlattices represent a new important class of nanostructure. Describe how such superlattices can be made from component quantum dots and other nanoparticles. See, for example, Shevchenko et al. (2006) and Talapin et al. (2009). Also consider the article by Kalsin et al. (2006).

13.21 Proteins

In tandem, biological scaffolds can be used to create hierarchical nanostructures. As examples, see Douglas et al. (2002) and Sun et al. (2007).

13.22 Manipulation

Along with being able to manufacture semiconductor and metal nanostructures, the ability to manipulate them is key to potential device applications. See the articles by Kim et al. (2001) and Ryan et al. (2006) describing the use of the Langmuir–Blodgett technique to align either metal or semiconductor nanorods.

13.23 Manipulation

It is also of interest to align and rationally position nanowires. Explore the manipulation of nanowires using the following techniques:

(a) Dielectrophoresis (Boote and Evans 2005)

(b) Directed flow (Messer et al. 2000; Huang et al. 2001)

(c) Langmuir–Blodgett assembly (Yang 2003; Tao et al. 2003; Whang et al. 2003)

(d) Bubble films (Yu et al. 2007)

13.8 REFERENCES

Barber DJ, Freestone IC (1990) An investigation of the origin of the colour of the Lycurgus cup by analytical transmission electron microscopy. *Archaeometry* **32**, 33.

Bawendi MG, Wilson WL, Rothberg L, et al. (1990) Electronic structure and photo-excited carrier dynamics in nanometer-size CdSe clusters. *Phys. Rev. Lett.* **65**, 1623.

Beaulac R, Archer PI, Ochsenbein ST, Gamelin DR (2008) Mn^{2+}-doped CdSe quantum dots: new inorganic materials for spin-electronics and spin-photonics. *Adv. Funct. Mater.* **18**, 3873.

Bimberg D, Grundmann M, Ledentsov NN (2001) *Quantum Dot Heterostructures.* Wiley, New York.

Blakemore R (1975) Magnetotactic bacteria. *Science* **190**, 377.

Boote JJ, Evans SD (2005) Dielectrophoretic manipulation and electrical characterization of gold nanowires. *Nanotechnology.* **16**, 1500.

Brennan JG, Siegrist T, Carroll PJ, et al. (1989) The preparation of large semiconductor clusters via the pyrolysis of a molecular precursor. *J. Am. Chem. Soc.* **111**, 4141.

Bruchez M Jr, Moronne M, Gin P, et al. (1998) Semiconductor nanocrystals as fluorescent biological labels. *Science* **281**, 2013.

Brust M, Walker M, Bethell D, et al. (1994) Synthesis of thiol-derivatised gold nanoparticles in a two-phase liquid–liquid system. *Chem. Soc. Chem. Commun.* 801.

Buriak JM (2004) Chemistry with nanoscale perfection. *Science* **304**, 692.

Chan WCW, Nie S (1998) Quantum dot bioconjugates for ultrasensitive nonisotopic detection. *Science* **281**, 2016.

Chemseddine A, Weller H (1993) Highly monodisperse quantum sized CdS particles by size-selective precipitation. *Ber. Bunsenges. Phys. Chem.* **97**, 636.

Dabbousi BO, Rodriguez-Viejo J, Mikulec FV, et al. (1997) (CdSe)ZnS core–shell quantum dots: synthesis and characterization of a size series of highly luminescent nanocrystallites. *J. Phys. Chem. B* **101**, 9463.

Dameron CT, Reese RN, Mehra RK, et al. (1989) Biosynthesis of cadmium sulphide quantum semiconductor crystallites. *Nature* **338**, 596.

Douglas T, Strable E, Willits D, et al. (2002) Protein engineering of a viral cage for constrained nanomaterials synthesis. *Adv. Mater.* **14**, 415.

Eisenstein M (2005) Bacteria find work as amateur chemists. *Nature Meth.* **2**, 6.

Faraday M (1857) The Bakerian Lecture: Experimental relations of gold (and other metals) to light. *Phil. Trans. R. Soc. Lond.* **147**, 145.

Fiedler I, Bayard MA (1986) Cadmium yellows, oranges and reds. In *Artists' Pigments. A Handbook of their History and Characteristics*, Volume 1. (ed. R. L. Feller). Oxford University Press, New York, p. 65.

Fredrickx P, Schryvers D, Janssens K (2002) Nanoscale morphology of a piece of ruby red Knuckel glass. *Phys. Chem. Glasses* **43**, 176

Freestone I, Meeks N, Sax M, Higgitt C (2007) The Lycurgus cup—a Roman nanotechnology. *Gold Bull.* **40**, 270.

Fojtik A, Weller H, Koch U, Henglein A (1984) Photo-chemistry of colloidal metal sulfides 8. Photo-physics of extremely small CdS particles: Q-state CdS and magic agglomeration numbers. *Ber. Bunsenges. Phys. Chem.* **88**, 969.

Gaponenko SV (1999) *Optical Properties of Semiconductor Nanocrystals.* Cambridge University Press, Cambridge, UK.

Green AA, Hersam MC (2007) Ultracentrifugation of single-walled nanotubes. *Mater. Today* **10**, 59.

Hanif KM, Meulenberg RW, Strouse GF (2002) Magnetic ordering in doped $Cd_{1-x}Co_xSe$ diluted magnetic quantum dots. *J. Am. Chem. Soc.* **124**, 11495.

Herron N, Calabrese JC, Farneth WE, Wang Y (1993) Crystal structure and optical properties of $Cd_{32}S_{14}(SC_6H_5)_{36} \cdot DMF_4$, a cluster with a 15 Angstrom CdS core. *Science* **259**, 1426.

Hersam MC (2008) Progress towards monodisperse single-walled carbon nanotubes. *Nature Nanotech.* **3**, 387.

Hines MA, Guyot-Sionnest P (1995) Synthesis and characterization of strongly luminescing ZnS-capped CdSe nanocrystals. *J. Phys. Chem. B* **100**, 468.

Huang Y, Duan X, Wei Q, Lieber CM (2001) Directed assembly of one-dimensional nanostructures into functional networks. *Science* **291**, 630.

Hulteen JC, Van Duyne RP (1995) Nanosphere lithography: a materials general fabrication process for periodic particle array surfaces. *J. Vac. Sci. Technol. A* **13**, 1553.

Jose-Yacaman M, Rendon L, Arenas J, Serra Puche MC (1996) Maya blue paint: a ancient nanostructured material. *Science* **273**, 223.

Kalsin AM, Fialkowski M, Paszewsi M, et al. (2006) Electrostatic self-assembly of binary nanoparticle crystals with a diamond-like lattice. *Science* **312**, 420.

Kalyanasundaram K, Borgarello E, Duonghong D, Grätzel M (1981) Cleavage of water by visible light irradiation of colloidal CdS solutions; inhibition of photocorrosion by RuO_2. *Angew. Chem. Int. Ed. Engl.* **20**, 987.

Kasuya A, Sivamohan R, Barnakov YA, et al. (2004) Ultra-stable nanoparticles of CdSe revealed from mass spectrometry. *Nature Mat.* **3**, 99.

Kim F, Kwan S, Akana J, Yang P (2001) Langmuir–Blodgett nanorod assembly. *J. Am. Chem. Soc.* **123**, 4360.

Kraus T, Malaquin L, Schmid H, et al. (2007) Nanoparticle printing with single-particle resolution. *Nature Nanotech.* **2**, 570.

Kumar A, Whitesides GM (1993) Features of gold having micrometer to centimeter dimensions can be formed through a combination of stamping with an elastomeric stamp and an alkanethiol ink followed by chemical etching. *Appl. Phys. Lett.* **63**, 2002.

Kuno M (1998) Band edge spectroscopy of CdSe quantum dots. PhD Thesis, MIT.

Kuno M (2008) An overview of solution-based semiconductor nanowires: synthesis and optical studies. *Phys. Chem. Chem. Phys.* **10**, 620.

Lam DM, Rossiter BY (1991) Chromoskedasic painting. *Sci. Am.* **265**, (November), 80.

LaMer VK, Dinegar RH (1950) Theory, production and mechanism of formation of monodispersed hydrosols. *J. Am. Chem. Soc.* **72**, 4847.

Lu W, Lieber CM (2006) Semiconductor nanowires. *J. Phys. D: Appl. Phys.* **39**, R387.

Maitra A (1984) Determination of size parameters of water–aerosol OT–oil reverse micelles from their nuclear magnetic resonance data. *J. Phys. Chem.* **88**, 5122.

Manna L, Milliron DJ, Meisel A, et al. (2003) Controlled growth of tetrapod-branched inorganic nanocrystals. *Nature Mater.* **2**, 382.

Mao C, Flynn CE, Hayhurst A, et al. (2003) Viral assembly of oriented quantum dot nanowires. *Proc. Natl Acad. Sci. USA* **100**, 6946.

Mao C, Solis DJ, Reiss BD, et al. (2004) Virus-based toolkit for the directed synthesis of magnetic and semiconducting nanowires. *Science* **303**, 213.

Matijevic E (1993) Preparation and properties of uniform size colloids. *Chem. Mater.* **5**, 412.

Messer B, Song JH, Yang P (2000) Microchannel networks for nanowire patterning. *J. Am. Chem. Soc.* **122**, 10232.

Mikulec FV (1999) Semiconductor nanocrystal colloids: manganese doped cadmium selenide, (core)shell composites for biological labeling and highly fluorescent cadmium telluride. PhD Thesis, MIT.

Murray CB, Norris DJ, Bawendi MG (1993) Synthesis and characterization of nearly monodisperse CdE (E = S,Se,Te) semiconductor nanocrystallites. *J. Am. Chem. Soc.* **115**, 8706.

Norris DJ, Efros AL, Erwin SC (2008) Doped nanocrystals. *Science* **319**, 1776.

Ouyang J, Badruz Zaman Md, Yan FJ, et al. (2008) Multiple families of magic-sized CdSe nanocrystals with strong bandgap photoluminescence via noninjection one-pot syntheses. *J. Phys. Chem. C* **112**, 13805.

Pearson DH, Tonucci RJ (1995) Nanochannel glass replica membranes. *Science* **270**, 68.

Peng X, Schlamp MC, Kadavanich AV, Alivisatos AP (1997) Epitaxial growth of highly luminescent CdSe/CdS core/shell nanocrystals with photostability and electronic accessibility. *J. Am. Chem. Soc.* **119**, 7019.

Peng X, Wickham J, Alivisatos AP (1998) Kinetics of II–VI and III–V colloidal semiconductor nanocrystal growth: focusing of size distributions. *J. Am. Chem. Soc.* **120**, 5343.

Peng X, Manna L, Yang W, et al. (1999) Shape control of CdSe nanocrystals. *Nature* **404**, 59.

Peng ZA, Peng X (2001) Formation of high-quality CdTe, CdSe, and CdS nanocrystals using CdO as precursor. *J. Am. Chem. Soc.* **123**, 183.

Penn RL, Banfield JF (1998) Imperfect oriented attachment: dislocation generation in defect-free nanocrystals. *Science* **281**, 969.

Perkowitz S (1998) *Empire of Light. A History of Discovery in Science and Art.* Joseph Henry Press, Washington, DC.

Pileni MP (2003) The role of soft colloidal templates in controlling the size and shape of inorganic nanocrystals. *Nature Mat.* **2**, 145.

Piner RD, Zhu J, Xu F, et al. (1999) Dip pen nanolithography. *Science* **283**, 661.

Reiss P, Bleuse J, Pron A (2002) Highly luminescent CdSe/ZnSe core/shell nanocrystals of low size dispersion. Nano Lett. **2**, 781.

Robinson RD, Sadtler B, Demchenko DO, et al. (2007) Spontaneous superlattice formation in nanorods through partial cation exchange. *Science* **317**, 355.

Rossetti R, Brus L (1982) Electron–hole recombination emission as a probe of surface chemistry in aqueous CdS colloids. *J. Phys. Chem.* **86**, 4470.

Rossetti R, Hull R, Gibson JM, Brus LE (1985) Excited electronic states and optical spectra of ZnS and CdS crystallites in the $\simeq$15 to 50 Å size range: evolution from molecular to bulk semiconducting properties. *J. Chem. Phys.* **82**, 552.

Ryan KM, Mastroianni A, Stancil KA, et al. (2006) Electric-field-assisted assembly of perpendicularly oriented nanorod superlattices. *Nano Lett.* **6**, 1479.

Salaita K, Wang Y, Mirkin CA (2007) Applications of dip-pen nanolithography. *Nature Nanotech.* **2**, 145.

Santhanam V, Andres RP (2004) Microcontact printing of uniform nanoparticle arrays. *Nano Lett.* **4**, 41.

Schalm O, Van der Linden V, Frederickx P, et al. (2009) Enamels in stained glass windows: preparation, chemical composition, microstructure and causes of deterioration. *Spectrochim. Acta B* **64**, 812.

Shevchenko EV, Talapin DV, Kotov NA, et al. (2006) Structural diversity in binary nanoparticle superlattices. *Nature* **439**, 55.

Smith EF (1943) M. Carey Lea, chemist. *J. Chem. Educ.* **20**, 577.

Son DH, Hughes SM, Yin Y, Alivisatos AP (2004) Cation exchange reactions in ionic nanocrystals. *Science* **306**, 1009.

Steigerwald ML, Alivisatos AP, Gibson JM, et al. (1988) Surface derivatization and isolation of semiconductor cluster molecules. *J. Am. Chem. Soc.* **110**, 3046.

Sugimoto T (1987) Preparation of monodispersed colloidal particles. *Adv. Coll. Interf. Sci.* **28**, 65.

Sun J, DuFort C, Daniel MC, et al. (2007) Core-controlled polymorphism in virus-like particles. *Proc. Natl Acad. Sci. USA* **104**, 1354.

Sweeney RY, Mao C, Gao X, et al. (2004) Bacterial biosynthesis of cadmium sulfide nanocrystals. *Chem. Biol.* **11**, 1553.

Talapin DV, Shevchenko EV, Bodnarchuk MI, et al. (2009) Quasicrystalline order in self-assembled binary nanoparticle superlattices. *Nature* **461**, 964.

Tang Z, Kotov NA, Giersig M (2002) Spontaneous organization of single CdTe nanoparticles into luminescent nanowires. *Science* **297**, 237.

Tao A, Kim F, Hess C, et al. (2003) Langmuir–Blodgett silver nanowire monolayers for molecular sensing using surface-enhanced Raman spectroscopy. *Nano Lett.* **3**, 1229.

Turkevich J, Stevenson PL, Hillier J (1951) A study in the nucleation and growth processes in the synthesis of colloidal gold. *J. Discuss. Faraday Soc.* **11**, 55.

Tweney RD (2006) Discovering discovery: how Faraday found the first metallic colloid. *Persp. Sci.* **14**, 97.

Wagner RS, Ellis WC (1964) Vapor–liquid–solid mechanism of single crystal growth. *Appl. Phys. Lett.* **4**, 89.

Wang F, Dong A, Sun J, et al. (2006) Solution–liquid–solid growth of semiconductor nanowires. *Inorg. Chem.* **45**, 7511.

Whang D, Jin S, Wu Y, Lieber CM (2003) Large-scale hierarchical organization of nanowire arrays for integrated nanosystems. *Nano Lett.* **3**, 1255.

Wu Y, Yan H, Huang M, et al. (2002) Inorganic semiconductor nanowires: rational growth, assembly, and novel properties. *Chem. Eur. J.* **8**, 1260.

Yang P (2003) Wires on water. *Nature* **425**, 243.

Yin AJ, Li J, Jian W, et al. (2001) Fabrication of highly ordered metallic nanowire arrays by electrodeposition. *Appl. Phys. Lett.* **79**, 1039.

Yin Y, Rioux RM, Erdonmez CK, et al. (2004) Formation of hollow nanocrystals through the nanoscale Kirkendall effect. *Science* **304**, 711.

Yu WW, Qu L, Guo W, Peng X (2003) Experimental determination of the extinction coefficient of CdTe, CdSe, and CdS nanocrystals. *Chem. Mater.* **15**, 2854.

Yu G, Cao A, Lieber CM (2007) Large-area blown bubble films of aligned nanowires and carbon nanotubes. *Nature Nanotech.* **2**, 372.

Zhang H, Chung S-W, Mirkin CA (2003) Fabrication of sub-50-nm solid-state nanostructures on the basis of dip-pen nanolithography. *Nano Lett.* **3**, 43.

Characterization

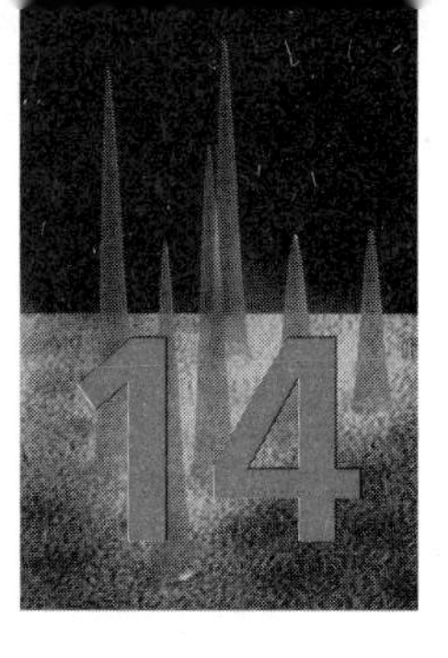

14.1 INTRODUCTION

In this chapter, we describe the various characterization techniques commonly used to study nanoscale materials. We focus on nanowires and quantum dots, since significant work has recently been devoted to these systems. Furthermore, we emphasize chemically synthesized nanostructures, since these materials are common within the nanoscience literature. In addition, they have numerous potential uses that will be described in Chapter 15.

Instead of just listing the various approaches, we will sort them based on use. Namely, there are techniques that focus on the core of these materials. Others primarily study their surfaces. We therefore begin this chapter with a brief overview about the anatomy of chemically synthesized quantum dots and nanowires.

14.2 ANATOMY OF CHEMICALLY SYNTHESIZED NANOSTRUCTURES

Figure 14.1 shows a schematic depicting the overall appearance of solution-synthesized nanocrystals and nanowires. In either case, there are two main parts: the core and the outer, chemically stabilized, surface. Recall that the core can be made out of a variety of materials and that we have focused primarily on metals and semiconductors in this text.

Recall also that the core dictates the optical and electrical properties of nanostructures. This has been described in Chapters 6–12. Namely, the core confines electrons to the physical dimensions of the material, leading to well-known confinement effects in semiconductors. In the case of colloidal quantum dots and metal nanoparticles, the overall core morphology is spherical. For nanowires, cylindrical geometries are common. However, cases exist where cubes, rods, and other nonspherical structures have been made using analogous solution chemistries.

A second important aspect of the nanocrystal/nanowire is its outer organic shell. Specifically, in the solution-phase synthesis of colloidal quantum dots and nanowires, some method of stabilizing resulting structures is needed to prevent them from aggregating or clumping. Numerous strategies have therefore been developed to passivate these materials. For some systems, the stabilizing layer consists of simple organic surfactants that provide steric stabilization (**Figure 14.1**). Nanostructures in close proximity to each other therefore encounter a repulsive potential, stemming from the presence of these surface-bound molecules.

Alternatively, nanostructures can be electrostatically stabilized by relying on the Coulomb repulsion of like charges to prevent particle agglomeration. In fact, chemically synthesized gold nanoparticles are

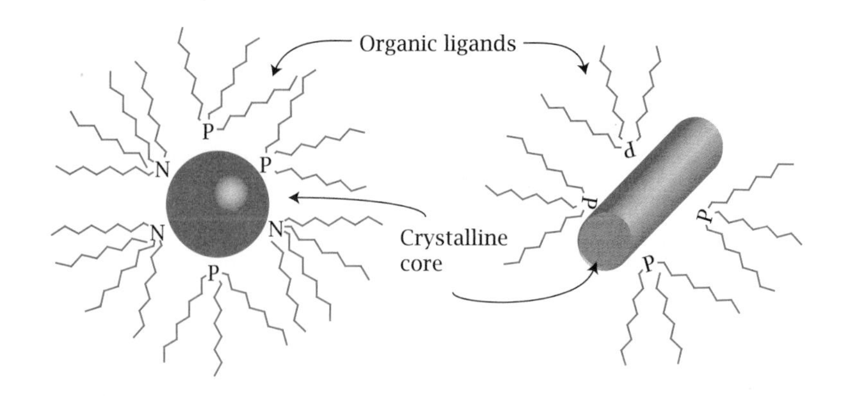

Figure 14.1 Anatomy of chemically synthesized semiconductor/metal nanoparticles and nanowires, sterically stabilized with organic ligands. Not to scale.

often passivated this way. Namely, they possess negatively charged citrate groups on their surface, which repel other negatively charged species.

In general, the structure of the organic passivating ligand has a head group that sticks to the nanocrystal/nanowire surface via dative or covalent bonds. The surfactant also possesses a "tail" that points into the surrounding medium. This tail is important because its hydrophobic or hydrophilic nature dictates whether passivated nanostructures can be suspended in water or in organic solvents. For many nanocrystals/nanowires, the surfactant tail groups are aliphatic. As a consequence, ligand exchange chemistries are needed to disperse them in hydrophilic solvents. This is especially important for their use in biological applications (Chapter 15).

The surfactant molecules are also important because they provide electronic stabilization of nanostructures. Specifically, they satisfy many of the dangling bonds on the nanocrystal/nanowire surface, resulting from their abrupt termination. These dangling bonds often lead to defect-related contributions to the nanostructure emission and also promote the rapid nonradiative relaxation of excitations owing to surface trapping of photogenerated carriers. As a consequence, passivating the surface of low-dimensional materials is a critical aspect of their development.

14.3 SIZING NANOSTRUCTURES

Perhaps the most important task in synthesizing nanoscale materials involves controlling their size. To an equal extent, controlling their shape is paramount. Both are critical parameters, since nanoscale semiconductors and metals exhibit size- as well as shape-dependent optical/electrical properties. We have described such properties in Chapters 3, 7, and 12.

The next most important task after making these materials involves determining their actual size and/or shape. A number of techniques therefore exist for determining the dimensions of nanoscale systems. In what follows, we briefly describe these approaches, since they routinely appear in the literature. Note that we cannot cover all approaches. Instead, we focus on a few important techniques. Where possible, we illustrate some of their salient features through back-of-the-envelope calculations.

14.3.1 Transmission Electron Microscopy (TEM)

A common approach for characterizing nanoscale materials involves imaging them using transmission electron microscopy (TEM). TEM is analogous to conventional light microscopy in that one obtains pictures or micrographs of a specimen of interest. However, instead of photons, electrons are used to construct the images.

In the TEM experiment, a thin or diluted sample is bombarded under high vacuum with a focused electron beam. Electromagnetic lenses steer the beam and focus it onto the specimen. Transmitted electrons then form contrast patterns that create images of the sample. In this regard, thicker regions of the specimen occlude more of the incident beam than thinner regions, causing intensity variations that help define the image. Note, though, that absorption is generally small in these experiments, since samples are inherently thin. As a consequence, most micrographs are actually constructed from scattered electrons that arise from interactions with the material. This scattering can be either elastic or inelastic and, in many cases, it is the elastically scattered electrons that are used to construct the micrographs. The transmitted electron beam image is then magnified onto a detector or phosphorescent screen to yield a picture of the specimen.

To illustrate the use of the TEM in imaging and sizing nanostructures, **Figure 14.2(a–d)** show a series of micrographs of chemically synthesized gold nanoparticles. They have been produced using the Turkevich approach (Chapter 13), yielding samples with sizes of about 14 nm. From the micrographs, the (effective) spherical shape of the individual particles is apparent. However, a closer examination at higher magnification reveals that the crystals exhibit many facets (**Figure 4.12**).

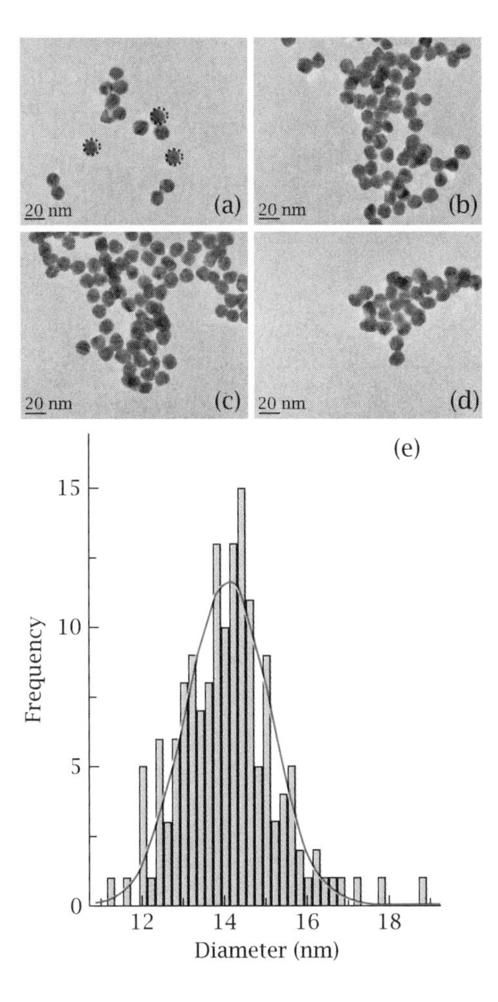

Figure 14.2 (a–d) Transmission electron microscope (TEM) images of chemically synthesized gold nanoparticles. Three particles in (a) have been highlighted. (e) Sizing histogram.

Sizing is then conducted by comparing the dimensions of each nanoparticle with a calibration marker. Occasionally, the process is automated. However, more often than not, it is done by hand.

What ultimately results is a histogram of sizes. An example is shown in **Figure 14.2(e)**. By fitting the histogram with a function (e.g., a Gaussian), the mean particle diameter and accompanying standard deviation can be obtained. For the gold nanoparticles shown in **Figure 14.2**, the mean diameter is 14 nm, with a standard deviation of 1 nm. The estimated ensemble size distribution is then 7%, which is typical of many colloidal nanocrystal preparations. Other nanostructures can be imaged this way, examples are shown in **Figures 1.1** and **4.2–4.12**.

Resolution of the TEM

A critical feature of the TEM is its inherently high degree of spatial resolution. Specifically, the classical diffraction limit of electromagnetic waves is roughly half their wavelength. As a consequence, a beam of electrons resolves much finer things than photons.

To illustrate, assume that the TEM has an accelerating voltage V that ranges from 100 to 300 kV. Electrons extracted from a filament or other source then acquire a kinetic energy

$$E = eV,$$

and since $E = \frac{1}{2}m_0 v^2$,

$$\frac{1}{2}m_0 v^2 = eV.$$

The resulting electron velocity in the microscope is then

$$v = \sqrt{\frac{2eV}{m_0}}. \tag{14.1}$$

Next, given an electron momentum $p = mv$, we find an expression that relates p to the microscope's accelerating voltage:

$$p = \sqrt{2m_0 eV}.$$

At this point, the de Broglie relationship between particles and waves, $\lambda = h/p$ (Equation 3.1), yields

$$\lambda = \frac{h}{\sqrt{2m_0 eV}}. \tag{14.2}$$

The expression can be written differently to make the connection between λ and the TEM accelerating voltage more apparent. Namely,

$$\lambda(\text{nm}) = \frac{1.225(\text{nm})}{\sqrt{V}}, \tag{14.3}$$

where V is in volts. The reader may verify this result as an exercise.

EXAMPLE 14.1

Wavelength of an Electron in a TEM To illustrate Equations 14.1 and 14.3, consider a TEM that accelerates electrons with a 100 keV potential difference. Equation 14.3 then shows that their resulting wavelength $\lambda = 0.0039$ nm. The corresponding velocity $v = 1.88 \times 10^8$ m/s. Similarly, if we have a TEM that uses a 200 keV accelerating voltage, the resulting electron wavelength is $\lambda = 0.0027$ nm. The corresponding velocity is $v = 2.65 \times 10^8$ m/s. Notice that these values are close to the speed of light. As a consequence, we should really be using relativistic expressions. In fact, the relativistic wavelength is

$$\lambda = \frac{h}{\sqrt{2m_0 eV\left(1 + eV/2m_0c^2\right)}}. \tag{14.4}$$

However, to keep things simple, let us continue using the nonrelativistic expressions above.

Next, the diffraction limit is defined as

$$l = \frac{0.612\lambda}{\text{NA}}, \tag{14.5}$$

which is the length below which we cannot distinguish two objects next to each other. The relationship is called the Abbé equation. In it, $\text{NA} = n_m \sin\theta$ is the numerical aperture of the imaging lens, n_m is the refractive index of the medium, and θ is the half-angle of the impinging rays from normal incidence.

The diffraction limit is roughly half the wavelength of light (i.e., in the best case, $n_m = 1$ and $\theta = \frac{1}{2}\pi$). As a consequence, $l \simeq \frac{1}{2}\lambda$, which means that we have a best case resolution of approximately 200 nm for visible light. This implies that light microscopy cannot directly resolve nanostructures.

By contrast, we have just seen that electron wavelengths are much smaller. As a consequence, we can obtain significantly higher-resolution images if we use a TEM. To illustrate, inserting the electron wavelength (Equation 14.3) into Equation 14.5 gives

$$l(\text{nm}) = \frac{(0.612)(1.225\text{nm})}{\sqrt{V} n_m \sin\theta},$$

and since TEM experiments are conducted in vacuum, $n_m = 1$. Furthermore, since θ is generally small, the trigonometric small-angle relationship for $\sin\theta$ can be invoked to obtain $\sin\theta \simeq \theta$. A critical length scale is therefore

$$l(\text{nm}) \simeq \frac{0.750}{\theta\sqrt{V}} \tag{14.6}$$

and is the theoretical TEM resolution when θ is expressed in radians and V is in volts. So, for a 100 keV system, the theoretical diffraction-limited resolution is $l \simeq 0.237$ nm (2.37 Å), assuming $\theta \simeq 0.01$ rad. This value is comparable to the interatomic spacings first discussed in Chapter 2. Likewise, for a 200 keV system, the diffraction limited resolution is $l \simeq 0.168$ nm, or 1.68 Å. In either case, the TEM can be used to image nanostructures. Note that other factors limit the practical resolution of the instrument and are discussed in Williams and Carter (1996).

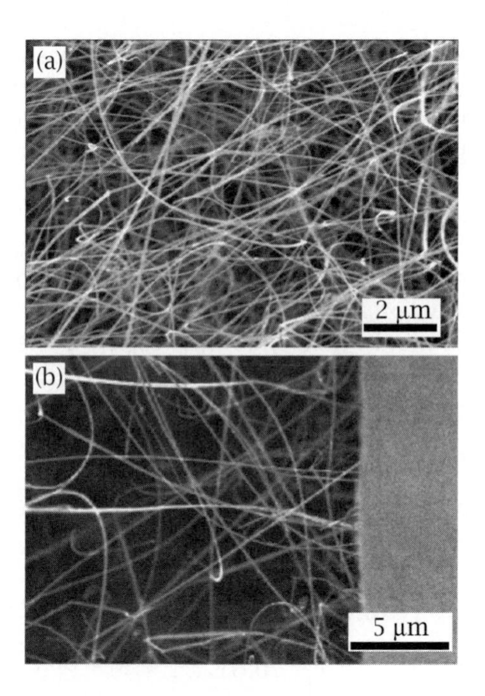

Figure 14.3 Scanning electron microscope (SEM) images of CdSe nanowires grown off a substrate: (a) top-down view; (b) side view.

14.3.2 Scanning Electron Microscopy (SEM)

A related form of electron microscopy is scanning electron microscopy (SEM). In the SEM, an electron beam is focused to a tight spot and is raster-scanned over the sample. When incident electrons interact with it, a number of effects take place. An example is the scattering of electrons off the specimen.

Some of these scattering events are inelastic in nature. As a consequence, energy is lost to the material. This can then lead to the ejection of secondary electrons from elements that are present. These low-energy electrons are highly localized to the region directly under the electron beam and hence can be used to create images of the sample. In tandem, elemental analyses of the specimen can be conducted by analyzing the X-ray energies originating from other inelastic scattering events. This will be seen later on when we discuss a technique called energy-dispersive X-ray spectroscopy.

Figure 14.3 shows example SEM images of CdSe nanowires grown off a substrate. By tilting the sample, different views are accessible. In particular, **Figure 14.3(a)** shows a top-down view of the specimen, whereas **Figure 14.3(b)** shows a side view, clearly revealing the underlying substrate. The appeal of the SEM technique is that it provides a three-dimensional rendition of the material being studied. This is because it possesses a large depth of field. Objects in the background therefore remain in focus and, as a consequence, sample morphologies are straightforward to determine. As with the TEM, the SEM's resolution is much higher than that of conventional light microscopy. Nanostructures are therefore readily sized this way by comparing their images against a calibration marker.

14.3.3 Atomic Force Microscopy (AFM)

Apart from the above electron microscopies, a separate class of scanning probe techniques exist that can be used to determine the size and morphology of nanoscale materials. One such approach is called atomic force microscopy (AFM). The basic idea behind the technique is that a sharp tip can be raster-scanned over a sample of interest to map out its

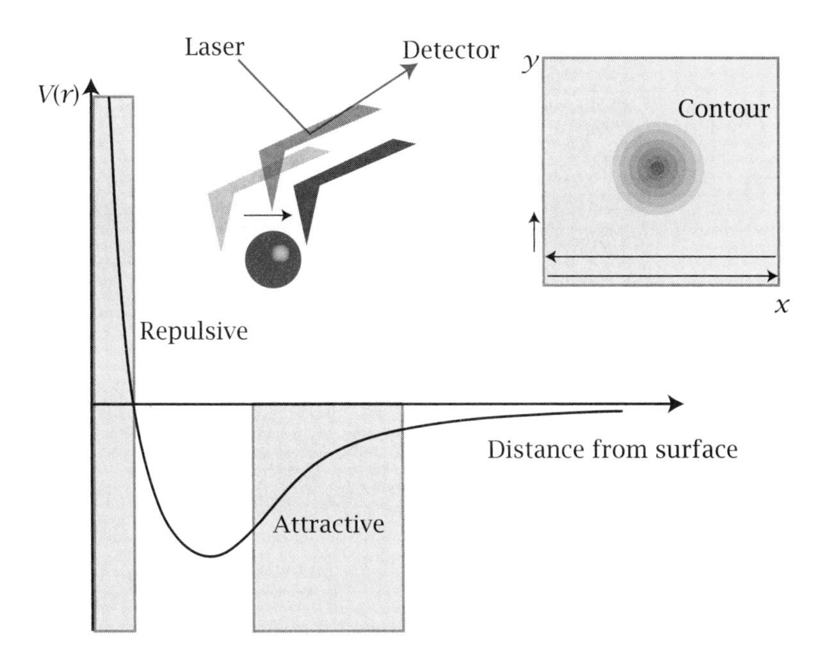

Figure 14.4 Schematic of the atomic force microscopy (AFM) technique as well as of the raster scanning required to construct an image of the sample. For illustration purposes, a sphere is depicted. The bottom shows the potential energy profile, stemming from the van der Waals interaction between the tip and the sample (or substrate).

topography. This is akin to a stylus from an old record player passing over the bumps and grooves in a vinyl record. Changes in the tip height as it passes over features are then recorded and are used to reconstruct images of the sample. This is outlined schematically in **Figure 14.4**, where the contours of a sphere are mapped out by raster-scanning a sharp tip.

The AFM technique, however, is more complicated. This is because, when the tip is brought close to the sample, interactive forces exist, which can be either repulsive or attractive. An important force involved in the operation of the AFM is the van der Waals force. Its associated potential energy curve is shown at the bottom of **Figure 14.4**. Attractive and repulsive parts of the potential therefore exist and depend on how close the tip is to the sample or to the substrate.

As a consequence, the AFM operates in multiple modes. For example, the AFM can function in the attractive part of the van der Waals potential, where the tip remains further away from the sample. This is called the "noncontact" mode of operation. Alternatively, it can work in the repulsive part of the potential where the tip is brought very close to the sample. This is called the "contact" mode of operation. Other modes such as "intermittent" contact (also called the "tapping" mode) exist; the reader may refer to Meyer et al. (2004) for more details.

In nearly all cases, the tip height over the sample is kept constant using a feedback mechanism, based on the deflection of a laser off the back of the cantilever onto a position-sensitive photodiode (**Figure 14.4**). When the cantilever bends, the position of the laser spot changes, allowing the feedback electronics to record tip height variations. In tandem, the tip is moved closer to or further away from the substrate to compensate for the sample's height. Since the AFM is very sensitive to tip height changes, nanostructures can be readily sized this way. We will see an illustration of this shortly.

A representative AFM image of a gold nanoparticle monolayer is shown in **Figure 14.5**. In it, individual particles can be seen as small round features. These are the same particles imaged by the TEM approach in **Figure 14.2**. Although the packing density looks high,

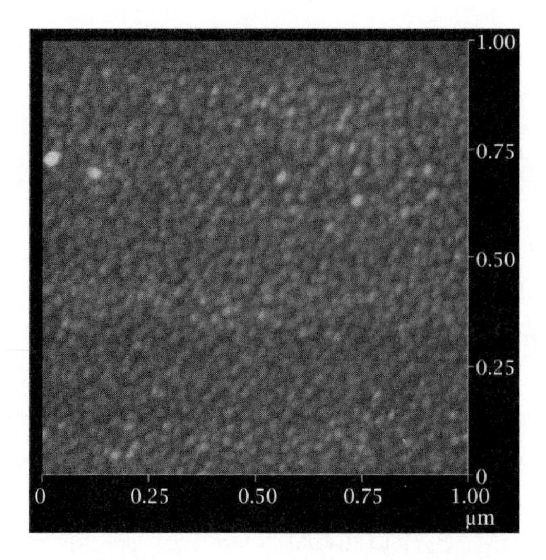

Figure 14.5 AFM image of a monolayer of gold nanoparticles.

in reality it is much smaller owing to a phenomenon called tip convolution (below). A separate AFM image of CdSe nanowires is shown in **Figure 14.6** and clearly reveals their overall wire-like morphology.

The primary disadvantage of the AFM technique is that the tip has a finite radius of curvature (i.e., it is not infinitely sharp). As a consequence, size measurements of nanoscale objects on the lateral scale are somewhat problematic because the recorded nanostructure width includes contributions from the tip's physical dimensions. This is called tip convolution and only by knowing the tip's exact physical shape can its contribution be deconvoluted (i.e., taken into account).

Additional issues include the fact that the AFM technique involves scanning. Data are therefore taken one point at a time over many points. As a consequence, the scan times needed to construct an image are of the order of minutes or tens of minutes, especially if large areas are sampled. Finally, additional feedback loops must be introduced if sample drift is to be minimized. Such drift occurs because of hysteresis in the piezoelectric devices used to position the tip and can also originate from thermal changes in the system. As a consequence, once a scan is repeated over a given area, samples are generally found to be shifted. This is problematic if one seeks to continuously monitor a specific nanostructure.

The AFM has other uses. For example, it has been employed to move individual nanostructures. It has also been used to scratch surfaces, creating nanoscale patterns from which other nanostructures can be

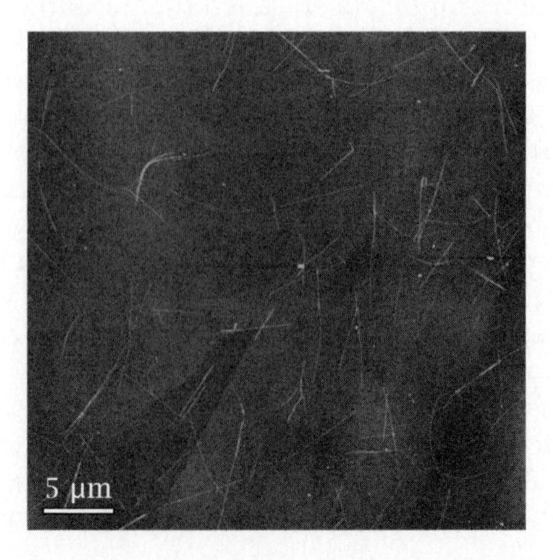

Figure 14.6 AFM image of CdSe nanowires deposited on a substrate.

made. In addition, variations of the technique exist. They operate on slightly different principles and extend the AFM's overall usefulness. For example, magnetic force microscopy (MFM) and electrostatic force microscopy (EFM) can be used to image magnetic domains as well as charges or potentials on surfaces. Additional information about these related techniques can be found in Meyer et al. (2004).

14.3.4 Illustration of the AFM Cantilever Sensitivity

To illustrate the AFM's sensitivity to nanometer-scale tip height changes, let us carry out a simple back-of-the-envelope calculation. Namely, we will evaluate the magnification of an idealized AFM cantilever when it encounters an object and bends upwards in response. In the example, we assume that the cantilever has a length l and that it sits at the origin of our coordinate system. We also assume that the macroscopic distance L between the cantilever and the detector, monitoring the laser, is large. We therefore have the situation outlined in **Figure 14.7**.

The incident laser reflects off the back of the cantilever with the same angle of incidence and reflection θ. Both are relative to the cantilever surface normal. Based on simple trigonometry, the location of the laser spot on the detector occurs at A. Namely,

$$\tan(90^\circ - \theta) = \cot\theta = \frac{A}{L},$$

giving

$$A = L\cot\theta.$$

Next, we assume that the tip encounters an object with a height a. In response, it bends upwards. For simplicity, let us model this as a change in the angle the cantilever makes relative to the original surface

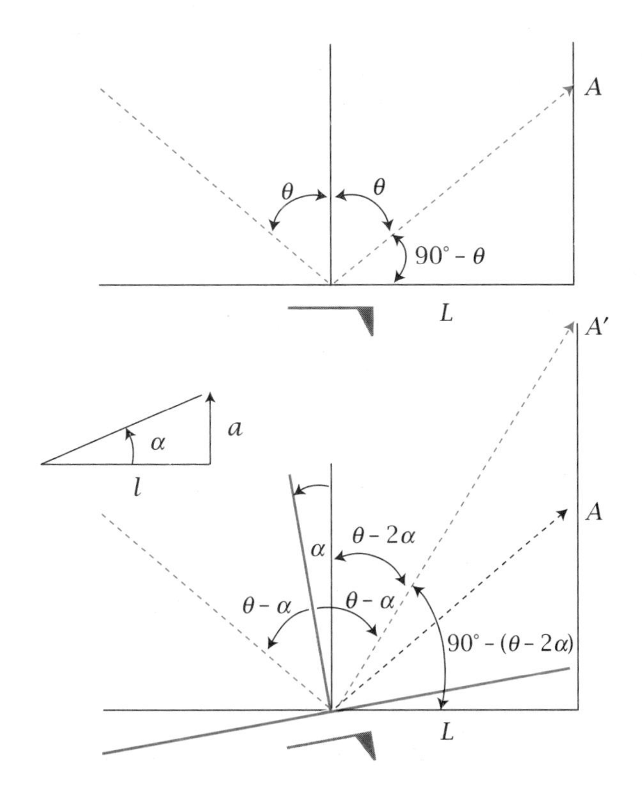

Figure 14.7 Angles used to illustrate AFM magnification and sensitivity.

normal. The angle therefore changes by α, as depicted in the inset in **Figure 14.7**. As a consequence,

$$\alpha = \tan^{-1}\left(\frac{a}{l}\right).$$

The laser's new angle of incidence onto the cantilever is then $\theta - \alpha$. Likewise, its angle of reflection is $\theta - \alpha$. Relative to the original surface normal, however, the angle of reflection is $\theta - 2\alpha$. The reader may verify this as an exercise. Thus, the location of the laser spot on the detector (assuming $L \gg l$ such that the cantilever height displacement can be ignored) is

$$\tan[90^\circ - (\theta - 2\alpha)] = \cot(\theta - 2\alpha) = \frac{A'}{L},$$

or

$$A' = L\cot(\theta - 2\alpha).$$

The overall displacement of the laser on the detector is then

$$\Delta A = L[\cot(\theta - 2\alpha) - \cot\theta].$$

EXAMPLE 14.2

Estimate of the AFM Cantilever Displacement We can now see how large this displacement is using actual numbers. Assume that

$$a = 1\,\text{nm},$$
$$\theta = 60^\circ,$$
$$l = 1\,\mu\text{m},$$
$$L = 1\,\text{cm}.$$

Inserting these values into the above expression then gives $\Delta A = 28\,\mu\text{m}$. Given that the original tip height displacement was 1 nm, this is an enormous magnification and illustrates how small tip height changes can readily be observed.

14.3.5 Scanning Tunneling Microscopy (STM)

A related approach to AFM is scanning tunneling microscopy (STM). STM was the original probe microscopy, developed by Gerd Binnig and Heinrich Rohrer at IBM Zurich in 1981. This ultimately led to their winning a Nobel Prize in 1986.

Like the AFM, the STM uses a sharp tip to image samples. It is raster-scanned over samples and data are taken one point at a time to create an image. However, unlike the AFM, STM exploits a different physical mechanism to sense tip height changes. Namely, instead of attractive and repulsive tip/sample forces that cause cantilever deflections, changes in a tunneling current are measured. In particular, the STM principle of operation is the tunneling of electrons from a conductive tip into a conductive substrate through a conductive sample of interest (notice the emphasis on conductive). Since the tunneling current depends exponentially on the tip-to-sample separation, the technique is exquisitely sensitive to sample height changes. We will illustrate this shortly. Nanostructures and even individual atoms can therefore be imaged.

A STM image of a CdSe nanowire is shown in **Figure 14.8**. This illustrates the technique's ability to image nanoscale structures and to

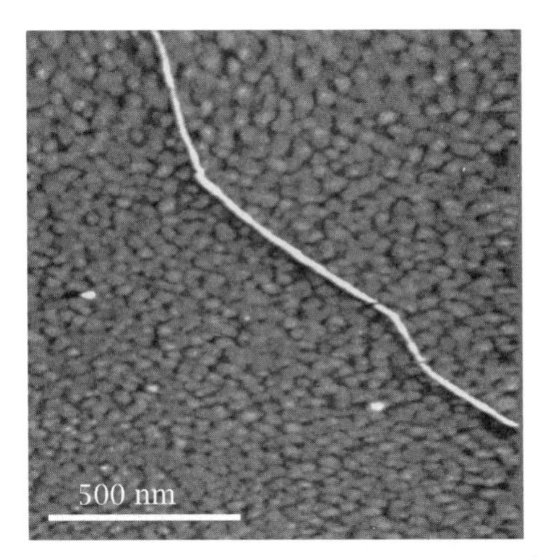

Figure 14.8 Scanning tunneling microscope (STM) image of a single CdSe nanowire deposited onto a conductive substrate.

reveal their size and/or morphology. Although the image is qualitatively similar to the AFM image shown in **Figure 14.6**, there are significant differences in its information content as well as in its interpretation. This is because the electrons that tunnel from the tip to the substrate ultimately pass through the nanowire. As a consequence, the wire's density of states (Chapter 9) plays an important role in determining the magnitude of the observed tunneling current. Thus, additional contributions to the apparent "height" of the sample exist.

Tunneling through a One-Dimensional Rectangular Barrier

We can illustrate the exquisite sensitivity of the STM to the tip-to-sample displacement as follows. This requires solving the quantum mechanical problem of a particle tunneling through a one-dimensional barrier. We saw some of this earlier when we outlined the finite particle-in-a-box problem in Chapter 8. We also touched on it in Chapter 10 when we solved the Kronig–Penney model, which yielded a simple description of bands.

Consider the situation shown in **Figure 14.9**. Assume that the barrier has a potential height V. Furthermore, assume that the electron energy E is smaller than V. In what follows, we sketch the solution to this problem. The reader may carry out an explicit derivation of these results as an exercise.

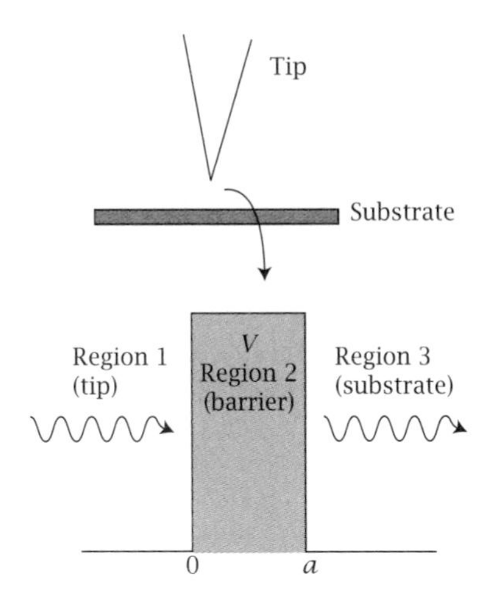

Figure 14.9 Quantum mechanical one-dimensional tunneling barrier in STM.

In the problem, we seek the wavefunctions in Regions 1, 2, and 3. Quantum mechanical matching conditions, first introduced in Chapter 6, apply at the boundaries. Namely,

$$\psi_1(0) = \psi_2(0),$$
$$\psi_1'(0) = \psi_2'(0)$$

and

$$\psi_2(a) = \psi_3(a),$$
$$\psi_2'(a) = \psi_3'(a).$$

To find ψ_1 in Region 1, we solve the Schrödinger equation with $V = 0$:

$$-\frac{\hbar^2}{2m_0}\frac{d^2}{dx^2}\psi_1(x) = E\psi_1(x).$$

This rearranges to

$$\frac{d^2\psi_1(x)}{dx^2} = -\frac{2m_0E}{\hbar^2}\psi_1(x)$$

and, as usual, we let $k^2 = 2m_0E/\hbar^2$ to obtain

$$\frac{d^2\psi_1(x)}{dx^2} + k^2\psi_1(x) = 0.$$

Its solutions are simple linear combinations of plane waves

$$\psi_1(x) = Ae^{ikx} + Be^{-ikx},$$

with A and B constant coefficients to be determined.

Next, in Region 3 we have the same situation. General solutions are

$$\psi_3(x) = Ee^{ikx} + Fe^{-ikx},$$

where E and F are two new constant coefficients to be determined. Notice that we can associate the first term (Ee^{ikx}) with a right-going traveling wave and the second (Fe^{-ikx}) with a left-going traveling wave. To see this, add a time dependence back into the solution by multiplying it with $e^{-i\omega t}$. The second term becomes $e^{-ikx-i\omega t} = e^{-i(kx+\omega t)}$. From our discussion of traveling waves in Chapter 5, we now keep its phase constant and let time increase to see which direction it is traveling in. Thus for $kx + \omega t$, if t increases x must decrease to keep the phase constant. As a consequence, this describes a left-going wave. The reader may likewise verify that the first term represents a right-going wave.

At this point, since there are no sources of particles on the right, nor are there any barriers off which reflections can occur, the left-going component must be zero. That is, any particle from Region 1 that penetrates the barrier and emerges into Region 3 continues moving towards the right (**Figure 14.9**). There is no component that returns towards the barrier. As a consequence, $F = 0$ and the resulting wavefunction in Region 3 is

$$\psi_3(x) = Ee^{ikx}.$$

Finally, in Region 2 (assume, for convenience, that the mass of the particle remains the same in the barrier),

$$-\frac{\hbar^2}{2m_0}\frac{d^2\psi_2(x)}{dx^2} + V\psi_2(x) = E\psi_2(x),$$

$$-\frac{\hbar^2}{2m_0}\frac{d^2\psi_2(x)}{dx^2} = (E - V)\psi_2(x),$$

$$\frac{d^2\psi_2(x)}{dx^2} = \frac{2m_0(V - E)}{\hbar^2}\psi_2(x).$$

We let $\beta^2 = 2m_0(V - E)/\hbar^2$, to obtain

$$\frac{d^2\psi_2(x)}{dx^2} - \beta^2\psi_2(x) = 0,$$

and we have already seen in Chapters 8 and 10 that general solutions to this equation are growing and decaying exponentials. Specifically,

$$\psi_2(x) = Ce^{\beta x} + De^{-\beta x},$$

with C and D constant coefficients to be determined. To summarize, we have the following wavefunctions in Regions 1, 2, and 3:

$$\psi_1(x) = Ae^{ikx} + Be^{-ikx},$$
$$\psi_2(x) = Ce^{\beta x} + De^{-\beta x},$$
$$\psi_3(x) = Ee^{ikx}.$$

We now define a transmission coefficient by dividing the transmitted flux in Region 3 by the incident flux in Region 1:

$$T = \frac{|E|^2}{|A|^2}.$$

We can rationalize this expression since the squares of the coefficients E and A represent the probability contributions of the transmitted and incident wavefunctions in Regions 1 and 3. Note that the actual form of the transmission coefficient is more complex, but reduces to the simple expression shown above under these circumstances.

To find an explicit ratio, we apply the quantum mechanical matching conditions shown earlier. Since $\psi_1(0) = \psi_2(0)$ and $\psi_1'(0) = \psi_2'(0)$,

$$A = \left(1 + \frac{\beta}{ik}\right)\frac{C}{2} + \left(1 - \frac{\beta}{ik}\right)\frac{D}{2}.$$

Next, C and D can be found in terms of E using the second set of matching conditions, $\psi_2(a) = \psi_3(a)$ and $\psi_2'(a) = \psi_3'(a)$. This gives

$$C = \left(1 + \frac{ik}{\beta}\right)\frac{E}{2}e^{ika}e^{-\beta a},$$
$$D = \left(1 - \frac{ik}{\beta}\right)\frac{E}{2}e^{ika}e^{\beta a}.$$

When inserted into A and consolidated (this is a bit of work, which the reader may do as an exercise), we find

$$A = \frac{E}{4\beta ik}e^{ika}\left[2(k^2 - \beta^2)\sinh\beta a + 4\beta ik\cosh\beta a\right].$$

Then,

$$|A|^2 = |E|^2\left[\frac{(k^2 - \beta^2)^2}{4\beta^2 k^2}\sinh^2\beta a + \cosh^2\beta a\right],$$

and thus

$$T = \frac{|E|^2}{|A|^2} = \frac{1}{\dfrac{(k^2 - \beta^2)^2}{4\beta^2 k^2}\sinh^2\beta a + \cosh^2\beta a}.$$

Since $\cosh^2\beta a = 1 + \sinh^2\beta a$, we alternatively write

$$T = \frac{1}{1 + \left[\dfrac{(k^2 - \beta^2)^2}{4\beta^2 k^2} + 1\right]\sinh^2\beta a}$$

whereupon expanding the term in square brackets gives

$$T = \frac{1}{1 + \dfrac{(\beta^2 + k^2)^2}{4\beta^2 k^2}\sinh^2\beta a}.$$

At this point we recall that $\beta = \sqrt{2m_0(V-E)/\hbar^2}$ and $k = \sqrt{2m_0E/\hbar^2}$ to obtain the tunneling probability commonly seen in many texts:

$$T = \frac{1}{1 + \dfrac{V^2}{4E(V-E)} \sinh^2 \beta a}. \tag{14.7}$$

If βa is large (i.e., the potential barrier is very large), we find an even simpler expression. Recall that when x is large, $\sinh x = \frac{1}{2}(e^x - e^{-x}) \simeq \frac{1}{2}e^x$. Thus, $\sinh^2 x \simeq \frac{1}{4}e^{2x}$ and

$$\begin{aligned} T &\simeq \frac{1}{1 + \dfrac{V^2}{4E(V-E)} \dfrac{e^{2\beta a}}{4}} \\ &\simeq \frac{16E(V-E)}{V^2} e^{-2\beta a} \\ &\simeq N e^{-2\beta a} \end{aligned}$$

with

$$N = \frac{16E(V-E)}{V^2}, \qquad \beta a = \sqrt{\frac{2m_0(V-E)}{\hbar^2}}\, a.$$

As a consequence, an expression for the transmission probability in this limit is

$$T \simeq N \exp\left[-2\sqrt{\frac{2m_0(V-E)}{\hbar^2}}\, a\right]. \tag{14.8}$$

One sees that it decays exponentially with the barrier width a, which represents our tip-to-sample separation. The above expression could also have been derived using the so-called WKB approximation in quantum mechanics. However, we did not discuss this in our brief overview of perturbation theory in Chapter 6. The interested reader may consult one of the quantum mechanics texts cited in the Further Reading in Chapter 6 for more details.

EXAMPLE 14.3

Illustration of the STM Current Tip-to-Substrate Separation Sensitivity To illustrate the tremendous sensitivity of the tunneling current to the tip-to-substrate separation, assume a barrier height of 3 eV along with a corresponding electron energy of 1 eV. If we further assume that T is proportional to the tunneling current in an actual experiment, we can see how dramatic a 1 nm change in the tip-to-substrate separation actually is. From Equation 14.8, we have

$$T_{old} = N \exp\left[-2\sqrt{\frac{2m_0(V-E)}{\hbar^2}}\, a_{old}\right],$$

$$T_{new} = N \exp\left[-2\sqrt{\frac{2m_0(V-E)}{\hbar^2}}\, a_{new}\right],$$

where a_{old} (a_{new}) is the old (new) displacement. Thus,

$$\frac{T_{new}}{T_{old}} = \exp\left[-2\sqrt{\frac{2m_0(V-E)}{\hbar^2}}\, \Delta a\right].$$

If the tip moves away from the substrate, $\Delta a = a_{new} - a_{old} = +1$ nm. We then find that

$$\frac{T_{new}}{T_{old}} = \exp\left[-2\sqrt{\frac{2(9.11\times10^{-31})(3-1)(1.602\times10^{-19})}{(1.054\times10^{-34})^2}} \times (1\times10^{-9})\right] = e^{-14.5},$$

or that

$$T_{new} = T_{old}(5\times10^{-7}).$$

A dramatic seven orders of magnitude drop of the tunneling current is apparent. This illustrates the great sensitivity of the STM experiment to the tip-to-substrate separation. Nanostructures can therefore be sized this way.

The STM has many uses in addition to sizing nanostructures. It has been employed to directly probe the density of states in colloidal quantum dots and in other systems. It has also been used to investigate the charging energies of small metal islands. This extension is called scanning tunneling spectroscopy (STS). Although not standard, STM has even been used to make nanostructures. Perhaps the most famous example of this is the IBM study leading to the creation of quantum corrals.

In all cases, a disadvantage of the STM is that it is a scanning technique and, just like AFM, can be slow. More importantly, it requires relatively conductive samples dispersed on a conductive substrate. This, in turn, imposes limitations on the materials that can be studied. As a consequence, the AFM technique described earlier was developed to circumvent this issue. More about the STM technique can be found in Meyer et al. (2004).

14.3.6 X-ray Powder Diffraction (XRD)

To conclude this section on nanostructure sizing, we discuss a third set of techniques that use scattered electromagnetic radiation to estimate the dimensions of nanostructures. However, unlike the electron microscopies seen earlier, the nanoparticles/nanowires are not directly imaged in these approaches. Instead, they are indirectly sized through their measured scattering response.

As one example, a common sizing technique for nanoscale materials involves powder X-ray diffraction (XRD) measurements. Recall that XRD is often used to determine the crystallographic phase and structure of a material. The underlying idea is that atomic planes in a crystal diffract incoming X-rays. Often, the intensity of these reflections and their exact angular positions are specific to a material. XRD patterns can therefore be used to "fingerprint" systems using tables of material-specific X-ray lines. The Joint Committee of Powder Diffraction Standards (JCPDS) is one such compilation commonly used in the literature.

XRD measurements entail depositing powders of nanocrystalline samples onto suitable substrates. These nanocrystal/nanowire powders are readily obtained by precipitation from solution (Chapter 13) and drying the recovered material under vacuum or under inert conditions. The sample is then exposed to monochromatic X-rays, whereupon the diffracted X-ray intensity is measured as a function of the incident angle. Because of the random orientation of different nanostructures in a specimen, all crystallographic planes are equally sampled. An exception to this occurs in the case of nanowires, leading to what is called crystallographic texture. Namely, the assumption that all crystallographic planes are equally sampled by incident X-rays in a randomly oriented powder is no longer necessarily true for nanowires. This is because when nanowires are deposited onto substrates, they tend to rest with their growth axes parallel to the substrate. As a consequence, certain crystallographic planes are under-sampled, since only crystal planes, parallel to the substrate, diffract strongly. X-ray reflections from nanowire samples may therefore exhibit significant distortions from their predicted (randomly oriented) powder patterns.

Origin of the X-Ray Reflection

The existence of characteristic X-ray reflections can be explained by a classical analysis of the interaction between X-rays and a crystalline sample. Consider the scenario shown in **Figure 14.10**. In the diagram, X-rays impinge upon the sample from the left. For illustration purposes, a pair of crystalline planes from the same family is shown.

To achieve constructive interference, the two outgoing rays must be in phase. This occurs when their pathlength difference Δl varies by an integer number of wavelengths. From a geometric standpoint, the total pathlength difference must therefore equal $2d\sin\theta$. As a consequence, constructive interference only occurs when

$$n\lambda = 2d\sin\theta. \tag{14.9}$$

In the expression, n is the "order" of the reflection and satisfies $n = 1, 2, 3\ldots.$ Equation 14.9 is called Bragg's law.

In a cubic lattice where a is the lattice constant, the spacing between planes is

$$d = \frac{a}{\sqrt{h^2 + k^2 + l^2}}, \tag{14.10}$$

where h, k, and l are Miller indices of a specific crystal plane. Other expressions for the interplanar spacing in different Bravais lattices (Chapter 2) exist and are listed in **Table 14.1**.

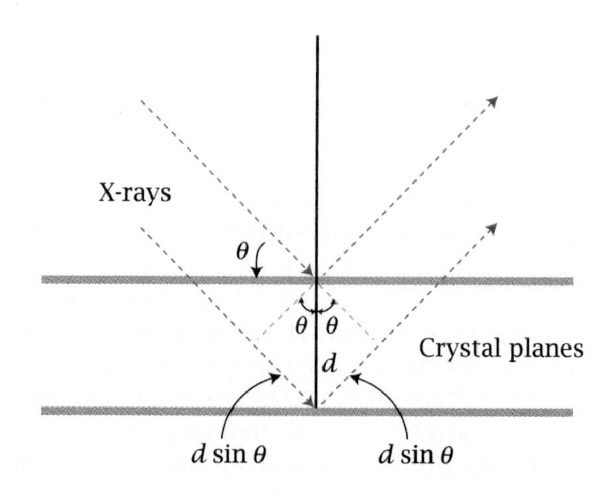

Figure 14.10 Relevant angles in a classic analysis of X-ray diffraction from a crystal.

Table 14.1 Interplanar spacing in different Bravais lattices

System	Angles	d_{hkl}
Cubic	$\alpha=\beta=\gamma=90^\circ$	$\dfrac{a}{\sqrt{h^2+k^2+l^2}}$
Tetragonal	$\alpha=\beta=\gamma=90^\circ$	$\dfrac{a}{\sqrt{h^2+k^2+\left(\frac{l}{c}\right)^2}}$
Orthorhombic	$\alpha=\beta=\gamma=90^\circ$	$\dfrac{b}{\sqrt{\left(\frac{h}{a}\right)^2+k^2+\left(\frac{l}{c}\right)^2}}$
Hexagonal	$\alpha=\beta=90^\circ, \gamma=120^\circ$	$\dfrac{a}{\sqrt{\frac{4}{3}(h^2+hk+k^2)+\left(\frac{l}{c}\right)^2}}$
Rhombohedral	$\alpha=\beta=\gamma\neq 90^\circ$	$\dfrac{a\sqrt{1+2\cos^3\alpha-3\cos^2\alpha}}{\sqrt{(h^2+k^2+l^2)\sin^2\alpha+2(hk+hl+kl)(\cos^2\alpha-\cos\alpha)}}$
Monoclinic	$\alpha=\gamma=90^\circ,\ \beta\neq 90^\circ$	$\dfrac{b}{\sqrt{\dfrac{\left(\frac{h}{a}\right)^2+\left(\frac{l}{c}\right)^2-\frac{2hl}{ac}\cos\beta}{\sin^2\beta}+k^2}}$
Triclinic	$\alpha\neq\beta\neq\gamma\neq 90^\circ$	$\dfrac{b}{\sqrt{\dfrac{\frac{h}{a}\begin{vmatrix}\frac{h}{a} & \cos\gamma & \cos\beta\\ k & 1 & \cos\alpha\\ \frac{l}{c} & \cos\alpha & 1\end{vmatrix}+k\begin{vmatrix}1 & \frac{h}{a} & \cos\beta\\ \cos\gamma & k & \cos\alpha\\ \cos\beta & \frac{l}{c} & 1\end{vmatrix}+\frac{l}{c}\begin{vmatrix}1 & \cos\gamma & \frac{h}{a}\\ \cos\gamma & 1 & k\\ \cos\beta & \cos\alpha & \frac{l}{c}\end{vmatrix}}{\begin{vmatrix}1 & \cos\gamma & \cos\beta\\ \cos\gamma & 1 & \cos\alpha\\ \cos\beta & \cos\alpha & 1\end{vmatrix}}}}$

For simplicity, let us focus on cubic systems such as the simple cubic, body-centered cubic, and face-centered cubic lattices. For first-order reflections, $n=1$ and we have from Bragg's law that

$$\lambda = 2d\sin\theta,$$

or

$$\sin\theta = \frac{\lambda}{2a}\sqrt{h^2+k^2+l^2}.$$

As a consequence, the following X-ray reflection angles are predicted:

$$\theta = \sin^{-1}\left[\frac{\lambda}{2a}\sqrt{h^2+k^2+l^2}\right]. \tag{14.11}$$

Notice that these angular positions are, on their own, not sufficient to reproduce an actual powder XRD pattern. What are also needed are the diffracted intensities. In what follows, we therefore show that some of these reflections exhibit no intensity at all owing to the destructive interference of scattered X-rays. Furthermore, in mixed-element crystals, variable reflection intensities occur due to the constructive and destructive interference of X-rays scattered off different elements.

The Structure Factor

To describe these X-ray intensities, we introduce what is called the structure factor $F(\theta)$. It is a measure of the scattering amplitude from a crystal's unit cell. The structure factor is defined as

$$F(\theta) = \sum_{i}^{n} f_i(\theta)e^{2\pi i(hx_i+ky_i+lz_i)}, \tag{14.12}$$

where $f(\theta)$ is an atomic scattering factor that varies from element to element, n is the number of atoms in the unit cell, and x_i, y_i, z_i represent their coordinates. Notice that $F(\theta)$ is the sum of all atomic scattering contributions in a unit cell, weighted by a phase factor to account for the different positions of individual atoms. Each contribution therefore possesses a slight phase shift relative to the others, leading to constructive and destructive interferences that affect the overall scattering intensity.

Next, $f(\theta)$ represents the amplitude of the scattered wave off a given atom. As a consequence, not only does it have an angular dependence, it also depends on the element's atomic number Z. In tandem, $f(\theta)$ is sensitive to the incident X-ray wavelength. However, since this will generally be fixed in an experiment, we will not worry about it.

Atomic scattering factors can be modeled using the following expression:

$$f(s) = Z - 41.78214s^2 \sum_{i=1}^{3,4} a_i e^{-b_i s^2}, \tag{14.13}$$

where Z is the element's atomic number and $s = \sin\theta/\lambda(\text{Å})$. The sum extends over 3 or 4 terms, depending on element. The coefficients a_i and b_i are obtained from tables of atomic scattering factor parameters; a large table can be found in De Graef and McHenry (2007).

To illustrate Equation 14.13, we find for the elements sulfur, selenium, and tellurium that

$$f_{\mathrm{S}}(\theta) = 16 - 41.78214s^2(1.659e^{-36.650s^2} + 2.386e^{-11.488s^2} + 0.790e^{-2.469s^2} + 0.321e^{-0.340s^2}),$$

$$f_{\mathrm{Se}}(\theta) = 34 - 41.78214s^2(2.298e^{-38.830s^2} + 2.854e^{-11.536s^2} + 1.456e^{-2.146s^2} + 0.590e^{-0.316s^2}),$$

$$f_{\mathrm{Te}}(\theta) = 52 - 41.78214s^2(4.785e^{-27.999s^2} + 3.688e^{-5.083s^2} + 1.500e^{-0.581s^2}).$$

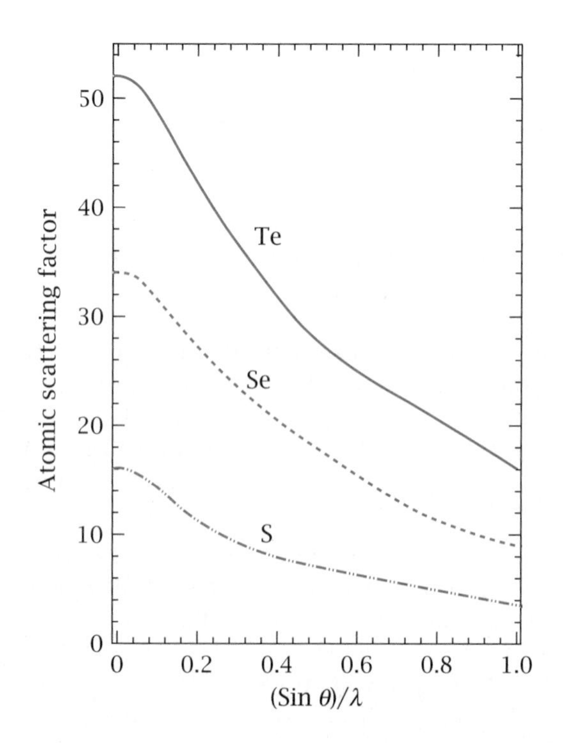

Figure 14.11 Atomic scattering factors for sulfur, selenium and tellurium.

These functions are plotted in **Figure 14.11**. In all cases, $f(\theta)$ decreases with increasing angle of incidence. Furthermore, the larger an element's atomic number, the larger $f(\theta)$ is at any given angle. Heavier elements, down the periodic table, therefore scatter better.

At this point, we can use our knowledge of $F(\theta)$ (Equation 14.12), $f(s)$ (Equation 14.13), and θ (Equation 14.11) to effectively construct a theoretical powder pattern for a given system.

Powder XRD Examples

Let us now show how one can effectively simulate XRD patterns of some representative systems. Note, though, that there exist other important correction factors that must be accounted for to model *actual* experimental powder patterns. The following calculations therefore give us a generic sense for how the reflection intensities of different lattices vary. For simplicity, let us again consider only cubic, single, and multielement crystals.

EXAMPLE 14.4

XRD Reflection Intensities for the Simple Cubic (SC) Lattice First, let us consider the SC lattice, where there exists only one atom per unit cell (Chapter 2). The location of the atom is at (0, 0, 0) within the unit cell's coordinate system. Using Equation 14.12, the structure factor is

$$F(\theta) = \sum_i^1 f(\theta) e^{2\pi i[h(0)+k(0)+l(0)]} = f(\theta)$$

and shows that $F(\theta)$ is identical to the atomic scattering factor $f(\theta)$. Since the scattering intensity is proportional to $|F(\theta)|^2$, we predict the existence of reflections from all possible planes, as summarized in **Table 14.2**. Their intensities, however, decrease towards larger angles given that $f(\theta)$ decreases with increasing θ. We have illustrated this in **Figure 14.11**.

EXAMPLE 14.5

XRD Reflection Intensities for the Body-Centered Cubic (BCC) Lattice Next, let us consider the single-element BCC lattice. From Chapter 2, we know that it has two atoms per unit cell. These atoms occur at (0, 0, 0) and $(\frac{1}{2}, \frac{1}{2}, \frac{1}{2})$, in units of the lattice constant. From Equation 14.12, the structure factor is then

$$\begin{aligned} F(\theta) &= \sum_i^2 f_i(\theta) e^{2\pi i(hx_i+ky_i+lz_i)} \\ &= f(\theta)\left(e^{2\pi i[h(0)+k(0)+l(0)]} + e^{2\pi i[h(1/2)+k(1/2)+l(1/2)]}\right) \\ &= f(\theta)\left(1 + e^{i\pi(h+k+l)}\right). \end{aligned}$$

We can leave $F(\theta)$ expressed this way. However, we can also rearrange it to

$$F(\theta) = f(\theta) e^{i(\pi/2)(h+k+l)}\left(e^{-i(\pi/2)(h+k+l)} + e^{i(\pi/2)(h+k+l)}\right),$$

whereupon the Euler relation can be invoked. As a consequence,

$$F(\theta) = 2f(\theta) e^{i(\pi/2)(h+k+l)} \cos\left[\frac{\pi}{2}(h+k+l)\right].$$

Table 14.2 X-ray diffraction reflections in single-element cubic systems (* denotes a nonzero intensity)

hkl	$h^2+k^2+l^2$	SC	BCC	FCC
{100}	1	*	–	–
{110}	2	*	*	–
{111}	3	*	–	*
{200}	4	*	*	*
{210}	5	*	–	–
{211}	6	*	*	–
–	–	–	–	–
{220}	8	*	*	*
{300} {221}	9	*	–	–
{310}	10	*	*	–
{311}	11	*	–	*
{222}	12	*	*	*
{320}	13	*	–	–
{321}	14	*	*	–
–	–	–	–	–
{400}	16	*	*	*
{410} {322}	17	*	–	–
{411} {330}	18	*	*	–
{331}	19	*	–	*
{420}	20	*	*	*

Then since the intensity $I(\theta)$ of the scattered X-rays is proportional to $|F(\theta)|^2$,

$$I(\theta) \propto 4|f(\theta)|^2 \cos^2\left[\frac{\pi}{2}(h+k+l)\right].$$

Notice that when the sum $h+k+l$ is even, the scattering intensity is nonzero. However, if $h+k+l$ is odd, $\cos^2[(\pi/2)(h+k+l)]$ is zero and hence $I(\theta)=0$. This arises from the destructive interference between scattered waves. In this way, a list of allowed and unallowed reflections for the BCC lattice can be compiled (**Table 14.2**).

EXAMPLE 14.6

XRD Reflection Intensities for the Face-Centered Cubic (FCC) Lattice We continue with a calculation of the single-element FCC structure factor and associated reflections. The FCC unit cell has four atoms per unit cell (Chapter 2). Constituent atoms are located at $(0,0,0)$, $\left(\frac{1}{2},\frac{1}{2},0\right)$, $\left(\frac{1}{2},0,\frac{1}{2}\right)$, and $\left(0,\frac{1}{2},\frac{1}{2}\right)$, in units of the lattice constant. As a consequence, we find from Equation 14.12 that

$$F(\theta) = \sum_i^4 f_i(\theta) e^{2\pi i(hx_i+ky_i+lz_i)}$$
$$= f(\theta)\left[1 + e^{i\pi(h+k)} + e^{i\pi(h+l)} + e^{i\pi(k+l)}\right].$$

The term in square brackets can alternatively be written as

$$\left(1+e^{i\pi(h+k)}\right)\left(1+e^{i\pi(h+l)}\right),$$

since $e^{2\pi ih} = 1$. The reader may verify this as an exercise. Thus,

$$F(\theta) = f(\theta)\left(1 + e^{i\pi(h+k)}\right)\left(1 + e^{i\pi(h+l)}\right)$$
$$= f(\theta)e^{i(\pi/2)(h+k)}\left[e^{-i(\pi/2)(h+k)} + e^{i(\pi/2)(h+k)}\right]$$
$$\times e^{i(\pi/2)(h+l)}\left[e^{-i(\pi/2)(h+l)} + e^{i(\pi/2)(h+l)}\right]$$
$$= f(\theta)2e^{i(\pi/2)(h+k)}\cos\left[\frac{\pi}{2}(h+k)\right]2e^{i(\pi/2)(h+l)}\cos\left[\frac{\pi}{2}(h+l)\right],$$

yielding

$$F(\theta) = 4f(\theta)e^{i(\pi/2)(2h+k+l)}\cos\left[\frac{\pi}{2}(h+k)\right]\cos\left[\frac{\pi}{2}(h+l)\right].$$

The scattered X-ray intensity is then

$$I(\theta) \propto 16|f(\theta)|^2\cos^2\left[\frac{\pi}{2}(h+k)\right]\cos^2\left[\frac{\pi}{2}(h+l)\right].$$

We see that only when h, k, l are all odd integers (or all even integers) is $I(\theta) \neq 0$. If any of the indices differ in parity, then $I(\theta) = 0$. The allowed and forbidden reflections of the FCC crystal structure can therefore be compiled (**Table 14.2**).

Mixed-Element Crystals

Finally, let us illustrate what happens when we consider mixed-element crystals. This is important, because many semiconductors are binary compounds. For simplicity, we carry out the above analysis for the CsCl and NaCl crystal structures first described in Chapter 2. The reader may conduct the analysis of the zinc blende structure as an exercise.

EXAMPLE 14.7

XRD Reflection Intensities for the CsCl Lattice Recall that the CsCl structure has two atoms per unit cell. The first element occurs at (0, 0, 0) while the second can be found at $\left(\frac{1}{2}, \frac{1}{2}, \frac{1}{2}\right)$ (Chapter 2). Using Equation 14.12, we then have the following structure factor:

$$F(\theta) = \sum_i^2 f_i(\theta)e^{2\pi i(hx_i+ky_i+lz_i)}$$
$$= f_{Cs}e^{2\pi i[h(0)+k(0)+l(0)]} + f_{Cl}e^{2\pi i(h/2+k/2+l/2)}$$
$$= f_{Cs} + f_{Cl}\cos\pi(h+k+l),$$

where f_{Cs} and f_{Cl} are the atomic scattering factors of Cs and Cl, respectively.

We see that if $h+k+l = 2n$ (an even number, $n = 1, 2, 3, \ldots$), then

$$F(\theta) = f_{Cs} + f_{Cl}.$$

Alternatively, if $h+k+l = 2n+1$ (an odd number), then

$$F(\theta) = f_{Cs} - f_{Cl}.$$

General X-ray diffraction intensities are therefore

$$I(\theta) \propto |f_{Cs} + f_{Cl}\cos[\pi(h+k+l)]|^2.$$

Limiting cases are

$$I(\theta) \propto |f_{Cs} + f_{Cl}|^2$$

if $h+k+l=2n$ and

$$I(\theta) \propto |f_{Cs} - f_{Cl}|^2$$

if $h+k+l=2n+1$.

EXAMPLE 14.8

XRD Reflection Intensities for the NaCl Lattice Finally, let us consider the NaCl crystal structure. It has eight atoms per unit cell. To simplify things, recall that the NaCl structure can be viewed as two interpenetrating single-element FCC lattices offset by $\left(\frac{1}{2}, \frac{1}{2}, \frac{1}{2}\right)$ (Chapter 2). Using our previous work on FCC lattices, we thus find the following structure factor:

$$\begin{aligned}
F(\theta) &= \sum_i^8 f_i(\theta) e^{2\pi i[hx_i+ky_i+lz_i]} \\
&= f_{Na}(e^{2\pi i(0)} + e^{2\pi i(h/2+k/2)} + e^{2\pi i(h/2+l/2)} + e^{2\pi i(k/2+l/2)}) \\
&\quad + f_{Cl}(e^{2\pi i(h/2+k/2+l/2)} + e^{2\pi i(h+k+l/2)} + e^{2\pi i(h+k/2+l)} \\
&\quad + e^{2\pi i(h/2+k+l)}) \\
&= f_{Na}(1 + e^{i\pi(h+k)} + e^{i\pi(h+l)} + e^{i\pi(k+l)}) \\
&\quad + f_{Cl}(e^{i\pi(h+k+l)} + e^{2\pi i(2h/2+2k/2+l/2)} + e^{2\pi i(2h/2+k/2+2l/2)} \\
&\quad + e^{2\pi i(h/2+2k/2+2l/2)}) \\
&= f_{Na}(1 + e^{i\pi(h+k)} + e^{i\pi(h+l)} + e^{i\pi(k+l)}) \\
&\quad + f_{Cl}e^{i\pi(h+k+l)}(1 + e^{i\pi(h+k)} + e^{i\pi(h+l)} + e^{i\pi(k+l)}) \\
&= (f_{Na} + f_{Cl}e^{i\pi(h+k+l)})[1 + e^{i\pi(h+k)} + e^{i\pi(h+l)} + e^{i\pi(k+l)}],
\end{aligned}$$

where f_{Na} and f_{Cl} are the atomic scattering factors of Na and Cl, respectively. Furthermore, we have already seen that the term in square brackets factorizes, giving

$$F(\theta) = (f_{Na} + f_{Cl}e^{i\pi(h+k+l)})(1 + e^{i\pi(h+k)})(1 + e^{i\pi(h+l)}),$$

which yields

$$\begin{aligned}
F(\theta) &= 4(f_{Na} + f_{Cl}e^{i\pi(h+k+l)})e^{i(\pi/2)(2h+k+l)} \\
&\quad \times \cos\left[\frac{\pi}{2}(h+k)\right]\cos\left[\frac{\pi}{2}(h+l)\right].
\end{aligned}$$

The scattered X-ray intensity is then

$$I(\theta) \propto 16\left|f_{Na} + f_{Cl}e^{i\pi(h+k+l)}\right|^2 \cos^2\left[\frac{\pi}{2}(h+k)\right]\cos^2\left[\frac{\pi}{2}(h+l)\right],$$

or

$$\begin{aligned}
I(\theta) &\propto 16\left\{f_{Na}^2 + f_{Cl}^2 + 2f_{Na}f_{Cl}\cos[\pi(h+k+l)]\right\}\cos^2\left[\frac{\pi}{2}(h+k)\right] \\
&\quad \times \cos^2\left[\frac{\pi}{2}(h+l)\right].
\end{aligned}$$

From this, we see that, if any index h, k, l possesses a different parity from the others, $I(\theta) = 0$. If all of the indices are even (i.e., $h + k + l = 2n$), then

$$I(\theta) \propto 16(f_{\text{Na}} + F_{\text{Cl}})^2.$$

Alternatively, if all of the indices h, k, l are odd (i.e., $h + k + l = 2n + 1$), then

$$I(\theta) \propto 16(f_{\text{Na}} - f_{\text{Cl}})^2.$$

The reader may prove these results as an exercise.

Summary

One sees from Equation 14.11 that the angles of the X-ray reflections depend on the lattice constant of the material. Furthermore, the pattern and intensities of the reflections are characteristic of the elements and of the crystal structure underlying the material. XRD can therefore be used to fingerprint materials. More about the technique can be found in De Graef and McHenry (2007), Klug and Alexander (1974), and Guinier (1963).

Sizing, the Scherrer Equation

At this point, having established the origin of both X-ray reflection angles and intensities, how can one obtain nanostructure sizing using XRD powder patterns? It turns out that one can show from Bragg's law that if the size of a crystal is finite, then broadening of the observed X-ray reflections occurs. In turn, by quantifying this broadening, we can determine the mean size of the crystals responsible for the scattering. This relationship between size and linewidth is referred to as the Scherrer equation:

$$D = \frac{K\lambda}{\beta \cos\theta}. \tag{14.14}$$

In this expression, D is a characteristic size of the crystal (in Å), K is a constant close to 1 ($K = 0.89$), λ is the wavelength of the incident X-rays (in Å), β is the breadth of the reflection (in rad), and θ is the Bragg angle.

In principle, β is the difference between the nanostructure's breadth and that of its bulk counterpart (i.e., $\beta = B - b$); b (B) is the bulk (nanomaterial) breadth. However, $\beta \simeq B$, and, as a consequence, the experimental linewidth is often used without reference to the bulk. In addition, β is evaluated from the experimental width in 2θ units (i.e., $\Delta(2\theta)$), after which it is converted to radians. Note that θ is half the reflection's experimental 2θ peak position.

EXAMPLE 14.9

Illustration of Nanoparticle Sizing using the Scherrer Equation To illustrate the Scherrer equation, consider the gold nanoparticle XRD powder pattern shown in **Figure 14.12(a)**. The TEM image in **Figure 14.12(b)** reveals that the particles have a mean diameter of 1.5 nm. From the experimental XRD powder pattern, we are therefore interested in sizing the particles to see if the Scherrer result corresponds to the TEM result.

The peak position of the reflection that we will use occurs at approximately 40°. Although the Scherrer equation is valid

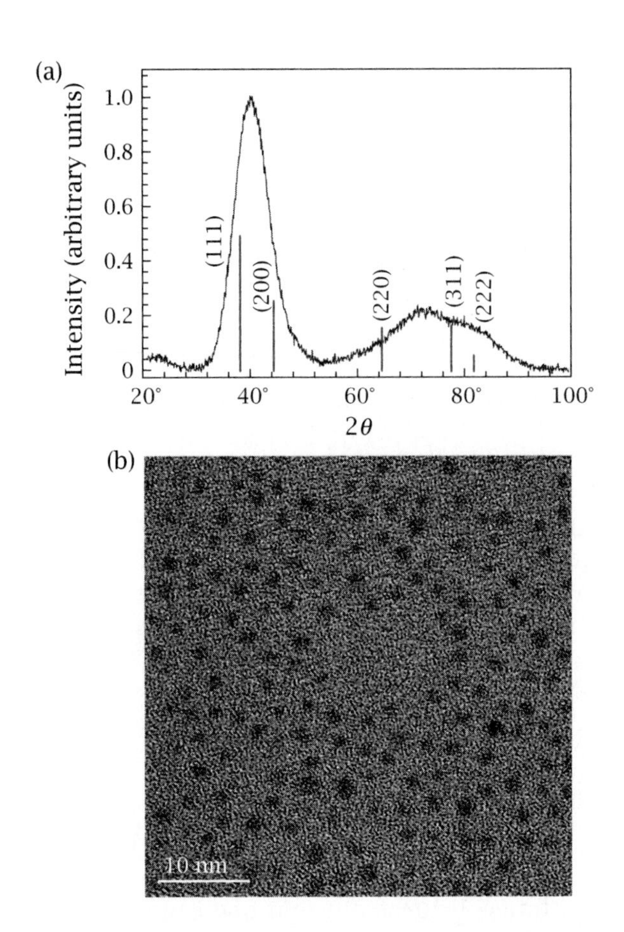

Figure 14.12 (a) Powder XRD pattern from an ensemble of small gold nanoparticles. (b) Corresponding TEM image.

for other reflections, it is best to avoid overlapping peaks. Half this peak position is then $\theta = 20°$. Next, its full width at half-maximum is 7.54° (in 2θ units). When converted to radians, $\beta = 0.1316$ rad. Inserting this into the Scherrer equation

$$D = \frac{(0.89)(1.54\,\text{Å})}{(0.1316)\cos(20°)}$$

then gives $D = 11.08$ Å, or $D = 1.11$ nm. We see that the result is comparable to the TEM mean diameter and confirms the usefulness of the Scherrer equation.

Note that complications exist when using XRD to size nanoscale materials. For example, amorphous specimens have large X-ray linewidths. Likewise, crystal strain contributes to line broadening. The choice of substrate is also important, since it may possess overlapping X-ray reflections. As a consequence, there are multiple reasons why X-ray reflections can appear broad. Finally, XRD experiments generally require a good deal of material to obtain powder patterns with large signal-to-noise ratios. Such large amounts may or may not be easy to obtain, and depend on the synthesis used to make the sample.

Derivation of the Scherrer Equation

The origin of the Scherrer equation and how the X-ray reflections of finite-sized crystals broaden begins with Bragg's law

$$\Delta l = n\lambda = 2d\sin\theta,$$

where Δl is the pathlength difference between two rays reflecting off two neighboring crystallographic planes. As we have seen earlier,

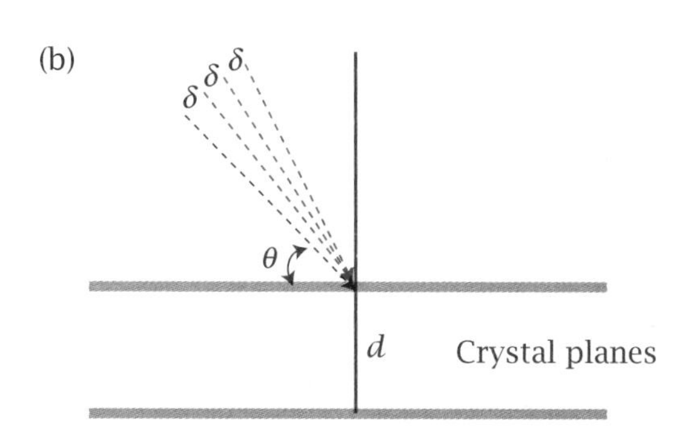

Figure 14.13 Variation of the incident X-ray angle used in deriving the Scherrer equation.

constructive interference between the rays therefore occurs when Δl (**Figure 14.10**) is an integer multiple of their wavelength.

What we want to do now is account for the diffracted X-ray intensity from a randomly oriented mass of the sample. We expect that due to the finite size of the crystal, a broadening of the X-ray reflection will occur. Conversely, in the case where we consider the bulk infinite-crystal limit, we should recover narrow linewidths. The following derivation follows Klug and Alexander (1974) and may be skipped if desired. First, let us assume a slight deviation δ of the incident X-ray angle on the finite-sized crystal, as illustrated in **Figure 14.13(a)**. This will ultimately account for its random orientation in an ensemble. To achieve constructive interference, the two sets of collinear rays must now satisfy the following equivalence:

$$\begin{aligned}\Delta l' &= 2d\sin(\theta+\delta)\\ &= 2d(\sin\theta\cos\delta+\cos\theta\sin\delta)\\ &= (2d\sin\theta)\cos\delta+2d\cos\theta\sin\delta\\ &= n\lambda\cos\delta+2d\cos\theta\sin\delta.\end{aligned}$$

If δ is small, we can invoke the small-angle approximation for $\sin\delta$ and $\cos\delta$ ($\cos\delta\simeq 1$ and $\sin\delta\simeq\delta$) to obtain

$$\Delta l' \simeq n\lambda+2d\delta\cos\theta.$$

The phase factor between X-rays at the new incident angle and those at the original angle is then

$$\begin{aligned}e^{i(2\pi/\lambda)\Delta l'} &= e^{i(2\pi/\lambda)(n\lambda+2d\delta\cos\theta)}\\ &= e^{i2\pi n}e^{i(2\pi/\lambda)(2d\delta\cos\theta)}\\ &= e^{i(4\pi/\lambda)d\delta\cos\theta},\end{aligned}$$

with a corresponding phase difference

$$\phi = \frac{4\pi d\delta}{\lambda}\cos\theta.$$

Next, it has been shown that if multiple waves of the same amplitude (say a) differ in phase by a uniform amount ϕ, then the amplitude A of the resulting scattered wave is (**Figure 14.13b**) (Schuster 1928)

$$A = an\frac{\sin\frac{1}{2}\phi}{\frac{1}{2}\phi},$$

where n is the number of waves present and $\frac{1}{2}\phi$ is half the phase difference between the first and last waves.

Since we have just found that the phase difference between the first and second waves is

$$\phi_{12} = \frac{4\pi\, d\delta}{\lambda}\cos\theta,$$

we can generalize this result to obtain the phase difference between other pairs of waves, assuming that their overall phase difference increases uniformly. The phase difference between the first and third waves is therefore

$$\phi_{13} = \left(\frac{4\pi\, d\delta}{\lambda}\cos\theta\right)(2),$$

and, by the same token, the phase difference between the first and nth waves is

$$\phi_{1n} = \left(\frac{4\pi\, d\delta}{\lambda}\cos\theta\right)(n).$$

As a consequence,

$$\frac{\phi_{\text{tot}}}{2} = \left(\frac{2\pi\, d\delta}{\lambda}\cos\theta\right)(n),$$

yielding

$$A = an\frac{\sin\frac{1}{2}\phi_{\text{tot}}}{\frac{1}{2}\phi_{\text{tot}}} = an\frac{\sin\left(\frac{2\pi\, nd\delta}{\lambda}\cos\theta\right)}{\frac{2\pi\, nd\delta}{\lambda}\cos\theta}.$$

If $\delta = 0$, then all n waves are in phase and $A_0 = an$. Thus,

$$A = A_0\frac{\sin\left(\frac{2\pi\, nd\delta}{\lambda}\cos\theta\right)}{\frac{2\pi\, nd\delta}{\lambda}\cos\theta}.$$

The scattered X-ray intensity is therefore

$$\begin{aligned} I &= |A|^2 \\ &= A_0^2\frac{\sin^2\left(\frac{2\pi\, nd\delta}{\lambda}\cos\theta\right)}{\left(\frac{2\pi\, nd\delta}{\lambda}\cos\theta\right)^2}, \end{aligned}$$

or

$$I = I_0\frac{\sin^2\left(\frac{2\pi\, nd\delta}{\lambda}\cos\theta\right)}{\left(\frac{2\pi\, nd\delta}{\lambda}\cos\theta\right)^2},$$

with $I_0 = A_0^2$. At the function's half-height, the corresponding value of $\frac{1}{2}\phi_{tot}$ is 1.4. Namely,

$$\frac{2\pi n d\delta}{\lambda}\cos\theta = \frac{\phi_{tot}}{2} = 1.4.$$

Solving for δ then gives

$$\delta = \frac{(1.4)\lambda}{2\pi(nd)\cos\theta}$$

and if the characteristic (finite) particle size is $D = nd$, we have

$$\delta = \frac{1.4\lambda}{2\pi D\cos\theta}.$$

Finally, if we multiply the result by two to account for both positive and negative deviations of the incident angle and then further multiply the result by two to define an angular spread in 2θ units, we obtain

$$\begin{aligned}\beta = 4\delta(\text{rad}) &= \frac{4(1.4)\lambda}{2\pi D\cos\theta}\\ &= \frac{0.89\lambda}{D\cos\theta}.\end{aligned}$$

Note that as $D \to \infty$, this breadth goes to zero and correctly accounts for the limiting scenario of a bulk crystal. At this point, we find the familiar Scherrer equation

$$D = \frac{0.89\lambda}{\beta\cos\theta},$$

with D the characteristic particle size, β the full width at half-maximum of the experimental data in 2θ units (expressed in radians), and θ half the experimental 2θ peak position.

14.3.7 Light scattering

Finally, before moving on to approaches for characterizing the core's composition, let us discuss two final scattering techniques used to size nanostructures. Here we refer to light scattering as opposed to X-ray scattering. Furthermore, whereas X-ray scattering entails scattering off individual atoms, light scattering involves, not only different frequencies, but also scattering off larger segments of the nanocrystal as well as the entire structure itself.

The phenomenon of light scattering by colloids dates back many centuries. The first to study this phenomenon was Faraday, using gold nanoparticles. Subsequent investigations by Tyndall revealed the Brownian motion of particles dispersed in a liquid. Rayleigh then conducted the first rigorous theoretical study of the effect in 1871, followed by Mie's seminal work in 1908.

Light scattering can be treated classically and requires no quantum mechanics. In fact, Rayleigh's study precedes its development. However, despite the depth of the resulting theory, the determination of a particle's size and shape, both of which affect the scattered light intensity, remains a complex problem. Recent developments in lasers, data acquisition, data analysis, and efficient numerical algorithms have made the subject more tractable. Using the technique, it is now possible to estimate not only the size of nanostructures but also their size distributions, their effective shapes, their diffusion coefficients, and even their surface potentials. More about light scattering can be found in Hiemenz and Rajagopalan (1997).

Why is Light Scattered?

As light propagates through matter, electrons in a material respond to its oscillating electric field. Depending on the material's polarizability (often denoted by α, the same letter used for the absorption coefficient described earlier in Chapter 5), the amplitude of the oscillating electrons is large or small. The larger α is, the larger the electron displacement, since electrons readily respond to the incident field. These oscillating electrons, in turn, emit light, since they represent a time-varying electric field. In the elastic case considered here, the scattered frequency is the same as that of the incident light, since no energy is lost to the material. The inelastic case leads to Raman scattering and is discussed in Chapter 15.

Rayleigh Scattering

Now, in the limit where particles are much smaller than the wavelength of the incident light, the scattering intensity can be readily described using Rayleigh's theory. To illustrate, let us describe the incident light as z-polarized and approaching the material along the x axis of a Cartesian coordinate system (**Figure 14.14a**). Its electric field is described by (Equation 11.8)

$$\mathbf{E} = E_0 \mathbf{e} \cos \omega t,$$

where E_0 is the field's amplitude, ω is its frequency, and $\mathbf{e}$ is a unit polarization vector ($\mathbf{e} = (0, 0, 1)$ in the case considered). As a consequence,

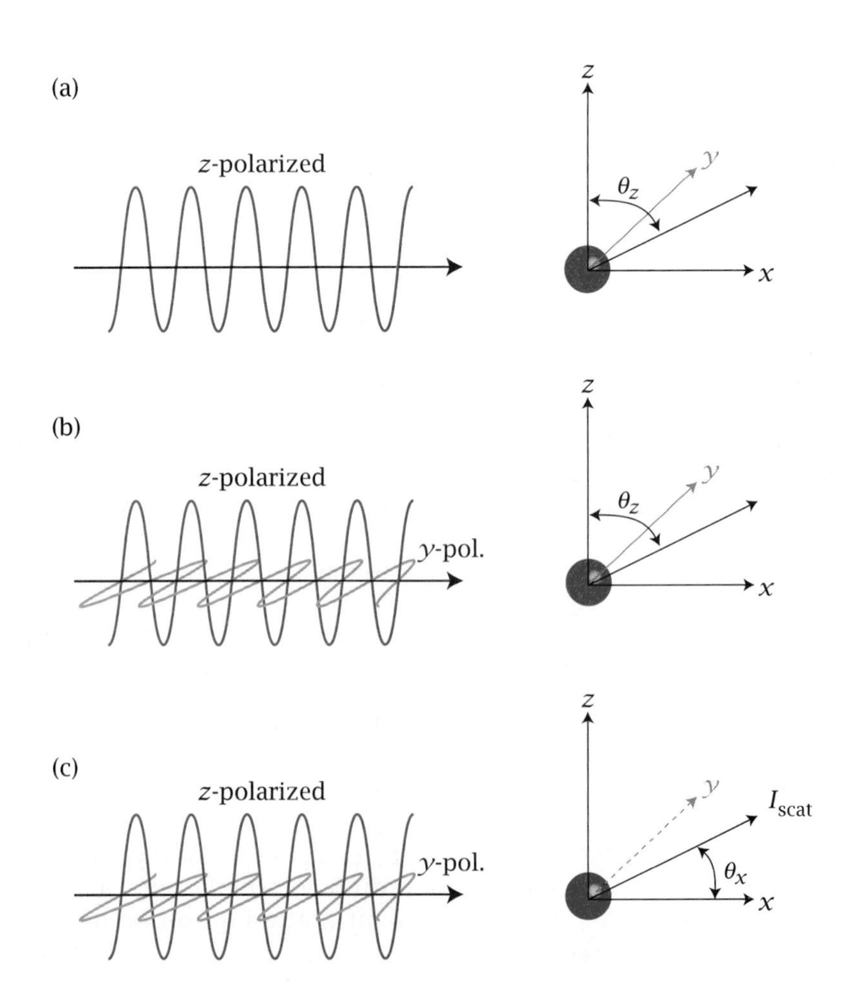

Figure 14.14 Illustration used to describe Rayleigh scattering.

$E_z = E_0 \cos \omega t$. Since $\omega = 2\pi\nu = 2\pi c/\lambda$, we find that

$$E_z = E_0 \cos\left(\frac{2\pi c}{\lambda} t\right).$$

The induced dipole μ in the system is then

$$\mu = \alpha E,$$

whereupon, using our previous expression for the electric field,

$$\mu = \alpha E_0 \cos\left(\frac{2\pi c}{\lambda} t\right). \tag{14.15}$$

Next, the oscillating dipole radiates. At a distance r from the particle and along an angle θ_z relative to the z axis (**Figure 14.14a**), the emitted electric field takes the form

$$E_{\text{scat}} = \frac{1}{4\pi\epsilon_0 c^2} \frac{d^2\mu}{dt^2} \left(\frac{\sin\theta_z}{r}\right). \tag{14.16}$$

This expression can be found in most electrodynamics texts. Implicit is the assumption that the medium is vacuum. However, if working in a different dielectric environment, we replace $1/4\pi\epsilon_0$ with $1/4\pi\epsilon\epsilon_0$.

Now, using Equations 14.15 and 14.16, we find that

$$E_{\text{scat}} = -\frac{\alpha E_0}{4\pi\epsilon_0} \frac{4\pi^2}{\lambda^2} \left(\frac{\sin\theta_z}{r}\right) \cos\left(\frac{2\pi c}{\lambda} t\right).$$

Since the scattered light intensity $I_{\text{scat}} = |E_{\text{scat}}|^2$, we have

$$I_{\text{scat}} = \frac{\alpha^2 I_0}{(4\pi\epsilon_0)^2} \left(\frac{16\pi^4}{\lambda^4}\right) \frac{\sin^2\theta_z}{r^2} \cos^2\left(\frac{2\pi c}{\lambda} t\right),$$

where $I_0 = E_0^2$. If we simply consider the magnitude of the scattered light intensity, then

$$I_{\text{scat}} = \frac{\alpha^2 I_0}{(4\pi\epsilon_0)^2} \left(\frac{16\pi^4}{\lambda^4}\right) \left(\frac{\sin^2\theta_z}{r^2}\right). \tag{14.17}$$

Equation 14.17 accounts for z-polarized light (**Figure 14.14a**). To model unpolarized light (**Figure 14.14b**), the orthogonal (y) component must also be considered. In this case,

$$I_{\text{scat, unpol}} = \tfrac{1}{2}(I_{\text{scat},z} + I_{\text{scat},y})$$

such that

$$I_{\text{scat, unpol}} = \frac{1}{2}\left[\frac{\alpha^2 I_0}{(4\pi\epsilon_0)^2} \left(\frac{16\pi^4}{\lambda^4}\right) \left(\frac{\sin^2\theta_z}{r^2}\right) + \frac{\alpha^2 I_0}{(4\pi\epsilon_0)^2} \left(\frac{16\pi^4}{\lambda^4}\right) \left(\frac{\sin^2\theta_y}{r^2}\right)\right].$$

This reduces to

$$I_{\text{scat, unpol}} \frac{\alpha^2 I_0}{(4\pi\epsilon_0)^2} \left(\frac{8\pi^4}{r^2\lambda^4}\right) (\sin^2\theta_z + \sin^2\theta_y),$$

and can be made simpler if we recall that the sum of the direction cosines squared is 1. Namely,

$$\cos^2\theta_z + \cos^2\theta_y + \cos^2\theta_x = 1. \tag{14.18}$$

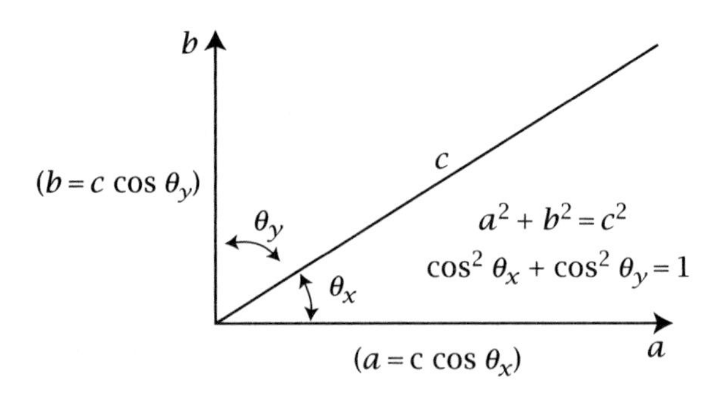

Figure 14.15 Direction cosines.

Equation 14.18 can be rationalized by recalling the Pythagorean theorem as illustrated in **Figure 14.15**. A simple extension to three dimensions then yields the above relationship. Since $\cos^2\theta_z + \cos^2\theta_y + \cos^2\theta_x = 1$, we find that

$$1 + \cos^2\theta_x = \sin^2\theta_z + \sin^2\theta_y,$$

through $\cos^2\theta_z = 1 - \sin^2\theta_z$ and $\cos^2\theta_y = 1 - \sin^2\theta_y$. This ultimately results in

$$I_{\text{scat, unpol}} = \frac{\alpha^2 I_0}{(4\pi\epsilon_0)^2}\left(\frac{8\pi^4}{r^2\lambda^4}\right)(1 + \cos^2\theta_x). \tag{14.19}$$

At this point, we invoke a relationship for spherical particles that relates their polarizability to their volume. This is called the Clausius–Mossotti relationship:

$$\alpha = \frac{3\epsilon_0}{N}\left(\frac{n_m^2 - 1}{n_m^2 + 2}\right). \tag{14.20}$$

Here n_m is the refractive index of the material and N is the number density of electrons (with units of number per unit volume). If we assume that $N = 1/V_{\text{tot}}$, where $V_{\text{tot}} = \frac{4}{3}\pi a^3$ is the volume of the particle and a its radius, then

$$\alpha = 3\epsilon_0\left(\frac{n_m^2 - 1}{n_m^2 + 2}\right)V_{\text{tot}}.$$

We thus obtain the following unpolarized-light, spherical-particle scattering intensity:

$$I_{\text{scat, unpol}} = I_0 a^6\left(\frac{n_m^2 - 1}{n_m^2 + 2}\right)^2\left(\frac{8\pi^4}{r^2\lambda^4}\right)(1 + \cos^2\theta_x). \tag{14.21}$$

From Equation 14.21, it is apparent that the scattering intensity depends on the sixth power of the particle's radius. Sizing can therefore be conducted provided that the observation angle θ_x is known (**Figure 14.14c**). In this regard, the intensity at multiple angles can be measured to ensure the validity of the extracted radius. Finally, the characteristic λ^{-4} dependence of Rayleigh scattering, often invoked to explain why the sky is blue, is evident.

A major limitation of the technique is that it requires the particle to be smaller than the wavelength of light. Furthermore, the model does not take into account shape effects. It also requires that the solutions be dilute, since multiple scattering events from different particles are not taken into account. Most of these limitations are thus dealt with using more advanced theories beyond the scope of this text.

Dynamic Light Scattering

Finally, apart from static light scattering experiments, a related (dynamic) approach can be used to size particles. Namely, we have just seen how the scattered light intensity from a dilute, *static* ensemble relates to a nanoparticle's size. However, in liquid suspensions, the particles are not stationary. They continuously move owing to Brownian motion. As a consequence, over a finite amount of time, the scattered light intensity changes. This may look like flickers of light to the eye. These intensity fluctuations are then exploited in dynamic light scattering to indirectly size scattering objects. This is done via a determination of their diffusion coefficient, which, in turn, relates to their physical dimensions.

Specifically, the "twinkling" of the scattered light intensity is sensed by a detector coupled to a time-correlated photon counter. A correlation function between the scattered light intensity at a time t and at a later time $t+\tau$ is then generated, where the intensity (or second-order) autocorrelation function is defined as

$$g_2(\tau) = \frac{\langle I(t)I(t+\tau)\rangle}{\langle I(t)\rangle^2}, \tag{14.22}$$

with

$$\langle I(t)I(t+\tau)\rangle = \lim_{\tau\to\infty} \frac{1}{\tau}\int_0^{\tau} I(t)I(t+\tau)\,dt,$$

$$\langle I(t)\rangle = \lim_{\tau\to\infty} \frac{1}{\tau}\int_0^{\tau} I(t)\,dt.$$

The underlying concept is that at short times, $g_2(\tau)$ is large because the particles have not moved much. However, as $\tau \to \infty$, it decays because the scattered light intensities from different particles become increasingly uncorrelated owing to their motion. It can be shown that, for a monodisperse system (i.e., where all the particles have the same size), the correlation function decays exponentially as

$$g_2(\tau) = 1 + e^{-2Dq^2\tau}, \tag{14.23}$$

with a characteristic time scale τ_c that relates to the particle's diffusion coefficient D:

$$\tau_c = \frac{1}{2Dq^2} \tag{14.24}$$

(note that q is not to be confused with a charge here). Namely,

$$|q| = \frac{2\pi}{\lambda}\left(2\sin\frac{\theta}{2}\right) = \frac{4\pi}{\lambda}\sin\frac{\theta}{2} \quad (q = k_i - k_{scat}),$$

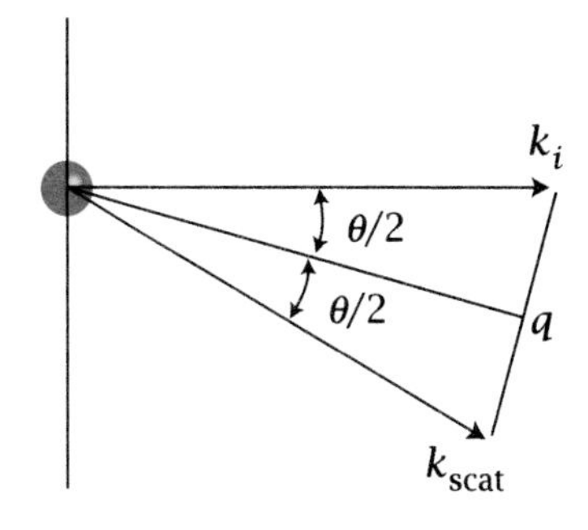

Figure 14.16 Scattering angle in dynamic light scattering.

where θ is the scattering angle between the incident and scattered light wavevectors (k_i and k_{scat}, **Figure 14.16**). Furthermore, when not working in vacuum, $\lambda \to \lambda/n_m$ where n_m is the medium's refractive index. As a consequence,

$$q = \frac{4\pi n_m}{\lambda}\sin\frac{\theta}{2}. \tag{14.25}$$

When dealing with spherical particles, the Stokes–Einstein relationship relates D to the particle radius a, as well as to the solvent viscosity η, through

$$D = \frac{kT}{6\pi\eta a}. \tag{14.26}$$

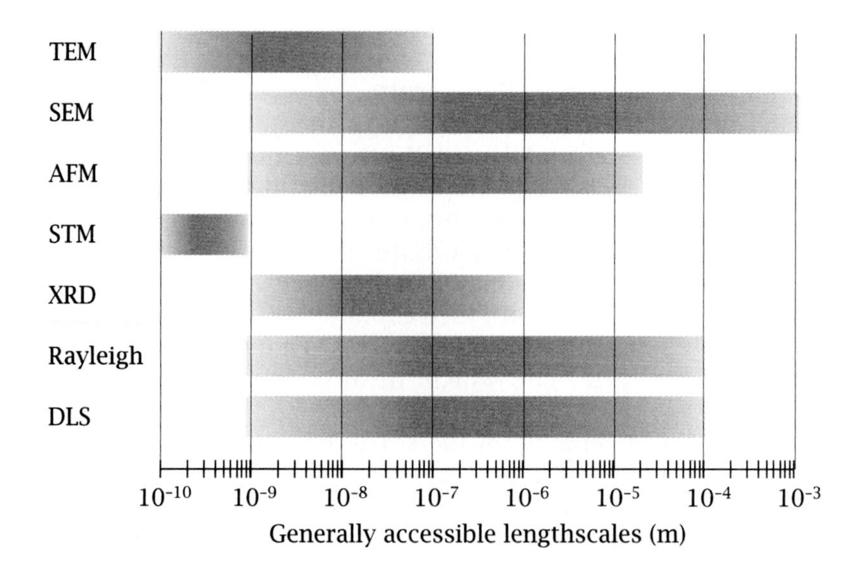

Figure 14.17 Illustration of accessible length scales for the various nanostructure sizing techniques discussed.

Defining $G(\tau) = g_2(\tau) - 1$ then gives

$$G(\tau) = e^{-2Dq^2\tau}, \tag{14.27}$$

whereupon plotting $\ln[G(\tau)]$ versus τ yields a straight line with a slope of $|m| = 2Dq^2 = 1/\tau_c$. Given

$$2Dq^2 = \frac{1}{\tau_c},$$

$$2\left(\frac{kT}{6\pi\eta a}\right)q^2 = \frac{1}{\tau_c},$$

we thus find a particle radius

$$a = \frac{kTq^2\tau_c}{3\pi\eta}. \tag{14.28}$$

Dynamic light scattering is therefore commonly used to determine the size of colloidal nanoparticles in solution. Provided with D for other geometries such as rods and cylinders, one can, in principle, obtain their characteristic dimensions as well. The ease of use of commercial dynamic light scattering instruments and the ability to size nanostructures in situ are therefore the approach's biggest advantages.

14.4 SUMMARY

Figure 14.17 summarizes the general length scales accessible to the nanostructure sizing techniques we have just discussed. As an exercise, the reader may think of the key advantages and disadvantages of each approach. Furthermore, the reader can think of defining aspects of each technique that result in their upper and lower sizing limits.

14.5 CHARACTERIZING THE CORE'S ELEMENTAL COMPOSITION

Finally, before moving on to the nanostructure's surface, we briefly discuss several techniques used to characterize its composition. We have already seen that powder X-ray diffraction can be used to identify the crystal lattice and even a material's composition based on its characteristic X-ray reflections and relative intensities. However, other more direct (chemical) approaches exist. They include techniques such as energy dispersive X-ray spectroscopy and inductively coupled plasma atomic emission spectroscopy.

14.5.1 Energy-Dispersive X-Ray Spectroscopy (EDXS)

Energy-dispersive X-ray spectroscopy (EDXS) is an X-ray technique based on the detection of characteristic X-rays emitted by materials when exposed to high-energy electrons. In the experiment, samples are bombarded with an electron beam. Constituent atoms of the material are then ionized, emitting X-rays. Physically, this entails the incident electrons ejecting inner-shell electrons out of elements that are present. Higher-energy, outer-shell electrons then relax to fill these previously occupied states, in turn, yielding X-rays.

Since energy differences between shells are characteristic of individual atoms, the detected X-rays can be associated with a specific element. In addition, the X-ray intensity gives information about an element's relative abundance. Furthermore, absolute quantities of a given element can be obtained if internal standards are used. **Figure 14.18** shows representative data from experiments conducted on ensembles of CdS, CdSe, and PbSe nanowires. The peaks match well to characteristic elemental lines from a computer database, allowing the material's composition to be determined.

14.5.2 Inductively Coupled Plasma Atomic Emission Spectroscopy (ICP-AES)

Along similar lines, inductively coupled plasma atomic emission spectroscopy (ICP-AES) can be used to determine a sample's composition. The technique involves digesting a material with an acid. It is then dispersed into the gas phase, where it is exposed to a plasma. Collisions with the plasma's electrons and ions further break down the sample and, in parallel, excite constituent atoms. This, in turn, leads to atomic transitions. As with EDXS, the atomic levels and their spacings are characteristic of specific elements. As a consequence, different atomic emission lines are observed and can be used to determine a sample's composition. Provided with an appropriate calibration of the emission intensity to the concentration of a given element, the method can be made quantitative.

14.6 CHARACTERIZING THE SURFACE

Finally, let us describe a way to characterize a nanostructure's surface. Our primary focus is on identifying the surface ligands seen in **Figure 14.1**. Since there are many direct and indirect approaches for this, let us just outline one familiar technique, Fourier transform infrared spectroscopy.

14.6.1 Fourier Transform Infrared Spectroscopy (FTIR)

Infrared spectroscopy is a useful approach for identifying surface chemical species. The underlying idea is identical to conventional absorption

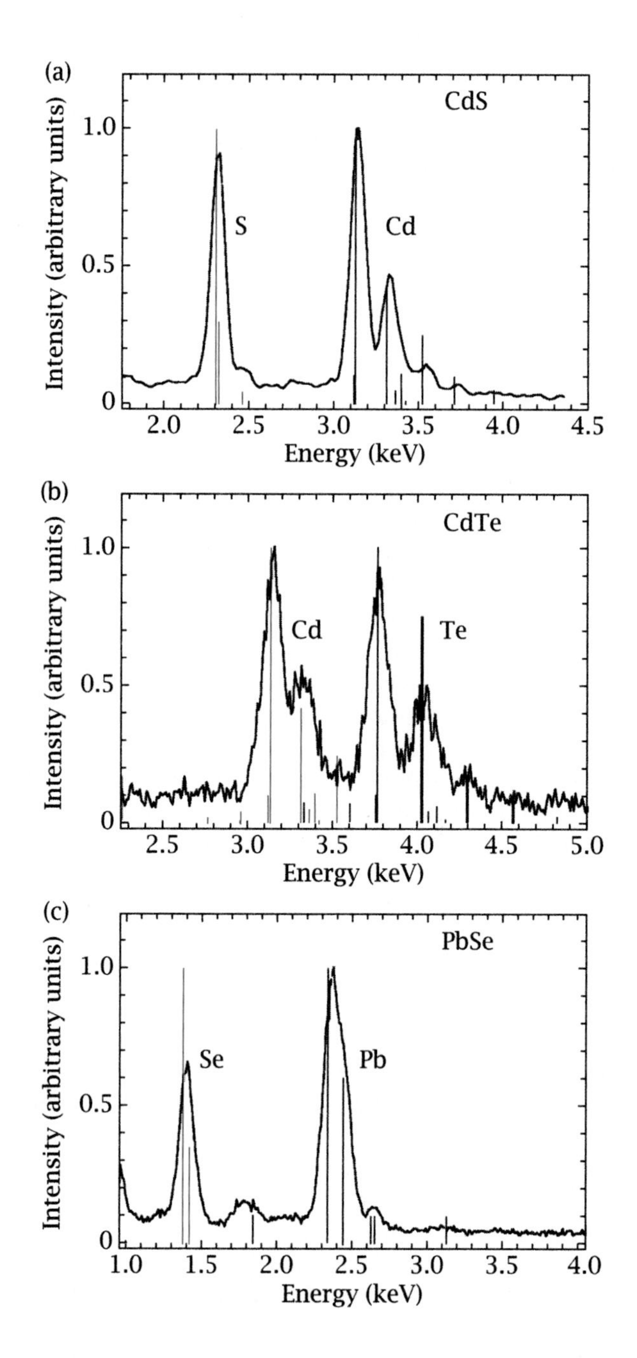

Figure 14.18 Energy-dispersive X-ray spectra (EDXS) of chemically synthesized semiconductor nanowires made of (a) CdS, (b) CdTe, and (c) PbSe. Vertical lines represent literature energies and intensities.

spectroscopy. However, infrared frequencies, as opposed to visible light, are used.

In this regard, many molecules have vibrational transitions in the red, beyond the visible. Depending on their atoms and their actual bonding geometry, one can find characteristic infrared transitions that subsequently identify a species of interest. For example, C=O vibrations occur at energies of approximately 1700 cm^{-1} (divide this number by 8 if you prefer vibrational energies in meV). Other characteristic energies are

- O–H vibrations, ~ 3600 cm^{-1}
- N–H vibrations, ~ 3300 cm^{-1}
- C≡N vibrations, ~ 2500 cm^{-1}
- C=O vibrations, ~ 1700 cm^{-1}

More extensive tables of common vibrational frequencies can be found in Drago (1992). These absorption energies can then be used to fingerprint molecules, especially if they possess many such vibrational "labels."

In practice, infrared absorption has been used to monitor the presence of a given species on a nanostructure's surface. Alternatively, in ligand exchange experiments (Chapter 13), it can establish the disappearance of the original ligand and the appearance of a new ligand. The ligand exchange efficiency can then be estimated.

Classical Description of Vibrations

To provide a semiquantitative description of these atomic vibrations, we revisit the following classical picture of a diatomic molecule whose atoms are connected by a spring with a force constant k_{Hooke} (with units of N/m). The reduced mass of the system is then μ, where $1/\mu = 1/m_a + 1/m_b$ and m_a (m_b) are the masses of the two atoms.

By Hooke's law, the restoring force on the system is

$$F = -k_{\mathrm{Hooke}}x,$$

where, for convenience, we have chosen the x axis of a Cartesian coordinate system to represent the molecule's vibrational coordinate. $x(t)$ represents the system's displacement from equilibrium and a negative sign accounts for the fact that the force is a restoring one.

Next, Newton's second law states that $F = \mu a = \mu d^2x(t)/dt^2$. As a consequence,

$$\mu\frac{d^2x(t)}{dt^2} = -k_{\mathrm{Hooke}}x(t).$$

We then find the time-dependent displacement $x(t)$ by rearranging the equation into a familiar form:

$$\frac{d^2x(t)}{dt^2} + \frac{k_{\mathrm{Hooke}}}{\mu}x(t) = 0.$$

Then letting $\omega^2 = k_{\mathrm{Hooke}}/\mu$, i.e.,

$$\omega = \sqrt{\frac{k_{\mathrm{Hooke}}}{\mu}}, \qquad (14.29)$$

gives

$$\frac{d^2x(t)}{dt^2} + \omega^2x(t) = 0.$$

We have previously seen (Chapter 7) that general solutions to this equation are linear combinations of plane waves:

$$x(t) = Ae^{i\omega t} + Be^{-i\omega t}.$$

We now apply the system's boundary conditions to determine the coefficients A and B. Namely, at $t = 0$, the displacement is simply the equilibrium displacement (i.e., $x(0) = x_0$). We therefore find that

$$x(0) = x_0 = A + B,$$

from which $B = x_0 - A$, giving

$$x(t) = Ae^{i\omega t} + (x_0 - A)e^{-i\omega t}.$$

Next, at $t = 0$, the velocity of the system is zero (i.e., $dx(0)/dt = 0$). Thus,

$$\begin{aligned}\frac{dx(0)}{dt} &= 0 = i\omega A e^{i\omega(0)} + (x_0 - A)(-i\omega)e^{-i\omega(0)} \\ &= i\omega A - i\omega(x_0 - A).\end{aligned}$$

This yields $A = \frac{1}{2}x_0$ and, since $B = x_0 - A$, we have $B = \frac{1}{2}x_0$. Putting everything together then gives

$$\begin{aligned}x(t) &= \tfrac{1}{2}x_0(e^{i\omega t} + e^{-i\omega t}) \\ &= x_0 \cos \omega t.\end{aligned}$$

We see that the system oscillates harmonically forever, since there are no damping forces built into the problem.

Recall now that the force is the negative gradient of the potential: $\mathbf{F} = -\nabla V$. For simplicity, if we consider the one-dimensional case,

$$F = -\frac{dV}{dx}.$$

Then $V = -\int F\, dx$ with $F = -k_{\text{Hooke}}x$ leads to

$$V(x) = \tfrac{1}{2}k_{\text{Hooke}}x^2,$$

which is the parabolic potential first seen in Chapter 8 when dealing with the harmonic oscillator problem.

A Quantum Description

In this regard, the functional form of $V(x)$ can be inserted into the Schrödinger equation to obtain the wavefunctions and energies of an analogous quantum harmonic oscillator. However, since we already did this when we solved the problem in Chapter 8, we will not do it again. Instead, recall that the resulting energies were quantized and had the form

$$E = \hbar\omega\left(n + \tfrac{1}{2}\right),$$

where $n = 1, 2, 3 \ldots$.

In the infrared absorption experiment, transitions between levels are probed. As a consequence, the energy difference between states is simply

$$\Delta E = \hbar\omega.$$

Then, given $\omega^2 = k_{\text{Hooke}}/\mu$ we can predict characteristic transitions in the experiment provided suitable force constants and reduced masses.

EXAMPLE 14.10

Estimate of the Characteristic Vibrational Frequency of a C–O Bond To illustrate, consider the characteristic vibrational frequency of a C–O bond. Assume that the force constant is $k_{\text{Hooke}} = 1870\,\text{N/m}$. Using atomic masses for C and O, we find a reduced mass of $\mu = 1.14 \times 10^{-26}\,\text{kg}$. The vibrational frequency $\omega = \sqrt{k_{\text{Hooke}}/\mu}$ is then $\omega = 4.05 \times 10^{14}\,\text{rad/s}$, with an associated linear vibrational frequency of $\nu = \omega/2\pi = 6.44 \times 10^{13}\,\text{Hz}$.

In terms of energy, it is common to express vibrational transitions using wavenumbers, defined as

$$\bar{\nu} = \frac{1}{\lambda} = \frac{\nu}{c}, \tag{14.30}$$

with units of cm^{-1}. So, for the above example, we find that the C–O vibration occurs at $\bar{\nu} = 2150\,\text{cm}^{-1}$.

14.7 OPTICAL CHARACTERIZATION

Finally, there are a number of ways to characterize nanoscale materials optically. Perhaps the most basic set of measurements involves acquiring their absorption and emission spectra. We have previously discussed the underlying physics behind the former in Chapters 5 and 12 when discussing the absorption cross section, the molar extinction coefficient, and the quantum mechanics of interband transitions.

Practically speaking, the absorption measurement is conducted by examining the transmission of light through a dilute specimen of interest. The sample is dispersed in a solvent. Alternatively, it can be a solid, provided that light passes through it. White light from a halogen lamp illuminates the sample, and the transmitted spectrum $I(\lambda)$ is measured as a function of wavelength. In turn, by knowing the incident spectrum $I_0(\lambda)$, we find the absorption of the sample as a function of wavelength using $A = \log(1/T) = \log[I_0(\lambda)/I(\lambda)]$.

Commercial instruments carry out the experiment by scanning the incident light across the visible range and, at each point, measuring the transmitted light intensity. Alternatively, the transmission at all colors is recorded simultaneously using a diode array. An example of absorption spectra for both colloidal CdSe quantum dots and a nanowire ensemble is shown in **Figure 14.19**. The size-dependent transitions, previously described in Chapter 12, are apparent.

Emission experiments involve exciting samples with a laser, or alternatively with a lamp, provided that suitable bandpass filters are used. Upon excitation, electrons are promoted from the material's valence band into its conduction band. Rapid electron/hole intraband relaxation occurs and once at their respective band edges they recombine, emitting light.

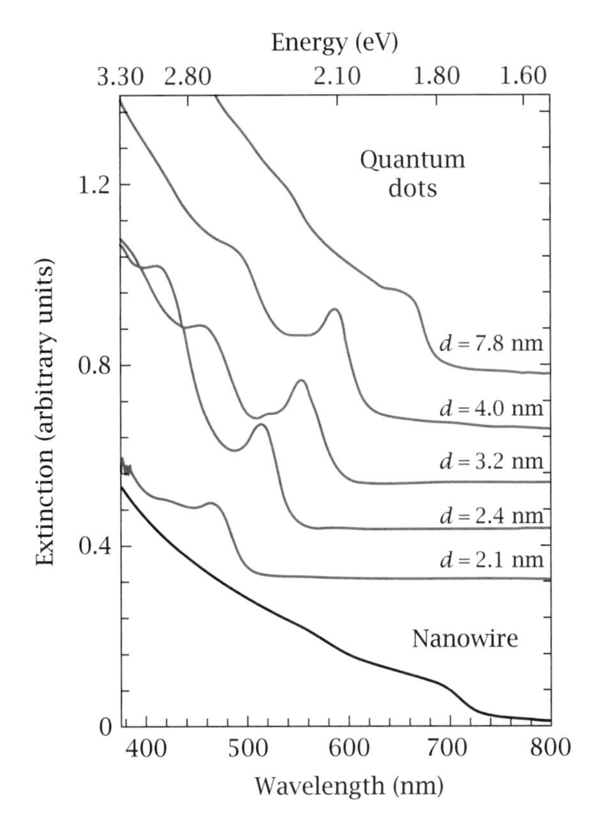

Figure 14.19 Linear extinction spectra of colloidal CdSe quantum dots and a nanowire ensemble.

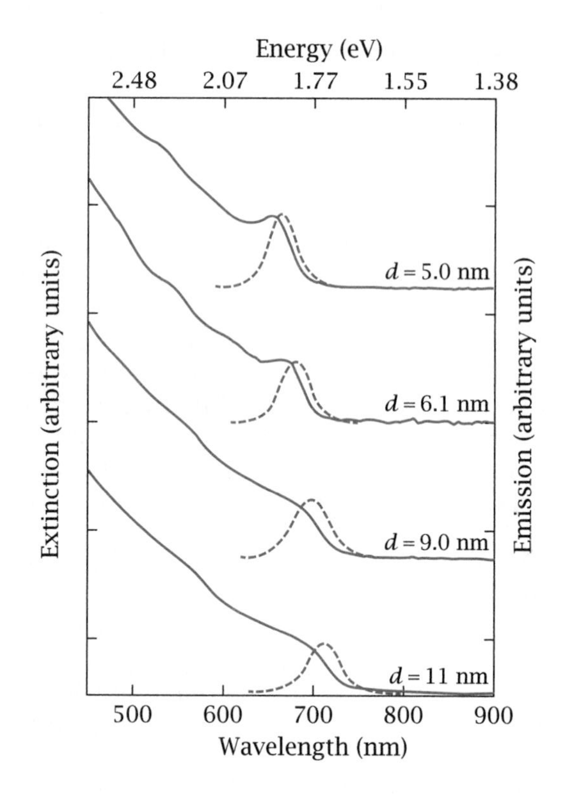

Figure 14.20 Linear extinction and emission spectra from a small size series of CdSe nanowires.

A lens then collects the emission with a filter to remove any unwanted laser light. Note that emission experiments work best when a significant spectral separation exists between where the sample is excited and where it emits. The filtered light is subsequently guided into a spectrometer, which disperses it into its component colors. This dispersed light is then imaged with a CCD detector, enabling the sample's spectrum to be obtained.

Figure 14.20 shows representative emission spectra from a small size series of solution-synthesized CdSe nanowires (dashed lines). The samples are excited at wavelengths below 500 nm, and the resulting emission beyond 600 nm is collected in the manner described above. Note that many other optical characterization approaches exist. These include Raman spectroscopy and even single-particle microscopies. However, given the breadth of these areas, we refer the reader to other more specialized texts for additional information about these subjects.

14.8 SUMMARY

With this, we conclude our discussion about the characterization of nanoscale materials. It has been by no means complete. However, we have seen several important approaches for determining the size and morphology of nanostructures. Some entail electron imaging. Others indirectly size these materials using their light or X-ray scattering response. Yet others measure their topography through scanning probe approaches. Various techniques also exist for determining the composition of nanostructures, as well as their surface passivation. Examples include energy-dispersive X-ray and infrared absorption spectroscopies. Common optical characterization techniques have also been mentioned, although our discussion was kept brief since we have already discussed the physics behind these approaches in previous chapters. We now describe uses of nanoscale materials in Chapter 15.

14.9 THOUGHT PROBLEMS

14.1 Electron tomography

In addition to the two-dimensional images of nanostructures seen in the main text, TEM can be used to obtain three-dimensional representations through what is called electron tomography. Read more about the technique in Midgley et al. (2001) and Ersen et al. (2007). An example of a recent application of the approach is described in Friedrich et al. (2009). Explain how the technique works and why it is useful.

14.2 Atom tomography

Atom probe tomography is another technique enabling the three-dimensional reconstruction of a specimen, this time atom by atom and with chemical specificity. Explore the technique and explain its underlying operating principle/s. Read more about its recent application in nanoscience research in Perea et al. (2009).

14.3 Scanning probe extensions

In addition to the AFM and STM techniques described in the main text, other variants of scanning probe microscopy exist. Examples include Kelvin probe microscopy and magnetic force microscopy. Read Pool (1990). Then follow up on the variants discussed in this article by finding recent examples of their use in nanoscience research.

14.4 Scanning tunneling spectroscopy (STS)

One particularly useful STM extension involves an approach called scanning tunneling spectroscopy (STS). Explore this technique and explain how it works. Then, to see how it has been used to resolve the discrete states of individual colloidal quantum dots, read Banin et al. (1999, 2003) and Bakkers et al. (2001). The application of the technique towards nanowires can be seen in Ma et al. (2003) and Mikkelsen et al. (2004).

14.5 CNT tips for AFM

Read more about using carbon nanotubes to improve the imaging resolution of AFMs in Wilson and Macpherson (2009). Assess for yourself the impact that such higher-resolution tips have had in nanoscience research.

14.6 AFM extension

We have seen that AFM can be used to characterize the size and shape of nanoscale objects. Find other examples in the literature that describe uses of the AFM to better understand nanomaterials.

14.7 Small-angle X-ray scattering (SAXS)

In addition to the X-ray diffraction technique discussed in the main text, there is an analogous characterization approach called small-angle X-ray scattering (SAXS). Read more about it in Narayanan et al. (2004) and Biswas et al. (2008). Then explain any underlying differences in information content between the two X-ray diffraction techniques.

14.8 Nuclear magnetic resonance spectroscopy (NMR)

Nanostructures can be characterized using magnetic resonance techniques such as nuclear magnetic resonance (NMR) spectroscopy. Read more about NMR in Becerra et al. (1994), Tomaselli et al. (1999), and Berrettini et al. (2004). Are there any potential limitations of the technique?

14.9 Electron paramagnetic resonance spectroscopy (EPR)

Another magnetic resonance technique, useful in studying doped nanostructures, involves electron paramagnetic resonance (EPR) spectroscopy. Read more about the technique in Murase et al. (1999) and Norris et al. (2001). Explain key differences between EPR and NMR.

14.10 X-ray photoelectron spectroscopy (XPS)

Along with NMR, nanostructure surfaces can be studied using a separate analytical technique called X-ray photoelectron spectroscopy (XPS). An early application of this, focused on understanding the surface of colloidal CdSe quantum dots, is described by Bowen Katari et al. (1994). Explain how XPS works and the origin of its chemical sensitivity.

14.11 Extended X-ray absorption fine structure (EXAFS)

Although not a common characterization technique, owing to its larger infrastructure requirements, extended X-ray absorption fine structure (EXAFS) has been used to study nanostructures. Read more about the technique and how it has been used to study nanoscale materials in Davis et al. (1997) and Sun et al. (2009). Explain how EXAFS works and illustrate how it differs from XPS as well as from more conventional X-ray diffraction techniques.

14.12 Thermal gravimetric analysis (TGA)

Thermal gravimetric analysis (TGA) is a technique based on measuring the weight loss of a sample upon heating. Investigate how this approach has been used in nanoscience research. As an example, see Kuno et al. (1997).

14.13 Time-of-flight mass spectroscopy (TOF-MS)

Along these lines, nanostructure masses can be measured using a technique called time-of-flight mass spectroscopy. Read more about it within the context of gold nanoparticles in Whetten et al. (1996) and Arnold and Reilly (1998). Does the approach have any mass limitations?

14.14 Transient differential absorption spectroscopy

We did not spend much time discussing the optical characterization of nanostructures. However, this is a large field. Here we briefly mention a few techniques often seen in the literature. One involves measuring the absorption of nanostructures as well as their dynamics by essentially monitoring their absorption as a function of time. Read more about transient differential absorption spectroscopy in the review articles by Burda and El-Sayed (2000) and Hartland (2010). Explain how the approach works.

14.15 Single-particle absorption microscopy

Single-particle microscopy is one way to characterize nanoscale materials. It circumvents the inability of current syntheses to make all nanostructures identical in terms of their size and shape. One area of interest centers on measuring the absorption of individual nanostructures. A number of ways exist for doing this. Read Arbouet et al. (2004) and Billaud et al. (2010) and then explain how the direct absorption technique works. An alternative single-particle absorption technique involves what is called photothermal heterodyne imaging. Read Berciaud et al. (2005) and Giblin et al. (2010) and explain how it works.

14.16 Single-particle emission microscopy

In addition to single nanostructure absorption measurements, it is common to look at their emission. Read the descriptions of two early studies probing the emission of individual colloidal CdSe quantum dots in Blanton et al. (1996) and Empedocles et al. (1996). Summarize for yourself what such measurements have taught us about nanostructures that could not be learned using more conventional ensemble measurements.

14.17 Near-field scanning optical microscopy (NSOM)

A related single-particle imaging technique is near-field scanning optical microscopy (NSOM). This is an optical technique that enables us to resolve things below the diffraction limit. Read more about it in Betzig et al. (1991). Examples of recent implementations of the technology are described in Johnson et al. (2001) and Gu et al. (2005). Explain how the technique works and why one would use this over the more conventional single-particle optical microscopies described above. Conversely, explain conditions where one might want to use NSOM in preference to the other approaches.

14.18 Time-correlated single photon counting (TCSPC)

Along with measuring the absorption and emission of individual nanostructures, one can also measure their excited-state lifetimes using a technique called time-correlated single photon counting (TCSPC). Read more about it and explain how the technique works. Examples of its use in nanostructure research can be found in Schlegel et al. (2002) and Fisher et al. (2004). Note that excited-state lifetimes can alternatively be measured using another approach called the Hanbury Brown and Twiss experiment. Explain how this technique works and how it differs from the above TCSPC measurement. As an example, see Michler et al. (2000).

14.19 Terahertz spectroscopy

Finally, a more recent optical characterization technique involves using radiation from a different part of the spectrum. Read more about it in Beard et al. (2002) and Wang et al. (2006) and explain what information can be learned from the approach

14.20 Other techniques

Find an example in the literature that uses another technique to characterize nanoscale materials. Explain how it works and what information it provides.

14.10 REFERENCES

Arbouet A, Christofilos D, Del Fatti N, et al. (2004) Direct measurement of the single-metal-cluster optical absorption. *Phys. Rev. Lett.* **93**, 127401-1.

Arnold RJ, Reilly JP (1998) High-resolution time-of-flight mass spectra of alkanethiolate-coated gold nanocrystals. *J. Am. Chem. Soc.* **120**, 1528.

Banin U, Cao YW, Katz D, Millo O (1999) Identification of atomic-like electronic states in indium arsenide nanocrystal quantum dots. *Nature* **400**, 542.

Banin U, Millo O (2003) Tunneling and optical spectroscopy of semiconductor nanocrystals. *Annu. Rev. Phys. Chem.* **54**, 465.

Bakkers EPAM, Hens Z, Zunger A, et al. (2001) Shell-tunneling spectroscopy of the single-particle energy levels of insulating quantum dots. *Nano Lett.* **1**, 551.

Beard MC, Turner GM, Schmuttenmaer CA (2002) Terahertz spectroscopy. *J. Phys. Chem. B* **106**, 7146.

Becerra LR, Murray CB, Griffin RG, Bawendi MG (1994) Investigation of the surface morphology of capped CdSe nanocrystallites by ^{31}P nuclear magnetic resonance. *J. Chem. Phys.* **100**, 3297.

Berciaud S, Cognet L, Lounis B (2005) Photothermal absorption spectroscopy of individual semiconductor nanocrystals. *Nano Lett.* **5**, 2160.

Berrettini MG, Braun G, Hu JG, Strouse GF (2004) NMR analysis of surfaces and interfaces in 2-nm CdSe. *J. Am. Chem. Soc.* **126**, 7063.

Betzig E, Trautman JK, Harris TD, et al. (1991) Breaking the diffraction barrier: optical microscopy on a nanometric scale. *Science* **251**, 1468.

Billaud P, Marhaba S, Grillet N, et al. (2010) Absolute optical extinction measurements of single nano-objects by spatial modulation spectroscopy using a white lamp. *Rev. Sci. Instrum.* **81**, 043101-1.

Biswas K, Varghese N, Rao CNR (2008) Growth kinetics of gold nanocrystals: a combined small-angle X-ray scattering and calorimetric study. *Small* **4**, 649.

Blanton SA, Hines MA, Guyot-Sionnest P (1996) Photoluminescence wandering in single CdSe nanocrystals. *Appl. Phys. Lett.* **69**, 3905.

Bowen Katari JE, Colvin VL, Alivisatos AP (1994) X-ray photoelectron spectroscopy of CdSe nanocrystals with applications to studies of the nanocrystal surface. *J. Phys. Chem.* **98**, 4109.

Burda C, El-Sayed MA (2000) High-density femtosecond transient absorption spectroscopy of semiconductor nanoparticles. A tool to investigate surface quality. *Pure Appl. Chem.* **72**, 165.

Davis SR, Chadwick AV, Wright JD (1997) A combined EXAFS and diffraction study of pure and doped nanocrystalline tin oxide. *J. Phys. Chem. B* **101**, 9901.

De Graef M, McHenry ME (2007) *Structure of Materials: An Introduction to Crystallography, Diffraction and Symmetry.* Cambridge University Press, Cambridge, UK.

Drago RS (1992) *Physical Methods for Chemists*, 2nd edn. WB Saunders, Orlando, FL.

Empedocles SA, Norris DJ, Bawendi MG (1996) Photoluminescence spectroscopy of single CdSe nanocrystallite quantum dots. *Phys. Rev. Lett.* **77**, 3873.

Ersen O, Hirlimann C, Drillon M, et al. (2007) 3D-TEM characterization of nanometric objects. *Solid State Sci.* **9**, 1088.

Fisher BR, Eisler HJ, Stott NE, Bawendi MG (2004) Emission intensity dependence and single-exponential behavior in single colloidal quantum dot fluorescence lifetimes. *J. Phys. Chem. B* **108**, 143.

Friedrich H, Gommes CJ, Overgaag K, et al. (2009) Quantitative structural analysis of binary nanocrystal superlattices by electron tomography. *Nano Lett.* **9**, 2719.

Giblin J, Syed M, Banning MT, et al. (2010) Experimental determination of single CdSe nanowire absorption cross sections through photothermal imaging. *ACS Nano* **4**, 358.

Gu Y, Kwak E-S, Lensch JL, et al. (2005) Near-field scanning photocurrent microscopy of a nanowire photodetector. *Appl. Phys. Lett.* **87**, 043111-1.

Guinier A (1963) *X-Ray Diffraction: In Crystals, Imperfect Crystals, and Amorphous Bodies.* WH Freeman, San Francisco (reprinted 1994 by Dover Publications, New York).

Hartland GV (2010) Ultrafast studies of single semiconductor and metal nanostructures through transient absorption microscopy. *Chem. Sci.* **1**, 303.

Hiemenz PC, Rajagopalan R (1997) *Principles of Colloid and Surface Chemistry*, 3rd edn. Marcel Dekker, New York.

Johnson JC, Yan H, Schaller RD, et al. (2001) Single nanowire lasers. *J. Phys. Chem. B* **105**, 11387.

Klug HP, Alexander LE (1974) *X-Ray Diffraction Procedures for Polycrystalline and Amorphous Materials*, 2nd edn. Wiley, New York.

Kuno M, Lee JK, Dabbousi BO, et al. (1997) The band edge luminescence of surface modified CdSe nanocrystallites: probing the luminescing state. *J. Chem. Phys.* **106**, 9869.

Ma DDD, Lee CS, Au FCK, et al. (2003) Small-diameter silicon nanowire surfaces. *Science* **299**, 1874.

Meyer E, Hug HJ, Bennewitz R (2004) *Scanning Probe Microscopy: The Lab on a Tip.* Springer-Verlag, Berlin.

Michler P, Imamoglu A, Mason MD, et al. (2000) Quantum correlation among photons from a single quantum dot at room temperature. *Science* **406**, 968.

Midgley PA, Weyland M, Thomas JM, Johnson BFG (2001) Z-contrast tomography: a technique in three-dimensional nanostructural analysis based on Rutherford scattering. *Chem. Commun.* 907.

Mikkelsen A, Skold N, Ouattara L, et al. (2004) Direct imaging of the atomic structure inside a nanowire by scanning tunneling microscopy. *Nature Mater.* **3**, 519.

Murase N, Jagannathan R, Kanematsu Y, et al. (1999) Fluorescence and EPR characteristics of Mn^{2+}-doped ZnS nanocrystals prepared by aqueous colloidal method. *J. Phys. Chem. B* **103**, 754.

Narayanan S, Wang J, Lin XM (2004) Dynamical self-assembly of nanocrystal superlattices during colloidal droplet evaporation by in situ small angle X-ray scattering. *Phys. Rev. Lett.* **93**, 135503-1.

Norris DJ, Yao N, Charnock FT, Kennedy TA (2001) High-quality manganese-doped ZnSe nanocrystals. *Nano Lett.* **1**, 3.

Perea DE, Hemesath ER, Schwalbach EJ, et al. (2009) Direct measurement of dopant distribution in an individual vapour–liquid–solid nanowire. *Nature Nanotech.* **4**, 315.

Pool R (1900) The children of the STM. *Science* **247**, 634.

Schlegel G, Bohnenberger J, Potapova I, Mews A (2002) Fluorescence decay time of single semiconductor nanocrystals. *Phys. Rev. Lett.* **88**, 137401-1.

Schuster A (1928) *An Introduction to the Theory of Optics*, 3rd edn. Arnold, London.

Sun ZH, Oyanagi H, Uehara M (2009) In-situ EXAFS study of nucleation process of CdSe nanocrystals. *J. Phys. Conf. Ser.* **190**, 012120.

Tomaselli M, Yarger JL, Bruchez M Jr, et al. (1999) NMR study of InP quantum dots: surface structure and size effects. *J. Chem. Phys.* **110**, 8861.

Wang F, Shan J, Islam MA, et al. (2006) Exciton polarizability in semiconductor nanocrystals. *Nature Mater* **5**, 861.

Whetten RL, Khoury JT, Alvarez MM, et al. (1996) Nanocrystal gold molecules. *Adv. Mater.* **8**, 428.

Williams DB, Carter CB (1996) *Transmission Electron Microscopy.* Plenum Press, New York.

Wilson NR, Macpherson JV (2009) Carbon nanotube tips for atomic force microscopy. *Nature Nanotech.* **4**, 483.

Applications

15.1 INTRODUCTION

As described in Chapter 1, nanostructures, whether they be quantum wells, nanowires, or quantum dots, all possess interesting size-dependent optical and electrical properties. The study of these intrinsic features is the realm of nanoscience. However, at the end of the day, we expect that some of this acquired knowledge (funded largely through your tax dollars) will be put to good use in developing the next generation of consumer products.

So how exactly are today's nanotechnologists trying to harness the potential of nano? In what follows, we will see examples of areas being pursued for applications. However, this chapter cannot truly be comprehensive, since a vast number of nano applications exist. Some exploit the mechanical and/or structural properties of nanostructures. As an illustration, it is not uncommon these days to find baseball or softball bats that contain carbon nanotube composites to make them lighter and stronger. The same is true with tennis rackets. Other applications focus on the improved catalytic properties of nanomaterials.

To stay consistent with our discussion in earlier chapters, we focus on applications that utilize the size-dependent optical and electrical responses of nanostructures. Where possible, we illustrate the underlying concept(s) behind these applications in a semiquantitative fashion. Let us therefore touch on some applications of quantum dots and nanowires, both as current commercial products and as future ones. Furthermore, given that the area of quantum wells is fairly mature, we defer to other references for more information about their use.

15.2 QUANTUM DOTS

In the realm of colloidal semiconductor quantum dots, the following applications have been proposed:

- Quantum dots as labels for biological imaging applications
- Quantum dots as optical filters and as active elements in light-emitting diodes
- Quantum dots as sensitizers for photovoltaic applications

Brief descriptions of each and the reasons why nanocrystals have distinct advantages over conventional solutions are presented below.

15.2.1 Biological Labeling

Conventional biological labeling is currently carried out using fluorescent organic dyes or in some cases with radioactive sources. In the case of organic fluorophores such as tetramethylrhodamine, the molecules are covalently attached to a biological specimen of interest using specific linker chemistries.

Organic dyes, however, possess several drawbacks. Namely, they suffer from an effect called photobleaching. This is a phenomenon where, after exposure to incident light for a modest amount of time, the molecules undergo photochemistry, which renders them nonfluorescent. The dyes are said to "fade." This makes labeling and tracking experiments difficult because of the finite observation window one has before their emission disappears. The phenomenon becomes especially problematic when dealing with discrete numbers of dye molecules in more sensitive single-molecule tracking/labeling experiments. As a general rule of thumb, individual organic dye molecules absorb and emit roughly 10^6 photons before succumbing to photobleaching. Under typical single-molecule imaging conditions, this means a few seconds worth of data.

An additional hindrance is that organic dyes tend to have fairly discrete absorption spectra. Thus, from dye to dye, their absorption wavelengths (energies) vary dramatically. This makes multicolor labeling experiments difficult, because each dye requires its own excitation frequency. Furthermore, proper filtering of the monitored emission becomes difficult because of the multiple colors present simultaneously. Finally, different dyes may require different functionalization chemistries to link them to a specimen. This can, in turn, make the labeling process difficult.

By contrast, colloidal quantum dots such as CdSe possess narrow emission spectra (see **Figure 1.3**). Furthermore, because of quantum confinement effects, different-sized nanocrystals emit different colors. As a consequence, it is possible to use one synthesis to make a variety of materials that emit across the entire visible spectrum. This reduces production costs and also helps simplify subsequent conjugation chemistries, since one is always dealing with the same material.

More importantly, because of their density of states (Chapter 9), the dots all absorb in the blue. As a consequence, multiple color labeling experiments are possible using a single excitation source. To illustrate, CdSe, a popular quantum dot label, can be excited at 488 nm (a common argon-ion laser line) and, depending on size, will emit anywhere from 500 to 650 nm.

At the same time, we have seen in Chapter 5 that dots have order-of-magnitude larger absorption cross sections than organic fluorophores. This means that they absorb light more efficiently. Hence, less quantum dot label or lower light intensity can be used in subsequent imaging experiments.

Finally, nanocrystals are made of inorganic compounds. As such, they are more robust than organic fluorophores when it comes to photobleaching. Dots have been observed to absorb and emit over 10^8 photons before experiencing irreversible photodarkening. This is two orders of magnitude more photons from a single emitter, meaning that nanocrystal-labeled ensembles practically do not fade.

So what is the catch? First, as-made CdSe quantum dots generally have emission quantum yields of 5–10%. Although this is sizable, a problem arises when trying to modify their surfaces to make them biologically compatible. In this regard, as-made nanocrystals typically have aliphatic organic ligands on their surfaces (Chapter 13). As a consequence, ligand exchange chemistries are needed to make the dots water-soluble.

The problem is that, after ligand exchange, the quantum yields of these particles drop dramatically. Presumably, this is due to their surfaces being damaged during the exchange process. Thus, the use of quantum dots for biological labels only became viable after the advent of surface overcoating techniques, which we described in Chapter 13.

In this approach, a shell of a wider-gap semiconductor is layered over the original nanocrystal surface. The coating then acts to passivate surface defects, which limit the particle's quantum yield. Following the procedure, near-unity values are seen at room temperature. Subsequent functionalization chemistries make the overcoated dots water-soluble and yield particles with quantum yields that readily exceed 10%. These materials are then useful for labeling applications.

Another problem with quantum dots is that their surface chemistry is still (relatively speaking) in its infancy. There is still much to be understood before both facile and general chemistries can be used to attach dots to specific sites on proteins or cells or on other biological specimens. This is an area where organic dyes still prevail. Furthermore, semiconductor quantum dots, though nanometer-sized, may be too large for some labeling experiments. Namely, their size could influence the behavior of a biological sample. To illustrate, imagine trying to watch a quantum-dot-labeled protein fold in real time. It might be reasonable to ask whether the dot affects the folding and unfolding pathways being explored. In this regard, the reader may consider the typical size of a protein and then compare it with the size of the nanocrystals first discussed in Chapter 4. Likewise, there may be cellular regions or membranes where the particles simply cannot penetrate owing to size restrictions.

In parallel, there are potential toxicity effects that arise from using particles containing heavy metals. As a consequence, motivation exists for further refinements in the general as well as surface-specific chemistries of colloidal nanocrystals. A good overview of the area can be found in Resch-Genger et al. (2008).

Finally, there is one additional undesirable feature of colloidal quantum dots. Namely, they tend to "blink" when imaged at the individual particle level. Blinking or fluorescence intermittency is a phenomenon where, under continuous excitation, the dots occasionally cease emitting light and become nonemissive. After some time, the dots turn "on" again and emit light. There is no way to predict when the dots will turn on or off. This is, on one hand, a fascinating scientific phenomenon, but, on the other, a nuisance when it comes to imaging. Imagine trying to follow a fluorescently labeled species that randomly turns on and off.

Blinking was discovered over 15 years ago by Nirmal and co-workers at Bell Laboratories. Despite a decade of research into its origins, it is still not fully understood. Cichos et al. (2007), Frantsuzov et al. (2008), and Stefani et al. (2009) provide good overviews of the current state of research in this area.

Although the following list is certain to change as companies undergo their natural boom-and-bust cycles, the following are some current vendors of commercial quantum dot biological labels:

- Invitrogen (under the Qdot name)
- Ocean Nanotech
- CrystalPlex
- PlasmaChem

15.2.2 Lighting and Possibly Lasing

Lighting has not changed much since the light bulb was invented by Edison and others close to a hundred years ago. More efficient fluorescent lighting has since been developed, but suffers from flicker and color purity issues. Recently, solid state light-emitting diodes (LEDs) have come on the market and are poised to revolutionize the lighting industry. LEDs exhibit tremendous brightness, consume little power, come in different colors, and emit little or no heat. They are already used in traffic lights and even in museums to illuminate paintings.

A major goal of the lighting industry is to replace all incandescent and fluorescent light bulbs with white light LEDs. Furthermore, brighter, more efficient, flat panel displays will emerge from this technology. Already, next-generation LED screens are competing with current flat panel liquid crystal displays (LCDs) in electronics superstores.

Quantum dots may eventually find a role in this area, provided that they can be made to electroluminesce efficiently. That is, by passing an electrical current through an array of dots, electrons and holes can be injected into their conduction and valence bands (note the informal use of this terminology), from where they can recombine by emitting light. By exploiting the nanocrystal's size-dependent emission, different colors can be obtained using the same material. At the same time, an array of dots with different sizes can be mixed to produce white light. This utilization of quantum dot electroluminescence, however, is still in its infancy.

One of the first applications of quantum dot products in this direction entails using highly luminescent quantum dots to absorb light from existing LEDs and to re-emit it at different frequencies. As an example, a single GaN-based blue LED can be used to obtain different colors across the visible by simply coating its surface with a film of quantum dots of a given size. A similar application entails achieving better color balance in existing white light LEDs by "re-shaping" their spectral output. Namely, a quantum dot filter element can be placed in front of an existing white light LED to absorb a fraction of its incident light. Depending on the composition of the particles in the filter, the resulting transmitted as well as re-emitted light can then be better balanced to soften the harshness of a standard white light LED. In turn, its output can be made warmer to more closely resemble popular incandescent lighting. An example of such a device, now being marketed by QD Vision, is shown in **Figures 15.1** and **15.2**.

Finally, we mention that the size-dependent emission from quantum dots can be exploited for possible lasing applications. In this regard, the development of low-dimensional nanostructures such as quantum dots and nanowires opens the door to lasers that are more efficient than their bulk counterparts. This is because their density of states is increasingly peaked near the band edge (Chapter 9). As a consequence, low-dimensional structures should, in principle, lead to devices with lower lasing thresholds and larger gain efficiencies. However, as with their electroluminescence, quantum dot lasing is still in its research phase.

15.2.3 Photovoltaics

Renewable energy has been an area of interest to many since the 1973 OPEC oil embargo. The field, however, has since undergone numerous growth-and-bust cycles. It is currently on the upswing, as evidenced by increased funding in the area, and a renewed sense of urgency to move away from fossil fuels. Solar is one facet of renewable energy, with wind and geothermal being others. The underlying motivation is to take advantage of the Sun's abundant energy by converting it into usable forms, much like photosynthesis in plants. What is needed, though, is

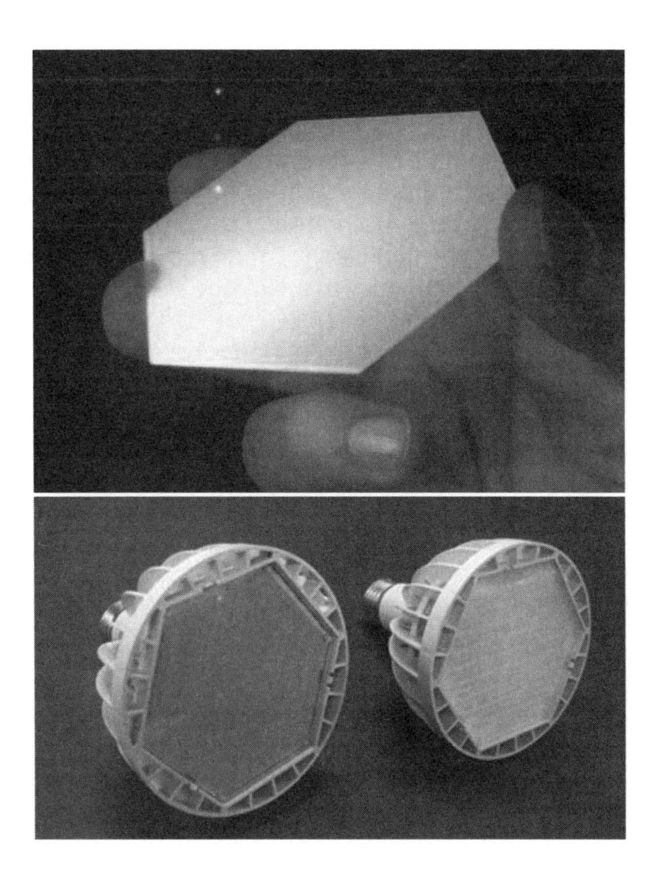

Figure 15.1 Illustration of QD Vision's first Quantum Light Optic product. This optic is placed in front of an LED lamp or fixture. Quantum Light materials absorb some of the cool blue LED light and efficiently re-emit it as warm red light. This balances the lighting color spectrum, creating a pleasing incandescent-quality light with significantly higher efficiency. Higher efficiency translates into the need for fewer LEDs to achieve a given brightness level, which translates to a lower cost to make the LED lamp or fixture. This picture shows Nexxus Lighting's Array Quantum LED R30 lamp with (left) and without (right) QD Vision's Quantum Light optic. Quantum Light-enhanced products demonstrate high color quality (with a color rendering index CRI ≥90) and efficiency, together with energy savings of up to 80% and 50 000 hours of product life. Images courtesy of QD Vision.

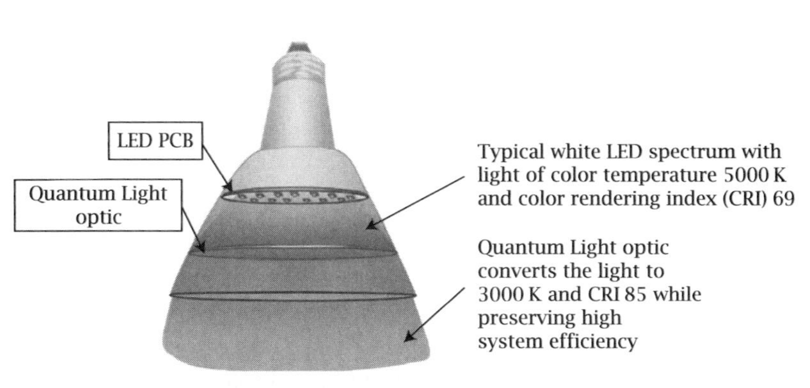

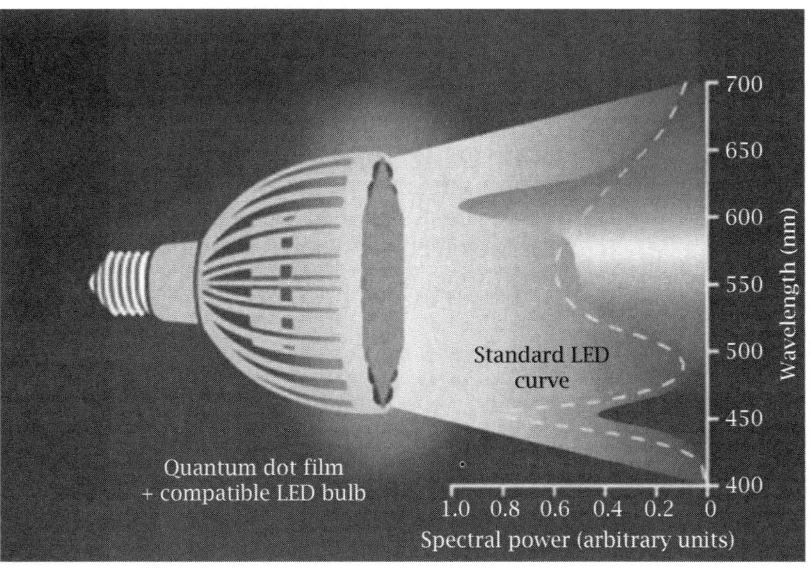

Figure 15.2 QD Vision Quantum Light optic converts "cold" LED light to "warm" white light while preserving system efficiency. Images courtesy of QD Vision.

an active material or system like chlorophyll that can absorb solar radiation and provide efficient charge separation and/or storage.

Most low-cost commercial solar cells are made using polycrystalline silicon. On exposure to light, electrons and holes are created in the material by transitions from its valence band into its conduction band. By design, the cell is made to prevent electrons and holes from recombining quickly. These separated carriers then move to electrodes, where they become the basis of an electrical current.

Typical device efficiencies are in the region of 15%. Although good by today's standards, most of the collected energy is actually wasted. As a consequence, improved solar cells use single-crystal silicon as well as multiple heterojunctions to achieve efficiencies that exceed 30%. Some have even reached 45%. However, this has come at the cost of being expensive and impractical for widespread commercial use. As of now, there is still more "bang for your buck" by burning fossil fuels.

The Dye-Sensitized Grätzel Cell

Because of the unique optical and electrical properties of nanoscale materials, there has been longstanding interest in applying nanocrystals and other nanostructures to the problem of solar energy conversion. Two decades ago, an important breakthrough was made in the area by Michael Grätzel and his co-worker Brian O'Regan (O'Regan and Grätzel 1991). What they found was that one could use sintered titanium dioxide (TiO_2) nanoparticles as substrates onto which ruthenium-based organic dyes could be used as "sensitizers" to absorb visible light.

Using these materials, Grätzel and O'Regan came up with a surprisingly efficient low-cost solar cell that is described in more detail below. As a brief aside, TiO_2 is ubiquitous in modern life. It can be found in foods and other common household products. To illustrate, the white powder in powdered donuts contains this material. TiO_2 is even found in toothpaste.

The basic construction of the Grätzel cell is as follows. TiO_2 nanoparticles are deposited onto a conductive transparent electrode such as indium tin oxide (ITO). The nanoparticles have a large surface-to-volume ratio (Chapter 2) and, as a consequence, represent a large surface area onto which dye molecules can physisorb. This is important for subsequent electron transfer events, since many sensitizers can be in close proximity to the charge accepting TiO_2 surface.

At the same time, the TiO_2 particles are sintered together to create a conductive pathway for electrons to the underlying ITO. The porous nanoparticle network also enables an iodine-based redox couple to remain in intimate contact with the sensitizer. This is important because the electrolyte removes holes from the dye (i.e., the dye is reduced) and allows it to subsequently participate in light absorption. A counter-electrode, in contact with the electrolyte, then completes the circuit. A schematic of the Grätzel cell is shown in **Figure 15.3**.

How Does It Work?

The Grätzel cell has the following operating principles. When sunlight is absorbed by the sensitizer, an electronic transition occurs to its excited state. The process is denoted by an upward arrow in **Figure 15.4**. If we wanted to, we could model this transition using our results from Chapters 5 and 11.

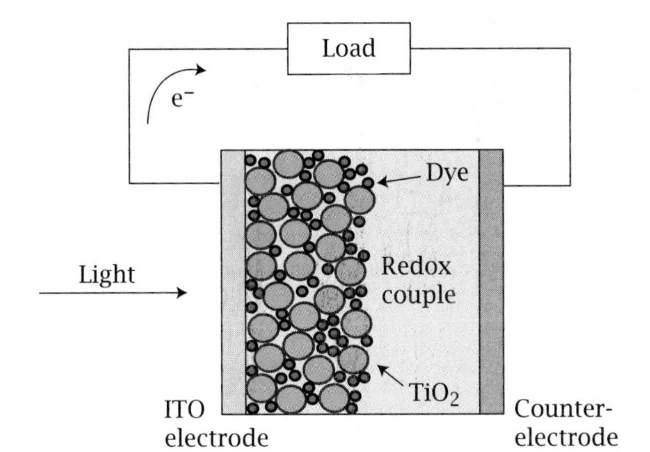

Figure 15.3 Illustration of the Grätzel cell.

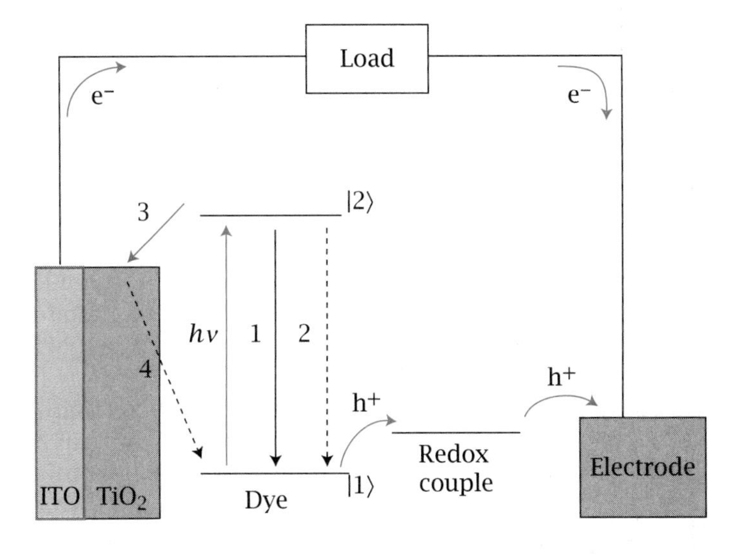

Figure 15.4 Energy diagram for the different processes involved in the Grätzel cell. Process 1 denotes radiative relaxation to the ground state. Process 2 represents nonradiative relaxation to the ground state. Process 3 is the desired electron transfer step. Process 4 is an undesired back electron transfer to the oxidized dye.

Next, while in the excited state, the sensitizer undergoes either radiative or nonradiative relaxation to the initial ground state (processes 1 and 2 in **Figure 15.4**). The former results in the emission of light, the latter heat. In the presence of TiO_2, however, another relaxation pathway opens up for the excited electron (process 3). This stems from possible charge transfer into the underlying TiO_2. Provided that this pathway has a rate competitive with the former two processes, an electron can be extracted from the sensitizer. Note that an undesired back electron transfer process exists, which reduces the carrier concentration in the electrode (process 4).

Once in the TiO_2, the extracted charge moves through the interconnected nanocrystal layer to reach the ITO. From there, it migrates into the rest of the circuit. In parallel, an iodine-based redox couple in solution extracts the hole from the sensitizer and reduces it. The oxidized iodine species then diffuses back to the counter-electrode, giving up the extracted hole. This, in turn, regenerates the couple, completes the circuit and creates an electrical current that can be used to power a load. The resulting cell has external power conversion efficiencies of the order of 10%. This is remarkable given the simplicity of the device. More importantly, the cell is competitive with commercial silicon solar cells, making it possible to conceive of low-cost, high-efficiency photovoltaics made of nanostructured materials. The Further

Reading contains more information about dye-sensitized solar cells as well as possible experiments that the reader can conduct.

An Extension: Quantum Dot Solar Cells

Colloidal quantum dots have recently received a lot of attention for potential use in such devices as replacements for the organic sensitizer. Among reasons for this, quantum dots have size-tunable absorption and emission spectra. As a consequence, different-sized particles can be used to absorb light from the ultraviolet into the infrared. These size-dependent properties can subsequently be exploited to make solar cells perfectly matched to the solar spectrum. In tandem, photovoltaics, reaching much of the currently unexploited infrared part of the solar spectrum, are also possible.

Additional benefits of using quantum dot sensitizers include their large absorption cross sections, which are an order of magnitude larger than those of organic dyes (**Table 5.2**). Colloidal nanocrystals are also solution processable. They can be ink-jetted, electrosprayed, or even spray-coated onto large surface areas. All of these properties then make them compatible with low-cost roll-to-roll processing.

Finally, there exist potential carrier multiplication effects in colloidal nanocrystals. Although still widely debated, the phenomenon could, in principle, lead to intriguing possibilities for improving solar cell efficiencies. Namely, generating multiple carriers with a single photon would dramatically improve device power conversion efficiencies. Hope therefore exists that quantum dots and other similar materials will lead to next-generation solar cells that are both low in cost and high in efficiency.

Many current quantum dot solar cells therefore bootstrap onto the original Grätzel cell, replacing the organic dye with colloidal nanocrystals. At the same time, other nanocrystal photovoltaic paradigms have been explored. As an example, mixtures of quantum dots with conductive polymers have been used to create what are referred to as bulk heterojunction solar cells. In addition, all-inorganic devices exist that use mixtures of nanocrystals and other nanostructures to harvest light. More information about Grätzel-cell-based quantum dot solar cells and their variants can be found in Kamat (2008).

15.2.4 The Ideal Diode Model

To model the operation of a solar cell and, more importantly, to estimate its efficiency, we introduce what is referred to as the ideal diode model. **Figure 15.5(a)** shows a generalized scheme of the device. It consists of a diode, an external current source (this is the light-generated current in our device), and a resistor that represents our load.

In practice, a slightly more complicated equivalent circuit accounts for imperfections in the cell. An example is illustrated in **Figure 15.5(b)**. The only difference is the addition of shunt and series resistances to model intrinsic power losses. Let us ignore these complications in what follows.

Given the diode in the model, an equation describing the current running through it was derived by Shockley a number of years ago. Note that Schockley was apparently a pretty colorful character. Along with John Bardeen and Walter Brattain, he developed the first transistor at Bell Laboratories in 1947. All three later won the Nobel Prize in physics (1956) for their work. A terrific account of the birth of the transistor

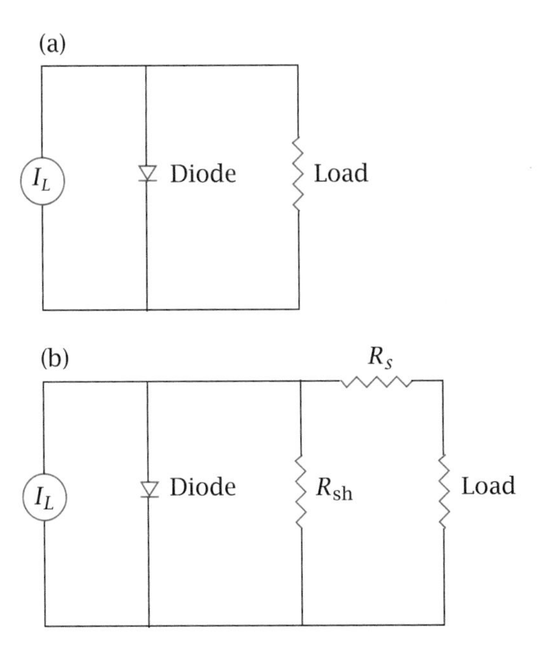

Figure 15.5 Illustration of the ideal diode solar cell model. (a) Idealized version. (b) More realistic version with shunt (R_{sh}), and series (R_s), resistances.

and a real behind-the-scenes description of how science was (and still is) conducted can be found in Riordan and Hoddeson (1997).

Now, when Schockley's equation is modified to describe the current in a solar cell, the following expression arises (the difference is just a negative sign to reflect that a negative current implies power generation):

$$I = I_0(1 - e^{qV/\gamma kT}).$$

Here I is the current, I_0 is called the diode reverse saturation current, V is the magnitude of the potential across the diode, and γ is a small fudge factor absent in ideal diodes but included here to account for the nonideality of actual cells. It takes values between 2 and 3, with $\gamma = 1$ being ideal. **Figure 15.6** shows the appearance of I when plotted against V.

Next, when an independent, light-generated, current I_L is added to the expression, the cell's current becomes

$$I = I_0(1 - e^{qV/\gamma kT}) + I_L. \tag{15.1}$$

Its appearance is illustrated in **Figure 15.6** and we see that I is simply offset from its original value by I_L.

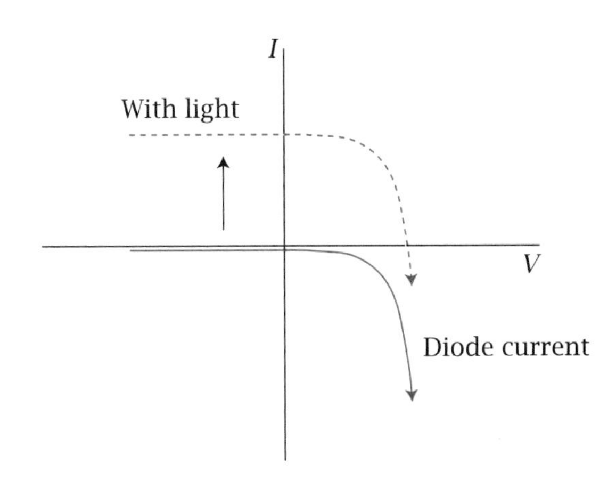

Figure 15.6 Illustration of the diode current I with (dashed line) and without (solid line) light.

Let us now use Equation 15.1 to derive several important properties of the idealized solar cell. Namely, we will find expressions for its open-circuit voltage V_{OC} as well as its short-circuit current I_{SC}. Both are important parameters that relate to its efficiency. In addition, we will derive relationships for the device's power curve, its maximum power generating voltage V_{max}, the corresponding maximum power current I_{max}, and the overall power conversion efficiency η. The latter requires use of the cell's so-called fill factor FF, which we will introduce shortly.

The Open-Circuit Voltage V_{OC}

To obtain the open-circuit voltage from Equation 15.1, we set $I = 0$. As the name implies, V_{OC} is the potential under which no current flows through the device. We thus find that

$$0 = I_0(1 - e^{qV_{OC}/\gamma kT}) + I_L,$$

which rearranges to

$$e^{qV_{OC}/\gamma kT} = 1 + \frac{I_L}{I_0}.$$

At this point, taking the natural logarithm of both sides gives

$$\ln(e^{qV_{OC}/\gamma kT}) = \ln\left(1 + \frac{I_L}{I_0}\right)$$

$$\frac{qV_{OC}}{\gamma kT} = \ln\left(1 + \frac{I_L}{I_0}\right),$$

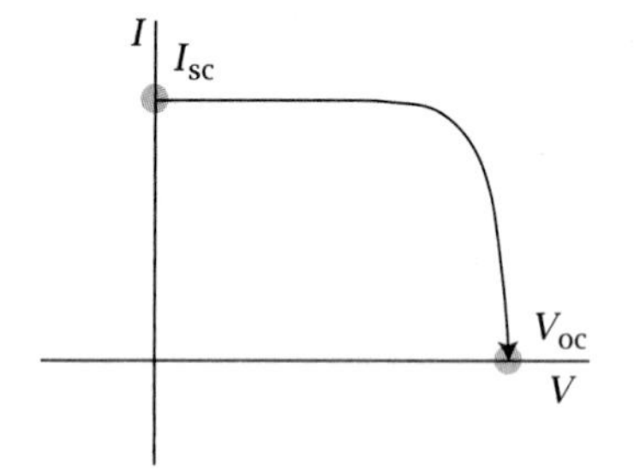

Figure 15.7 Illustration of important points in the diode current.

and from here we find the desired open-circuit voltage

$$V_{OC} = \frac{\gamma kT}{q} \ln\left(1 + \frac{I_L}{I_0}\right). \quad (15.2)$$

The position of V_{OC} in an I–V plot can be seen in **Figure 15.7**.

The Short-Circuit Current I_{SC}

Next, to obtain the short-circuit current, we set $V = 0$ in Equation 15.1. This gives

$$I_{SC} = I_L \quad (15.3)$$

and represents the maximum current generated in the device. I_{SC} will be important as a measure of the cell's efficiency when discussing its external efficiency (below). The position of I_{SC} in a I–V plot is also shown in **Figure 15.7**.

The Power Curve

Finally, the power P of the cell is simply the product of its current and the potential energy difference: $P = IV$. The units are watts when I is in amps and V is in volts. From Equation 15.1, we find that

$$P = I_0 V\left(1 - e^{qV/\gamma kT}\right) + I_L V. \quad (15.4)$$

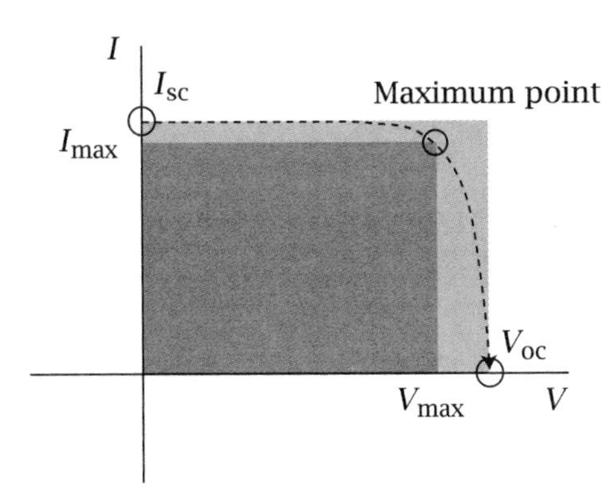

Figure 15.8 Illustration of the diode current maximum power point and the associated area, representing the maximum power output of the cell. For comparison purposes, the ideal area, defined by the open-circuit voltage and short-circuit current, is illustrated.

The voltage and current associated with the maximum power output of the cell, V_{max} and I_{max} (**Figure 15.8**), are found by taking the derivative of Equation 15.4 and setting it to zero. We thus have

$$\frac{dP}{dV} = 0 = (I_0 + I_L) - I_0 e^{qV_{max}/\gamma kT}\left(1 + \frac{qV_{max}}{\gamma kT}\right),$$

whereupon solving for the roots of

$$\left(1 + \frac{I_L}{I_0}\right) - e^{qV_{max}/\gamma kT}\left(1 + \frac{qV_{max}}{\gamma kT}\right) = 0$$

yields V_{max}. This is done numerically with a computer. Equation 15.1 can subsequently be used to find I_{max}.

Fit to Real Data

At this point, we can fit experimental data to the above ideal diode model. This is readily done in the limit where $e^{qV/\gamma kT} \gg 1$. In this case, Equation 15.1 becomes

$$I \simeq -I_0 e^{qV/\gamma kT} + I_L.$$

Consolidating terms then gives $I_L - I \simeq I_0 e^{qV/\gamma kT}$, and taking the natural logarithm of both sides yields

$$\ln(I_L - I) \simeq \frac{qV}{\gamma kT} + \ln I_0. \tag{15.5}$$

Plotting $\ln(I_L - I)$ versus V therefore results in a straight line with a slope of $q/\gamma kT$ and an intercept $\ln I_0$. In turn, the experimental data are straightforward to fit using this expression.

The Power Conversion Efficiency and Fill Factor

Having established V_{max}, I_{max}, V_{oc}, and I_{sc} as well as the cell's output power, two important parameters that characterize the solar cell's performance are its power conversion efficiency η and its fill factor (FF). The former is basically the ratio of the cell's output power P_{out} to that of the incident light, P_{in}:

$$\eta = \frac{P_{out}}{P_{in}} = \frac{I_{max}V_{max}}{P_{in}} = \frac{I_{sc}V_{oc}\mathrm{FF}}{P_{in}}. \tag{15.6}$$

FF is a unitless number between 0 and 1 and also reflects the cell's efficiency. It is defined as the ratio of areas specified by I_{max} and V_{max}

Table 15.1 Nanostructured solar cell efficiencies

System types	AM 1.5, 1 sun power conversion efficiency (%)	References
Photoelectrochemical systems		
CdSe quantum dots/TiO_2	2.9	Lee et al. (2008a)
CdS/CdSe quantum dots/TiO_2	2.8	Niitsoo et al. (2006)
Ru dye/ZnO nanowire	1.5	Law et al. (2005)
CdSe quantum dots/TiO_2	1.0	Lee et al. (2008b)
CdSe quantum dots/ZnO nanowire	0.4	Leschkies et al. (2007)
InAs quantum dots/TiO_2	0.3	Yu et al. (2006)
Bulk heterojunction systems		
CdSe tetrapods/PPV	1.8	Sun et al. (2003)
CdSe nanorods/P3HT	1.7	Huyhn et al. (2002)
CdS nanowire/MEH-PPV	1.6	Lee et al. (2009)
ZnO nanowire/P3HT	0.6	Unalan et al. (2008)
CdSe quantum dots/MEH-PPV	0.1	Greenham et al. (1996)
PbSe quantum dots/P3HT	0.04–0.14	Jiang et al. (2007)
All-inorganic systems		
CdSe nanorod/CdTe nanorod	2.9	Gur et al. (2005)
PbS quantum dot	1.8	Johnston et al. (2008)
PbSe quantum dot	1.1	Koleilat et al. (2008)
PbS quantum dot	0.5	Klem et al. (2007)
Si nanowire	0.1	Tsakalakos et al. (2007)

as well as I_{sc} and V_{oc}:

$$FF = \frac{I_{max} V_{max}}{I_{sc} V_{oc}}. \tag{15.7}$$

This is illustrated in **Figure 15.8** as a ratio of two shaded rectangular areas. Basically, FF is the ratio of the cell's maximum operating power output to the idealized output, as defined by V_{oc} and I_{sc}.

Note that in experiments that measure η, simulated sunlight is used to obtain the (practical) device power conversion efficiency. In this regard, we have seen that the original Grätzel cell has power conversion efficiencies of the order of 10%. More recent quantum-dot-based devices, however, have efficiencies of the order of a few percent. This shows that such nanostructured photovoltaics are still in their early stages of development. **Table 15.1** provides a short compilation of efficiencies, obtained from the literature.

Incident-Photon-to-Current Conversion Efficiency (IPCE)

Finally, there are times when, in addition to reporting the cell's broadband FF and power conversion efficiency, we seek to find its efficiency at individual wavelengths. This leads to a different experiment involving the cell's external efficiency or incident-photon-to-current conversion efficiency (IPCE).

In the experiment, an "action" spectrum is taken by measuring the device's photocurrent when excited at different wavelengths across the visible. **Figure 15.9** shows the response of several CdSe quantum dot photoelectrochemical solar cells. One sees that, depending on the quantum dot diameter, the response has different absorption edges,

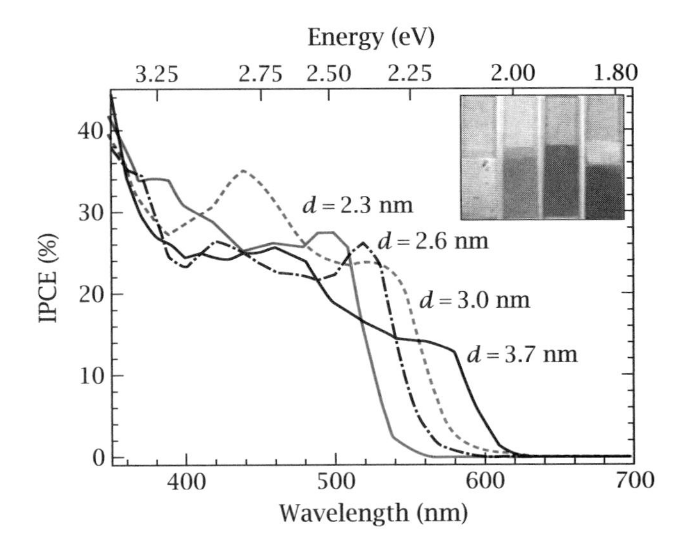

Figure 15.9 Incident-photon-to-current conversion efficiency (IPCE) photocurrent action spectra of CdSe quantum dot photoelectrochemical solar cells made using different-sized quantum dots. The inset is a photograph of the different electrodes, consisting of a quantum dot/TiO_2 layer deposited over ITO. Courtesy of K Tvrdy.

which shift to the blue with decreasing particle size. Furthermore, the response clearly mimics the linear absorption of the particles.

At each wavelength, the IPCE is defined as the ratio of the number of carriers measured for a given amount of time, n_{carrier}, to the number of incident photons on the device during that same period, n_{photon}:

$$\text{IPCE} = \frac{n_{\text{carrier}}}{n_{\text{photon}}}.$$

Implicit in the calculation is that every photon creates a carrier.

The maximum number of charges measured is then the product of the device's short circuit current density j_{sc} (with units of $\text{A/cm}^2 = \text{C s}^{-1}\,\text{cm}^{-2}$), a conversion factor $1/(1.602 \times 10^{-19})$, and the cell's area A_{cell} (cm^2)

$$n_{\text{carrier}} = j_{\text{sc}} \left(\frac{1}{1.602 \times 10^{-19}} \right) A_{\text{cell}}.$$

Likewise, the number of incident photons on the cell during that time is the incident light intensity I (with units of $\text{W/cm}^2 = \text{J s}^{-1}\,\text{cm}^{-2}$) divided by the photon's energy at a specific wavelength, $[1240/\lambda(\text{nm})] \times 1.602 \times 10^{-19}$ (J/photon) times the detector area A_{detector} (cm^2). Note that the first part of the conversion factor, $1240/\lambda$, is simply Equation 5.27, which describes how one converts from wavelength to eV. The second factor, 1.602×10^{-19}, converts eV to J. We thus find that

$$n_{\text{photon}} = I \frac{\lambda}{1240 \times (1.602 \times 10^{-19})} A_{\text{detector}}.$$

The ratio $n_{\text{carrier}}/n_{\text{photon}}$ is then

$$\text{IPCE} = \frac{j_{\text{sc}} \frac{1}{1.602 \times 10^{-19}} A_{\text{cell}}}{I \frac{\lambda}{1240 \times (1.602 \times 10^{-19})} A_{\text{detector}}} = \left(\frac{1240}{\lambda} \right) \left(\frac{j_{\text{sc}}}{I} \right) \left(\frac{A_{\text{cell}}}{A_{\text{detector}}} \right),$$

and if we assume that $A_{cell} = A_{detector}$, we obtain the expression for IPCE that is commonly seen in the literature:

$$\mathrm{IPCE}(\lambda) = \left[\frac{1240}{\lambda(\mathrm{nm})}\right]\left[\frac{j_{sc}(\mathrm{A/cm^2})}{I(\mathrm{W/cm^2})}\right]. \tag{15.8}$$

Clearly, the closer to unity this value is, the more efficient is the cell at that particular wavelength. Note, however, that the external efficiency should not be confused with the power conversion efficiency. This is because the former does not take into account the full broadband solar spectrum. Instead, it represents the device's response at a specific wavelength. Furthermore, there are other more subtle distinctions between the two that we will not discuss.

15.3 METAL NANOSTRUCTURES

In the area of metal nanostructures, an exciting area of current nanoscience and nanotechnology involves the nano/bio interface. Namely, there exists tremendous interest in applying nanoscale materials in addressing biological problems. One of the most advanced areas in this field employs metal nanoparticles to carry out biological sensing studies.

Motivating much of this are commercial interests in developing low-cost medical diagnostic kits. Other reasons exist. The anthrax attacks after 9/11 clearly illustrated the need for sensitive biosensors that could be readily deployed. Numerous researchers have therefore taken advantage of the relatively facile surface functionalization chemistries of colloidal nanostructures to make low-cost biosensors. In what follows, we highlight a few of these applications within the context of metal nanostructures.

15.3.1 Contrast Agents for Biological Labeling

Early uses of metal nanoparticles have focused on applying them as labels in biological imaging studies. There are a number of reasons for this. First, the surface chemistry of gold nanoparticles is relatively easy. For example, well-known gold–thiol chemistry can be exploited to attach different chemical moieties to a nanoparticle surface (Chapter 13). This can subsequently be used to functionalize nanoparticles in such a way as to impart biological recognition capabilities.

Next, when samples are imaged using electron microscopy (e.g., transmission electron microscopy (Chapter 14)), the electron-dense gold particles are clearly visible against a much weaker background from the specimen. This strong contrast then enables one to readily determine the location of a species of interest, given the recognition capabilities of the functionalized particles. At the same time, different-sized gold nanocrystals can be used to stain different regions of the same specimen, providing multiplexed labeling capabilities.

Gold nanoparticles can also be imaged optically. This is because they scatter light efficiently. We have already discussed light scattering in Chapter 14 and will discuss more about the plasmon resonance of these gold nanoparticles shortly. Thus, metal nanoparticles can also be used as optical microscopy labels in addition to electron microscopy stains. Sperling et al. (2008) provide an overview of the field.

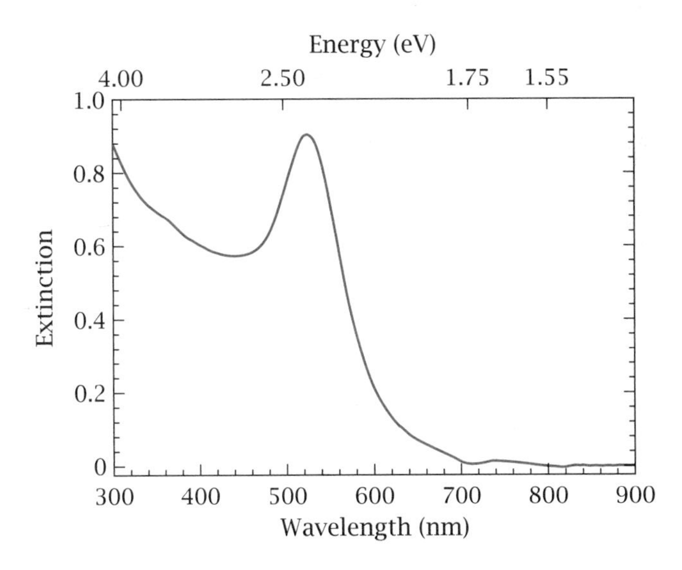

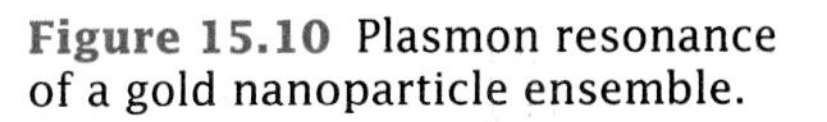

Figure 15.10 Plasmon resonance of a gold nanoparticle ensemble.

15.3.2 Plasmon Resonance Sensors

Metals are typically characterized by the presence of free electrons. In nanoscale metals such as gold sols, the collective excitation of these unbound charges leads to strong light absorption, referred to as a plasmon resonance. This resonance has a spectral position and linewidth that depends on the particle's size as well as its shape. An illustration of the plasmon resonance of gold nanoparticles is shown in **Figure 15.10**. The peak near 520 nm is evident and is characteristic of gold particles with diameters of approximately 15 nm.

Mie was the first to model the extinction due to small spherical particles (the extinction is simply the sum of any absorbed and scattered light). By solving Maxwell's equations for a polarizable sphere interacting with plane wave electromagnetic radiation, he was able to derive expressions for the absorption and scattering of light by particles with sizes smaller than its wavelength. Under dilute conditions where the scattering is small, Mie showed that the particle's plasmon absorption had a Lorentzian lineshape.

Furthermore, the corresponding frequency-dependent absorption cross section was found to be

$$\sigma(\omega) = \frac{9\omega V \epsilon_m^{3/2}}{c} \frac{\epsilon_2(\omega)}{[\epsilon_1(\omega) + 2\epsilon_m]^2 + [\epsilon_2(\omega)]^2}, \tag{15.9}$$

where V is the nanoparticle volume, ϵ_1 (ϵ_2) is the real (imaginary) part of the material's dielectric constant, and ϵ_m is the dielectric constant of the surrounding medium. Notice that the plasmon resonance peaks when the denominator $[\epsilon_1(\omega) + 2\epsilon_m]^2 + [\epsilon_2(\omega)]^2$ is minimized. Under these conditions, $\epsilon_1(\omega) = -2\epsilon_m$, showing that both the cross section and the resonance's peak position depend on the environment's dielectric constant. This sensitivity can therefore be used as the basis of a plasmon resonance sensor as we will see.

EXAMPLE 15.1

Estimate of the Absorption Cross Section of a Gold Nanoparticle To illustrate the use of Equation 15.9 let us calculate the cross section of a representative 16 nm diameter gold nanoparticle, analogous to those shown in **Figures 4.12** and **14.2**. The result can then be compared with the cross sections of other materials previously listed in **Table 5.2**. Towards this end, the optical constants (n and k) of bulk gold at 520 nm

(2.38 eV) are taken from Johnson and Christy (1972). The reader may rationalize this choice as an exercise. Furthermore, from Chapter 5, ϵ_1 and ϵ_2 are related to n and k through (Equation 5.14)

$$\epsilon_1 = n^2 - k^2$$

and (Equation 5.15)

$$\epsilon_2 = 2nk.$$

When all are introduced into Equation 15.9, we obtain $\sigma(520\,\text{nm}) = 6.1 \times 10^{-12}\,\text{cm}^2$. This is a large number and shows that metal nanoparticles absorb strongly. The estimated σ-value is also in good agreement with Demers et al. (2000), who determined an experimental molar extinction coefficient $\epsilon = 4.2 \times 10^8\,\text{M}^{-1}\,\text{cm}^{-1}$ at 520 nm. To make the agreement more explicit, Equation 5.6 from Chapter 5 can be used to convert ϵ to an absorption cross section $\sigma(520\,\text{nm}) = 1.61 \times 10^{-12}\,\text{cm}^2$.

A Plasmon Resonance Sensor

We have seen that the peak of the nanoparticle plasmon absorption depends on the dielectric constant of the surrounding medium. We now show this explicitly by finding how the frequency associated with the particle's peak absorption depends on ϵ_m. Recall from Chapter 3 that the real part of a metal's dielectric constant is (Equation 3.41)

$$\epsilon_1 = 1 - \frac{\omega_p^2}{\omega^2 + \gamma^2},$$

where ω_p is the bulk plasma frequency $\omega_p^2 = nq^2/\epsilon_0 m$ and γ is a damping frequency with values of the order of $10^{15}\ \text{s}^{-1}$. Alternatively,

$$\epsilon_1 \simeq \epsilon_\infty - \frac{\omega_p^2}{\omega^2 + \gamma^2},$$

since $\epsilon_\infty \to 1$ at high frequencies; note that ϵ_∞ is the metal's high-frequency dielectric constant. Furthermore, given that $\omega \gg 10^{15}$ rad/s in the denominator, γ^2 can be ignored because it takes smaller values of order $2\pi \times 10^{15}$ rad/s. This leaves us with

$$\epsilon_1 \simeq \epsilon_\infty - \frac{\omega_p^2}{\omega^2}.$$

At this point, we have seen that the peak of the plasmon resonance occurs when $\epsilon_1 = -2\epsilon_m$. Inserting the above expression for ϵ_1 into this leads to the following equivalence:

$$\epsilon_\infty - \frac{\omega_p^2}{\omega^2} = -2\epsilon_m,$$

$$\epsilon_\infty + 2\epsilon_m = \frac{\omega_p^2}{\omega^2}.$$

From this, we obtain

$$\omega^2 = \frac{\omega_p^2}{\epsilon_\infty + 2\epsilon_m}, \tag{15.10}$$

which provides a direct relationship between the resonance's peak frequency and the dielectric constant of the surrounding medium. We can find an even more convenient expression if we express everything in terms of wavelength. Recall that $\omega = 2\pi\nu = 2\pi c/\lambda$. As a consequence,

$$\frac{1}{\lambda^2} = \frac{1}{\lambda_p^2(\epsilon_\infty + 2\epsilon_m)},$$

or

$$\lambda^2 = \lambda_p^2(\epsilon_\infty + 2\epsilon_m), \tag{15.11}$$

where $\lambda_p^2 = (2\pi c/\omega_p)^2$. There is therefore a linear relationship between λ^2 (the square of the peak plasmon wavelength) and ϵ_m (the medium's dielectric constant). This dependence can subsequently be used as the basis of an analytical technique to detect biological moieties attached to nanoparticle surfaces. Namely, when a species of interest adsorbs onto gold nanoparticles, their plasmon resonance will shift. These changes can then be detected through simple absorption measurements.

Other sensing approaches based on changes in a metal nanoparticle's plasmon resonance exist. In this regard, the plasmon resonances of metal nanostructures with other sizes/morphologies differ from the spherical particle solution shown earlier (Equation 15.9). They also peak in different parts of the spectrum. As an example, metal nanorods, with aspect ratios greater that 1, have plasmon resonances in the red beyond 600 nm. The surface chemistry of colloidal nanoparticles can therefore be exploited to detect a species of interest by inducing them to aggregate. This is because the plasmon resonances of nanoparticles in close proximity to each other interact. In effect, the overall nanoparticle geometry changes, leading to dramatic shifts in the observed plasmon resonance relative to that of the original (isolated) particle.

In practice, such an approach entails using the analyte to neutralize surface charges that electrostatically stabilize the particles (Chapter 13). Alternatively, it could mean forming covalent/dative bonds between the analyte and multiple particles, causing them to flocculate. In all cases, a dramatic spectral shift results, enabling one to sense things. Elghanian et al. (1997) provide more information about the effect. As a final note, some commercial home pregnancy test kits already use this phenomenon as the basis of their color change.

15.3.3 Metal Nanoparticle Photothermal Therapy

Another emerging use of metal nanoparticles entails photothermal therapy. This is an approach that uses heat to destroy cancer cells. The underlying idea is that a sensitizer can be localized on or within a tumor. Upon absorbing light, any excess energy due to the nonradiative relaxation of excitations causes temperature changes in the immediate environment of the particle. On sustained exposure to light, the tumor heats up and is eventually destroyed.

The reason why metal nanoparticles such as gold have found widespread use in this area is because they possess large absorption cross sections (Equation 15.9). We have already seen σ values that are an order of magnitude larger than those of organic sensitizers. The particles therefore absorb strongly, allowing low laser powers to be used. Furthermore, because the particles do not readily emit light, any excess energy that is absorbed is ultimately released as heat into the local environment. Their facile surface chemistries also allow them to be functionalized to impart biorecognition capabilities. We have already seen this ability exploited in electron microscopy staining applications. More about uses of metal nanoparticles in photothermal therapy can be found in Lal et al. (2008) and Huang et al. (2006).

15.3.4 Surface-Enhanced Raman Scattering

Finally, an important technique for identifying chemical species is Raman spectroscopy. This is a complementary approach to the infrared absorption spectroscopy discussed earlier in Chapter 14. Raman spectroscopy entails looking at the scattered light from an analyte when illuminated with a laser. Within this light, there exist multiple frequencies due to elastic and inelastic scattering events. The dominant Rayleigh-scattered component possesses the same frequency as the incident light (Chapter 14). However, this is not of interest in the Raman experiment. Instead, the power of the technique comes from the other frequencies buried within the scattered light, which contains information about the sample's vibrational modes.

The existence of such inelastically scattered light can be rationalized through the following back-of-the-envelope estimate. Namely, we will describe the response of a molecule, having a single vibrational mode (ω_{vib}), to a time-varying electromagnetic field. Much of this has already been foreshadowed by our discussion of light scattering in Chapter 14, and the parallels will be evident.

In the Raman experiment, the induced dipole of the material, interacting with light, is

$$\mu = \alpha \mathbf{E},$$

where α is the polarizability of the molecule and $\mathbf{E}$ is the incident light's electric field. The latter can be written as (Equation 11.8)

$$\mathbf{E} = E_0 \mathbf{e} \cos \omega_0 t,$$

where $\mathbf{e}$ is a unit polarization vector. By analogy to our prior discussion on infrared vibrations (Chapter 14), the system's associated time-dependent displacement is

$$x(t) = x_0 \cos \omega_{vib} t.$$

For simplicity, we have assumed the x direction of a Cartesian coordinate system as our vibrational coordinate, which, in turn, means that we are dealing with x-polarized light: $E_x = E_0 \cos \omega_0 t$.

Next, if the polarizability of the molecule changes due to interactions with the light's electric field, the following series expansion can be invoked:

$$\begin{aligned} \alpha &= \alpha_0 + \left(\frac{\partial \alpha}{\partial x}\right) x + \cdots \\ &= \alpha_0 + \left(\frac{\partial \alpha}{\partial x}\right) (x_0 \cos \omega_{vib} t) + \cdots . \end{aligned}$$

Keeping only the first two terms of the expansion then gives us the following time-varying polarizability:

$$\alpha \simeq \alpha_0 + \left(\frac{\partial \alpha}{\partial x}\right) (x_0 \cos \omega_{vib} t).$$

Note that in our previous discussion of Rayleigh scattering, we considered only the first term.

Inserting everything into μ yields

$$\begin{aligned} \mu &\simeq \left[\alpha_0 + \left(\frac{\partial \alpha}{\partial x}\right) x_0 \cos \omega_{vib} t\right] (E_0 \cos \omega_0 t) \\ &\simeq \alpha_0 E_0 \cos \omega_0 t + \left(\frac{\partial \alpha}{\partial x}\right) x_0 \cos \omega_0 t \cos \omega_{vib} t, \end{aligned}$$

whereupon invoking the trigonometric identity $\cos a \cos b = \frac{1}{2}[\cos(a - b)t + \cos(a + b)t]$ gives

$$\mu(t) \simeq \alpha_0 E_0 \cos \omega_0 t + \left(\frac{\partial \alpha}{\partial x}\right) \frac{x_0}{2} [\cos(\omega_0 - \omega_{vib})t - \cos(\omega_0 + \omega_{vib})t] .$$

From this, it is apparent that the sample's dipole moment oscillates with time. Furthermore, since a time-varying (oscillating) dipole radiates light, three different frequencies arise due to the three harmonic terms present. The first, at ω_0, accounts for Rayleigh scattering (Chapter 14). The second, at a lower frequency $\omega_0 - \omega_{vib}$, accounts for what is called Stokes Raman scattering. The third, at a higher frequency, $\omega_0 + \omega_{vib}$, accounts for anti-Stokes Raman scattering. The chemical specificity of the Raman technique therefore arises from having scattered light with frequency shifts (away from the Rayleigh line) that reflect the molecule's vibrational frequency.

The Stokes Raman line turns out to have a stronger intensity than the anti-Stokes line. This is due to differences in the thermal population of states, as dictated by the Boltzmann distribution. However, the important thing to note is that both lines are much weaker than the Rayleigh line. As a consequence, the sensitivity of the Raman experiment is not intrinsically high. Samples must therefore be highly concentrated for a signal to be seen. Alternatively, the incident laser intensity must be large. In situations focused on detecting trace amounts of a material, improvements to the Raman approach are therefore necessary.

Towards this end, it was discovered in the late 1970s that roughened metal surfaces led to large enhancements in the Raman scattering efficiency. Initially, the explanation for the effect was that roughened surfaces possessed more surface area onto which analytes could adsorb (Fleischmann et al. 1974). This would, in turn, boost their local concentration, leading to larger Raman signals. However, it soon became apparent that another effect could better explain the phenomenon (Jeanmaire and Van Duyne 1977; Albrecht and Creighton 1977). Namely, metal nanostructures scatter strongly in the vicinity of a plasmon resonance.

To illustrate, a qualitative description for the magnitude of the field radiated by a metal particle, E_{np} is

$$E_{np} = fE_0,$$

where E_0 is the electric field of the incident light and f is a local, near-field enhancement factor. Molecules adsorbed onto metal nanoparticles therefore experience an enhanced light intensity, which consequently increases their Raman scattering magnitude. Next, the Raman-scattered light is itself enhanced by the metal nanoparticle. The overall magnitude of the resulting (surface-enhanced) scattered electric field is therefore

$$\begin{aligned} E_{SERS} &\sim f' E_{np} \\ &\sim f'(fE_0), \end{aligned}$$

where f' is the field enhancement factor at the Raman-shifted frequency $\omega_0 - \omega_{vib}$. The intensity of the resulting scattered light is then the square of the electric field:

$$\begin{aligned} I_{SERS} &\sim |ff'|^2 |E_0|^2 \\ &\sim |ff'|^2 I_0 . \end{aligned}$$

with $I_0 = |E_0|^2$. If the enhancement factor at $\omega_0 - \omega_{vib}$ is essentially the same as that at ω_0 (i.e., $f \simeq f'$), the surface-enhanced Raman scattering intensity becomes

$$I_{SERS} \sim |f|^4 I_0. \tag{15.12}$$

This shows that the intensity of an analyte's Raman-scattered light in the immediate vicinity of a metal nanoparticle is enhanced by a factor raised to the fourth power. Such electromagnetic enhancement factors are quite large and can easily be on the order of 10^6–10^8. In addition, enhancement factors as large as 10^{14} have been observed, enabling the Raman spectroscopy of individual molecules (Nie and Emory 1997; and Kneipp et al. 1997).

Because of these dramatic enhancement capabilities, there has been great interest in developing metal nanoparticles for Raman sensing applications. The realization that gaps between metal nanoparticles also lead to similarly large enhancement factors has motivated the subsequent development of architectures that contain nanoscale junctions. More about SERS and the use of metal nanostructures can be found in Kneipp et al. (2006).

15.3.5 Single-Electron Transistors

We have seen in Chapter 3 that small metal particles have sizable capacitances. Depending on their diameter, charging energies as large as 0.24 eV are possible. Researchers therefore realized early on that if one could make small enough nanocrystals, their capacitance would be large enough to store discrete charges at room temperature. In turn, this could be used as the basis of a single-electron transistor (SET) in next-generation electronics. Such SETs would then operate by turning on and off upon the addition of individual charges. More information about SETs and their operating principle can be found in Feldheim and Keating (1998).

15.4 NANOWIRES

Having seen a few of the many applications of semiconductor/metal nanoparticles, we note that the commercial use of nanowires is not presently so advanced. This is because the historical development of zero-dimensional structures precedes that of nanowires, and only recently have high-quality one-dimensional structures been made through the development of VLS and other one-dimensional growth strategies (Chapter 13).

The move towards applications, however, has been fast. The key selling point of nanowires is that, in addition to having a unique density of states, which is peaked near the band edge (Chapter 9), nanowires are (as their name suggests) wires. This means that electrical connections to the outside world are much easier to achieve than with other systems like quantum dots.

To date, nanowires have been used in a variety of proof-of-concept demonstrations. From the electronics standpoint, they have been used as the basis of nanoscale field effect transistors and logic elements. From the optoelectronic standpoint, wires have been used in prototype photodetectors, waveguides, and light-emitting diodes, and even as lasing elements. More recently, there has been interest in using one-dimensional structures for possible photovoltaic applications.

On the biological side of things, nanowires have been used as sensors for different compounds or gases adsorbed to their surface. This is done

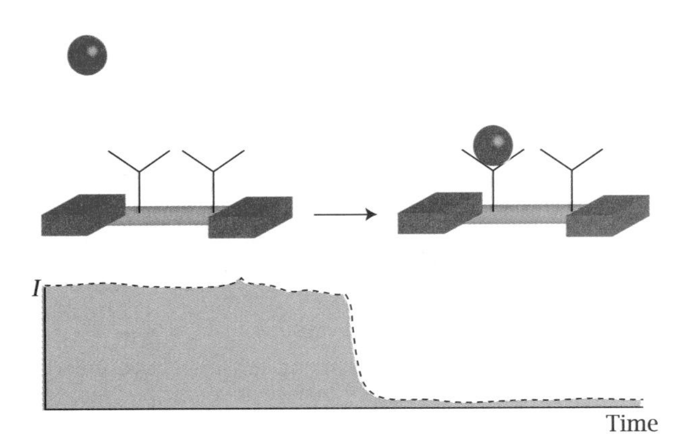

Figure 15.11 Illustration of a nanowire sensor, based on conductivity changes that follow a binding event.

by monitoring changes in their conductivity following binding events. To illustrate, **Figure 15.11** qualitatively depicts the change in an electrical current running through a wire upon a binding event. The effect stems from variations in the wire's surface potential and can, in turn, be used to sense biological events. More information about these sorts of devices as well as general overviews about nanowire applications can be found in the Further Reading.

15.5 SUMMARY

With this, we end our introduction to nanoscience. We have seen in the past chapters that crystalline nanostructures exhibit size-dependent properties. Furthermore, various length scales have been defined to determine where size-dependent effects take place. One of the most interesting aspects of these low-dimensional systems has been their unique optical properties. Quantum mechanical models have therefore been introduced to describe their linear absorption as well as the underlying physics behind their optical transitions. In tandem, we have seen how the density of states of wells, wires, and dots changes with dimensionality. With the last three chapters, we have come to the logical conclusion of our brief survey and have focused on the synthesis, characterization, and application of these materials.

15.6 THOUGHT PROBLEMS

15.1 Barcodes

Read the article about so-called nanobarcodes by Nicewarner-Pena et al. (1994). Explain their use and how the technology works.

15.2 Fluorescent beads

Apart from their direct use as fluorescent probes, colloidal quantum dots have also been incorporated into silica or polystyrene beads for analogous biological imaging applications. Read Han et al. (2001) and explain how the dots were incorporated into the beads and what the advantages of such a system are.

15.3 Magnetic beads for separation

Although we have not discussed the magnetic properties of nanostructures, the colloidal synthesis of magnetic nanoparticles employs many of the same strategies used to synthesize the emissive semiconductor systems described in the main text. One application of these magnetic particles involves their use in separations. See, for example, Grass et al. (2007). Explain the concept and find other examples of this technology.

15.4 MRI contrast agents

Along the same lines, magnetic nanoparticles can also be used as contrast agents in magnetic resonance imaging (MRI). Read Na et al. (2009) and explain how magnetic particles function in this application.

15.5 Ferrofluids

Find uses of ferrofluids in common everyday applications. Now explore the types of magnetic nanostructures employed in this regard. Example syntheses of such materials are described by Berger et al. (1999) and Sun and Zeng (2002).

15.6 Security

Researchers have proposed using colloidal quantum dots, nanoparticles and other nanostructures as elements in anti-counterfeiting measures. Investigate this area and explain how such a nano-enabled technology would work.

15.7 Molecular rulers

Measuring distances on the nanometer scale is an important task for a number of scientific investigations. Provide a few examples. More recently, colloidal nanoparticles have been used as molecular rulers. Read Sonnichsen et al. (2005) and explain how the technology works.

15.8 Photonic bandgaps

Arrays of close-packed quantum dots may have uses as photonic band-gap materials. Read Vlasov et al. (1999), and explain the interest in photonic bandgap systems, using this and other articles you find in the literature.

15.9 Single-electron transistors

Single-electron transistors have been built with colloidal quantum dots and with metal nanoparticles. We alluded to this in the main text. Read more about this technology in Klein et al. (1997) and Feldheim and Keating (1998).

15.10 Solar cells

We discussed the use of low-dimensional systems in next-generation solar photovoltaics. Read more about this area and, in particular, the articles about quantum dot solar cells by Huynh et al. (2002), Gur et al. (2005), and McDonald et al. (2005). Then see the articles about analogous nanowire based systems by Law et al. (2005), Tian et al. (2007), and Fan et al. (2009). Are there advantages of nanowire systems over comparable quantum dot devices? What limits the overall power conversion efficiencies of these solar cells?

15.11 Photodetectors

As with solar photovoltaics, there also exists strong interest in creating nanoscale photodetectors. Read more about these proof-of-concept devices in Qi et al. (2005), Oertel et al. (2005), and Konstantatos et al. (2006) for quantum dots and in Hayden et al. (2006) and Pettersson et al. (2006) for nanowires. As before, are there advantages of nanowire systems over comparable quantum dot devices?

15.12 Light-emitting diodes

Another area of interest, described in the main text, has to do with next-generation lighting and displays. Read more about this area in Qian et al. (2005) Huang et al. (2005), and Caruge et al. (2008). Summarize what improvements must be made before this technology becomes competitive with current lighting and/or display technologies.

15.13 Transistors

Explain how nanowire field effect transistors work. See Cui et al. (2003) and Xiang et al. (2006). Describe the different approaches and underlying strategies for creating these transistors.

15.14 Nanowire sensors

Many nanowire transistor schemes have also been employed towards developing nanowire sensors. Read the articles by Favier et al. (2001) on the creation of gas sensors and Patolsky and Lieber (2005) and Zheng et al. (2005) on more biologically oriented applications.

15.15 Quantum dot sensors

Colloidal quantum dot sensors have also been made. For an example within a biological context, see Willard and Van Orden (2003) and Medintz et al. (2003).

15.16 Thermoelectrics

Although we did not discuss the thermal properties of nanostructures, read Dresselhaus et al. (2007), Hochbaum et al. (2008), and Boukai et al. (2008), and summarize for yourself the interest in low-dimensional systems for potential thermoelectrics.

15.17 An extension

Explain potential applications of nanocrystal superlattices. See Shevchenko et al. (2006) and then see if you can devise an extension of this work with an applied bent.

15.18 An extension

Explain how nanowires have been assembled to make circuits. See Huang et al. (2001). Now see if you can come up with an alternative way of assembling these materials.

15.19 Literature

Find other applications of colloidal quantum dots, metal nanoparticles or nanowires that we have not discussed here or in the main text.

15.20 An application

Having seen the above applications of low-dimensional materials, attempt to devise an original application of wells, wires, or dots (or any other nano system). Describe your application, state why it has advantages over bulk analogues, and explain how you intend to make such a device.

15.7 REFERENCES

Albrecht MG, Creighton JA (1977) Anomalously intense Raman spectra of pyridine at a silver electrode. *J. Am. Chem. Soc.* **99**, 5215.

Berger P, Adelman NB, Beckman KJ, et al. (1999) Preparation and properties of an aqueous ferrofluid. *J. Chem. Ed.* **76**, 943.

Boukai AI, Bunimovich Y, Tahir-Kheli J, et al. (2008) Silicon nanowires as efficient thermoelectric materials. *Nature* **451**, 168.

Caruge JM, Halpert JE, Wood V, et al. (2008) Colloidal quantum-dot light-emitting diodes with metal-oxide charge transport layers. *Nature Photonics* **2**, 247.

Cichos F, von Borczyskowski C, Orrit M (2007) Power-law intermittency of single emitters. *Curr. Opin. Colloid Interface Sci.* **12**, 272.

Cui Y, Zhong Z, Wang D, et al. (2003) High performance silicon nanowire field effect transistors. *Nano Lett.* **3**, 149.

Demers LM, Mirkin CA, Mucic RC, et al. (2000) A fluorescence-based method for determining the surface coverage and hybridization efficiency of thiol-capped oligonucleotides bound to gold thin films and nanoparticles. *Anal. Chem.* **72**, 5535.

Dresselhaus MS, Chen G, Tang MY, et al. (2007) New directions for low-dimensional thermoelectric materials. *Adv. Mater.* **19**, 1.

Elghanian R, Storhoff JJ, Mucic RC, et al. (1997) Selective colorimetric detection of polynucleotides based on the distance dependent optical properties of gold nanoparticles. *Science* **277**, 1078.

Fan Z, Razavi H, Do J, et al. (2009) Three-dimensional nanopillar-array photovoltaics on low-cost and flexible substrates. *Nature Mater.* **8**, 648.

Favier F, Walter EC, Zach MP, et al. (2001) Hydrogen sensors and switches from electrodeposited palladium mesowire arrays. *Science* **293**, 2227.

Feldheim DL, Keating CD (1998) Self-assembly of single electron transistors and related devices. *Chem. Soc. Rev.* **27**, 1.

Fleischmann M, Hendra PJ, McQuillan AJ (1974) Raman spectra of pyridine adsorbed at a silver electrode. *Chem. Phys. Lett.* **26**, 163.

Frantsuzov P, Kuno M, Janko B, Marcus RA (2008) Universal emission intermittency in quantum dots, nanorods and nanowires. *Nature Phys.* **4**, 519.

Grass RN, Athanassiou EK, Stark WJ (2007) Covalently functionalized cobalt nanoparticles as a platform for magnetic separations in organic synthesis. *Angew. Chem. Int. Ed.* **46**, 4909.

Greenham NC, Peng X, Alivisatos AP (1996) Charge separation and transport in conjugated polymer/semiconductor-nanocrystal composites studied by photoluminescence quenching and photoconductivity. *Phys. Rev. B* **54**, 17628.

Gur I, Fromer NA, Geier ML, Alivisatos AP (2005) Air-stable all-inorganic nanocrystal solar cells processed from solution. *Science* **310**, 462.

Han M, Gao X, Su JZ, Nie S (2001) Quantum-dot-tagged microbeads for multiplexed optical coding of biomolecules. *Nature Biotech.* **19**, 631.

Hayden O, Agarwal R, Lieber CM (2006) Nanoscale avalanche photodiodes for highly sensitive and spatially resolved photon detection. *Nature Mater.* **5**, 352.

Hochbaum AI, Chen R, Delgado RD, et al. (2008) Enhanced thermoelectric performance of rough silicon nanowires. *Nature* **451**, 163.

Huang Y, Duan X, Wei Q, Lieber CM (2001) Directed assembly of one-dimensional nanostructures into functional networks. *Science* **291**, 630.

Huang Y, Duan X, Lieber CM (2005) Nanowires for integrated multicolor nanophotonics. *Small* **1**, 142.

Huang X, Jain PK, El-Sayed IV, El-Sayed MA (2006) Determination of the minimum temperature required for selective photothermal destruction of cancer cells with the use of immunotargeted gold nanoparticles. *Photochem. Photobiol.* **82**, 412.

Huynh WU, Dittmer JJ, Alivisatos AP (2002) Hybrid nanorod–polymer solar cells. *Science* **295**, 2425.

Jeanmaire DL, Van Duyne RP (1977) Surface Raman spectroelectrochemistry. *J. Electroanal. Chem.* **84**, 1.

Jiang X, Schaller RD, Lee SB, et al. (2007) PbSe nanocrystal/conducting polymer solar cells with an infrared response to 2 micron. *J. Mater. Res.* **22**, 2204.

Johnson PB, Christy RW (1972) Optical constants of the noble metals. *Phys. Rev. B* **6**, 4370.

Johnston KW, Pattantyus-Abraham AG, Clifford JP, et al. (2008) Schottky-quantum dot photovoltaics for efficient infrared power conversion. *Appl. Phys. Lett.* **92**, 151115.

Kamat PV (2008) Quantum dot solar cells. Semiconductor nanocrystals as light harvesters. *J. Phys. Chem. C* **112**, 18737.

Klein DL, Roth R, Limb AKL, et al. (1997) A single-electron transistor made from a cadmium selenide nanocrystal. *Nature* **389**, 699.

Klem EJD, MacNeil DD, Cyr PW, et al. (2007) Efficient solution-processed infrared photovoltaic cells: planarized all-inorganic bulk heterojunctions devices via inter-quantum-dot bridging during growth from solution. *Appl. Phys. Lett.* **90**, 183113.

Kneipp K, Wang Y, Kneipp H, et al. (1997) Single molecule detection using surface-enhanced Raman scattering (SERS). *Phys. Rev. Lett.* **78**, 1667.

Kneipp K, Moskovits M, Kneipp H (eds) (2006) Surface-Enhanced Raman Scattering: Physics and Applications. Springer-Verlag, Berlin.

Koleilat GI, Levina L, Shukla H, et al. (2008) Efficient, stable infrared photovoltaics based on solution-cast colloidal quantum dots. *ACS Nano*, **2**, 833.

Konstantatos G, Howard I, Fischer A, et al. (2006) Ultrasensitive solution-cast quantum dot photodetectors. *Nature* **442**, 180.

Lal S, Clare SE, Halas NJ (2008) Nanoshell-enabled photothermal cancer therapy: impending clinical impact. *Acc. Chem. Res.* **41**, 1842.

Law M, Greene LE, Johnson JC, et al. (2005) Nanowire dye-sensitized solar cells. *Nature Mater.* **4**, 455.

Lee YL, Huang BM, Chien HT (2008a) Highly efficient CdSe-sensitized TiO photoelectrode for quantum-dot-sensitized solar cell applications. *Chem. Mater.* **20**, 6903.

Lee HJ, Yum JH, Leventis HC, et al. (2008b) CdSe quantum dot-sensitized solar cells exceeding efficiency 1% at full-sun intensity. *J. Phys. Chem. C* **112**, 11600.

Lee JC, Lee W, Han SH, et al. (2009) Synthesis of hybrid solar cells using CdS nanowire array grown on conductive glass substrates. *Electrochem. Commun.* **11**, 231.

Leschkies KS, Divakar R, Basu J, et al. (2007) Photosensitization of ZnO nanowires with CdSe quantum dots for photovoltaic devices. *Nano Lett.* **7**, 1793.

McDonald SA, Konstantatos G, Zhang S, et al. (2005) Solution-processed PbS quantum dot infrared photodetectors and photovoltaics. *Nature Mater.* **4**, 138.

Medintz IL, Clapp AR, Mattoussi H, et al. (2003) Self-assembled nanoscale biosensors based on quantum dot FRET donors. *Nature Mater.* **2**, 630.

Na HB, Song IC, Hyeon T (2009) Inorganic nanoparticles for MRI contrast agents. *Adv. Mater.* **21**, 2133.

Nicewarner-Pena SR, Freeman RG, Reiss BD, et al. (1994) Submicrometer metallic barcodes. *Science* **294**, 137.

Nie S, Emory SR (1997) Probing single molecules and single nanoparticles by surface-enhanced Raman scattering. *Science* **275**, 1102.

Niitsoo O, Sarkar SK, Pejoux C, et al. (2006) Chemical bath deposited CdS/CdSe-sensitized porous TiO_2 solar cells. *J. Photochem. Photobiol. A* **181**, 306.

Oertel DC, Bawendi MG, Arango AC, Bulovic V (2005) Photodetectors based on treated CdSe quantum-dot films. *Appl. Phys. Lett.* **87**, 213505-1.

O'Regan B, Grätzel M (1991) A low-cost, high efficiency solar cell based on dye-sensitized colloidal TiO_2 films. *Nature* **353**, 737.

Patolsky F, Lieber CM (2005) Nanowire nanosensors. *Mater. Today* **8**, 20.

Pettersson H, Tragardh T, Persson AI, et al. (2006) Infrared photodetectors in heterostructure nanowires. *Nano Lett.* **6**, 229.

Qi D, Fischbein M, Drndic M, Selmic S (2005) Efficient polymer–nanocrystal quantum-dot photodetectors. *Appl. Phys. Lett.* **86**, 093103-1.

Qian F, Gradecak S, Li Y, et al. (2005) Core/multishell nanowire heterostructures as multicolor, high-efficiency light-emitting diodes. *Nano Lett.* **5**, 2287.

Resch-Genger U, Grabolle M, Cavaliere-Jaricot S, et al. (2008) Quantum dots versus organic dyes as fluorescent labels. *Nature Meth.* **5**, 763.

Riordan M, Hoddeson L (1997) *Crystal Fire: The Birth of the Information Age.* WW Norton, New York.

Shevchenko EV, Talapin DV, Kotov NA, et al. (2006) Structural diversity in binary nanoparticle superlattices. *Nature* **439**, 55

Sonnichsen C, Reinhard BM, Liphardt J, Alivisatos AP (2005) A molecular ruler based on plasmon coupling of single gold and silver nanoparticles. *Nature Biotech.* **23**, 741.

Sperling RA, Gil PR, Zhang F, et al. (2008) Biological applications of gold nanoparticles. *Chem. Soc. Rev.* **37**, 1896.

Stefani FD, Hoogenboom JP, Barkai E (2009) Beyond quantum jumps: blinking nanoscale light emitters. *Phys. Today* **62**, 34.

Sun S, Zeng H (2002) Size-controlled synthesis of magnetite nanoparticles. *J. Am. Chem. Soc.* **124**, 8204.

Sun B, Marx E, Greenham NC (2003) Photovoltaic devices using blends of branched CdSe nanoparticles and conjugated polymers. *Nano Lett.* **3**, 961.

Tian B, Zheng X, Kempa TJ, et al. (2007) Coaxial silicon nanowires as solar cells and nanoelectronic power sources. *Nature* **449**, 885.

Tsakalakos L, Balch J, Fronheiser J, et al. (2007) Silicon nanowire solar cells. *Appl. Phys. Lett.* **91**, 233117.

Unalan HE, Hiralal P, Kuo D, et al. (2008) Flexible organic photovoltaics from zinc oxide nanowires grown on transparent and conducting single walled carbon nanotube thin films. *J. Mater. Chem.* **18**, 5909.

Vlasov YA, Yao N, Norris DJ (1999) Synthesis of photonic crystals for optical wavelengths from semiconductor quantum dots. *Adv. Mater.* **11**, 165.

Willard DM, Van Orden A (2003) Quantum dots: resonant energy-transfer sensor. *Nature Mat.* **2**, 575.

Xiang J, Lu W, Hu Y, et al. (2006) Ge/Si nanowire heterostructures as high performance field-effect transistors. *Nature* **441**, 489.

Yu P, Zhu K, Norman AG, et al. (2006) Nanocrystalline TiO_2 solar cells sensitized with InAs quantum dots. *J. Phys. Chem. B* **110**, 25451.

Zheng G, Patolsky F, Cui Y, et al. (2005) Multiplexed electrical detection of cancer markers with nanowire sensor arrays. *Nature Biotech.* **23**, 1294.

15.8 FURTHER READING

Dye-Sensitized Solar Cells

Meyer GJ (1997) Efficient light-to-energy electrical conversion: nanocrystalline TiO_2 films modified with inorganic sensitizers. *J. Chem. Ed.* **74**, 652.

Siemsen F, Bunk A, Fischer K, et al. (1998) Solar energy from spinach and toothpaste: fabrication of a solar cell in schools. *Eur. J. Phys.* **19**, 51.

Smestad GP, Grätzel M (1998) Demonstrating electron transfer and nanotechnology: a natural dye-sensitized nanocrystalline energy converter. *J. Chem. Ed.* **75**, 752.

Nanowire Applications

Agarwal R, Lieber CM (2006) Semiconductor nanowires: optics and optoelectronics. *Appl. Phys.* **A85**, 209.

Li Y, Qian F, Xiang J, Lieber CM (2006) Nanowire electronic and optoelectronic devices. *Mater. Today* **9**, 18.

Patolsky F, Zheng G, Lieber CM (2006) Nanowire sensors for medicine and the life sciences. *Future Med.* **1**, 51.

Appendix: Useful Constants

$$
\begin{aligned}
q &= 1.602 \times 10^{-19}\,\text{C} \\
m_0 &= 9.11 \times 10^{-31}\,\text{kg} \\
a_0 &= 5.29 \times 10^{-11}\,\text{m} \\
\\
h &= 6.62 \times 10^{-34}\,\text{J s} \\
\hbar &= 1.054 \times 10^{-34}\,\text{J s} \\
\\
N_a &= 6.022 \times 10^{23} \\
k &= 1.38 \times 10^{-23}\,\text{J/K} \\
\\
c &= 3 \times 10^{8}\,\text{m/s} \\
\epsilon_0 &= 8.854 \times 10^{-12}\,\text{F/m} \\
\mu_0 &= 4\pi \times 10^{-7}\,\text{H/m} \\
1\,\text{eV} &= 1.602 \times 10^{-19}\,\text{J}
\end{aligned}
$$

Index